Climate Change

WHAT THE SCIENCE TELLS US

Second Edition

Charles Fletcher
University of Hawai'i

EDITORIAL DIRECTOR	Veronica Visentin
EXECUTIVE EDITOR	Glenn Wilson
PROJECT MANAGER	Jennifer Yee
SENIOR EDITORIAL MANAGER	Leah Michael
EDITORIAL MANAGER	Judy Howarth
CONTENT MANAGEMENT DIRECTOR	Lisa Wojcik
CONTENT MANAGER	Nichole Urban
SENIOR CONTENT SPECIALIST	Nicole Repasky
PRODUCTION EDITOR	Kavitha Balasundaram
COVER PHOTO CREDIT	© Nickolay Lamm / Climate Central

This book was set in 10/12pt NewBaskervilleStd-Roman by SPi Global and printed and bound by Quad/Graphics.

Founded in 1807, John Wiley & Sons, Inc. has been a valued source of knowledge and understanding for more than 200 years, helping people around the world meet their needs and fulfill their aspirations. Our company is built on a foundation of principles that include responsibility to the communities we serve and where we live and work. In 2008, we launched a Corporate Citizenship Initiative, a global effort to address the environmental, social, economic, and ethical challenges we face in our business. Among the issues we are addressing are carbon impact, paper specifications and procurement, ethical conduct within our business and among our vendors, and community and charitable support. For more information, please visit our website: www.wiley.com/go/citizenship.

Evaluation copies are provided to qualified academics and professionals for review purposes only, for use in their courses during the next academic year. These copies are licensed and may not be sold or transferred to a third party. Upon completion of the review period, please return the evaluation copy to Wiley. Return instructions and a free of charge return shipping label are available at: www.wiley.com/go/returnlabel. If you have chosen to adopt this textbook for use in your course, please accept this book as your complimentary desk copy. Outside of the United States, please contact your local sales representative.

ISBN: 978-1-118-79306-0 (PBK)

ISBN: 978-1-119-39941-4 (EVALC)

Library of Congress Cataloging in Publication Data:
Names: Fletcher, Charles H., author.
Title: Climate Change : What the Science Tells Us / Charles Fletcher
 (University of Hawaii).
Description: Second edition. | Hoboken, NJ : Wiley, [2019] | Includes index.
 | Identifiers: LCCN 2018034907 (print) | LCCN 2018037715 (ebook) | ISBN
 9781119399483 (Adobe PDF) | ISBN 9781119399391 (ePub) | ISBN 9781118793060
 (pbk.)
Subjects: LCSH: Climatic changes. | Global warming.
Classification: LCC QC903 (ebook) | LCC QC903 .F64 2019 (print) | DDC
 551.6—dc23
LC record available at https://lccn.loc.gov/2018034907

The inside back cover will contain printing identification and country of origin if omitted from this page. In addition, if the ISBN on the back cover differs from the ISBN on this page, the one on the back cover is correct.

To Ruth.
May our children and grandchildren thrive and find happiness in a bright new world.

CONTENTS

As I write this, humans are engaged in a dangerous global experiment that threatens the very foundation of our society. We are surrounded by evidence that Earth's surface is reacting to the unbridled release of greenhouse gases much as originally calculated in the 19th century by Swedish physicist Arrhenius Svante.[1] Since those early days, a solid body of careful observations and modeling unmistakably tells us that these gases are raising Earth's surface temperature . . . with devastating consequences.

Heat waves,[2] extreme weather,[3] animal and plant extinctions,[4] and wildfires[5] are on the rise. Worldwide, glaciers are melting,[6] the oceans are warmer,[7] and sea level is rising.[8] Unexpected effects are emerging such as political turmoil,[9] ocean anoxia,[10] and declining nutritional value of our food.[11]

Global warming is generating dangerous climate change and making parts of Earth unlivable. The resulting tide of refugees strains geopolitical relationships and is giving rise to radical nationalism. If you pay attention to geopolitical developments in North Africa, the Middle East, and Europe, you can glimpse our future.

Recognizing the signs, governments everywhere have made significant commitments to fight global warming. Trailblazing corporations are engaged in providing low-carbon services and products. Communities are reimagining their futures as more sustainable and resilient. There is abundant evidence that we are in the midst of a rapid global transformation[12] not only to renewable energy, but also to a cleaner, healthier human society.

It is easy to be hopeful that humanity is on a forward march to stop global warming as envisioned in the 2015 Paris Accord.[13]

However, among climate researchers, it is widely accepted that we cannot stop the warming at 1.5°C, and there is great uncertainty about halting at 2°C.[14]

After 3 years of stable emissions, 2017 witnessed the return of increasing carbon dioxide output.[15] India, China, and other developing nations are rising to join the middle class and the demand for new energy is nearly insatiable. Despite rapid deployment of renewables, most of this demand is being met with fossil fuels.

Energy economists are predicting a rise in carbon dioxide emissions over the next two decades. The remaining carbon budget that will fully commit us to dangerously warm climate is being rapidly spent.

Global commitments to decrease greenhouse gas emissions are only one-third of what is needed to stop warming at 2°C.[16] Over 15,000 scientists published[17] a "Warning to Humanity" that said

---

[1] Svante Arrhenius, https://en.wikipedia.org/wiki/Svante_Arrhenius

[2] Mora, C., et al. (2017) Global risk of deadly heat. *Nature Climate Change*; DOI: 10.1038/NCLIMATE3322

[3] Fischer, E.M. and Knutti, R. (2015) Anthropogenic contribution to global occurrence of heavy-precipitation and high temperature extremes. *Nature Climate Change*, v.5, June, p. 560–565.

[4] Wiens, J.J. (2016) Climate-related local extinctions are already widespread among plant and animal species, *PLOS Biology*, 14(12), e2001104, doi: 10.1371/journal.pbio.2001104

[5] Abatzoglou, J.T. and Williams, A.P. (2016) Impact of anthropogenic climate change on wildfire across western U.S. forests. *PNAS*; 201607171 doi: 10.1073/pnas.1607171113.

[6] Radić, V. and Hock, R. (2011) Regionally differentiated contribution of mountain glaciers and ice caps to future sea-level rise, *Nature Geoscience*, 4, 91-94, doi: 10.1038/ngeo1052

[7] Cheng, L., et al. (2017) Improved estimates of ocean heat content from 1960 to 2015, *Science Advances* 10 Mar., v. 3, no. 3, e1601545, DOI: 10.1126/sciadv.1601545

[8] Dangendorf, S., et al. (2017) Reassessment of 20th Century global mean sea level rise, *PNAS*, doi: 10.1073/pnas.1616007114

[9] Kelley, C.P., et al. (2015) Climate change in the Fertile Crescent and implications of the recent Syrian drought. *PNAS*, 112, 3241–3246.

[10] Schmidtko, S., et al. (2017) Decline in global oceanic oxygen content during the past five decades, *Nature*, 542, 335–339, 16 February, doi:10.1038/nature21399

[11] Chuang, Z., et al. (2017) Temperature increase reduces global yields of major crops in four independent estimates, *PNAS*, 114 (35) 9326–9331; doi:10.1073/pnas.1701762114

[12] A world in transformation: World Energy Outlook 2017, https://www.iea.org/newsroom/news/2017/november/a-world-in-transformation-world-energy-outlook-2017.html

[13] The Paris Agreement, http://unfccc.int/paris_agreement/items/9485.php

[14] Raftery, A.E., et al. (2017) Less than 2°C warming by 2100 unlikely, *Nature Climate Change*, 31 July, DOI: 10.1038/NCLIMATE3352

[15] CarbonBrief, https://www.carbonbrief.org/analysis-global-co2-emissions-set-to-rise-2-percent-in-2017-following-three-year-plateau

[16] UN News Centre, http://www.un.org/apps/news/story.asp?NewsID=57999#.Wk14eLaZOL4

[17] Ripple, W.J., et al. (2017) World Scientists' Warning to Humanity: A Second Notice. *BioScience*, DOI: 10.1093/biosci/bix125, http://scientists.forestry.oregonstate.edu/sites/sw/files/Ripple_et_al_warning_2017.pdf

humans have pushed Earth's ecosystems to their breaking point and are well on the way to ruining the planet.

A large body of research[18] reveals that humanity is using Earth's natural capital at an unsustainable pace, plants and animals are experiencing disruptions that scale from the gene to the ecosystem,[19] and the very foundation of the natural world is at risk. It is said that we are in the midst of Earth's sixth mass extinction.[20]

One point of view indicates that we are on our way to creating a safe and clean future.

Another suggests that the end of this century will be a dark and dire time.[21] How are we to assimilate these opposing visions of the future?

We must each come to our own conclusions regarding the future of humanity.

Let us do it, at least, on the basis of scientifically rigorous information that is uncontaminated by dogma or loyalty.

The world is changing so rapidly, and in such complex and often unpredictable ways, that it is critically important for every member of the human community to have a working knowledge of climate change.

Global warming threatens the stability of civilization and the security of human communities. The one most powerful tool that we have in overcoming this global threat is education that leads to positive action.

The rise of a sustainable Earth begins in the classroom.

This second edition of *Climate Change: What the Science Tells Us* is completely rewritten and updated. It has several features:

- A new Table of Contents organized to build your understanding of Earth's climate system step by step.
- Abundant use of footnotes to identify peer-reviewed scientific research, informative videos, empirical datasets, and websites with more information.
- All new chapters cover the fundamentals of Earth's climate system as well as the latest research on the impacts of climate change.
- Chapter 1 presents the hard truths about climate change: Earth's surface is already warming in ways that are dangerous for humanity, ecosystems are suffering, and projections for the future are grim.
- Chapter 2 introduces you to the processes behind radiative equilibrium and the greenhouse effect. The atmospheric energy budget and sources of heat-trapping greenhouse gases are reviewed.
- Chapter 3 presents Earth's climate system including the atmosphere, the ocean, and continental climate zones. Special focus is given to extreme weather and its relationship to changes in the jet stream, how the ocean is changing in alarming ways, Arctic amplification, and global desertification.
- Chapter 4 dissects climate denier arguments that continue to sow confusion throughout the United States. The carbon cycle is explained, and our understanding of paleoclimate is presented including orbital parameters, the conveyor-belt hypothesis, and the bipolar seesaw.
- Chapter 5 delves into the global phenomenon of sea-level rise with detailed descriptions of the changing cryosphere, ocean heat content, and the impacts of sea-level rise on coastal communities.
- Chapter 6 examines global climate modeling, the CMIP project, IPCC-AR5 projections and findings, and climate phenomena that introduce spatial and temporal variability.
- Chapter 7 reviews climate change impacts to social and economic sectors in the United States with new findings from IPCC-AR5, and the Fourth National Climate Assessment.
- Chapter 8 is about dangerous climate, and walks you through the research describing a world that has warmed 3°C.
- Throughout the book, I've highlighted special topics in a feature called "Spotlight on Climate Change." I've made sure to include key issues such as the 2015 Paris Agreement, community resilience, the earliest climate scientists, research groups that model climate, and other important subjects.
- Everywhere, I've tried to use a familiar, jargon-free voice that delivers accurate and scientifically valid findings described in easy to understand language.
- The end of each chapter has several levels of resources for instructors to use in class or as homework: a list of high-quality animations and videos available online, comprehension questions, longer more thought-provoking questions called "Thinking Critically," solo and group activities for students as homework or in the classroom, and key terms.
- Additional resources, including a color image gallery, are available on the instructor companion website at: (www.wiley.com/go/fletcher/climatechange2e)

---

[18] Rothman, D.H. (2017) Thresholds of catastrophe in the Earth system. *Science Advances*, DOI: 10.1126/sciadv.1700906

[19] Scheffers, B.R., et al. (2016) The broad footprint of climate change from genes to biomes to people. *Science*, November, DOI: 10.1126/science.aaf7671

[20] Ceballos, G., et al. (2017) Biological annihilation via the ongoing sixth mass extinction signaled by vertebrate population losses and declines, *PNAS*, 114 (30) E6089–E6096; doi:10.1073/pnas.1704949114

[21] The Uninhabitable Earth, http://nymag.com/daily/intelligencer/2017/07/climate-change-earth-too-hot-for-humans.html?utm_campaign=nym&utm_medium=s1&utm_source=fb

CHAPTER 1

Evidence of Climate Change

Climate change enhances the conditions under which wildfires occur. Studies indicate that global warming is causing earlier snowmelt in the spring and summer, which leads to hot, dry conditions persisting for longer periods of the year. Warmer temperatures and drier conditions increase the chances of a fire starting and help a burning fire spread. In the western United States, anthropogenic climate change has doubled the area affected by forest fires over the past 30 years. The number of acres of forest burning yearly in large western fires ballooned ninefold from 1984 to 2015 and is expected to increase exponentially in coming decades.[3]

David R. Frazier Photolibrary, Inc. / Alamy Stock Photo

LEARNING OBJECTIVE

Since modern measurements began in 1880, Earth's global mean surface temperature has increased approximately 1.1°C (1.8°F).[1] Called **global warming**, this is caused by the accumulation of heat-trapping gases in the atmosphere that are produced when we burn coal, natural gas, and oil for energy. Deforestation and large-scale animal farming also produce these gases (called **greenhouse gases**).[2] As a consequence of global warming, scientists have observed widespread **climate change**, which in turn is causing significant changes to Earth's environments and ecosystems. Some changes are so dramatic that they are considered dangerous, and continued warming constitutes a global crisis that threatens human well-being, the health of the oceans and land-based ecosystems, and the very sustainability of Planet Earth.

1.1 Chapter Summary

Climate change is a result of global warming, a genuine phenomenon about which there is no significant debate within the scientific community.[4] Rather, scientists debate the questions "How sensitive is climate to greenhouse gas buildup?" and "What will climate change look like regionally and locally?"

There is abundant, convincing, and reproducible scientific evidence that the increase in Earth's surface temperature is having measurable impacts on human communities and natural environments: glaciers are melting; spring is

---

[1] NASA discussion: http://earthobservatory.nasa.gov/IOTD/view.php?id=89469.

[2] Greenhouse gases are heat-trapping gases in the atmosphere that raise the temperature. The higher temperature is known as the "greenhouse effect." The natural greenhouse effect has made Earth livable, as without it the average temperature would be below freezing. However, every day, humans add huge quantities of greenhouse gases to the atmosphere, and the greenhouse effect has become amplified. This anthropogenic greenhouse effect is changing Earth in dangerous and unsustainable ways.

[3] Abatzoglou, J.T., Williams, A.P. (2016) Impact of anthropogenic climate change on wildfire across western U.S. forests. *Proceedings of the National Academy of Sciences*; 201607171 DOI: 10.1073/pnas.1607171113.

[4] Union of Concerned Scientists, *Scientific Consensus on Global Warming*. http://www.ucsusa.org/global_warming/science_and_impacts/science/scientific-consensus-on.html#.WKiN1hiZOV4.

coming earlier; the tropics are expanding; sea level is rising; the global water cycle is amplified; storms are more intense and last longer; ecosystems are shifting, and plants and animals are experiencing local extinction[5]; atmospheric and oceanic circulation are changing; the ocean is turning acidic and anoxic; drought, flooding, and other types of extreme weather are more common; heat waves kill thousands more people each year than historically; tropical diseases are spreading; food is more expensive; and insured losses have tripled.

These and many other observations document that Earth's surface, which has always sustained humanity, is rapidly changing in dangerous and expensive ways. This is the result of global warming caused by human release of greenhouse gases into the atmosphere; a phenomenon known as anthropogenic climate change.[6]

Climate change threatens all forms of life on this planet and raises the question "How will the human socio-economic framework change in response to climate change?"

In this chapter, you will learn the following:

- Human activities have caused levels of the heat-trapping gas carbon dioxide and other greenhouse gases to increase in the atmosphere.
- Humans are now manipulating global climate more rapidly than any geological analog other than a large extraterrestrial impact, leaving us with no natural guidelines on which to base predictions for how climate change will affect the world in coming years.
- In over 140 years of observations, increases in Earth's surface temperature are setting new records every few years with record highs far outpacing record lows.
- Excess heat in the atmosphere is changing ecosystems, weather patterns, and other climate-dependent aspects of Earth's surface.
- A number of natural processes cause variability in Earth's climate—volcanic eruptions, the El Niño Southern Oscillation, variations in the Sun, changes in ocean circulation, and others. In recent decades, their net effect has been to cool the climate and thus do not account for observed rapid warming.
- If global warming continues at its current rate, Earth will be increasingly characterized by more abnormally hot days and nights; fewer cold days and nights; more frequent and severe droughts, hurricanes, and cold-season storms; a decrease in glaciers and ice sheets; erosion and flooding of coastal areas; and other effects capable of displacing large portions of the human population.
- All levels of the ocean are growing hotter, more acidic, and anoxic. These effects severely stress marine ecosystems.

- Leading research centers at universities, government offices, and independent institutions conduct scientific investigations and publish their results in peer-reviewed, critically evaluated journals and reports. The great majority of these are credible, unbiased,[7] and abundant sources of information about climate change.
- Certain media sources that report on climate change are not credible as they seek to generate controversy through misinformation rather than report the facts about climate change.
- The Intergovernmental Panel on Climate Change (IPCC) releases special reports on topics relevant to the implementation of the United Nations Framework Convention on Climate Change, an international treaty that acknowledges the reality of harmful climate change.
- Certain "human fingerprints" on the climate system confirm that humans are the cause of global warming.

1.2 Introduction

Earth's climate has always been restless, influenced by massive volcanic eruptions, extraterrestrial impacts, weathering mountain ranges, changes in Earth's orientation to the Sun, variations in solar output, the El Niño Southern Oscillation (ENSO), shifts in ocean circulation and heat storage, and others. Some of these natural phenomena cause short-term climate variability (e.g., ENSO),[8] and some are capable of causing global climate to change in dramatic, long-term, and profound ways (e.g., extraterrestrial impacts). These causes of climate change are intensively studied and, to this day, have yet to yield all of their secrets.

Modern climate change, however, does not fit natural history; in the past half-century, the rate and extent of climate change have been nothing short of astonishing. In fact, natural processes today have the net effect of cooling the climate,[9] yet anthropogenic warming far outweighs their

---

[5] Local extinction occurs when plants and animals are no longer found in their natural location.

[6] VIDEO: National Geographic, http://video.nationalgeographic.com/video/news/101-videos/151201-climate-change-bill-nye-news.

[7] Harlos, C., et al. (2016) No evidence of publication bias in climate change science. *Climatic Change*, 140 (3–4): 375 DOI: 10.1007/s10584-016-1880-1

[8] Climate variability refers to short-term (months to decades) climate processes that occur at the same time that there is long-term (decades to centuries) climate change. Individual volcanic eruptions, El Niño and La Niña events, and multiyear to decadal redistribution of heat in the oceans are examples. Climate variability is an important field of study in order to improve understanding of how short-term climate variations influence and interact with long-term climate change. Such studies help clarify the characteristics of long-term changes in climate, as well as how short-term variability is changing as a result. An example of the study of climate variability is the analysis of Pacific sea-level changes published by Hamlington, B. D., et al. (2016) An ongoing shift in Pacific Ocean sea level, *J. Geophys. Res. Oceans*, 121, 5084–5097, DOI:10.1002/2016JC011815.

[9] Foster, G., and Rahmstorf, S. (2011) Global temperature evolution 1979–2010, *Environmental Research Letters*, 6(4). See also, https://skepticalscience.com/foster-and-rahmstorf-measure-global-warming-signal.html.

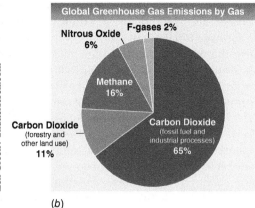

(a)   (b)

FIGURE 1.1 (a) Human communities have been using the atmosphere like an open sewer for over a century as we spew heat-trapping gases into the air. (b) Carbon dioxide, methane, nitrous oxide, and fluorinated gases (F-gases) are the key greenhouse gases emitted by human activities. Carbon dioxide is released to the atmosphere by burning oil and coal for energy and by deforestation. Methane is released through concentrated animal feeding operations, waste management, and biomass burning. Nitrous oxide is released when we use fertilizers and burn biomass. Fluorinated gases are used in refrigeration and a variety of consumer products. All of these gases are collecting in the atmosphere in ever-greater quantities, and the heat they trap is changing the climate.

Source: https://www.epa.gov/ghgemissions/global-greenhouse-gas-emissions-data.[10]

influence. Around the world, ice is melting, seas are rising, local plant and animal extinctions cripple ecosystems, and record-setting storms rake the land and sea. Despite extensive research by thousands of scientists, no known natural process accounts for these extraordinary events.

1.2.1 The Greenhouse Effect

Yet an obvious culprit is well known; humans have been treating the atmosphere like an open sewer. For over a century, we have spewed industrial exhausts of **carbon dioxide** (CO_2), **methane** (CH_4), **nitrous oxide** (N_2O), and various **fluorinated gases** into the air as if it had infinite capacity to absorb these pollutants without negative effect. Simultaneously, we have **deforested** wide tracts of the planet that were effective at storing carbon dioxide and developed **concentrated animal farming operations** (CAFO) that emit massive quantities of methane.

These gases have a special property; they trap heat that comes off Earth's surface after it has been warmed by the Sun. The more of these greenhouse gases there are in the atmosphere, the more heat is trapped and the hotter the air becomes. This is analogous to continuously adding blankets to your bed to capture your body heat and stay warm while you sleep, at some point you will add too many blankets and grow uncomfortably hot. Heat-trapping gases operate like heat-trapping blankets. Thus, Earth's surface temperature rises in what is called the **greenhouse effect**. There is a natural greenhouse effect that allows the surface

to be warmer than it would be otherwise (and support life). However, the human release of these same gases has created an enhanced greenhouse effect that is producing global warming, which in turn is causing climate change. (**Figure 1.1**).

1.2.2 Early Climate Science

Nearly two centuries of research underpin the modern science of climate change. Physicists in the 19th century realized that direct heating by the Sun did not sufficiently account for Earth's surface temperature. Careful measurement revealed that once sunlight warms Earth's surface, heat emitted back into the air by the land and the oceans can be trapped by water vapor and carbon dioxide and several other naturally occurring gases.[11] In recognition of this process, the term *greenhouse effect* was coined at the beginning of the 20th century. The greenhouse effect is a natural process that warms Earth's surface approximately 33 degrees Celsius (60 degrees Fahrenheit) more than it would otherwise be, allowing life as we know it on Earth.

Nineteenth-century scientists even predicted that burning coal and other carbon-rich fuels had the potential to double the amount of carbon dioxide in the air, trapping more heat and changing the climate. To conclude that modern climate change is a consequence of greenhouse

[10] Greenhouse Gas Emissions Data: https://www.epa.gov/ghgemissions/global-greenhouse-gas-emissions-data.

[11] See a simple and effective illustration of the Greenhouse Effect by the Australian Department of the Environment and Energy: http://www.environment.gov.au/climate-change/climate-science-data/climate-science/greenhouse-effect

gas accumulation is obvious; in fact, among scientists, this has been a well-documented, widely accepted conclusion for over 100 years.[12]

Writing in the 1820s, French mathematician and physicist **Jean-Baptiste Joseph Fourier** (1768–1830) was the first to calculate that incoming solar radiation alone did not sufficiently account for Earth's surface temperature. Although he ultimately favored interstellar radiation as the source of the additional warmth, he did consider the possibility that the atmosphere acts as an insulator of some kind. Today, this is widely recognized as the first proposal of the greenhouse effect, although Fourier never called it that.[13]

French physicist **Claude Servais Mathias Pouillet** (1790–1868) expanded Fourier's ideas and developed the first real mathematical treatment of Earth's radiation budget. He speculated that water vapor and carbon dioxide trapped infrared (IR) radiation (heat) and therefore could account for the excess warmth identified by Fourier. However, there was no experimental proof that these gases absorbed heat.

The missing observations were provided by Irish physicist **John Tyndall** (1820–1893). At the time, early geologists were presenting evidence that Earth's climate had recently emerged from a so-called *ice age*. Tyndall speculated that variations in carbon dioxide and water vapor might account for the changes in air temperature that would characterize an ice age (and the present *interglacial*). He provided the first measurements that these (and other) gases trapped heat, and he correctly concluded that water vapor was the principal gas controlling air temperature. Tyndall was the first to prove the existence of the greenhouse effect, although by his time, it was widely surmised.

Swedish scientist **Svante August Arrhenius** (1859–1927) was the first to use basic principles of chemistry to estimate the increase in Earth's surface temperature based on the growth of carbon dioxide. He concluded that humanity's burning of coal and oil was of such magnitude that it was capable of changing the air temperature, leading to global warming.[14] He concluded that this warming would be strong enough to prevent the world from entering a new ice age; a good thing, in his opinion, as a warmer Earth would be needed to feed the rapidly increasing human population (**Figure 1.2**).

Guy Stewart Callendar (1897–1964) was an English engineer who compiled temperature measurements from the 19th and 20th centuries. He correlated these with contemporaneous measurements of atmospheric carbon dioxide concentrations. Callendar concluded that over the previous

Photogravure Meisenbach Riffarth & Co. Leipzig. - Zeitschrift für Physikalische Chemie, Band 69, von 1909. - Scanned, image processed and uploaded by Kuebi = Armin Kübelbeck

FIGURE 1.2 Svante August Arrhenius (1859–1927) was the first to use the basic principles of physical chemistry to calculate the extent to which increases in atmospheric carbon dioxide would raise Earth's surface temperature.[15]

Source: https://en.wikipedia.org/wiki/Svante_Arrhenius#/media/File:Svante_Arrhenius_01.jpg.

half-century, global land temperatures had increased. This, he concluded, could be explained as an effect of the increase in carbon dioxide. His estimates have been shown as remarkably accurate, especially as they were performed without the aid of a computer. However, his findings were met with skepticism at the time.

Charles David Keeling (1928–2005) was an American chemist who developed the first instrument to measure carbon dioxide in atmospheric samples. In 1958, he established an atmospheric monitoring base on Mauna Loa, Hawaii. By 1960, he had found that levels of carbon dioxide in the atmosphere rise and fall each year as plants, through photosynthesis and respiration, take up the gas in spring and summer and release it in fall and winter. In Keeling's words, "We were witnessing for the first-time nature's withdrawing CO_2 from the air for plant growth during summer and returning it each succeeding winter."[16]

Keeling's measurements collected at Mauna Loa show a steady increase in mean atmospheric carbon dioxide concentration from about 315 parts per million (ppm) in 1958 to 411 ppm as of June 2018. This increase is due to the combustion of fossil fuels and has been accelerating in recent years (over the past decade the rate of increase is about

---

[12] VIDEO: What we knew in '82, https://www.youtube.com/watch?v=OmpiuuBy-4s.

[13] Who first coined the phrase "greenhouse effect?" Origin of the term is not clear, but an easy to read discussion of this question can be found here: http://www.easterbrook.ca/steve/2015/08/who-first-coined-the-term-greenhouse-effect/

[14] Arrhenius, S. (1896) On the Influence of Carbonic Acid in the Air on the Temperature of the Ground, *Philosophical Magazine and Journal of Science*, Series 5, v. 41, April p. 237–276. http://www.rsc.org/images/Arrhenius1896_tcm18-173546.pdf

[15] Wikipedia: Svante Arrhenius: https://en.wikipedia.org/wiki/Svante_Arrhenius

[16] The History of the Keeling Curve; https://scripps.ucsd.edu/programs/keelingcurve/2013/04/03/the-history-of-the-keeling-curve/

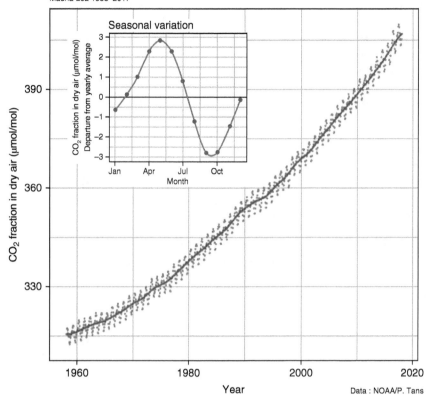

Monthly mean CO_2 concentration
Mauna Loa 1958–2017

FIGURE 1.3 The Keeling Curve is a measurement of the concentration of carbon dioxide in the atmosphere made on the slopes of Hawaii's Mauna Loa volcano since 1958. It is the longest-running such measurement in the world. Inset—the seasonal concentration of carbon dioxide in the atmosphere rises in the winter and falls in the summer. This is because of photosynthesis where plants accumulate carbon in the spring and summer when they're active and release carbon back to the air in the fall and winter.

Source: https://en.wikipedia.org/wiki/Keeling_Curve.

2 ppm per year).[17] Since carbon dioxide is a greenhouse gas, this has significant implications for global warming.[18] By the early 1970s, this curve was getting serious attention and played a key role in launching a government research program into the effect of rising CO_2 on climate. Since then, the rise has been relentless and shows a remarkably constant relationship with fossil-fuel burning based on the simple premise that 57 percent of fossil-fuel emissions remain airborne.[19]

Now the longest continuous record of atmospheric carbon dioxide in the world, the Keeling Curve, has become an iconic symbol of both the value of long-term monitoring of natural systems and the powerful impact human activities have had on the planet (**Figure 1.3**).

1.2.3 Climate Change in Every Corner of Our Lives

The 2017 Fourth National Climate Assessment, Climate Science Special Report states:

> . . . "it is extremely likely that human influence has been the dominant cause of the observed warming since the mid-20th

century. For the warming over the last century, there is no convincing alternative explanation supported by the extent of the observational evidence."[20]

It is natural for all forms of life to change their environment, and humans are no exception. In fact, the human population is now of such size (almost 8 billion people), and technological sophistication that change is evident in many global systems.

For instance, humans affect the extent and health of the global forest,[21] global marine ecosystem,[22] global water quality and availability,[23] global sedimentation,[24] global reefs,[25]

---

[17] NOAA Earth System Research Laboratory: https://www.esrl.noaa.gov/gmd/ccgg/trends/gr.html

[18] Notes from Wikipedia: https://en.wikipedia.org/wiki/Keeling_Curve

[19] The History of the Keeling Curve.

[20] Wuebbles, D.J., et al. (2017) Executive summary. In: Climate Science Special Report: Fourth National Climate Assessment, Volume I [Wuebbles, D.J., et al. (eds.)]. U.S. Global Change Research Program, Washington, DC, USA, pp. 12–34, DOI: 10.7930/J0DJ5CTG.

[21] Keenan, R.J., et al. (2015) Dynamics of global forest area: Results from the FAO global forest resources assessment. *Forest Ecology and Management,* 352, 7 Sept., pp. 9–20: https://dx.doi.org/10.1016/j.foreco.2015.06.014

[22] McCauley, D.J., et al. (2015) Marine defaunation: Animal loss in the global ocean, *Science,* 347(6219), 16, Jan, DOI: 10.1126/science.1255641

[23] Vorosmarty, C.J., et al. (2010) Global threats to human water security and river biodiversity, *Nature,* 467, 30 Sept., 555–561, DOI: 10.1038/nature09440

[24] Syvitski, J.P.M., et al. (2005) Impact of humans on the flux of terrestrial sediment to the global coastal ocean, *Science,* 307(5720), 15 Apr., 376–380: DOI: 10.1126/science.1109454

[25] Heron, S.F., et al. (2016) Warming trends and bleaching stress of the world's coral reefs 1985–2012, *Scientific Reports,* 6, 38402, DOI: 10.1038/srep38402

global river discharge[26] (an estimated 36,000 dams interrupt the flow of nearly all of Earth's river systems), global precipitation[27] (storms and drought), and others. Accepting that climate is affected by human activities makes sense because it accounts most elegantly for the global phenomena scientists have observed over the past half-century (and, it has been suggested, much longer[28]).

Because the cause of modern climate change is linked directly to human industrial activities and government policies,[29] when we talk about the kinds of measures we could take to protect our children and ourselves from its worst effects, the discussion inevitably turns to jobs, taxes, government policies, human livelihoods, and personal choices. Unfortunately, when the discourse veers down these paths, the bright line around the science of climate change is blurred by political opinion, personal worldview, and individual beliefs.

Climate has even become political dogma[30]; some assume that if you vote Democratic, you accept climate theory, if you vote Republican, you do not. Today in the United States, politicians filter their messages to appeal to their largest voter base.[31] Climate change can also have religious connotations, "It is the height of human arrogance to think that we could control God's creation," is an oft-stated opinion.

It has been observed by social scientists that those with opinions at the extreme ends of the spectrum of views on climate change (i.e., deniers versus alarmists), have the loudest voices, while those in the middle with more moderate views don't need to shout as much.

Researchers[32] have categorized "Global warming's six Americas." Ranked from highest belief in global warming to lowest belief in global warming, these six (and

their proportion of the U.S. adult population) are: Alarmed (18%), Concerned (33%), Cautious, (19%), Disengaged (12%), Doubtful (11%), and Dismissive (7%). Essentially, the American public does not speak with a single voice on the issue of global warming—they fall into six distinct populations.

Notably, the same study also found that the American public is almost unanimous on the positive value of sustainable energy, and they widely support investments in alternative energy. The alarmed and the dismissive alike endorse the idea of investing more resources into research on renewable energy. This informative revelation can be used to tailor public education efforts and develop strategies for how to best mobilize U.S. citizens to reduce the worst impacts of global warming. See the on this phenomenon.[33]

Climate change enters the discussion of what to teach in schools, what is polite conversation, what kind of car to buy, the design of our buildings and cities, our source of electricity, and more. There are many examples of what is now known as the "climate debate." The irony? That among mainstream scientists, there is no climate debate.

To paraphrase the National Research Council in their 2011 report, *America's Climate Choices*,[34] it is "settled fact" that the climate system is warming and much of this warming is very likely due to human activities. Why use the phrase "very likely?" Because volcanic eruptions, El Niño, and variations in the Sun's energy also affect global temperature, but these have been intensely studied, and research indicates that their net influence on global climate has been to cool down Earth, but the global warming trend has been strong enough to overpower them.[35]

Climate does change naturally. See the box *Spotlight on Climate Change*, "Ice Ages and Interglacials."

1.3 Climate Change is Real and it is Dangerous

Earth's surface temperature is currently the warmest since modern record-keeping began in about 1880, and each of the last three decades has been successively warmer than any preceding decade since 1850 (**Figure 1.5**). According to the National Aeronautics and Space Administration (NASA) and the National Oceanic and Atmospheric Administration (NOAA), the average surface temperature of the ocean and land combined has risen about

[26] van Vliet, M.T.H., et al. (2013) Global river discharge and water temperature under climate change, *Global Environmental Change*, 23(2), Apr., 450–464: https://dx.doi.org/10.1016/j.gloenvcha.2012.11.002

[27] Medvigy, D. and Beaulieu, C. (2011) Trends in daily solar radiation and precipitation coefficients of variation since 1984, *Journal of Climate*, 25(4), 1330–1339, DOI: 10.1175/2011JCLI4115.1

[28] Ruddiman, W.F. (2003) The anthropogenic greenhouse era began thousands of years ago, *Climatic Change*, 61, 261–293.

[29] Heede, R. and Oreskes, N. (2016) Potential emissions of CO_2 and methane from proved reserves of fossil fuels: An alternative analysis, *Global Environmental Change*, 36, 12–20, DOI: 10.1016/j.gloenvcha.2015.10.005. See also Heede, R. (2013) Tracing anthropogenic CO_2 and methane emissions to fossil fuel and cement producers, 1854–2010, *Climatic Change*, DOI: 10.1007/s10584-013-0986-y

[30] Pew Research Center, http://www.pewinternet.org/2016/10/04/public-views-on-climate-change-and-climate-scientists/. See also "No Green Tea: What Americans Think about Climate Change, by Political Allegiance," *The Economist*, 2011, http://www.economist.com/blogs/dailychart/2011/09/american-public-opinion-and-climate-change

[31] Miami Herald, March 8, 2015: "In Florida, officials ban term 'climate change'", http://www.miamiherald.com/news/state/florida/article12983720.html

[32] Yale Program on Climate Change Communication: http://climatecommunication.yale.edu/about/projects/global-warmings-six-americas/

[33] VIDEO: Global Warmings Six Americas, http://climatecommunication.yale.edu/news-events/global-warmings-six-americas/

[34] National Research Council, 2010, *America's Climate Choices: Panel on Advancing the Science of Climate Change*, Washington, D.C., National Academies Press, pp. 21–22, http://www.nap.edu/catalog.php?record_id=12782

[35] Foster, G. and Rahmstorf, S. (2011)

Box 1.1 | Spotlight on Climate Change

Ice Ages and Interglacials

Over the past 500,000 years or so, Earth's history has been characterized by natural swings in global climate, from extreme states of cold called **ice ages**, to warm periods (such as humanity has enjoyed for the past 10,000 years) called **interglacials** (**Figure 1.4**). The most recent of these was the transition, beginning about 18,000 years ago and lasting 8,000 to 10,000 years, from the end of the last ice age to the beginning of the current interglacial. This entailed worldwide warming of 6 to 10 degrees Celsius (10.8 to 18 degrees Fahrenheit). Notably, the temperature rise over the past century has been around 10 times faster than this natural rate of warming.[36]

Repeating cycles of warm–cold–warm, known as **interglacial cycles**, occur in a regular rhythm approximately every 100,000 years. These cycles are guided by the amount of sunlight that reaches the Arctic due to oscillations in Earth's tilt and orbit around the Sun. True to this timing, Earth's surface temperature has been gently cooling over the past 10,000 years as it heads into the next ice age. However, because of modern global warming, humans have hijacked climate with greenhouse gases. Now, calculations indicate that we will delay the next ice age approximately 50,000 years because of the dramatic temperature changes associated with anthropogenic global warming.[37]

Ice ages are characterized by the growth of massive **continental ice sheets** reaching across North America and northern Europe. At their maximum, these glaciers were over 4 km (2.5 mi) thick. Accompanying the spread of ice sheets was the dramatic expansion of glaciers located in mountain valleys (known as **alpine glaciers**), many of which expanded out of their valleys, coalesced with neighboring glaciers, and grew into thick icecaps that covered vast areas of mountainous territory.

The formation of all this ice required an immense reservoir of water, and the oceans were the obvious source. Evaporation from the oceans fed snow precipitation in the cold climate, and glaciers expanded around the world. As a result, sea level fell by as much as 130 m (426 ft), exposing shallow seafloor around the continents.

In several places, newly exposed continental shelves connected adjacent lands that were previously separated by water. These "land bridges" allowed early communities of humans and animals to migrate to new lands. If you had been alive during the last ice age, you could have walked on the exposed seafloor between Siberia and Alaska (a land bridge presumed to have aided the first peoples of the Americas), from France into England and from there to Ireland, and from Malaysia across Indonesia and on to Borneo. In many places, today's shore is tens or even hundreds of kilometers from where it existed during the last ice age.

We currently live in an interglacial called the **Holocene Epoch**, which began about 10,000 years ago. The last ice age began approximately 75,000 years ago and culminated between 20,000 and 30,000 years ago. For approximately 50,000 years, the last ice age dominated Earth's climate. It was a time of great changes to the landscape, as glaciers expanded and contracted, leaving myriad erosional and depositional **glacial landforms**.

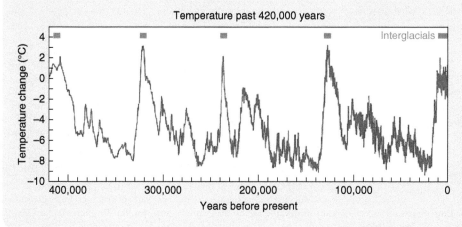

Temperature past 420,000 years

FIGURE 1.4 Recent geologic history is characterized by a series of ice ages and interglacials that occur on roughly 100,000-year cycles. In **Chapter 4**, we discuss in detail what causes interglacial cycles and how scientists study other forms of natural climate change.

Source: N. Hulbirt; Data: SkepticalScience.com.

[36] NASA: "How is todays warming different from the past?" https://earthobservatory.nasa.gov/Features/GlobalWarming/page3.php

[37] Ganopolski, A., et al. (2016) Critical insolation–CO$_2$ relation for diagnosing past and future glacial inception. *Nature*, 529(7585), 200, DOI: 10.1038/nature16494

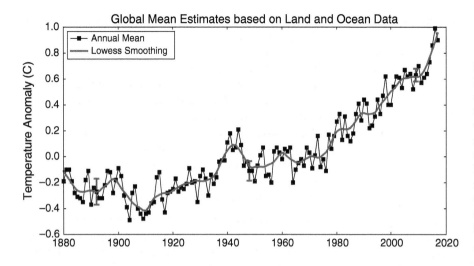

FIGURE 1.5 Land–ocean temperature anomaly, 1880 to 2017, relative to the average temperature over the period 1951–1980. In 2017, the surface temperature (land and ocean) were the second warmest that they have been since modern record-keeping began in 1880. 2017 set a record as the warmest year that was not boosted by an El Niño. Individual data points show global annual mean surface temperature anomaly. Running through these data is a 5 year trend line ("Lowess Smoothing"). A bracket at about 1945 shows the 95% confidence interval.

Source: https://data.giss.nasa.gov/gistemp/graphs/.

1.1 degrees Celsius (2.0 degrees Fahrenheit) over the past 136 years.[38] This period is now the warmest in the history of modern civilization.[39] Increased anthropogenic greenhouse gas emissions into the atmosphere drive this change.[40] We explore greenhouse gases more fully in **Chapter 2**.

Relative to the period of 1986–2005, Earth's surface temperature is projected to rise another 0.3 to 4.8 degrees Celsius (0.5 to 8.6 degrees Fahrenheit) before the end of the 21st century, depending on the rate and amount of greenhouse gases that are released to the atmosphere.[41] The projected temperature rise makes it very likely that heat waves will occur more often and last longer, extreme precipitation events will become more intense and frequent in many regions, the ocean will continue to warm and acidify, and global mean sea level will rise.

While these temperature changes may appear to be small and not very serious, minor changes in the average surface temperature of the planet can translate to large and potentially dangerous shifts in climate and weather. As this chapter will show, the evidence is clear. Rising global temperatures are already accompanied by dangerous changes in surface conditions on Earth.

For instance, the planet's oceans and glaciers have already experienced some big changes—oceans are warming and becoming more acidic as they absorb some of the excess carbon dioxide that we have dumped into the atmosphere, ice caps are melting, and sea levels are rising.

As these and other changes become more pronounced in coming decades, they will present costly, daunting challenges to our society and our environment.[42]

1.3.1 Earth Resources Are More Depleted Every Day

In his book *Earth in Mind*,[43] David Orr writes that on a typical day, we lose about 300 square km (116 sq. mi) of rainforest to logging (one acre per second), 186 square km (72 sq. mi) of land to encroaching deserts, and numerous species to extinction. Other sources tell us that in a day, the world's human population increases by more than 200,000,[44] we add 100 million tons[45] of carbon dioxide to the atmosphere, and we burn an average of 84.4 million barrels of oil (1000 barrels *per second*[46]). By the end of the day, Earth's freshwater, soil, and ocean are more acidic,[47]

[38] NASA: https://www.nasa.gov/press-release/nasa-noaa-data-show-2016-warmest-year-on-record-globally

[39] Wuebbles et al. (2017)

[40] Learn about Greenhouse Gases from the U.S. Environmental Protection Agency, https://www3.epa.gov/climatechange/kids/basics/today/greenhouse-gases.html

[41] IPCC, 2014: *Climate Change 2014: Synthesis Report. Contribution of Working Groups I, II and III to the Fifth Assessment Report of the Intergovernmental Panel on Climate Change* [Core Writing Team, R.K. Pachauri and L.A. Meyer (eds.)]. IPCC, Geneva, Switzerland, 151 pp.

[42] Henson, S.A., et al. (2017) Rapid emergence of climate change in environmental drivers of marine ecosystems, *Nature Communications*, 8, 14682, DOI: 10.1038/ncomms14682

[43] Orr, D.W. (2004) *Earth in Mind*, Washington, D.C., Island Press.

[44] Answers.com, "How Much Does World Population Increase Each Day?" http://wiki.answers.com/Q/How_much_does_world_population_increase_each_day

[45] CO$_2$Now.org: "What the World Needs to Watch," http://CO$_2$now.org/Current-CO$_2$/CO$_2$-Now/

[46] Tertzakian, P. (2006) *A Thousand Barrels a Second: The Coming Oil Break Point*, New York, McGraw-Hill.

[47] The U.S. Geological Survey has found that mining and burning coal, mining and smelting metal ores, and use of nitrogen fertilizer are the major causes of chemical oxidation processes that generate acid in the Earth-surface environment. These widespread activities have increased carbon dioxide in the atmosphere and resulted in: increased acidity of oceans; increased acidity of freshwater bodies and soils because of acid rain; increased acidity of freshwater streams and groundwater due to drainage from mines; and increased acidity of soils due to added nitrogen to crop lands. In Rice, K. and Herman, J. (2012) Acidification of Earth: an assessment across mechanisms and scales, *Applied Geochemistry*, 27(1), 1–14.

its natural resources more depleted, and its temperature a little hotter.[48]

These unrelenting impacts to Earth's ecosystems and natural resources have led researchers to conclude that our planet is perched on the edge of a tipping point, a planetary-scale critical transition resulting from human influence.[49] Scientists are warning that human population growth, widespread destruction of natural ecosystems, and climate change are pushing Earth's ecosystems and resources toward irreversible change.[50, 51]

Toward the end of 2017, an astonishing 15,000 scientists joined together as coauthors issuing a dire warning to humanity.[52] They reported that humans have pushed Earth's ecosystems to their breaking point and are well on the way to *ruining the planet*. This letter follows an earlier warning[53] that was issued 25 years earlier in 1992 from the Union of Concerned Sciences. The environmental impacts listed in that earlier warning included stratospheric ozone depletion, air and water pollution, the collapse of fisheries and loss of soil productivity, deforestation, species loss, and catastrophic global climate change caused by the burning of fossil fuels.

Global climate change leads the new letter's list of planetary threats. Global average temperatures have risen by more than 0.5 degree Celsius (0.9 degree Fahrenheit) since 1992, and annual carbon dioxide emissions have increased by 62 percent. But climate change is not the only problem the world faces.

Access to freshwater has declined, as has the amount of forestland and the number of wild-caught fish (a marker of the health of global fisheries). The number of ocean dead zones has increased. The human population grew by a massive 2 billion, while the populations of all other mammals, reptiles, amphibians, and fish have declined by nearly 30 percent. The 2017 warning points to several negative environmental trends, including:

- A 26% reduction in the amount of freshwater available per capita
- A drop in the harvest of wild-caught fish, despite an increase in fishing effort
- A 75% increase in the number of ocean dead zones
- A loss of nearly 300 million acres of forestland, much of it converted for agricultural uses
- Continuing significant increases in global carbon emissions and average temperatures
- A 53% rise in human population
- A collective 29% reduction in the numbers of mammals, reptiles, amphibians, birds, and fish

The warning came with steps to reverse negative trends, but the authors suggest that it may take a groundswell of public pressure to convince political leaders to take the right corrective actions. Such activities could include establishing more terrestrial and marine reserves, strengthening enforcement of anti-poaching laws and restraints on wildlife trade, expanding family planning and educational programs for women, promoting a dietary shift toward plant-based foods, and massively adopting renewable energy and other "green" technologies.

1.3.2 National Academy of Sciences

Orr was not merely speculating. According to the U.S. National Academy of Sciences, it is "settled fact" that the Earth system is warming, and there is 90–99 percent probability that humans are the cause.[54]

> "Some scientific conclusions or theories have been so thoroughly examined and tested, and supported by so many independent observations and results, that their likelihood of subsequently being found to be wrong is vanishingly small. Such conclusions and theories are then regarded as *settled facts*. This is the case for the conclusions that the Earth system is warming and that much of this warming is very likely due to human activities."[55]

This quotation, published in 2011 by a panel of scientists convened by the U.S. National Academy of Sciences, is included in a set of five volumes collectively called *America's Climate Choices*. The panel was compelled to conclude that "There is a strong, credible body of scientific evidence showing that climate change is occurring, is caused largely by human activities, and poses significant risks for a broad range of human and natural systems."[56]

[48] Largest Natural Disaster in US History Declared Today: http://www.examiner.com/article/largest-natural-disaster-u-s-declared-today. See also USDA Announces Streamlined Disaster Designation Process with Lower Emergency Loan Rates and Greater CRP Flexibility in Disaster Areas: https://www.usda.gov/wps/portal/usda/usdahome?contentidonly=true&contentid=2012/07/0228

[49] Barnosky, A.D., et al. (2012) Approaching a state shift in Earth's biosphere, *Nature*, 486, 7 June, 52–58.

[50] Biologists Say Half of All Species Could be Extinct by End of the Century: https://www.theguardian.com/environment/2017/feb/25/half-all-species-extinct-end-century-vatican-conference

[51] VIDEO: James Hansen on Ice Sheets, https://www.youtube.com/watch?v=Ykn8_ayFqNI&t=25s

[52] Ripple, W.J., et al. (2017) World Scientists' Warning to Humanity: A Second Notice. *BioScience*, DOI: 10.1093/biosci/bix125, http://scientists.forestry.oregonstate.edu/sites/sw/files/Ripple_et_al_warning_2017.pdf

[53] World Scientists Warning to Humanity (1992) over 1700 authors (w/ 104 Nobel Laureates), Union of Concerned Scientists: http://www.ucsusa.org/sites/default/files/attach/2017/11/World%20Scientists%27%20Warning%20to%20Humanity%201992.pdf

[54] National Research Council (2011) *Advancing the Science of Climate Change*, Washington, D.C., National Academies Press, 21–22, http://www.nap.edu/catalog.php?record_id=12782#toc

[55] National Research Council (2011)

[56] National Research Council (2011)

According to a joint publication[57] by the British Royal Society and the U.S. National Academy, rigorous analysis of all data and lines of evidence shows that most of the observed global warming over the past 50 years or so cannot be explained by natural causes and instead requires a significant role for the influence of human activities.

Global warming causes climate change. Warming is a consequence of deforestation (**Figure 1.6**), industrial agriculture, manufacturing, transportation, and other human activities that increase the concentration of heat-trapping greenhouse gases in the atmosphere. Actions such as burning oil and coal[58] release these gases to the atmosphere in quantities that have increased with the rise of the industrial age (and, as one respected climatologist proposes, since humans first domesticated animals and cleared land for farms beginning 8,000 years ago[59]).

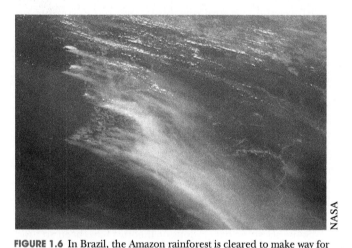

NASA

FIGURE 1.6 In Brazil, the Amazon rainforest is cleared to make way for farming. Across the middle and top of the photo, taken by astronauts aboard the International Space Station, light land areas reveal the extent of deforestation. Brush and debris are burned, releasing their carbon to the atmosphere and forever destroying the natural ability of the rainforest to store carbon and counteract global warming.[60]

Source: https://earthobservatory.nasa.gov/IOTD/view.php?id=84403.

1.3.3 Climate Change Evidence

There is abundant, convincing, and reproducible scientific evidence that the increase in Earth's surface temperature has measurable impacts on human communities and natural ecosystems. Scientists accept this as fact.

Consequently, research efforts center on questions such as: Is climate changing faster than anticipated? How is climate change affecting ecosystems, the weather, the ocean, and Earth's cryosphere? What are the best ways to predict future impacts of climate change?[61]

- In September 2016, the global atmospheric carbon dioxide concentration permanently passed 400 parts per million (ppm[62]),[63] growing at an average annual rate of about 2 ppm, more than twice the growth rate during the 1990s.[64] Carbon dioxide in the atmosphere is now rising faster than it has in hundreds of thousands of years.[65] This concentration is the highest since the **Miocene Epoch** (15 million years ago),[66] when sea level is estimated to have been 25 to 40 m (82 to 131 ft) higher and global temperature 3 to 6°C (5 to 10°F) warmer than current temperatures.

- Today's release of planet-warming carbon dioxide is about ten times faster than the most rapid event of any time in at least the past 66 million years, when an asteroid impact killed the dinosaurs. Richard Zeebe and colleagues[67] arrived at this conclusion when they compared the rate of carbon released (less than approximately 1.1 trillion kilograms of carbon per year) during the **Paleocene–Eocene Thermal Maximum** (PETM), a major warming episode 56 million years ago, to the present rate of carbon released by industrial activities (approximately 10 trillion kilograms of carbon per year). Global climate then was

[57] U.S. National Academy of Sciences http://nas-sites.org/americasclimate choices/events/a-discussion-on-climate-change-evidence-and-causes/

[58] Called "fossil fuels" because coal is made from fossil plants, and oil is made of fossil marine algae.

[59] William Ruddiman has proposed the "anthropogenic hypothesis." It is built on ice core data and calculations of the Earth–Sun orbital geometry, suggesting that the relatively warm climate of the past several thousand years is unnatural and should instead have been characterized by cooling. Ruddiman proposes that through the production of excess methane and carbon dioxide, human agricultural practices took control of Earth's climate as early as 5,000 to 8,000 years ago. Ruddiman (2003); Ruddiman, W.F. (2005) Cold climate during the closest stage 11 analog to recent millennia, *Quaternary Science Reviews* 24, 1111–1121; and Ruddiman, W.F. (2005) *Plows, Plagues, and Petroleum: How Humans Took Control of Climate,* Princeton, N.J., Princeton University Press.

[60] VIDEO: Deforestation, https://www.nytimes.com/2017/02/24/business/ energy-environment/deforestation-brazil-bolivia-south-america.html?_r=1

[61] Kerr, R.A. (2009) Amid worrisome signs of warming, climate fatigue sets in, *Science*, 326, 926–928.

[62] The abbreviation "ppm" means parts per million. It is a measurement of abundance (or concentration) the same way that "per cent" means parts per hundred. In this case ppm means molecules of CO_2 per million molecules of air.

[63] Climate Central, http://www.climatecentral.org/news/world-passes-400-ppm-threshold-permanently-20738

[64] Kerr (2009)

[65] The annual growth rate of atmospheric carbon dioxide measured at NOAA's Mauna Loa Observatory in Hawaii jumped by 3.05 ppm during 2015, the largest year-to-year increase in 56 years of monitoring. In another first, 2015 was the fourth consecutive year that CO_2 grew more than 2 ppm, said Pieter Tans, lead scientist of NOAA's Global Greenhouse Gas Reference Network. "Carbon dioxide levels are increasing faster than they have in hundreds of thousands of years," Tans said. "Its explosive compared to natural processes."

[66] Tripati, A.K., et al. (2009) Coupling of CO_2 and ice sheet stability over major climate transitions of the last 20 million years, *Science*, 326(5958), 1394–1397, http://www.sciencemag.org/cgi/content/abstract/1178296

[67] Zeebe, R.E., et al. (2016) Anthropogenic carbon release rate unprecedented during the past 66 million years, *Nature Geoscience*, DOI: 10.1038/ngeo2681

more than 8 degrees Celsius (14.4 degrees Fahrenheit) warmer than current temperatures. There are various hypotheses as to what caused the PETM including a comet impact, degassing of CO_2 from a massive magma body, changes in ocean circulation, extensive volcanic activity, and release of methane from frozen seafloor sediments. Regardless of the cause, Zeebe's calculations indicate that humans are now manipulating global climate more rapidly than any geological analog other than a large extraterrestrial impact, leaving us with no natural guidelines on which to base predictions for how climate change will affect the world in coming years.[68]

- In 2014, the average temperature on Earth's surface was the hottest in the modern era. The year 2015 exceeded that by 20 percent, a record-setting increase never before seen. In 2016, the temperature was boosted approximately 0.12 degrees Celsius (0.2 degrees Fahrenheit) by a strong El Niño (we discuss the El Niño Southern Oscillation, or ENSO, in more detail in **Chapter 5**) and set yet another new record.[69] Overall, Earth's average surface temperature has increased approximately 1.0°C (1.8°F) since modern record-keeping began in 1880. According to NASA, this change is driven by increased carbon dioxide and other greenhouse gas emissions into the atmosphere.[70]
- Overall, the global annual temperature has increased at an average rate of 0.07 degrees Celsius (0.13 degrees Fahrenheit) per decade since 1880 and at an average rate of 0.2 degrees Celsius (0.36 degrees Fahrenheit) per decade since 1970.[71]

The years 1998 and 2005 set records for warmth. In 2008, however, global mean temperature dropped, returning to temperatures not seen since the mid-1990s (although 2008 was, nonetheless, the ninth warmest year on record at the time). A graph of annual temperatures from 1998 to 2013 looked as if global warming had stopped when, in fact, average annual global temperature over the period still had a positive trend. Nevertheless, the drop in temperature from 2005 to 2008 influenced national attitudes and a trend of "global cooling" was reported in some media. However, mistaking *short-term climate variability* (year-to-year changes in temperature) for *long-term trends* (climate change) is a fundamental error.

By 2009, global temperature was on the upswing again, and it tied with a cluster of years—1998, 2002, 2003, 2006, and 2007—as the second warmest year on record. December 2009 also marked the end of the warmest decade with a warming trend over the previous 30 years of 0.2 degrees

Celsius (0.36 degrees Fahrenheit) per decade.[72] Typical of the history now established, subsequent years displayed a high degree of variability: 2010 tied for the warmest year on record; 2011 ranked as ninth warmest year; 2012 replaced 2011 as ninth warmest year; and 2013 tied with 2009 and 2006 for seventh warmest year.

However, the years 2014, 2015, and 2016 all, in turn, outstripped each other for warmest years on the modern record. Earth's surface temperature had ended a 16 years-long period of variability and entered a phase of fearsome warming.

Three record-setting years in a row is unprecedented in the 140-year-long modern history of global climate. Temperatures in 2016 not only set an annual record, but January through May and July through September (8 out of 12 months) were the warmest on record for those respective months. October and November were the second warmest of those months, just behind records set in 2015.[73]

Earth's global surface temperatures in 2017 ranked as the second warmest since 1880. Continuing the planet's long-term warming trend, globally averaged temperatures in 2017 were 0.90 degrees Celsius (1.62 degrees Fahrenheit) warmer than the 1951–1980 mean, according to scientists at NASA's Goddard Institute for Space Studies (GISS) in New York.[74] That is second only to global temperatures in 2016.

Scientists do not expect global warming to be expressed as a smooth annual rise in average atmospheric temperature from one year to the next. They understand that short-term climate processes (e.g., the El Niño Southern Oscillation, the sunspot cycle, volcanic eruptions) dominate year-to-year temperatures[75] and that it does not get warmer everywhere at the same time. Similar to the rise in stock market value since the 1970s, climate is taking a bumpy ride of ups and downs as it undergoes a long-term increase in global surface temperature.

In the same way, scientists do not take every snowstorm and cool day as evidence that global warming is not a looming issue or that it does not exist at all. In fact, anthropogenic climate change is responsible for record-setting snowstorms and cold snaps through what is known as Arctic amplification, the rapid warming taking place in the Arctic, which is three times faster than the global average. We explain this linkage of Arctic amplification to extreme winter events in **Chapter 3**.

On the contrary, scientists recognize that global warming is a "noisy" process that requires analysis of both short-term events and the long-term trends. By 2009, studies emerged that supported this view. In a blind test, the Associated Press gave prominent statisticians global temperature data

---

[68] Mashable http://mashable.com/2016/03/21/CO_2-fastest-66-million-years/#SqheckQsyuqN

[69] NASA: http://climate.nasa.gov/news/2537/nasa-noaa-data-show-2016-warmest-year-on-record-globally/

[70] NASA: https://www.giss.nasa.gov/research/news/20170118/

[71] NOAA: https://www.ncdc.noaa.gov/sotc/global/201613

[72] NASA: https://www.giss.nasa.gov/research/news/20100121/

[73] NASA: http://earthobservatory.nasa.gov/IOTD/view.php?id=89469

[74] NASA: https://climate.nasa.gov/news/2671/long-term-warming-trend-continued-in-2017-nasa-noaa/

[75] Hansen, J., et al. (2010) Global surface temperature change, *Reviews of Geophysics* 48, RG4004, DOI: 10.1029/2010RG000345

without identifying its source or what the numbers represented. The statisticians rejected global cooling because they recognized that short-term statistical variability does not represent a long-term trend.[76] Also, the U.S. National Climate Data Center published peer-reviewed research[77] reporting that climate history since the 1970s reveals many episodes when the average temperature of the atmosphere temporarily stopped rising, and even reversed its upward climb, but that strong net warming over the entire period is indisputable.

1.4 The Earth System is Changing

From soil at the equator to ice at the poles, the circulation of heat through Earth's atmosphere and oceans links the planet's living organisms and environments.[78] Certain global processes connect the poles and tropics, deserts and forests, continents and oceans even though Earth is 40,075 km (24,901 mi) in circumference and has a surface area of 509,600,000 square km (196,757,000 sq. mi). This vast area is characterized chiefly by mixing of the atmosphere and ocean, the water cycle, seasonal heating and cooling, and more.

Global warming causes changes to these processes on the scale of the whole Earth. Detailed analysis of ice core records of climate over the past 20,000 years reveals that today's changes in climate are unique over that entire period; there is no "natural process" that can explain today's warming.[79]

1.4.1 Irreversible Change

Earth is not an unchanging ball of rock hurtling through space. Energy from within and without alters it. For example, heat diffuses upward from the **core** through the **mantle**, Earth's thickest layer, causing rock in the mantle to plastically flow and migrate. As heat moves through the **crust**, the outermost layer, it drives **plate tectonics** and causes **volcanism**.[80] Heat also arrives from the Sun. As this heat circulates through the atmosphere and oceans,

and is carried by ocean and air currents around the planet, it influences Earth's weather and climate.

These processes make Earth dynamic and cause it to constantly change on long and short timescales, and it has been this way, in various forms, throughout its 4.6 billion-year history. For most of that history, natural processes have controlled those changes, and many of them have been enormous (e.g., the collision of continents driven by plate tectonics; the increasing diversity of living forms; the massive extinctions that characterize the fossil record). The natural processes that cause global climate change include plate tectonics, volcanic eruptions, extraterrestrial impacts, and variations in Earth's orbit (which we study in **Chapter 4**). Human activities can also cause global climate change.

On modern Earth, human activities have indeed caused global changes in land use, air and water quality, and the abundance of natural resources,[81] particularly over the past two centuries. There is scientific consensus that human activities are altering Earth's climate, largely owing to increasing levels of the heat-trapping gas carbon dioxide and other greenhouse gases (**Figure 1.7**).

Because it can reside in the atmosphere for more than 1,000 years,[84] carbon dioxide is the most dangerous greenhouse gas. It is released when we burn fossil fuels, sources of energy provided by burning fossil carbon, such as petroleum (fossil marine algae) and coal (fossil continental wetland plants).

A study by NOAA[85] concluded that climate change is largely irreversible for the next 1,000 years because of the long lifetime of CO_2 in the atmosphere. As a result, at higher levels of CO_2 (450 to 600 ppm), sea-level rise, changes in rainfall, severe weather events, and other consequences of global warming will come to permanently characterize the planet's surface.

[76] Borenstein, S., Statisticians Reject Global Cooling, Associated Press, October 26, 2009.

[77] Easterling, D. and Wehner, M. (2009) Is the climate warming or cooling? *Geophysical Research Letters*, 36, L08706. NASA, "The Ups and Downs of Global Warming," http://climate.nasa.gov/news/index.cfm?FuseAction=ShowNews&NewsID=175

[78] VIDEO: The Ocean, A Driving Force for Weather and Heat, https://www.youtube.com/watch?v=6vgvTeuoDWY

[79] Björck, S. (2011) Current global warming appears anomalous in relation to the climate of the last 20,000 years, *Climate Research* 48(1), 5, DOI: 10.3354/cr00873

[80] VIDEO: Plate Tectonics, https://www.youtube.com/watch?v=Xzpk9110Lyw

[81] According to the 2016 Edition of the National Footprint Accounts, humanity demanded the resources and services of 1.6 planets in 2016; such demand has increased over 2.5 times since 1961. This situation, in which total demand for ecological goods and services exceeds the available supply for a given location, is known as "overshoot". **Global overshoot** indicates that stocks of ecological capital may be depleting, and/or that waste is accumulating, beyond Earth's capacity to sustain a healthy ecosystem. It means that we are borrowing from the future to live today. Global Footprint Network, http://www.footprintnetwork.org/en/index.php/GFN/page/at_a_glance/

[82] Dai, A. (2011) Characteristics and trends in various forms of the Palmer drought severity index during 1900–2008, *Journal of Geophysical Research*, 116, D12115, DOI: 10.1029/2010JD015541. See also Dai, A. (2010) Drought under global warming: a review, *Wiley Interdisc. Rev. Clim*, 2, 45–65.

[83] Ault, T.R., et al. (2015) Relative impacts of mitigation, temperature, and precipitation on 21st-century megadrought risk in the American Southwest, *Science Advances*, 2(10), e1600873, DOI: 10.1126/sciadv.1600873

[84] Solomon, S., et al. (2009) Irreversible climate change due to carbon dioxide emissions, *Proceedings of the National Academy of Science* 106, 1704–1709, DOI: 10.1073/pnas.-9128211-6

[85] Solomon et al. (2009)

FIGURE 1.7 The global percentage of land area in drought has increased 10% as a result of climate change.[82] Droughts in the Southwest and Central Plains of the United States in the second half of the 21st century could be drier and longer than anything humans have seen in those regions in the past 1,000 years.[83]

1.4.2 Observed Impacts

Changes in precipitation (rain and snowfall), the source of our drinking water, cause of floods, and the crucial factor governing the health of critical ecosystems that provide us with natural resources, are of special concern to humanity. Studies[86] document that global warming directly influences precipitation because the water-holding capacity of air increases by about 7 percent for each 1 degree Celsius (1.8 degrees Fahrenheit) of warming. Thus, storms that are provided with more moisture produce more-extreme precipitation events.[87] Warmer air also results in greater evaporation that dries Earth's surface, increasing the intensity and duration of drought. Global warming is producing a world that is drier, yet, ironically, prone to greater flooding.

Because warming is producing only modest changes in winds, generalized precipitation patterns do not shift their location much, and thus, wet areas are becoming wetter and dry areas are becoming drier.[88] Globally, a warmer atmosphere produces more rainfall instead of snow, and winter snowpack develops later and melts earlier. This increases runoff in late winter and early spring, raising the risk of flooding (**Figure 1.8**) and extending the duration and intensity of summer drought. Farmers, communities, and government agencies responsible for public safety and health all find it challenging to adapt to this new pattern.

Simultaneously there are climate processes related to global warming that increase extreme winter weather. For instance, on the eastern seaboard of the United States, the warmer water of the Gulf Stream provides additional water vapor that can produce heavier snowfall during winter storms (nor'easters). As cold air masses from the north-central United States collide with the increased moisture off the warmer Atlantic, some record-setting snowstorms have occurred. Similarly, among the snow belts along the Great Lakes, warmer winters are keeping more of the lake surfaces ice-free, promoting evaporation and heavier snows downwind of Lakes Erie, Ontario, and Michigan particularly.

These phenomena may lead students to say "How can Earth be warming when it's snowing more in winter?" The answer lies with paying attention to the details of Earth's weather systems and knowing how warmer air influences climate patterns in ways that are scientifically consistent. In **Chapter 3**, we introduce the role of global warming in altering the Polar Jet Stream and how this also leads to increased extreme weather in the Northern Hemisphere, including more heat waves, and record-setting cold events.

Many studies indicate that the climate change observed during the 20th and early 21st centuries is due to a combination of changes in solar radiation, volcanic activity, land use, and increases in atmospheric greenhouse gases. Of these, greenhouse gases are the dominant long-term influence, and they are causing the lower atmosphere, the air closest to Earth, to warm. This excess heat is causing dramatic changes in ecosystems, weather patterns, and

[86] Trenberth, K. (2011) Changes in precipitation with climate change, *Climate Research*, DOI: 10.3354/cr00953

[87] NOAA: https://www.climate.gov/news-features/featured-images/heavy-downpours-more-intense-frequent-warmer-world

[88] Durack, P., et al. (2012) Ocean salinities reveal strong global water cycle intensification during 1950 to 2000, *Science*, 336(6080), 455–458, DOI: 10.1126/science.1212222

FIGURE 1.8 Tropical storm Harvey hit Houston, Texas in August, 2017, and set a major record: the most rainfall ever from a single storm in the continental United States. Global warming is changing precipitation patterns.[89] These changes pose an extraordinary challenge to the natural environment as well as human communities. Heavy downpours are more intense and frequent, and the global occurrence of extreme rainfall has increased 12%.[90]

Joe Raedle / Staff / Getty Images

Source: GettyImages_840239148.0

other climate-dependent aspects of Earth's surface. These changes include the following.

1. Carbon dioxide levels in the air have passed 400 ppm compared to a natural level of 280 ppm—an increase of over 40%. This is the highest level in millions of years.[91]
2. Today, release of planet-warming carbon dioxide is ten times faster than the most rapid event in the past 66 million years, when an asteroid impact killed the dinosaurs.[92]
3. Humans are causing the climate to change 170 times faster than natural forces.[93]
4. Global temperature has risen approximately 1°C (1.8°F) from the late 19th Century.[94]
5. The likely range of global temperature increase is 2.0–4.9°C, with median 3.2°C (5.76°F) and a 5% (1%) chance that it will be less than 2°C (1.5°C).[95]

6. The last time it was this warm was during the last interglacial (ca. 125,000 years ago) when global mean sea level was 6.6 m (20 ft) above present.[96]
7. Atmospheric humidity is rising.[97]
8. The global water cycle has accelerated.[98]
9. Air temperature over the oceans is rising.[99]
10. Greenland is losing ~286 billion tons of ice annually, Antarctica is losing ~159 tons, and alpine glaciers are losing over 200 billion tons of ice annually.[100, 101, 102]
11. The west Antarctic ice sheet is in "unstoppable" collapse.[103]

[89] Lehmann, J., et al. (2015) Increased record-breaking precipitation events under global warming, *Climatic Change*, DOI: 10.1007/s10584-015-1434-y

[90] NOAA: https://www.climate.gov/news-features/featured-images/heavy-downpours-more-intense-frequent-warmer-world

[91] Tripati et al. (2009)

[92] Zeebe et al. (2016)

[93] Gaffney, O., and Steffen, W. (2017) The Anthropocene equation, *The Anthropocene Review*, https:// dx.doi.org/10.1177%2F2053019616688022

[94] Haustein, K. et al. (2017) A global warming index. *Nature Scientific Reports*, DOI:10.1038/s41598-017-14828-5

[95] Raftery, A.E., et al. (2017) Less than 2°C warming by 2100 unlikely, *Nature Climate Change*, 7, 637–641, DOI:10.1038/nclimate3352.

[96] Hoffman, J.S., et al. (2017) Regional and global sea surface temperatures during the last interglaciation, *Science*, 355(6322), 276–279, DOI: 10.1126/science.aai8464

[97] Willett, K., et al. (2007) Attribution of observed surface humidity changes to human influence, *Nature*, 449, 710–712, DOI: 10.1038/nature06207

[98] Durack et al. (2012)

[99] NOAA National Climatic Data Center, "State of the Climate: Global Analysis for May 2011," http://www.ncdc.noaa.gov/sotc/global/

[100] Radić, V. and Hock, R. (2011) Regionally differentiated contribution of mountain glaciers and ice caps to future sea-level rise, *Nature Geoscience*, 4, 91–94, DOI: 10.1038/ngeo1052

[101] Zemp, M., et al. (2015) Historically unprecedented global glacier decline in the early 21st century. *Journal of Glaciology*, 61(228): 745 DOI: 10.3189/2015JoG15J017

[102] Rignot, E., et al. (2011) Acceleration of the contribution of the Greenland and Antarctic ice sheets to sea level rise, *Geophysical Research Letters*, 38, L05503, DOI: 10.1029/2011GL046583

[103] Joughlin, I., et al. (2014) Marine ice sheet collapse potentially underway for the Thwaites Glacier Basin, West Antarctica, *Science*, May 12. See also Rignot, E., et al. (2014) Widespread, rapid grounding line retreat of Pine Island, Thwaites, Smith and Kohler glaciers, West Antarctica from 1992 to 2011, *Geophysical Research Letters*, 3502–3509, DOI:10.1002/2014GL060140.

12. Glaciers worldwide are melting.[104]
13. Melting on Greenland and Antarctica has accelerated.[105]
14. Cloud cover over Greenland is decreasing at about 0.9% per year. Each 1% of decrease drives an additional 27±13 billion tons of ice melt each year.[106]
15. Alpine glaciers have shrunk to their lowest levels in 120 years and are wasting two times faster than they did in the period 1901–1950, three times faster than they did in 1851–1900, and four times faster than they did 1800–1850.[107]
16. Continental ice sheets are shrinking.[108]
17. Arctic sea ice is shrinking as a result of global warming.[109]
18. Winter Arctic sea ice was the lowest on record in 2017.[110]
19. In the Arctic, average surface air temperature for the year ending September 2016 was the highest since 1900, and new monthly record highs were recorded for January, February, October, and November 2016.[111]
20. Rapid warming in the Arctic is causing the jet stream to slow and develop large planetary waves.[112]
21. Regions of Earth where water is frozen for at least 1 month each year are shrinking with impacts on related ecosystems.[113]
22. In North America, spring snow cover extent in the Arctic was the lowest in the satellite record, which started in 1967.[114]
23. Extreme warm events in winter are much more prevalent than cold events.[115]
24. Snow cover is shrinking.[116]
25. The southern boundary of Northern Hemisphere permafrost is retreating poleward.[117]
26. Large parts of permafrost in northwest Canada are slumping and disintegrating into running water. Similar large-scale landscape changes are evident across the Arctic including in Alaska, Siberia, and Scandinavia.[118]
27. Tree lines are shifting poleward and to higher elevations.[119]
28. Spring is coming sooner to some plant species in the Arctic while other species are delaying their emergence amid warm winters. The changes are associated with diminishing sea ice.[120]
29. Air temperature over land is rising.[121]
30. The global percentage of land area in drought has increased about 10%.[122]
31. The global occurrence of extreme rainfall has increased 12%.[123]
32. Heavy downpours are more intense and frequent.[124]
33. Extreme weather events are more frequent.[125]
34. 0.5°C of global warming has been enough to increase heat waves and heavy rains in many regions of the planet.[126]
35. Storm tracks are shifting poleward.[127]
36. The number of weather disasters is up 14% since 1995–2004 and has doubled since 1985–1994.[128]

[104] Radić and Hock (2011)

[105] B. Wouters, et al. (2013) Limits in detecting acceleration of ice sheet mass loss due to climate variability. *Nature Geoscience* 6, 613–616.

[106] Hofer, S., et al. (2017) Decreasing cloud cover drives the recent mass loss on the Greenland ice sheet. *Science Advances*, 28 June, v. 3, no. 6, e1700584, DOI: 10.1126/sciadv.1700584

[107] Zemp et al. (2015)

[108] Rignot et al. (2011)

[109] Arctic Report Card (2016) http://www.arctic.noaa.gov/Report-Card/Report-Card-2016. See also Serreze, M., et al. (2007) Perspectives on the Arctic's shrinking sea-ice cover, *Science* 315, 1533–1536.

[110] See: http://nsidc.org/arcticseaicenews/

[111] See: http://www.arctic.noaa.gov/Report-Card/Report-Card-2016

[112] Francis, J., and Skific, N. (2015) Evidence linking rapid Arctic warming to mid-latitude weather patterns, *Phil. Trans. R. Soc.* 373, 20140170, https://dx.doi.org/10.1098/rsta.2014.0170 See: https://insideclimatenews.org/news/08062016/greenland-arctic-record-melt-jet-stream-wobbly-global-warming-climate-change

[113] Fountain, A., et al. (2012) The disappearing cryosphere: Impacts and ecosystem responses to rapid cryosphere loss, *BioScience* 62(4), 405–415, DOI: 10.1525/bio.2012.62.4.11

[114] Arctic Report Card (2016)

[115] Guirguis, K., et al. (2011) Recent warm and cold daily winter temperature extremes in the northern hemisphere, *Geophysical Research Letters*, 38, L17701, DOI: 10.1029/2011GL048762

[116] Déry, S. J. and Brown, R.D. (2007) Recent northern hemisphere snow cover extent trends and implications for the snow albedo feedback, *Geophysical Research Letters*, 34, L22504

[117] Thibault, S. and Payette, S. (2009) Recent permafrost degradation in bogs of the James Bay area, Northern Quebec, Canada, *Permafrost and Periglacial Processes*, 20(4), 383, DOI: 10.1002/ppp.660.

[118] See: https://insideclimatenews.org/news/27022017/global-warming-permafrost-study-melt-canada-siberia. See also Kokelj, S.V., et al. (2017) Climate-driven thaw of permafrost preserved glacial landscapes, northwestern Canada, *Geology*, first published online February 2017, DOI: 10.1130/G38626.1

[119] Beck, P.S.A., et al. (2011) Changes in forest productivity across Alaska consistent with biome shift, *Ecology Letters*, DOI: 10.1111/j.1461-0248.2011.01598.x

[120] Post, E., et al. (2016) Highly individualistic rates of plant phenological advance associated with Arctic sea ice dynamics, *Biology Letters*, 12(12), 20160332, DOI: 10.1098/rsbl.2016.0332

[121] Menne, M.J., et al. (2010) On the reliability of the U.S. surface temperature record, *Journal of Geophysical Research* 115, D11108, DOI: 10.1029/2009JD013094

[122] Dai (2011)

[123] Lehmann et al. (2015)

[124] NOAA: https://www.climate.gov/news-features/featured-images/heavy-downpours-more-intense-frequent-warmer-world

[125] Medvigy and Beaulieu (2011)

[126] Schleussner, C-F, et al. (2017) In the observational record half a degree matters, *Nature Climate Change*. DOI: 10.1038/nclimate3320

[127] Bender, F. A-M, et al. (2012) Changes in extratropical storm track cloudiness 1983–2008: Observational support for a poleward shift, *Climate Dynamics* 38, 2037–2053, DOI: 10.1007/s0038-011-1065-6

[128] Centre for Research on the Epidemiology of Disasters, UN International Strategy for Disaster Reduction: http://reliefweb.int/report/world/human-cost-weather-related-disasters-1995-2015

37. In Australia, record-setting hot days outnumber record-setting cold days by a factor of 12 to 1.[129]

38. Extreme heat waves are projected to cover double the amount of global land by 2020 and quadruple by 2040, regardless of future emissions trends.[130]

39. New records continue to be set for warm temperature extremes. For instance, in the United States during February, 2017, there were 3,146 record highs set compared to only 27 record lows, a ratio of 116 to 1.[131]

40. Nine of the ten deadliest heat waves have occurred since 2000, causing 128,885 deaths around the world.[132]

41. Nearly, one-third of the world's population is now exposed to climatic conditions that produce deadly heat waves.[133]

42. If global temperatures rise 2°C (3.6°F), the combined effect of heat and humidity will turn summer into one long heat wave. Temperature will exceed 104°F every year in many parts of Asia, Australia, Northern Africa, South and North America.[134]

43. If global temperatures rise 4°C (7.2°F), a new "super-heatwave" will appear with temperatures peaking at above 131°F making large parts of the planet unlivable including densely populated areas such as the U.S. east coast, coastal China, large parts of India and South America.[135]

44. Aridification will emerge over about 20% to 30% of the world's land surface by the time global mean temperature change reaches 2°C.[136]

45. Drought severity has been increasing across the Mediterranean, southern Africa, and the eastern coast of Australia over the course of the 20th century, while semiarid areas of Mexico, Brazil, southern Africa, and Australia have encountered desertification for some time as the world has warmed.[137]

46. Global warming has reached at least 1°C (1.8°F) above the natural background. The last time it was this warm was during the last interglacial (ca. 125,000 years ago)[138] when global mean sea level was approximately 6.6 m (20 ft) above the present.[139]

47. Global wind speed has accelerated.[140]

48. Spring is coming earlier.[141]

49. Warmer winters with less snow have resulted in a longer lag time between spring events and a more protracted vernal window (the transition from winter to spring).[142]

50. Plants are leafing out and blooming earlier each year.[143]

51. Climate-related local extinctions have already occurred in hundreds of species, including 47% of 976 species surveyed.[144]

52. Plant and animal extinctions, already widespread, are projected to increase from twofold to fivefold in the coming decades.[145]

53. Rising CO_2 decreases the nutrient and protein content of wheat, leading to a 15% decline in yield by mid-century.[146]

[129] Lewis, S.C. and King, A.D. (2015) Dramatically increased rate of observed hot record breaking in recent Australian temperatures, *Geophys. Res. Lett.*, 42, 7776–7784, DOI: 10.1002/2015GL065793. See also Meehl, G., et al. (2009) The relative increase of record high maximum temperatures compared to record low minimum temperatures in the US, *Geophysical Research Letters*, 36, L23701, DOI: 10.1029/2009GL040736

[130] Coumou, D. and Robinson, A. (2013) Historic and future increase in the global land area affected by monthly heat extremes, *Environmental Research Letters*, 8(3), 034018, DOI: 10.1088/1748-9326/8/3/034018

[131] Climate Central.org: http://www.climatecentral.org/news/record-high-temperature-february-21186

[132] Vaidyanathan, G. (2015) Killer heat grows hotter around the world, *Scientific American*, August 6, https://www.scientificamerican.com/article/killer-heat-grows-hotter-around-the-world/

[133] Mora, C., et al. (2017) Global risk of deadly heat. *Nature Climate Change*, DOI: 10.1038/NCLIMATE3322

[134] Russo, S., et al. (2017) Humid heat waves at different warming levels. *Scientific Reports*, 7 (1) DOI: 10.1038/s41598-017-07536-7

[135] Russo et al. (2017)

[136] Park, C.-E., et al. (2018) Keeping global warming within 1.5°C constrains emergence of aridification. *Nature Climate Change*, DOI: 10.1038/s41558-017-0034-4

[137] Park et al. (2018)

[138] Hoffman, J.S., et al. (2017) Regional and global sea surface temperatures during the last interglaciation, *Science*, 355(6322), 276–279, DOI: 10.1126/science.aai8464

[139] Kopp, R.E., et al. (2009) Probabilistic assessment of sea level during the last interglacial stage, *Nature*, 462, 863–867, DOI: 10.1038/nature08686

[140] Young, I.R., et al. (2011) Global trends in wind speed and wave height, *Science*, 332(6028), 451–455, DOI: 10.1126/science.1197219

[141] "The U.S. Geological Survey hails an early spring and ties it to climate change": http://www.chron.com/news/houston-weather/article/The-U-S-Geological-Survey-hails-an-early-spring-10958042.php. See also Kahru, M., et al. (2010) Are phytoplankton blooms occurring earlier in the Arctic? *Global Change Biology*, DOI: 10.1111/j.1365-2486.2010.02312.x

[142] Contosta, A.R., et al. (2017) A longer vernal window: the role of winter coldness and snowpack in driving spring transitions and lags. *Global Change Biology*; 23 (4): 1610 DOI: 10.1111/gcb.13517

[143] Wolkovich, E., et al. (2012) Warming experiments underpredict plant phenological responses to climate change, *Nature*, 485(7399), 494–497, DOI: 10.1038/nature11014

[144] Wiens, J.J. (2016) Climate-related local extinctions are already widespread among plant and animal species, *PLOS Biology*, 14(12), e2001104, DOI: 10.1371/journal.pbio.2001104

[145] Wiens (2016)

[146] Myers, S.S., et al. (2014) Increasing CO_2 threatens human nutrition, *Nature*, 510, 139–142, DOI: 10.1038/nature13179. See also Feng, Z., et al. (2015) Constraints to nitrogen acquisition of terrestrial plants under elevated CO_2, *Global Change Biology*, 21(8), 3152–3168, DOI: 10.1111/gcb.12938

54. By 2050, climate change will lead to per-person reductions of 3% in global food availability, 4% in fruit and vegetable consumption, and 0.7% in red meat consumption. These changes will be associated with 529,000 climate-related deaths worldwide.[147]

55. Without changes to policy and improvements to technology, food productivity in 2050 could look like it did in 1980 because, at present rates of innovation, new technologies won't be able to keep up with the damage caused by climate change in major growing regions.[148]

56. Certain groups of Americans—including children, elders, the sick, and the poor—are most likely to be harmed by climate change.[149]

57. Climate change is harming human health now. These harms include heat-related illness, worsening chronic illnesses, injuries and deaths from dangerous weather events, infectious diseases spread by mosquitoes and ticks, illnesses from contaminated food and water, and mental health problems.[150]

58. The lower atmosphere (troposphere) is warming.[151]

59. The tropics have expanded.[152]

60. Species are migrating poleward and to higher elevations.[153]

61. Atmospheric humidity is rising.[154]

62. The global water cycle has accelerated.[155]

63. Air temperature over the oceans is rising.[156]

64. Sea surface temperature is rising.[157]

65. The oceans are warming rapidly.[158]

66. Sea level is rising, and the rate of rise has accelerated.[159]

67. Today, global mean sea level is rising three times faster than it was in the 20th century.[160]

68. Between 1993 and 2014, the rate of global mean sea-level rise increased 50% with the contribution from melting of the Greenland Ice Sheet rising from 5% in 1993 to 25% in 2014.[161]

69. We have already committed to a long-term future sea level 1.3 to 1.9 m (4.3 to 6.2 ft) higher than today and are adding about 0.32 m/decade (1 ft/decade) to the total: ten times the rate of observed contemporary sea-level rise.[162]

70. Over 90% of the heat trapped by greenhouse gases since the 1970s has been absorbed by the oceans, and today, the oceans absorb heat at twice the rate they did only 18 years ago.[163]

71. The world's oceans have warmed at twice the rate of previous decades and the extra heat has reached deeper waters.[164, 165]

72. Sea surface temperatures have increased in areas of tropical cyclone genesis suggesting a connection with strengthened storminess.[166]

[147] Springmann, M., et al. (2016) Global and regional health effects of future food production under climate change: a modeling study, *The Lancet*, March 2, DOI: 10.1016/S0140-6736(15)01156-3

[148] Liang, X.Z., et al. (2017) Determining climate effects on US total agricultural productivity, *PNAS*, www.pnas.org/cgi/doi/10.1073/pnas. 1615922114

[149] Medical Alert! Climate Change is Harming Our Health, Medical Society Consortium on Climate and Health, 24p. https://medsocietiesforclimatehealth.org/wp-content/uploads/2017/03/medical_alert.pdf

[150] U.S. Global Change Research Program (2016) The Impacts of Climate Change on Human Health in the United States: A Scientific Assessment. Crimmins, A.J., et al. Global Change Research Program, Washington, DC, 312 pp. http://dx.doi. org/10.7930/J0R49NQX

[151] Thorne, P.W., et al. (2010) Tropospheric temperature trends: History of an ongoing controversy, *Wiley Interdisciplinary Reviews: Climate Change*, DOI: 10.1002/wcc.80

[152] Seidel, D.J., et al. (2008) Widening of the tropical belt in a changing climate, *Nature Geoscience*, 1, 21–24, DOI: 10.1038/ngeo.2007.38

[153] Loarie, S.R., et al. (2009) The velocity of climate change, *Nature*, 462, 1052–1055.

[154] Willett, K., et al. (2007) Attribution of observed surface humidity changes to human influence, *Nature*, 449, 710–712, DOI: 10.1038/nature06207

[155] Durack et al. (2012)

[156] NOAA National Climatic Data Center, "State of the Climate: Global Analysis for May 2011," published online June 2011, http://www.ncdc.noaa.gov/sotc/global/

[157] Levitus, S., et al. (2009) Global ocean heat content 1955–2008 in light of recently revealed instrumentation problems, *Geophys. Res. Lett.*, 36, L07608, DOI:10.1029/2008GL037155.

[158] Wang, G., et al. (2017) Consensuses and discrepancies of basin-scale ocean heat content changes in different ocean analyses, *Climate Dynamics.* DOI: 10.1007/s00382-017-3751-5

[159] Nerem, R.S., et al. (2018) Climate-change–driven accelerated sea-level rise detected in the altimeter era. *Proceedings of the National Academy of Science*, DOI: 10.1073/pnas.1717312115

[160] Dangendorf, S, et al. (2017) Reassessment of 20th Century global mean sea level rise, *Proceedings of the National Academy of Sciences*, DOI: 10.1073/pnas.1616007114

[161] Chen, X., et al. (2017) The increasing rate of global mean sea-level rise during 1993–2014, *Nature Climate Change*, DOI: 10.1038/nclimate3325

[162] Strauss, B.H. (2015) Rapid accumulation of committed sea level rise from global warming, *PNAS*, 110(34), 13699–13700.

[163] Cheng L., et al. (2015) Global upper ocean heat content estimation: recent progress and the remaining challenges. *Atmospheric and Oceanic Science Letters*, 8. DOI:10.3878/AOSL20150031. See also Glecker, P.J., et al. (2016) Industrial era global ocean heat uptake doubles in recent decades. *Nature Climate Change.* DOI:10.1038/nclimate2915

[164] Cheng, L., et al. (2017) Improved estimates of ocean heat content from 1960 to 2015, *Science Advances* 10 Mar., v. 3, no. 3, e1601545, DOI: 10.1126/sciadv.1601545

[165] Song, Y.T. and Colberg, F. (2011) Deep ocean warming assessed from altimeters, gravity recovery and climate experiment, in situ measurements, and a non-Boussinesq ocean general circulation model, *Journal of Geophysical Research* 116, C02020, DOI: 10.1029/2010JC006601. See also Volkov, D.L., et al. (2017) Decade-long deep-ocean warming detected in the subtropical South Pacific, *Geophysical Research Letters*, DOI: 10.1002/2016GL071661

[166] Defforge, C.L. and Merlis, T.M. (2017) Observed warming trend in sea surface temperature at tropical cyclone genesis, *Geophysical Research Letters*, DOI: 10.1002/2016GL071045

73. Oxygen levels in the ocean have declined by 2% over the past five decades because of global warming, probably causing habitat loss for many fish and invertebrate species.[167]

74. Marine ecosystems can take thousands, rather than hundreds, of years to recover from climate-related upheavals.[168]

75. The world's richest areas for marine biodiversity are also those areas mostly affected by both climate change and industrial fishing.[169]

76. The Atlantic Meridional Overturning Circulation has decreased 20%. The North Atlantic has the coldest water in 100 years of observations.[170]

77. The number of coral reefs impacted by bleaching has tripled over the period 1985–2012.[171, 172]

78. By 2050, more than 98% of coral reefs will be afflicted by bleaching-level thermal stress each year.[173]

79. Scientists have concluded that once seas are hot enough for long enough, nothing can protect coral reefs. The only hope for securing a future for coral reefs is urgent and rapid action to reduce global warming.[174]

80. Average pH of ocean water fell from 8.21 to 8.10, a 30% increase in acidity. Ocean water is more acidic from dissolved CO_2, which is negatively affecting marine organisms.[175]

81. Marine ecosystems are under extreme stress.[176]

82. Production of oxygen by photosynthetic marine algae is threatened at higher temperatures.[177]

83. Extreme weather is increasing.[178]

84. Global warming is changing life on Earth on a global scale.[179]

85. There is only a 5% chance of avoiding dangerous global warming of 2°C (3.6°F).[180]

1.4.3 U.S. Global Change Research Program

How will all these dramatic and overwhelmingly negative changes play out in a future characterized by continued global warming? This question has been at the root of much of the research being conducted by climate scientists in recent years. For instance, in a report produced by the U.S. Global Change Research Program,[181] a combined effort of more than a dozen government science agencies, researchers found the following.

- In the future, abnormally hot days and nights and heat waves are very likely to become more common.
- Cold days and cold nights are very likely to become much less common. The number of days with frost is very likely to decrease.
- Future sea ice extent will continue to decrease and could even disappear entirely in the Arctic Ocean in summer in coming decades. Sea ice loss has increased coastal erosion in Arctic Alaska and Canada because of increased exposure of the coastline to wave action. It has disrupted long-standing ecosystem processes such as predator–prey relationships, mating, seasonal migrations, forage and overall food availability, and prevalence of pests, diseases, and other health and mortality factors.
- Future precipitation is likely to be less frequent but more intense, and precipitation extremes are very likely to increase.
- Future droughts are likely to become more frequent and severe in some regions (e.g., U.S. Southwest, Mexico), leading to a greater need to respond to reduced water supplies, increased wildfires, and various ecological impacts.
- Future hurricanes in the North Atlantic and North Pacific are likely to have increased rainfall and wind speeds; for each 1°C (1.8°F) increase in tropical

[167] Schmidtko, S., et al. (2017) Decline in global oceanic oxygen content during the past five decades, *Nature*, 542, 335–339, 16 February, DOI: 10.1038/nature21399

[168] Moffitt, S.E., et al. (2015) Response of seafloor ecosystems to abrupt global climate change. *PNAS*, DOI: 10.1073/pnas.1417130112

[169] Ramírez, F., et al. (2017) Climate impacts on global hot spots of marine biodiversity. *Science Advances*, 3 (2): e1601198 DOI: 10.1126/sciadv.1601198

[170] Rahmstorf, S., et al. (2015) Exceptional twentieth-century slowdown in Atlantic Ocean overturning circulation. *Nature Climate Change*. DOI: 10.1038/nclimate2554

[171] Heron et al. (2016)

[172] VIDEO: Saving Corals, http://films.economist.com/blancpain-ocean?utm_source=Saving%20Coral&utm_medium=Editors%20 Picks&utm_content=Blancpain%20Ocean.

[173] Heron et al. (2016)

[174] Hughes, T.P., et al. (2017) Global warming and recurrent mass bleaching of corals. *Nature*, 543 (7645): 373 DOI: 10.1038/nature21707

[175] Barton, A., et al. (2012) The Pacific oyster, *Crassostrea gigas*, shows negative correlation to naturally elevated carbon dioxide levels: Implications for near-term ocean acidification effects, *Limnology and Oceanography* 57(3), 698–710, DOI: 10.4319/lo.2012.57.3.0698.

[176] McCauley, D.J., et al. (2015) Marine defaunation: Animal loss in the global ocean, *Science*, 347(6219), 16, Jan, DOI: 10.1126/science.1255641. Henson, S.A., et al. (2017) Rapid emergence of climate change in environmental drivers of marine ecosystems, *Nature Communications*, 8, 14682, DOI: 10.1038/ncomms14682.

[177] Sekerci, Y. and Petrovskll (2015) Mathematical modeling of Plankton-Oxygen dynamics under the climate change. *Bulletin of Mathematical Biology*: DOI: 10.1007/sl11538-015-0126-0

[178] O'Gorman, P. (2010) Understanding the Varied Response of the Extratropical Storm Tracks to Climate Change, *Proceedings of the National Academy of Sciences* 107, no. 45: 19176–19180, DOI: 10.1073/pnas.1011547107. See also T. R. Karl, et al. (eds.), *Weather and Climate Extremes in a Changing Climate*. Report by U.S. Climate Change Science Program and Subcommittee on Global Change Research. (Washington, D.C., Department of Commerce, NOAA National Climatic Data Center, 2008); http://www.climatescience .gov/Library/sap/sap3-3/final-report/default.htm

[179] Rosenzweig, C., et al. (2008) Attributing Physical and Biological Impacts to Anthropogenic Climate Change, *Nature* 453, no. 7193: 353–357.

[180] Raferty et al. (2017)

[181] Karl, T.R., et al. (2008)

sea-surface temperatures, rainfall rates will increase by 6% to 18% and wind speeds of the strongest hurricanes will increase by 1% to 8%.

- In the future, strong cold-season storms in both the Atlantic and Pacific are likely to be more frequent, with stronger winds and more extreme wave heights.

Climate change has already transformed our planet. Air temperatures have risen and, as a result, heat waves and drought are more common. Storms have increased in frequency and intensity, seasons have shifted, and the ranges of plant and animal life have moved. The glaciers are melting, global sea level is rising and accelerating, and the temperature of the oceans has increased. Climate change is rapidly altering the lands and waters we depend on for survival, and the cause is the buildup of greenhouse gases produced by human activities.

During the summer of 2012, following the warmest spring on record, the average temperature of the continental United States was 2.5 degrees Celsius (4.5 degrees Fahrenheit) above average. As a result, the 12 months ending June 30, 2012, were the warmest 12-month period on record for the United States. This unusual heat wave, more intense than any on record, led to the declaration of the largest natural disaster in the history of the United States.[182] Fifty-six percent of the continental United States was enveloped in intense drought conditions and the U.S. Department of Agriculture declared a nationwide state of emergency.[183]

If we don't act to lessen the cause of global warming and adapt our socioeconomic systems to the new reality that has emerged, we may leave our children—and all living things—with a world characterized by the most dangerous consequences of climate change.

1.5 Reliable Sources of Climate Change Information

"Global warming" refers to an increase in the average temperature of Earth's surface, including the air, land, and oceans. Global warming is not a political position. It is a scientific certainty that has been verified by independent studies of literally thousands of scientists.[184]

A 2009 study[185] of scientific consensus on global warming published by the American Geophysical Union[186] concludes:

> The debate on the authenticity of global warming and the role played by human activity is largely nonexistent among those who understand the nuances and scientific basis of long-term climate processes. The challenge, rather, appears to be how to effectively communicate this fact to policy makers and to a public that continues to mistakenly perceive debate among scientists.

Multiple studies published in peer-reviewed scientific journals[187] show that 97 percent or more of actively publishing climate scientists agree that: *Climate-warming trends over the past century are very likely due to human activities.* In addition, most of the leading scientific organizations worldwide have issued public statements endorsing this position.[188]

The release of greenhouse gases is a result of burning fossil fuels (coal, natural gas, petroleum), industrial agriculture, and deforestation.[189] This has led to their accumulation in the atmosphere in concentrations that have not been seen in millions of years of geologic history. These are heat-trapping gases, and consequently, the temperature of the atmosphere and the oceans has been rising for over a century at an accelerating pace.[190]

---

[182] News article http://www.examiner.com/article/largest-natural-disaster-u-s-declared-today

[183] USDA Announces Streamlined Disaster Designation Process with Lower Emergency Loan Rates and Greater CRP Flexibility in Disaster Areas: http://www.usda.gov/wps/portal/usda/usdahome?contentid=2012/07/0228.xml&navid=NEWS_RELEASE&navtype=RT&parentnav=LATEST_RELEASES&edeployment_action=retrievecontent

[184] Harlos et al. (2016)

[185] Doran, P.T., and Zimmerman, M.K. (2009) Examining the Scientific Consensus on Climate Change, *Eos Transactions American Geophysical Union*, v. 90 Iss. 3, 22, DOI: 10.1029/2009EO030002.

[186] The American Geophysical Union is a prominent international scientific organization of 60,000 researchers, teachers, and students in 139 countries. You can read their position statement about human impacts on climate change here: http://sciencepolicy.agu.org/files/2013/07/AGU-Climate-Change-Position-Statement_August-2013.pdf

[187] Cook, J., et al (2013) Quantifying the consensus on anthropogenic global warming in the scientific literature, *Environmental Research Letters* Vol. 8 No. 2, (June); Quotation from p.3 "Among abstracts that expressed a position on AGW [Anthropogenic, or human-caused, Global Warming], 97.1% endorsed the scientific consensus. Among scientists who expressed a position on AGW in their abstract, 98.4% endorsed the consensus." See also Anderegg, W.R.L. (2010) Expert Credibility in Climate Change, *Proceedings of the National Academy of Sciences*, v. 107 No. 27, 12107–12109 (21 June); DOI: 10.1073/pnas.1003187107. See also Doran, P.T., and Zimmerman, M.K. (2009). See also Oreskes, N. (2004) Beyond the Ivory Tower: The Scientific Consensus on Climate Change, *Science*, v. 306 no. 5702, p. 1686 (3 Dec.); DOI: 10.1126/science.1103618

[188] NASA: "Scientific Consensus: Earth's Climate is Warming" http://climate.nasa.gov/scientific-consensus/

[189] US Environmental Protection Agency: http://www3.epa.gov/climatechange/

[190] U.S. National Academies of Science: http://nas-sites.org/americasclimatechoices/sample-page/panel-reports/87-2/. See also, IPCC, 2013: *Climate Change 2013: The Physical Science Basis. Contribution of Working Group I to the Fifth Assessment Report of the Intergovernmental Panel on Climate Change* [Stocker, T., et al. (eds.)]. Cambridge University Press, Cambridge, United Kingdom and New York, NY, USA, 1535 pp.

1.5.1 Climate Data

Several groups collect data on global temperature. In the U.S., climate data are collected, maintained, and analyzed by the NASA Goddard Institute of Space Studies (GISS)[191] and the National Oceanographic and Atmospheric Administration (NOAA).[192, 193] Because climate knows no boundaries, both these organizations work closely with governments and researchers worldwide (**Figure 1.9**). In the United Kingdom, the Met Office Hadley Center (UKMET)[194] is the foremost climate change research center, with responsibility to collect and analyze global climate information. The Japan Meteorological Agency[195] provides weather observation and forecasting and climate change and global environmental tracking services. These units along with researchers at universities,[196] government offices, and institutions around the world conduct scientific investigations and publish their results in peer-reviewed journals and reports.

1.5.2 Media

One would expect that mainstream media accounts of science would be generally reliable, but this is not always the case. Headlines are often conceived to sell controversy, not communicate fact. It is important to read past the headlines and filter personal opinion from scientific observation. Other sources of information include websites,[197] institutional reports, and newsletters.[198]

1.5.3 Scientific Peer Review

The peer-review process, while not perfect, is the best available system for assessing the accuracy of scientific findings and ensuring that a rigorous standard applies to the work of those who report on the results of their research. Typically, peer review begins when a scientist sends a manuscript describing the results of a research project to the editor of a scientific journal and requests publication. The editor reviews the work and, if it is deemed appropriate for the journal, sends it to specialists in the field to get their opinions on its quality. The specialists return their review within a few weeks and recommend publication, revision, or rejection. It is common for different reviewers to have conflicting opinions, and it is up to the editor to sort this out. On the basis of these comments, the editor makes a decision to publish the piece, reject it, or request revisions from the author subject to further review.

Peer-reviewed research forms the basis of improving our understanding of the details of climate change. What are the characteristics of changing air temperature? How rapidly is the ocean warming and how is this affecting

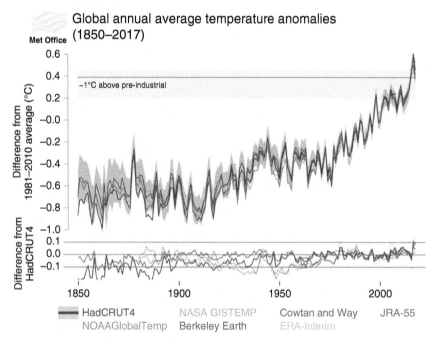

FIGURE 1.9 Global annual average temperature anomalies (°C, relative to the long-term average for 1981–2010). Six datasets are shown as indicated in the legend. Gray shading, 95% uncertainty. The horizontal gray line with yellow shading indicates approximate point at which temperatures exceed 1°C above "preindustrial" levels. At the top, the horizontal line with the grey band is the difference from the 1850–1900 average. (bottom) differences of each data set from HadCRUT4 (the Met Office data set) on an expanded scale.

Source: https://www.metoffice.gov.uk/research/news/2018/global-surface-temperatures-in-2017

---

[191] NASA: http://www.giss.nasa.gov/
[192] NOAA: http://www.noaa.gov/climate.html
[193] VIDEO: Climate Change Visualization 1880–2016, https://climate.nasa.gov/climate_resources/139/
[194] MET Office: http://www.metoffice.gov.uk/
[195] Japan Meteorological Agency: http://www.jma.go.jp/jma/indexe.html
[196] Berkeley Earth Group: http://berkeleyearth.org

[197] For instance, *Science Daily:* http://www.sciencedaily.com/. See also *Science News:* http://www.sciencenews.org/view/home
[198] For instance, the Pew Center on Global Climate Change: http://www.pewclimate.org/ Also the Union of Concerned Scientists: http://www.ucsusa.org/. See also US National Academy of Sciences: http://americasclimatechoices.org/

marine ecosystems? Are there shifts in precipitation patterns, global winds, snow cover, and storminess? These and other questions drive the engine of climate research so that scientists are constantly building knowledge.

A study of 120 research articles published in the field of climate research between 1997 and 2013 found that climate researchers do not conceal uncomfortable facts that could potentially disprove climate change. Study authors concluded, "It was gratifying to see that the scientific method is robust. It is important to show that we can trust the results of climate research, even if more work is needed about how those results are reported."[199]

1.5.4 Climate Literacy

Established in 1989 under the Executive Office of the President, the U.S. Global Change Research Program (USGCRP)[200] coordinates and integrates the climate change activities of 13 federal departments and agencies. The program is a ready source of peer-reviewed summaries on the subject of climate change and its impacts in the United States and the world.

The USGCRP has produced a short guide for educators to promote climate literacy among individuals and communities: *The Climate Literacy Guide.*[201] This guide provides a summary of essential principles underlying how Earth's climate system works and how climate change is occurring. The guide lists seven principles:

1. The Sun is the primary source of energy for Earth's climate system.
2. Climate is regulated by complex interactions among components of the Earth system.
3. Life on Earth depends on, is shaped by, and affects climate.
4. Climate varies over space and time through both natural and human-made processes.
5. Understanding of the climate system is improved through observations, theoretical studies, and modeling.
6. Human activities are affecting the climate system.
7. Climate change will have consequences for the Earth system and human lives.

This and the following chapters expand on many of these principles.

1.5.5 IPCC Assessments

Because the public and their leaders and decision-makers may not keep up on the latest scientific research, it is important to provide summaries of improving climate knowledge to stakeholders on a regular basis. This is a key role of the Intergovernmental Panel on Climate Change (IPCC). The IPCC is an international organization under the joint auspices of the United Nations Environmental Program and the World Meteorological Organization.[202] The IPCC produces global assessments of climate change every 5 to 7 years representing the status of professional understanding.

Past IPCC reports have been published in 1990, 1995, 2001, 2007, and 2013. The next report, Assessment Report 6 (AR6), is scheduled for release as a series of special studies. In 2018, a special study reported on the impacts of global warming of 1.5 degrees Celsius (2.7 degrees Fahrenheit) above preindustrial levels and related global greenhouse gas emission pathways.

The IPCC also has reports on: climate change and oceans and the cryosphere; climate change, desertification, land degradation, sustainable land management, and food security; and greenhouse gas fluxes in terrestrial ecosystems. These are part of the AR6 cycle. The main AR6 report, released in three working group contributions in 2020/2021 and a Synthesis Report in 2022, can be accessed at their website.[203]

The IPCC does not carry out original research, nor does it do the work of monitoring climate or related phenomena. Its primary role is publishing special reports that assess the state of climate science on topics relevant to the implementation of the United Nations Framework Convention on Climate Change (UNFCCC). The UNFCCC is an international treaty that acknowledges the possibility of harmful climate change.[204] The IPCC is organized in three working groups. Working Group I reports on the physical science basis of climate change. Working Group II reports on climate change impacts, adaptation, and vulnerability. Working Group III reports on mitigation of climate change.

Knowledge about global warming can be acquired from IPCC assessment reports. For instance, Assessment Report 5 [AR5] was published in 2013 and 2014[205] and provides an advanced, detailed, and thorough review of global, regional, and local climate patterns and processes. The findings of AR5 are discussed in more detail in **Chapter 6**.

Knowledge can also be gained from peer-reviewed scientific literature published between IPCC assessments in reputable journals such as *Science,*[206] *Nature,*[207] *Nature Climate*

---

[199] Harlos et al. (2016)

[200] USGCRP http://www.globalchange.gov/

[201] USGCRP: "Climate Literacy: The Essential Principles of Climate Science," http://www.globalchange.gov/resources/educators/climate-literacy

[202] IPCC: http://www.ipcc.ch/

[203] IPCC http://wg1.ipcc.ch/AR6/AR6.html

[204] All countries are members of this international treaty, the United Nations Framework Convention on Climate Change. The UNFCCC is designed to consider what can be done to reduce global warming and to cope with whatever temperature increases are inevitable. See their homepage: http://unfccc.int/2860.php

[205] The 2013/2014 Fifth Assessment Report (AR5) consists of four elements, one from each of the three working groups and a Synthesis Report. All four reports, as well as past reports, can be found at: https://www.ipcc.ch/report/ar5/

[206] *Science* http://www.sciencemag.org/

[207] *Nature* http://www.nature.com/

Change,[208] and others. Certain organizations also provide daily climate news feeds to your digital device. Among those that I receive are: Sciencedaily,[209] The Daily Climate,[210] and Google-alerts on any content you wish[211] (I receive Google-alerts on "coastal erosion," "sea-level rise," and "ice collapse," which are areas of special interest that I have). Also, through your digital phone App store, you can find many climate news feeds.

1.5.6 Climate Change Impacts in the United States

The U.S. Global Change Research Program[212] publishes congressionally mandated reports that summarize the state of knowledge regarding climate impacts globally and across the United States. The National Climate Assessment is conducted every 4 years to assess and inform the Nation about the impacts of climate change. Teams of researchers from across the United States author the report. It is a very inclusive and collaborative process and provides a comprehensive assessment of national climate impacts.

In late 2017, the Fourth National Climate Assessment, NCA4, published a Climate Science Special Report[213] that provided reviews of the state of scientific understanding of a range of topics including sea-level rise, extreme weather, projections of climate change, ecosystem impacts, and others.

Key findings[214] from the NCA4 Executive Summary[215] are as follows:

1. The global, long-term, and unambiguous warming trend has continued during recent years. Since the last National Climate Assessment was published, 2014 became the warmest year on record globally; 2015 surpassed 2014 by a wide margin; and 2016 surpassed 2015. Sixteen of the warmest years on record for the globe occurred in the last 17 years (1998 was the exception).

2. Annual average temperature over the contiguous United States has increased by 1.8°F (1.0°C) for the period 1901–2016 and is projected to continue to rise (very high confidence).

3. There have been marked changes in temperature extremes across the contiguous United States. The number of high temperature records set in the past two decades far exceeds the number of low temperature records (very high confidence).

4. Heavy precipitation events in most parts of the United States have increased in both intensity and frequency since 1901 (high confidence). There are important regional differences in trends, with the largest increases occurring in the northeastern United States (high confidence).

5. Extreme temperatures in the contiguous United States are projected to increase even more than average temperatures (very high confidence).

6. Future decreases in surface soil moisture from human activities over most of the United States are likely as the climate warms under the higher scenarios (medium confidence).

7. The world's oceans have absorbed about 93% of the excess heat caused by greenhouse gas warming since the mid-20th century, making them warmer and altering global and regional climate feedbacks (very high confidence).

8. Global mean sea level has risen by about 7–8 inches (about 16–21 cm) since 1900, with about 3 in (about 7 cm) occurring since 1993 (very high confidence) (**Figure 1.10**).

9. The world's oceans are currently absorbing more than a quarter of the CO_2 emitted to the atmosphere annually from human activities, making them more acidic (very high confidence), with potential detrimental impacts to marine ecosystems.

FIGURE 1.10 As sea level rises, it erodes previously stable shorelines and accelerates erosion on shores that were already eroding.

Source: Dolan Eversole, University of Hawaii Sea Grant.

---

[208] *Nature Climate Change* http://www.nature.com/nclimate/index.html

[209] Science Daily https://www.sciencedaily.com/news/earth_climate/

[210] Daily Climate http://www.dailyclimate.org

[211] Google Alerts https://www.google.com/alerts

[212] USGCRP: http://www.globalchange.gov/

[213] USGCRP (2017) Climate Science Special Report: Fourth National Climate Assessment, Volume I, Wuebbles, D.J., et al. (eds.). U.S. Global Change Research Program, Washington, DC, USA, 470 pp, DOI: 10.7930/J0J964J6.

[214] Findings in NCA4 are evaluated through two metrics: Confidence in the validity of a finding based on the type, amount, quality, strength, and consistency of evidence (such as mechanistic understanding, theory, data, models, and expert judgment); the skill, range, and consistency of model projections; and the degree of agreement within the body of literature. Likelihood, or probability of an effect or impact occurring, is based on measures of uncertainty expressed probabilistically (based on the degree of understanding or knowledge, e.g., resulting from evaluating statistical analyses of observations or model results or on expert judgment).

[215] USGCRP (2017)

10. Annual average near-surface air temperatures across Alaska and the Arctic have increased over the last 50 years at a rate more than twice as fast as the global average temperature (very high confidence).

11. Since the early 1980s, annual average Arctic sea ice has decreased in extent between 3.5% to 4.1% per decade, has become thinner by between 4.3 and 7.5 ft, and is melting at least 15 more days each year. September sea ice extent has decreased between 10.7% and 15.9% per decade (very high confidence).

12. The observed increase in global carbon emissions over the past 15–20 years has been consistent with higher scenarios. In 2014 and 2015, emission growth rates slowed as economic growth became less carbon-intensive. Even if this slowing trend continues, however, it is not yet at a rate that would limit the increase in the global average temperature to well below 3.6°F (2°C) above preindustrial levels.

13. Choices made today will determine the magnitude of climate change risks beyond the next few decades.

14. Unanticipated and difficult or impossible-to-manage changes in the climate system are possible throughout the next century as critical thresholds are crossed and/or multiple climate-related extreme events occur simultaneously.

1.5.7 United Nations Framework Convention on Climate Change

In 1992, at an international conference in Rio de Janeiro, an environmental treaty was negotiated among the world's nations and entered into force in 1994. Known as the United Nations Framework Convention on Climate Change (UNFCCC),[216] the treaty objective was to stabilize greenhouse gas concentrations in the atmosphere at a level that would prevent dangerous human interference with the climate system. The UNFCCC has been signed by 198 parties (nations for the most part) and is broadly viewed as a lawful and legitimate agreement due to its global membership. Since 1995, there have been annual meetings, called Conference of the Parties (COPs), wherein member nations assess progress in dealing with climate change.

In 1997, under the auspices of the UNFCCC, representatives of the United States and 83 other nations met in Kyoto, Japan, to establish legally binding obligations for developed countries to reduce their greenhouse gas emissions. The result of their deliberations was the Kyoto Protocol,[217] an international agreement that carbon dioxide emissions should be regulated. The protocol proposed establishing carbon quotas for each country based on its population and level of industrialization. A carbon quota is a fixed permissible amount of CO_2 that a country is allowed to release to the atmosphere each year.

For example, the Kyoto Protocol required that by 2012, industrialized countries such as the United States reduce their carbon dioxide emissions to 7 percent below the levels measured in 1990 and by 20 percent by 2020. As the leading developed nation, the United States has historically produced the bulk of the world's carbon dioxide emissions. Then, in 2007, China's output surpassed that of the United States. Today, powerful economic growth in India is rapidly escalating emission production there as well.

The United States never signed the Kyoto Protocol because the U.S. Congress refused to endorse the treaty, yet it reached the 2012 goal nonetheless, due in part to an economic recession that began in 2008,[218] improved automobile fuel efficiency, and growth in the use of natural gas as an energy source, which results in less carbon release to the atmosphere than burning traditional coal or oil.

The COP21, which met in Paris in late 2015, produced a landmark covenant replacing the Kyoto Protocol. Known as the **Paris Agreement**,[219] it lays out global commitments on climate change reduction measures from 2020. The agreement entered into force with the joining of at least 55 countries that together represent at least 55 percent of global greenhouse gas emissions. On April 22, 2016 (Earth Day), 174 countries signed the agreement in New York and began the work of adopting it within their own legal systems (called "ratification"). The expected primary result of the agreement is to limit global warming to less than 2 degrees Celsius (3.6 degrees Fahrenheit), compared to preindustrial levels, and to reach zero net anthropogenic greenhouse gas emissions during the second half of the 21st century.

The Paris Agreement also contains language to pursue efforts to limit the temperature increase to 1.5 degrees Celsius (2.7 degrees Fahrenheit), which will require reaching zero emissions sometime between 2030 and 2050. The future of the UNFCCC, and the Paris Agreement, is to establish protocols for monitoring individual nations' emissions. Because it has been calculated that the reductions agreed to by member nations thus far are not sufficient to limit warming to 2 degrees Celsius, the UNFCCC also works to establish new agreements that continue the process of lowering future greenhouse gas emissions.

For more on the Paris Agreement, see the box *Spotlight on Climate Change*, "COP 21."

---

[216] UNFCCC: http://unfccc.int/2860.php
[217] UNFCCC: http://unfccc.int/kyoto_protocol/items/2830.php

[218] During periods of recession in the U.S. and elsewhere, greenhouse gas emissions tend to decrease because manufacturing and other industrial activities decrease.
[219] UNFCCC: http://unfccc.int/paris_agreement/items/9485.php

Box 1.2 | Spotlight on Climate Change

COP21

In late 2015, under the guidance of the United Nations, 195 nations pledged to stop global warming temperatures "well below" 2°C (3.6°F) and if possible, below 1.5°C (2.7°F). This landmark climate treaty was signed in Paris (**Figure 1.11**) at the Conference of Parties 21, or COP21, the 21st year of meetings designed to achieve such an agreement.

Hailed as a "turning point," a "victory for the planet," and "the end of the era of fossil fuels," the agreement is a major milestone in over two decades of effort to get the world to unify under a single plan to halt global warming.

But in truth, the decreased greenhouse gas emissions approved in Paris will fail to halt warming at the agreed target.[220] Global temperatures will sail past 2°C[221] and not likely stop until passing 3°C (5.4°F) in the second half of this century.[222] In fact, a global study[223] of gross domestic product (GDP) per capita and carbon intensity, concluded that the likely range of warming by the end of the century is 2.0 to 4.9°C (3.6 to 8.82°F) with a median temperature of 3.2°C (5.76°F) above the natural background. They calculate there is only a 5% chance that warming will be less than 2°C, and a 1% chance of it being less than 1.5°C.

Emissions that have already been discharged through 2015 have locked in 1.3 to 1.7 m (4.3 to 5.6 ft) of global sea-level rise[224] when the temperatures of the ocean, the air, and the melting ice all reach equilibrium. Deep cuts in greenhouse gas emissions of 40 to 70% by mid-century will be needed to avert the worst of global warming. These cuts will require a tripling or a quadrupling of the share of low-carbon energies including solar, wind, or nuclear power.[225]

In light of these hard truths, how should we interpret the victory in Paris? A giant shove in the right direction might be an apt assessment. Binding nearly 200 countries to a global framework is no small feat, but the world needs to continue to "up its game." Here are some critical steps that the global human community must take if COP21 is to achieve its goals.

1. Every nation must ratify its commitment. That is, the committed decrease in greenhouse gas emissions must take on legal status within each nation. At this writing, 175 of the 197 parties to the convention have ratified the agreement in their own nations.

2. Governments of the world need to end the practice of subsidizing the production of fossil fuels. These subsidies often take the form of tax relief for oil and coal companies and outright investments by nations in producing fossil fuels. Experts have estimated that over $600 billion in subsidies is provided every year to offset expenses in producing carbon fuels.

Courtesy of Benjamin Géminel

FIGURE 1.11 The Paris Treaty, signed in 2015, unified the world's nations under a single agreement to end global warming.

[220] Rogelj, J., et al. (2016) Paris Agreement climate proposals need a boost to keep warming well below 2°C. *Nature*, 534, pp. 631–639 (30 June), DOI:10.1038/nature18307

[221] UNEP (2016) The Emissions Gap Report 2016. United Nations Environment Program (UNEP), Nairobi. A digital copy of this report along with supporting appendices are available at http://uneplive.unep.org/theme/index/13#egr

[222] Gasser, T., et al. (2015) Negative emissions physically needed to keep global warming below 2°C. *Nature Communications* 6. 7958: http://www.nature.com/ncomms/2015/150803/ncomms8958/full/ncomms8958.html

[223] Raftery et al. (2017)

[224] Strauss, B.H. (2015) Rapid accumulation of committed sea level rise from global warming. *PNAS*, v. 110, no.34, p. 13699–13700.

[225] Reuters: http://www.reuters.com/article/us-climatechange-solutions-idUSKBN0G71SF20140807

3. Put a global tax on carbon pollution. By taxing carbon, we would be charging those who use carbon fuels for the damage they do to the environment and the threats to public health and safety, a justifiable cost that will decrease the use of fossil fuels and increase the production of clean energies.

4. Invest in greener technologies that can take the place of dirty carbon energy.
5. Step up global commitments to decreasing greenhouse gas production.

1.6 How Unusual is the Present Warming?

To identify the difference between the present warming and natural climate changes, it is useful to study climate in a longer geologic context. A number of studies have done this.

One[226] approach is to search the geologic record of the past several thousand years for simultaneous changes in the Northern and Southern Hemispheres ("global" warming) such as happened over the past century. Svante Björck, a climate researcher at Lund University in Sweden, used this approach and showed that simultaneous warming of the two hemispheres has not occurred in the past 20,000 years. This is as far back as it is possible to analyze with sufficient precision to compare with modern climate changes occurring at a rapid pace. His study concludes that what is happening today is unique from a historical geological perspective.

1.6.1 Unprecedented Warming

Several independent studies confirm that recent warming is unprecedented in both magnitude (the amount of warming) and speed (the rate of warming). A study[227] of North Atlantic currents flowing into the Arctic highlights the fact that the Arctic is responding more rapidly to global warming than most other areas on our planet. Researchers concluded that early 21st century temperatures of Atlantic water entering the Arctic Ocean are unprecedented over the past 2,000 years and are presumably linked to the Arctic amplification of global warming.

These facts have raised alarms among scientists, some of whom[228] have concluded that the Arctic Ocean is already suffering the effects of a dangerous climate change. Another study[229] concluded that 20th century warming of deep North Atlantic currents has had no equivalent during the last 1,000 years. Still another research effort[230] concluded that the past few decades have been characterized by a global temperature rise that is unprecedented in the context of the last 1,600 years.

Research[231] by the National Center for Atmospheric Research concluded that Arctic temperatures in the 1990s and 2000s reached their warmest level of any decade of the past 2,000 years. They found that the Arctic would be experiencing a long-term cooling trend (due to the nature of Earth's orbital configuration with the Sun) were it not for greenhouse gases that are overpowering natural climate patterns.

The aggregate conclusion of these independent studies is unmistakable. Present warming is unprecedented in recent geologic history, no natural mechanism can be identified accounting for modern climate change, and human greenhouse gas emissions have the obvious potential to be the cause of the present warming.

In 2006, the U.S. Congress asked the National Research Council (NRC) to study Earth's climate and report on the levels of warming in recent history. The NRC concluded[232] that Earth's average surface temperature today is the highest of the past 1,300 years. Their report states that Earth's surface warmed 0.6 degrees Celsius (1 degree Fahrenheit) during the 20th century and is projected to warm by an additional (approximately) 2–6 degrees Celsius (3.6 to 10.8 degrees Fahrenheit) during the 21st century. Global average temperature measurements by instruments indicate a near-level trend from 1856 to about 1910, a rise till 1945, a slight decline to about 1975, and a rise to the present. Global warming is also verified by several independent sources including the National Climatic Data Center,[233] NASA,[234] U.K. Met Office,[235] Japan Meteorological Agency,[236] the IPCC, and others.

[226] Björck (2011)
[227] Spielhagen, R., et al. (2011) Enhanced Modern Heat Transfer to the Arctic by Warm Atlantic Water, *Science* 331, no. 6016: 450–453, DOI: 10.1126/science.1197397
[228] Duarte, C., et al. (2012) Abrupt Climate Change in the Arctic, *Nature Climate Change* 2: 60–62, DOI: 10.1038/nclimate1386
[229] Thibodeau, B., et al. (2010) Twentieth-Century Warming in Deep Waters of the Gulf of St. Lawrence: A Unique Feature of the Last Millennium, *Geophysical Research Letters* 37: L17604, DOI: 10.1029/2010GL044771
[230] Kellerhals, T., et al. (2010) Ammonium Concentration in Ice Cores: A New Proxy for Regional Temperature Reconstruction? *Journal of Geophysical Research* 115: D16123, DOI: 10.1029/2009JD012603
[231] Kaufman, D.S., et al. (2009) Recent Warming Reverses Long-Term Arctic Cooling, *Science* 325: 1236–1239
[232] National Research Council, *Surface Temperature Reconstructions for the Last 2,000 Years* (Washington, D.C., National Academies Press, 2006), p. 29, http://www.nap.edu/catalog.php?record_id=11676
[233] National Climate Data Center http://www.ncdc.noaa.gov/oa/climate/globalwarming.html
[234] NASA http://data.giss.nasa.gov/gistemp/
[235] Climate Research Unit http://www.cru.uea.ac.uk/cru/info/warming/
[236] Japan Meteorological Agency http://www.jma.go.jp/jma/en/Activities/cc.html

1.7 Surface Temperature

How, when, and why heat is distributed across the planet surface, is critical to the existence of life in every region and environment on Earth. The total amount of heat, and its variation from place to place, drive global winds that circulate the atmosphere and control regional weather patterns, rainfall, growing seasons, and living conditions to which humans have adapted since civilization began.

Earth is the right distance from the Sun (about 148 million km; 92 million mi), has the right combination of gases in its atmosphere, and has water covering more than 70 percent of the planet's surface, which allow the origin and evolution

of life and the resources necessary to sustain life. As far as we know, no other planet in our solar system has the thermal, physical, and chemical conditions that allow life to exist. This is what makes our blue planet unique and habitable.

However, global warming threatens severe changes to aspects of this system, including the temperature regime under which human civilization has developed, the location and distribution of agriculture and other protein sources that sustain us, the natural ecosystems that provide important services, and the supply of water around which we have built communities. By studying climate change, we gain critical knowledge that will support efforts to adapt to, and mitigate the negative impacts of, global warming (**Figure 1.12**).

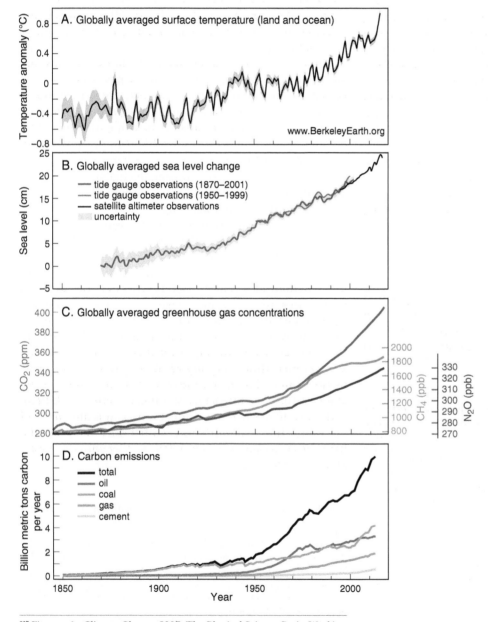

FIGURE 1.12 Observations of global climate change. (A.) Average annual Earth's surface temperature relative to the average over the period 1986–2005. (B.) Annual average global sea level. (C.) Atmospheric concentrations of the greenhouse gases carbon dioxide (CO_2, upper line) methane (CH_4, middle line) and nitrous oxide (N_2O, lower line) determined from ice core data (prior to the 1960s) and from direct atmospheric measurements. (D.) Global anthropogenic carbon emissions per year from burning of fossil fuel including gas and cement production. Cumulative annual emissions of carbon from these sources (upper line).[237]

Source: N. Hulbirt

[237] Figure *after* Climate Change 2007: The Physical Science Basis. Working Group I Contribution to the Fourth Assessment Report of the Intergovernmental Panel on Climate Change, Figure TS.6. Cambridge University Press.

1.7.1 Sensors

Several types of independent sensors have documented warming over the past decade. Weather balloon measurements have found that the global mean near-surface air temperature is warming by approximately 0.18 degrees Celsius (0.32 degrees Fahrenheit) per decade,[238] satellite measurements of the lower atmosphere show warming of 0.16 to 0.24 degrees Celsius (0.29 to 0.43 degrees Fahrenheit) per decade since 1982,[239] continental weather stations document warming of approximately 0.2 degrees Celsius (0.36 degrees Fahrenheit) per decade,[240] and ocean measurements using various types of sensors show persistent heating since 1970.[241]

The ocean has been a great ally to land dwelling ecosystems, absorbing over 90 percent of the excess heat trapped by greenhouse gases. However, the ocean's ability to store this heat is not limitless, and it has paid a price. Warming of the ocean is causing marine species to move to new, cooler locations, and natural relationships involving mating, predator–prey, interspecies contact, and seasonal migration are being affected.[242] The oceans are now taking up heat at twice the rate they were only two decades ago.[243]

Notably, consistent with theory, satellite records of warming in the layers of the atmosphere near Earth's surface are matched by simultaneous cooling in the higher layers of the atmosphere (**Figure 1.13**). This makes perfect sense given that greenhouse gases trap heat at Earth's surface and therefore limit heat flow into overlying layers.

1.7.2 Excess Heat

In a normal planetary atmosphere where the amount of heat arriving from the Sun equals the amount that radiates back out to space, the temperature would not change and global warming would not occur. We discuss this concept in **Chapter 2**. Temperature would be stable within a certain band of variability. However, with increasing greenhouse gases, the amount of heat radiating back out to space is less than the amount arriving from the Sun, and the difference is trapped near Earth's surface by carbon dioxide and other anthropogenic gases.

By dissecting the temperature record of the past 160 years, researchers[245] have been able to define the components of temperature change that are the result of volcanic eruptions, the El Niño Southern Oscillation, variations in the Sun's energy, and warming due to increasing greenhouse gases. Finding that more than 90 percent of the excess heat trapped by greenhouse gases has been absorbed by the oceans, the study concludes that since 1850 and 1950, approximately 75 and 100 percent, respectively, of the observed global warming is due to human influence.

In fact, it was determined that greenhouse gas emissions are responsible for 166 percent of the observed warming since 1950; that is, there would have been more greenhouse warming produced over the period, but it has been offset by aerosols (fine particles that reflect sunlight in the upper atmosphere, thus providing a cooling effect) produced by human manufacturing.

1.8 Human Fingerprints on Climate

The world is changing in significant ways that are most simply explained in the context of global warming. Taken together, the thousands of observations of changing ecosystems, environments, and natural processes that appear to be shifting in unusual fashion constitute a massive database pointing to the impacts of a warming world.[246] But what is the evidence that humans are the cause? Several principal observations define a "human fingerprint" on climate change that the majority of the scientific community find sufficient to support the hypothesis *Global warming is the result of industrial emissions of greenhouse gas.*[247]

[238] Angell, J.K. (2009) Global, Hemispheric, and Zonal Temperature Deviations Derived from Radiosonde Records, In *Trends Online: A Compendium of Data on Global Change.* (Oak Ridge, Tenn., Carbon Dioxide Information Analysis Center, Oak Ridge National Laboratory, U.S. Department of Energy), DOI: 10.3334/CDIAC/cli.005; http://cdiac.esd.ornl.gov/trends/temp/angell/angell.html

[239] Vinnikov, K.Y. and Grody, N.C. (2003) Global Warming Trend of Mean Tropospheric Temperature Observed by Satellites, *Science* 302, no. 5643: 269–272, DOI: 10.1126/science.1087910

[240] Hansen, J., et al. (2006) Global Temperature Change, *Proceedings of the National Academy of Sciences* 103: 14288–14293, DOI: 10.1073/pnas.0606291103, http://pubs.giss.nasa.gov/abstracts/2006/Hansen_etal_1.html

[241] Wang, G., et al. (2017) Consensuses and discrepancies of basin-scale ocean heat content changes in different ocean analyses, *Climate Dynamics.* DOI: 10.1007/s00382-017-3751-5. See also Levitus, S., et al. (2009)

[242] Doney, S.C., et al. (2012) Climate change impacts on marine ecosystems, *Annual Review of Marine Science*, v.4, 11-37, DOI: 10.1146/annurev-marine-041911-111611

[243] Glecker et al. (2016) See also Cheng et al. (2015)

[245] Huber, M. and R. Knutti (2011) Anthropogenic and Natural Warming Inferred from Changes in Earth's Energy Balance, *Nature Geoscience* 5: 31–36, DOI: 10.1038/ngeo1327

[246] Thorne, P.W., et al. (2010) Tropospheric temperature trends: history of an ongoing controversy. *Wiley Interdisciplinary Reviews: Climate Change*, 2010; DOI: 10.1002/wcc.80

[247] Skeptical Science "10 Indicators of a Human Fingerprint on Climate Change" at http://www.skepticalscience.com/10-Indicators-of-a-Human-Fingerprint-on-Climate-Change.html

Troposphere

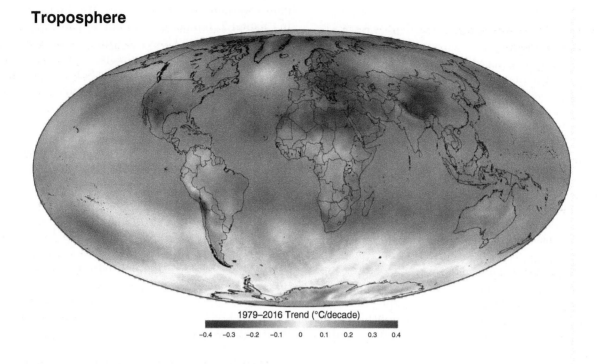

1979–2016 Trend (°C/decade)

−0.4 −0.3 −0.2 −0.1 0 0.1 0.2 0.3 0.4

Lower Stratosphere

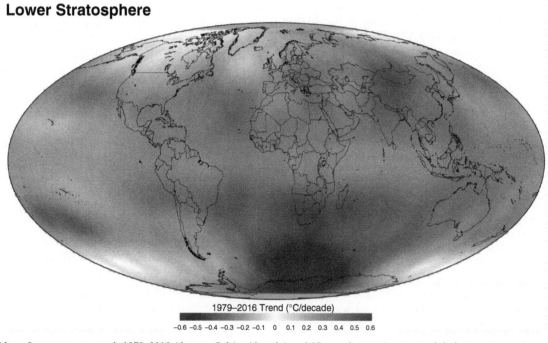

1979–2016 Trend (°C/decade)

−0.6 −0.5 −0.4 −0.3 −0.2 −0.1 0 0.1 0.2 0.3 0.4 0.5 0.6

FIGURE 1.13 Map of temperature trends 1979–2016 (degrees Celsius/decade) and 12 months running mean global temperature time series with respect to 1979–2016, in the lower troposphere (near surface) where temperature has risen in recent decades, and in the lower stratosphere (above the troposphere) where temperature has fallen because of heat trapping in the troposphere.[244]

Source: Wikipedia, https://en.wikipedia.org/wiki/File:RSS_troposphere_stratosphere_trend.png

[244] Mears, C.A., et al. (2011) Assessing Uncertainty in Estimates of Atmospheric Temperature Changes from MSU and AMSU using a Monte-Carlo Estimation Technique, *Journal of Geophysical Research* 116: D08112, DOI: 10.1029/2010JD014954

Fingerprint #1: There is more industrial carbon in the atmosphere

One clear human fingerprint is found in the type of carbon (C) released into the air by industrial emissions. Carbon, and other elements, occurs naturally in forms known as isotopes.[248] In the case of carbon, there are three forms: ^{12}C, ^{13}C, and ^{14}C. While engaging in photosynthesis, plants prefer to absorb CO_2 wherein the carbon is composed of the lighter form, ^{12}C. Petroleum is composed largely of fossilized marine algae, and coal is composed of fossilized terrestrial wetland plants; thus, both energy sources contain high abundances of light carbon (^{12}C) because they come from fossil plants. This can be measured using the ratio of the two types of carbon: $^{13}C/^{12}C$.

In plants, $^{13}C/^{12}C$ is relatively low (i.e., the amount of ^{12}C is high). Burning oil and coal releases this light carbon, which immediately combines with oxygen in the atmosphere to form $^{12}CO_2$; thus, we should be able to detect a decrease in $^{13}C/^{12}C$ in the atmosphere as more fossil fuel emissions accumulate.

Indeed, this is exactly what is found. Measurements of the ratio of $^{13}C/^{12}C$ in the air,[249] and in corals,[250] and in sponges[251] that take up atmospheric carbon (mixed in seawater) reveal a strong decrease in $^{13}C/^{12}C$ over the past 200 years, with a significant acceleration in the decrease since about 1960–1970. Thus, the growth of carbon dioxide in the atmosphere is wholly attributable to combustion of coal and petroleum; humans are raising the CO_2 level.

Another aspect of burning carbon is that when unburned carbon enters the air, it uses two oxygen atoms to form the molecule CO_2—it thus draws down the level of oxygen in the atmosphere. Measurements of oxygen not only show this decline, but the rate of decline is consistent with the rate of CO_2 increase.[252]

Fingerprint #2: Less heat is escaping to space

Direct evidence that more carbon dioxide causes warming is found in the fact that less heat is escaping into space. As discussed earlier, carbon dioxide traps IR radiation (proved by modern laboratory experiments[253] and as early as 1859 in laboratory measurements by John Tyndall[254]) from the planet surface that would otherwise escape to space. A decrease in the IR energy emitted by Earth from 1979 to 2016 has been detected by satellites[255] and has since been verified by additional measurements.[256] Because this heat is trapped in the lowest atmospheric layer, the troposphere, it is warming the air and Earth's surface.

Fingerprint #3: Oceans are warming from the top down

The oceans are warming[257] in the only manner possible under an enhanced greenhouse: from the top down. Measurements of ocean warming show that the water temperature has a depth profile that varies widely by ocean; natural internal climate variability, increased heat from Earth's mantle, or solar and volcanic forcing cannot explain it. The pattern of warming is complex, but it has been captured by sensors that depict the upper layer of the oceans (varying from 75 to 500 m [246 to 1,640 ft] depth) warming in a way that is consistent with models simulating human production of greenhouse gases.

Fingerprint #4: Nights are warming faster than days

During the day, sunlight heats the air. At night, the air cools by radiating heat out to space. Greenhouse gases trap part of this heat. Thus, in a situation where the Sun has not increased its output, but the greenhouse effect is increasing, nights will become warmer faster than days. That is, if global warming were caused by the Sun, we would expect to see that the days would warm faster than the nights. Observations[258] clearly show that nights are warming faster than days. Thus, the detailed pattern of global warming is consistent with an amplified greenhouse effect (resulting from growing emissions of heat-trapping gases).

Fingerprint #5: More heat is returning to Earth

Radiation works both ways. IR radiation can be measured moving upward from a warm Earth surface as well as moving

(Continued)

---

[248] Atoms of an element all have the same atomic number (the number of protons), but they may have different numbers of neutrons. The number of neutrons plus the number of protons is known as an atom's *mass number*. Atoms of the same element with differing mass numbers are known as *isotopes*. The mass number (identifying the isotope) is written as a superscript on the left side of an element's symbol.

[249] Manning, A., and Keeling, R. (2006) Global Oceanic and Land Biotic Carbon Sinks from the Scripps Atmospheric Oxygen Flask Sampling Network, *Tellus* 58: 95–116.

[250] Wei, G., et al. (2009) Evidence for Ocean Acidification in the Great Barrier Reef of Australia, *Geochemica, Cosmochemica Acta* 73: 2332–2346.

[251] Swart, P., et al. (2010) The ^{13}C Suess Effect in Scleractinian Corals Mirror Changes in the Anthropogenic CO_2 Inventory of the Surface Oceans, *Geophysical Research Letters* 37: L05604, DOI: 10.1029/2009GL041397

[252] Keeling, R.F., et al., Abstract: https://www.esrl.noaa.gov/gmd/icdc7/proceedings/abstracts/keeling.rFF328Oral.pdf

[253] Burch, D. (1970) Investigation of the Absorption of Infrared Radiation by Atmospheric Gases, *Semi-Annual Technical Report*, AFCRL, publication U-4784.

[254] NASA https://earthobservatory.nasa.gov/Features/Tyndall/

[255] J. Harries, et al (2001) Increases in Greenhouse Forcing Inferred from the Outgoing Longwave Radiation Spectra of the Earth in 1970 and 1997, *Nature*, 410: 355–357; DOI: 10.1038/35066553

[256] J. Griggs and J. Harries (2004) Comparison of Spectrally Resolved Outgoing Longwave Data between 1970 and Present, *Proceedings of SPIE*, 5543, 164. See also C. Chen, et al. (2007) Spectral Signatures of Climate Change in the Earth's Infrared Spectrum between 1970 and 2006. *European Organization for the Exploitation of Meteorological Satellites* (EUMETSAT), http://www.eumetsat.eu/Home/Main/Publications/Conference_and_Workshop_Proceedings/groups/cps/documents/document/pdf_conf_p50_s9_01_harries_v.pdf. Talk given to the 15th American Meteorological Society (AMS) Satellite Meteorology and Oceanography Conference, Amsterdam, September.

[257] T. Barnett, et al. (2005) Penetration of Human-Induced Warming into the World's Oceans, *Science* 309, no. 5732: 284–287.

[258] L. Alexander, et al. (2006) Global Observed Changes in Daily Climate Extremes of Temperature and Precipitation, *Journal of Geophysical Research* 111, DOI: 10.1029/2005JD006290

downward from a warm atmosphere. With an enhanced greenhouse effect, where the molecules of CO_2, CH_4, CFCs, and other greenhouse gases are reradiating heat (IR) in all directions, one would expect to observe an increase in downward IR radiation from the troposphere to the ground. As expected, this has been directly observed.[259] In fact, researchers state, "This experimental data should effectively end the argument by skeptics that no experimental evidence exists for the connection between greenhouse gas increases in the atmosphere and global warming." Thus, because of an amplified greenhouse effect due to industrial emissions, more heat is returning to Earth.

Fingerprint #6: Winter is warming faster than summer

If the Sun were causing global warming, you would continue to see a seasonal effect in the warming pattern. But that is not what we see. Data show that winter is warming faster than summer. That is, temperature is becoming more uniform throughout the year. One way to think about this is to realize that in an atmosphere that is uniformly warming under an amplified greenhouse effect, a cool winter would be more out of equilibrium with the rising temperature than summer. Thus, one would expect winter to warm faster than summer. This is exactly what has been observed.[260]

Fingerprint #7: The stratosphere is cooling

A corollary to fingerprint #2 (less heat is escaping to space) is that less heat is finding its way to the stratosphere. Because the amplified greenhouse effect is located in the troposphere near Earth's surface, industrial greenhouse gases below are trapping heat that would otherwise find its way to the atmospheric layers above. As a result, satellites and weather balloons are recording[261] cooling temperatures in the stratosphere simultaneous with warming in the troposphere (see Figure 1.13).

Fingerprint #8: Physical models require human greenhouse gas emissions

As we will see in **Chapter 6**, the fundamental laws of nature that explain the movement of heat, the behavior of molecules, and the physics of natural processes, can be programmed to build computer models of climate. These are basically larger and more complex versions of computer models that are used to predict the weather on the TV news every day. When these models attempt[262] to simulate the past century of warming using only natural factors (e.g., variations in sunlight, volcanic eruptions, and other natural climate processes), they instead predict *global cooling*. But when human emissions of greenhouse gases are introduced to the models along with the natural factors, they faithfully reproduce the observed temperature record: global warming. In fact, researchers[263] have found that computer models are growing in sophistication and accuracy to the point that they are approaching direct observation of the planet as a reliable source of information.

Fingerprint #9: Multiple and independent lines of evidence

Attributing global warming to industrial emissions is not only the consensus among the scientific community, it is the common explanation for multiple and independent lines of evidence. Direct observations show the following.

1. There is more industrial carbon in the atmosphere, and it is amplifying the greenhouse effect.
2. Less heat is escaping to space.
3. Oceans are warming from the top down.
4. Nights are warming faster than days.
5. More heat is returning to Earth's surface from the atmosphere.
6. Winter is warming faster than summer.
7. The stratosphere is cooling.
8. Physical laws of nature predict global warming consistent with observations.

Most of the scientific community finds the authenticity and agreement of these observations amply sufficient to support the hypothesis that *Global warming is the result of industrial emissions of greenhouse gases and other human activities.*

[259] Evans, W. and Puckrin, E. (2006) Measurements of the Radiative Surface Forcing of Climate, P1.7, *AMS 18th Conference on Climate Variability and Change.* See also Wang, K. and Liang, S. (2009) Global Atmospheric Downward Longwave radiation over Land Surface Under All-Sky Conditions from 1973 to 2008. *Journal of Geophysical Research* 114, D19.

[260] Braganza, K., et al. (2003) Indices of Global Climate Variability and Change: Part I. Variability and Correlation Structure, *Climate Dynamics* 20: 491–502. See also Braganza, K., et al. (2004) Simple Indices of Global Climate Variability and Change: Part II: Attribution of Climate Change during the Twentieth Century," *Climate Dynamics* 22: 823–838, DOI: 10.007/s00382-004-0413-1

[261] Jones, G., et al. (2003) Causes of Atmospheric Temperature Change 1960–2000: A Combined Attribution Analysis, *Geophysical Research Letters* 30: 1228. See also Mears, C. and Wentz, F. (2009) Construction of the remote sensing systems, v. 3.2 Atmospheric Temperature Records from the MSU and AMSU Microwave Sounders, *Journal of Atmospheric and Oceanic Technology* 26: 1040–1056.

[262] Solomon, S., et al. (eds.), *Contribution of Working Group I to the Fourth Assessment Report of the Intergovernmental Panel on Climate Change* (Cambridge, U.K., Cambridge University Press, 2007).

[263] Reichler, T. and Junsu, K. (2008) How Well Do Coupled Models Simulate Today's Climate? *Bulletin of the American Meteorological Society* 89: 303–311, DOI: https://dx.doi.org/10.1175/BAMS-89-3-303.

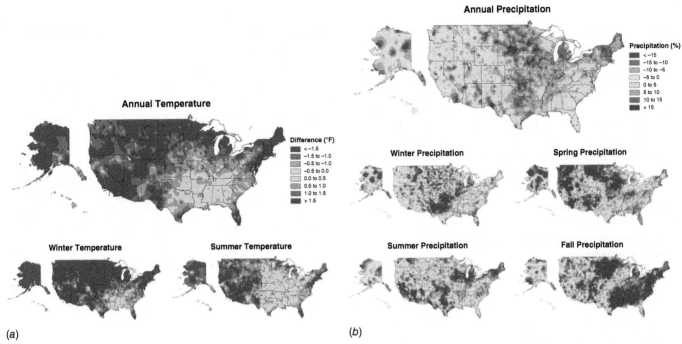

FIGURE 1.14 Temperature and precipitation change in the Unites States. (a) Observed[264] changes in annual, winter, and summer temperature (°F). Changes are the difference between the average for present day (1986-2016) and the average for the first half of the 20th century (1901–1960; 1925–1960 for Alaska and Hawaii). Annual average temperature over the contiguous United States has increased by 0.7°C (1.2°F). Surface and satellite data are consistent in their depiction of rapid warming since 1979. Paleo-temperature evidence shows that recent decades are the warmest of the past 1,500 years. (b) Annual and seasonal changes in precipitation over the United States.[265] Changes are the average for present-day (1986–2015) minus the average for the first half of the last century (1901–1960; 1925–1960 for Alaska and Hawaii). Annual precipitation has decreased in much of the West, Southwest, and Southeast and increased in most of the Northern and Southern Plains, Midwest, and Northeast. A national average increase of 4% in annual precipitation since 1901 mostly a result of large increases in the fall season.

Source: https://science2017.globalchange.gov

1.9 A Consistent Picture Emerges

In a warmer world caused by greenhouse gas emissions, one would expect to observe certain changes including melting glaciers, warming and acidifying oceans, sea-level rise, changes in ecosystems, plant and animal extinctions, new patterns in the weather, and long-term changes in temperature (**Figure 1.14a**) and precipitation (**Figure 1.14b**). In fact, these have all been observed.

For example, Greenland ice and Antarctic ice are melting at an accelerating rate.[266] Sensors on satellites have measured this over a sufficient period (nearly two decades) that we not only know melting is a persistent annual trend but also know that the rate of melting is *accelerating* from one year to the next. In fact, the melting has become so persistent that researchers now conclude that the melting of the West Antarctic Ice Sheet has become *irreversible*.[267]

The ocean is getting warmer and it is acidifying[268] as it mixes with an atmosphere that is enriched with excess carbon dioxide. The concentration of dissolved CO_2 in the ocean grows with each year. Not surprisingly, both warming and acidification are occurring at rates that are predicted by long-established chemical and physical theory.

Sea level is rising, and the rate of rise has accelerated.[269] This is an anticipated consequence of a warming world.[270]

[264] Vose, R.S., et al. (2017) Temperature changes in the United States. In: Climate Science Special Report: Fourth National Climate Assessment, Volume I [Wuebbles, D.J., et al. (eds.)]. U.S. Global Change Research Program, Washington, DC, USA, pp. 185–206, DOI: 10.7930/J0N29V45.

[265] Easterling, D.R., et al. (2017) Precipitation change in the United States. In: Climate Science Special Report: Fourth National Climate Assessment, Volume I [Wuebbles, D.J., et al. (eds.)]. U.S. Global Change Research Program, Washington, DC, USA, pp. 207–230, DOI: 10.7930/J0H993CC.

[266] News article http://www.colorado.edu/news/r/f595fae00e6b451d4016 ab9a43a049f8.html. See also Velicogna, I. (2009) Increasing Rates of Ice Mass Loss from the Greenland and Antarctic Ice Sheets Revealed by GRACE, *Geophysical Research Letters* 36: L19503, DOI: 10.1029/2009GL040222. See also Rignot et al. (2011) See also Hofer, S., et al. (2017) Decreasing cloud cover drives the recent mass loss on the Greenland ice sheet. *Science Advances*, 28 June, v. 3, no. 6, e1700584, DOI: 10.1126/sciadv.1700584

[267] Rignot et al. (2014)

[268] Pelejero, C., et al. (2010) Palaeo-Perspectives on Ocean Acidification, *Trends in Ecology and Evolution* 25, no. 6: 332–344, DOI: 10.1016/j .tree.2010.02.002. See also Murphy, D., et al. (2009) An Observationally Based Energy Balance for the Earth since 1950, *Journal of Geophysical Research* 114: D17107, DOI: 10.1029/2009JD012105.

[269] Nerem et al. (2018)

[270] Dangendorf et al. (2017)

In a warming world, it would be expected that the southern line of permanently frozen ground (permafrost) would begin to migrate to the north as warmer climate zones expand in the northern hemisphere. Indeed, this has already been observed in Canada.[271] Simultaneously, the boundaries of the tropics, defined by temperature, rainfall, wind, and ozone patterns, have shifted poleward by at least 2° latitude in the past 25 years.[272]

Excess heating of the tropics has sped up the rate of evaporation and atmospheric circulation so that surface winds have accelerated nearly around the entire planet.[273] These and many other phenomena stand in testimony to the reality of warming and, taken as a whole, are consistently in keeping with expectations of how a warmer atmosphere would change the world.

Global warming is also changing the weather. In the past decade, the United States experienced twice as many record daily high temperatures than record lows; that is, the hotter days are getting hotter and the colder days are getting hotter.[274] Throughout the spring, summer, and fall of every record-setting hot year, thousands of daily high temperature records are set across the United States, outnumbering the daily record lows by as much as 116 to 1[275] (February, 2017), which exceeds the 2 to 1 average of the previous decade.[276]

In the summer of 2012, the United States experienced a prolonged heat wave that severely impacted agricultural production that year. In fact, extreme weather events are expected to grow in frequency and magnitude as the world continues to warm.[277] Ecosystem changes are occurring as well. Mild winters in British Columbia allow an infestation of the boring mountain pine beetle,[278] and warming oceans have led to coral bleaching, a problem in some hot months that has reached epidemic proportions.[279] There are many other examples of observed changes around the world that are consistent with expected impacts of warming, and these are highlighted throughout this textbook.

---

[271] Thibault and Payette (2009)

[272] Seidel et al. (2008)

[273] Li, G. and Ren, B. (2012) Evidence for Strengthening of the Tropical Pacific Ocean Surface Wind Speed during 1979–2011, *Theoretical and Applied Climatology* 107, no. 1–2: 59–72, DOI:10.1007/s00704-011-0463-3. See also Young et al. (2011)

[274] Meehl, G.A., et al. (2009) The Relative Increase of Record High Maximum Temperatures Compared to Record Low Minimum Temperatures in the U.S., *Geophysical Research Letters* 36: L23701, DOI: 10.1029/2009GL040736.

[275] Climate Central.org: http://www.climatecentral.org/news/record-high-temperature-february-21186

[276] CapitalClimate: http://capitalclimate.blogspot.com/2010/10/endless-summer-xii-septembers.html

[277] Medvigy and Beulieu (2011)

[278] Kurz, W.A., et al. (2008) Mountain Pine Beetle and Forest Carbon Feedback to Climate Change, *Nature* 452: 987–990 DOI: 10.1038/nature06777.

[279] Mao-Jones, J., et al. (2010) How Microbial Community Composition Regulates Coral Disease Development, *PLoS Biology* 8, no. 3: e1000345, DOI: 10.1371/journal.pbio.1000345. See also Graham, N.A.J., et al. (2006) Dynamic Fragility of Oceanic Coral Reef Ecosystems, *Proceedings of the National Academy of Science* 103: 8425–8429.

1.9.1 Optimism about Climate Change?

There is hope that the mounting evidence of climate change, and that humans are the cause, may be having an impact on community attitudes across the globe.[280] The annual rate of global carbon dioxide emissions slowed[281] in 2014, 2015, and 2016. But in 2017 it surged upwards by 1.5 percent on increased emissions from developing nations (**Figure 1.15**).[282]

In the decade of the 2000s, global carbon dioxide emissions grew at an average annual rate of 3.5 percent greater than the previous year. Over the decade 2006–2015, the rate of increase slowed to 1.8 percent and slowed further

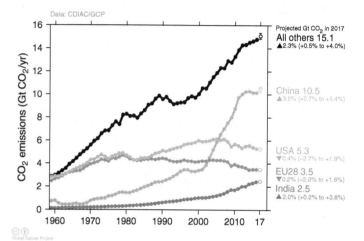

FIGURE 1.15 Global emissions from fossil fuels and industry rose by 1.5% in 2017.[283] Every year, the Global Carbon Project[284] publishes data on global carbon emissions. The United States is the world's second largest emitter; in 2017, its emissions fell by 0.4%, a smaller annual reduction than in recent years. In China, the world's largest emitter, emissions grew by over 3% after falling for 2 years. In India, the world's fourth largest emitter, emissions have grown about 6% per year over the past decade, and grew only 2% in 2017; but this reduction is not expected to persist. Europe, the third largest emitter, has had falling emissions over the past decade, but the rate of decrease has slowed. Over the past decade, emissions in a total of 22 countries have fallen while their economies have grown.[285]

Source: Global Carbon Project http://folk.uio.no/roberan/GCP2017.shtml

---

[280] VIDEO: The Case for Optimism on Climate Change, https://www.ted.com/talks/al_gore_the_case_for_optimism_on_climate_change

[281] CarbonBrief://www.carbonbrief.org/what-global-CO₂-emissions-2016-mean-climate-change. See also Peters, G., et al. (2012) Rapid Growth in CO₂ Emissions after the 2008–2009 Global Financial Crisis, *Nature Climate Change* 2: 2–4.

[282] Climate Central: http://www.climatecentral.org/news/americas-climate-pollution-falling-epa-21165

[283] International Energy Agency, Global Energy and CO₂ Status Report, 2017: https://www.iea.org/geco/

[284] Global Carbon Project: http://www.globalcarbonproject.org/index.htm

[285] VIDEO: Future Earth Global Carbon Budget, https://www.youtube.com/watch?v=KWKFxbA-T94

to essentially zero growth over the period 2014–2016. Remarkably, this stabilization in the rate of carbon dioxide growth occurred at the same time that the world economy grew at more than 3 percent per year—decoupling greenhouse gas emissions from economic growth for the first time.[286]

Slowing, and eventually stopping, the production of greenhouse gases from industrial activities requires the directed energies of all the world's economies. The recent slowdown in global missions suggests that efforts to move away from coal and oil, toward renewable energy, are beginning to have an effect (despite the uptick in 2017). To achieve falling greenhouse gas emissions, global businesses, nations, families, and individuals must be attentive to their carbon footprint; a measure of how much one's daily activities are responsible for raising greenhouse gases.

A number of short-lived greenhouse gases with atmospheric life times of under a few decades contribute significantly to the radiative forcing that drives climate change. Some experts argue that aggressively reducing the release of these "short-lived gases" should be an essential part of any climate mitigation strategy. The prime targets for this mitigation include methane, hydrofluorocarbons, black carbon, and ozone. However, one report[287] argues that the benefits of early short-lived gas mitigation have been greatly exaggerated. The reason is that CO_2, because of its long atmospheric residence time, causes nearly irreversible climate change persisting millennia after emissions cease.

For instance, for a sudden pulse of carbon dioxide in the atmosphere that raises the concentration to 1,250 parts per million over preindustrial levels (which were about 280 parts per million), about 900 parts per million of carbon dioxide will still be up there after 100 years. Indeed, concentrations will only decline to 675 parts per million over another 900 years.

Eventual mitigation of short-lived gases can make a useful contribution to halting global warming before it reaches catastrophic levels, but there is little to be gained by expending time and money on short-lived gases before stringent carbon dioxide controls are in place and have caused annual emissions to approach zero. Any earlier implementation of short-lived gas mitigation that replaces efforts to reduce CO_2 emissions will lead to a climate irreversibly warmer than would a strategy with an all-out focus on carbon dioxide. Short-lived gas mitigation does not buy time for implementing stringent controls on CO_2 emissions.

Thus, much of the carbon pollution emitted over the last few centuries is very much still with us. It's still determining our future even today, which means that the global warming that the world is currently experiencing will continue and accelerate as long as we persist in using fossil fuels. . . . and well beyond that.

See the box *Spotlight on Climate Change*, "Community Resilience."

Box 1.3 | Spotlight on Climate Change

Community Sustainability and Resilience

Over the past five decades, the average temperature of the atmosphere has increased at the fastest rate in recorded history. Under this trend, the average temperature could be 4.0 to 6.0°C (7.2 to 10.8°F) higher by the end of the century. In those conditions, cities will be exposed to heat waves, extreme weather, crippling summer temperatures, water shortages, drought, high-energy demand for air-conditioning, and food shortages.

Global warming is making life more dangerous. To adjust to this new reality, cities, towns, and suburbs can take steps[288] to increase their **sustainability** and **resilience** in the face of climate change (**Figure 1.16**).

Sustainable and resilient communities are characterized by:[289]

- **Local food production**—Local and regional food production helps to eliminate dependence on greenhouse gas intensive global industrial food systems.
- **Green LEED**[290] **building design**—LEED-certified buildings use less water and energy and produce less greenhouse gas emissions. Companies that specialize in LEED building and other sustainability initiatives can ensure energy efficiency and healthier buildings.
- **Renewable energy production**—Replacing electricity generated by fossil fuels with renewable

---

[286] Obama, B.H. (2017) The irreversible momentum of clean energy. *Science* 09 (Jan) aam6284 DOI: 10.1126/science.aam6284

[287] Pierrehumbert, R.T. (2014) Short-lived Pollution, *Annual Review of Earth and Planetary Sciences* 2014 42:1, 341–379

[288] Kaid Benfield, "Think Progress," http://thinkprogress.org/romm/2012/04/03/450059/nine-low-tech-steps-for-community-resilience-in-a-warming-climate/

[289] "Let's Go Solar": https://www.letsgosolar.com/consumer-education/sustainable-cities/

[290] LEED, or Leadership in Energy and Environmental Design, is the most widely used green building rating system in the world. Available for virtually all building, community and home project types, LEED provides a framework to create healthy, highly efficient and cost-saving green buildings. LEED certification is a globally recognized symbol of sustainability achievement.

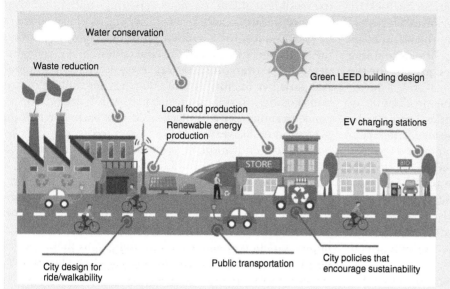

FIGURE 1.16 A sustainable community meets the needs of the present while ensuring sustainable resources and conditions for future generations and improves the quality of life and health in both the ecological and social environment.

Source: https://www.letsgosolar.com/consumer-education/sustainable-cities/

energy made by solar, wind, hydro, geothermal, and other sources of clean energy help curb greenhouse gas emissions.

Public transportation—Reducing the use of single-occupant vehicles as people make their daily commutes means less carbon emissions.

City design for ride/walkability—Cities that are designed to promote walking and bicycling not only produce less emissions but also encourage healthy living.

EV charging stations—Cities wanting to see more electric vehicles replacing traditional gasoline-powered cars need to develop the infrastructure to support those changes.

Water conservation—Water conservation helps residents save money, prevents water pollution in local lakes, rivers, and watersheds, and helps ensure availability of water for future generations.

Waste reduction—Prevention and reduction efforts help to prevent waste from ending up in the landfill where it contributes to climate change, leads to pollution, and uses natural resources and energy to manage.

City policies that encourage sustainability—(ban plastic grocery bags, provide space for community gardens, etc.). From purchasing procedures, to city planning and permitting, to energy efficiency and recycling programs, city policies can significantly impact the way local governments and their residents act when it comes to environmental issues.

Bring more vegetation into neighborhoods[291]—Green roofs, roadside plantings, vegetated swales, rain gardens, and other features improve storm water management, lower the temperature, and absorb carbon dioxide from the air.

Plant community gardens[292]—Community gardens such as urban orchards and vegetable patches help lower temperature, and growing food in neighborhoods reduces the number of driving errands in a community.

Drought-resistant landscaping—Landscaping that does not require frequent watering is a way to save water, because water shortages are likely to become more frequent with warmer temperatures.

Light-colored pavement, roofing, and other surfaces—Dark colors absorb heat, but light surfaces reflect sunlight and lower the planet's temperature. For instance, on the hottest day of the New York City summer in 2011, a white roof was found to be 23°C (42°F) cooler than a traditional black roof. This lowers electricity demand for air-conditioning, which in turn reduces carbon emissions from power plants.

Stop building on coastlines[293]—Sea-level rise is real. It is already accelerating and storms, tsunamis, high waves, and high winds cause more damage

---

[291] "Green Infrastructure," http://switchboard.nrdc.org/blogs/kbenfield/how_green_infrastructure_for_w.html

[292] "City Gardens that Respect the Urban Fabric," http://switchboard.nrdc.org/blogs/kbenfield/city_gardens_that_respect_the.html

[293] "Are You in the Zone? New Tool Helps Communities Prepare for Surging Seas," http://switchboard.nrdc.org/blogs/dlashof/are_you_in_the_zone_new_tool_h.html

when the ocean is higher. What used to be the storm of the century has now become the storm of the decade. Communities can adapt to sea-level rise, but planning needs to begin in advance of the problem.

Save older buildings—New construction generates heat, requires large volumes of water, disrupts vegetation, and adds to the carbon dioxide in the atmosphere.

Follow new "Original Green Building Practices" [294]—Especially in a warmer climate, it is important that buildings be constructed and sited to make best use of natural advantages. These practices include building front porches and planting deciduous trees on the south side where they provide shade in summer and allow sun in the winter. Plant evergreens on the side that will benefit by protection from winter winds.

Use close-to-the-source materials and a naturally insulating design. Place new buildings in walkable settings with everyday conveniences nearby.

Keep the community footprint small and well connected—One characteristic of urban sprawl is that it is vehicle dependent, and transportation is a major source of carbon dioxide. Walkable and bikeable destinations, effective mass transit, and small-scale commuting and errands promote a low carbon footprint among communities.

Update zoning and building codes—Updated codes promote resilience and a low carbon footprint.

As you read through the following chapters, keep in mind the steps that you can take to lower your contribution to global warming and to increase your own safety in a warming world.

Animations and Videos

1. Climate Change 101 With Bill Nye: National Geographic video. http://video.nationalgeographic.com/video/news/101-videos/151201-climate-change-bill-nye-news

2. Global Warming: What we knew in 82. https://www.youtube.com/watch?v=OmpiuuBy-4s

3. Global Warming's Six America's. http://climatecommunication.yale.edu/news-events/global-warmings-six-americas/

4. James Hansen in 2016 discussing growing concerns with global ice sheets. https://www.youtube.com/watch?v=Ykn8_ayFqNI&t=25s

5. Recent Acceleration in Deforestation. https://www.nytimes.com/2017/02/24/business/energy-environment/deforestation-brazil-bolivia-south-america.html?_r=1

6. The Ocean, A Driving Force for Weather and Heat. https://www.youtube.com/watch?v=6vgvTeuoDWY

7. Plate Tectonics. https://www.youtube.com/watch?v=Xzpk9110Lyw

8. Saving Corals. http://films.economist.com/blancpain-ocean?utm_source=Saving%20Coral&utm_medium=Editors%20Picks&utm_content=Blancpain%20Ocean

9. NASA, Climate Change Visualization 1880–2016. https://climate.nasa.gov/climate_resources/139/

10. The Case for Optimism on Climate Change. https://www.ted.com/talks/al_gore_the_case_for_optimism_on_climate_change

11. Future Earth Global Carbon Budget. https://www.youtube.com/watch?v=KWKFxbA-T94

Comprehension Questions

1. What is the relationship between global warming and climate change?

2. Describe some of the scientific evidence that increased surface temperature is having measurable impacts on human communities and natural ecosystems.

3. Describe the principle features of the graph showing global temperature anomalies (Figure 1.5).

4. What evidence supports the conclusion that humans are the primary cause of global warming?

5. Describe the primary human activities causing the problem of global warming and climate change.

6. If global warming is real, why is the stratosphere cooling?

7. Temperature records show that climate varies strongly from one year to the next. What does this mean in terms of interpreting the data for the presence or absence of global warming?

[294] Natural Resources Defense Council on Original Green. http://switchboard.nrdc.org/blogs/kbenfield/they_dont_makeem_like_they_use.html

8. What is the IPCC?

9. What can we expect the Earth to be like in the future if the climate crisis is not addressed?

10. Describe the findings of U.S. government agencies that study climate change.

Thinking Critically

1. Which aspect of climate change worries you the most? Why?

2. Suppose that a scientist reported that the climate has been cooling for several decades in one county in the central United States. What questions would you ask before accepting this information? And once you accept these data as true, what impact would they have on your understanding of global warming?

3. As mayor of a small town in Florida, what steps are you considering with regard to the problem of climate change?

4. Solar output over the period 2008–2010 was low, and scientists are predicting that this trend will continue for another decade or so before Sun's heat recovers to normal levels.

Describe the impact that low solar output could have on global warming both over the next decade and after.

5. As a homeowner planning on staying in your new home for at least 30 years, what proactive steps will you consider to adapt your house to climate change?

6. What steps would you like to see the U.S. President and Congress take with regard to climate change?

7. What effects could heat waves have on a large city?

8. What is the average rate of global warming?

9. Describe a study designed to test the theory that humans are causing global warming.

10. Why is the weather very likely to change as climate changes?

Activities

1. Explore the following websites. What evidence is provided by each of these that climate is changing and humans are the most important cause?

   (a) NOAA Centers for Environmental Information: https://www.ncdc.noaa.gov

   (b) NASA Vital Signs: https://climate.nasa.gov

   (c) U.S. Global Change Research Program: http://www.globalchange.gov

2. NASA uses satellites to gather information about the increase in Earth's temperature. Which climate aspects that satellites use for this purpose are mentioned in the NASA video "The Temperature Puzzle," https://www.youtube.com/watch?v=DjILZWW6Ko0?

3. Watch Al Gore, "The Case for Optimism on Climate Change," https://www.youtube.com/watch?v=u7E1v24Dllk.

   (a) List the main points and explain why these are reasons to be optimistic about the climate problem.

Key Terms

Acidic

Adaptation

Alpine glaciers

Anoxic

Anthropogenic climate change

Arctic amplification

Arrhenius, Svante August (1859–1927)

Callendar, Guy Stewart (1898–1964)

Carbon dioxide

Carbon footprint

Climate change

Climate variability

Concentrated animal farming operations

Continental ice sheets

Core

Crust

Deforested

El Niño

El Niño Southern Oscillation

Fluorinated gases

Fourier, Jean-Baptiste Joseph (1768–1830)

Glacial landforms

Global overshoot

Global warming

Greenhouse effect

Greenhouse gases

Holocene Epoch

Ice ages

Interglacials

Interglacial cycles

Intergovernmental Panel on Climate Change

Keeling Curve

Keeling, Charles David (1928–2005)

Local extinction

Mantle

Methane

Miocene Epoch

Mitigation

Nitrous oxide

Paleocene-Eocene Thermal Maximum

Paris agreement

Plate tectonics

Pouillet, Claude Servais Mathias (1790–1868)

Resilience

Sunspot cycle

Sustainability

Tyndall, John (1820–1893)

United Nations Framework Convention on Climate Change

Volcanism

CHAPTER 2

Radiative Equilibrium

NASA

A cross section of the atmosphere and the setting Sun taken by crew-members onboard the International Space Station. Ultimately, energy from the Sun, and how it heats Earth's atmosphere, land, and ocean, determines the climate.

LEARNING OBJECTIVE

The Sun is the principal source of energy creating Earth's climate. Solar energy drives photosynthesis, **evaporation** and rainfall, circulation in the ocean and the atmosphere, and weather patterns. If the amount of energy radiated by Earth into space does not equal the amount of energy arriving from the Sun, the planet will heat up or cool down accordingly. Once it enters Earth's climate system, solar energy is reflected, absorbed, and reradiated in complex ways. The greenhouse effect, wherein atmospheric gases such as carbon dioxide and methane trap infrared radiation from the surface, makes the troposphere warmer than it would be otherwise.

2.1 Chapter Summary

When the amount of sunlight that warms Earth equals the amount of energy that radiates away from Earth, the climate is stable.

In this chapter you will learn that:

- At Earth's average distance from the Sun (about 150 million km; 93 million mi), the average intensity of solar energy reaching the top of the atmosphere is about 1,361 W/m².
- Because only half of Earth receives sunlight at one time, and solar energy is reduced away from the equator, the amount of sunlight arriving at the top of Earth's atmosphere is only one-fourth of the total solar energy, or approximately 340.25 W/m² averaged over the entire planet.
- About 29% of the solar energy that arrives at the top of the atmosphere is reflected back to space by clouds, atmospheric particles, or bright ground surfaces like sea ice and snow. This energy plays no role in Earth's climate system.

- About 23% of incoming solar energy is absorbed in the atmosphere by water vapor, dust, and ozone.
- About 48% of incoming solar energy passes through the atmosphere and is absorbed by Earth's surface.
- Thus, about 71% of the total incoming solar energy (241.57 W/m²) is absorbed by the Earth system.
- In order for Earth's surface temperature to remain stable, an amount of energy must be radiated back to space that is equivalent to the incoming solar energy.
- As atoms and molecules on Earth absorb energy, they also radiate energy in the thermal infrared portion of the **electromagnetic spectrum**. This energy is heat.
- The **Stefan–Boltzmann Law** states that the amount of heat a surface radiates is proportional to the fourth power of its temperature. If temperature doubles, radiated energy increases by a factor of 16 (2 to the fourth power). This is **radiative cooling**.
- Energy leaves Earth's surface through evaporation, **convection/conduction**, and emission of thermal

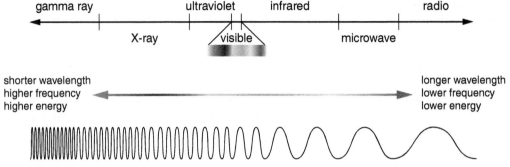

FIGURE 2.1 The electromagnetic spectrum describes the range of all types of energy. X-rays used in a hospital, light from a light bulb, and microwaves used in your microwave oven are all types of radiation that are found on the electromagnetic spectrum. The shorter (hotter) wavelength/higher frequency portion of the spectrum contains gamma rays, X-rays, and ultraviolet radiation. Infrared light, microwaves, and radio waves are all part of the longer (cooler) wavelength/lower frequency portion of the spectrum.

Source: NASA

infrared energy. These processes release heat to the atmosphere that joins heat already absorbed by clouds, water vapor, ozone, and **aerosols**.

- Water vapor, carbon dioxide, methane, nitrous oxide, and fluorine compounds are trace gases that absorb thermal infrared energy radiated by Earth's surface. These are called greenhouse gases.
- Greenhouse gas molecules radiate heat in all directions, including back to Earth's surface where it is absorbed. Absorbing this heat causes the surface to become warmer than it would be if it were heated only by direct sunlight. The extra heating is the **natural greenhouse effect**.
- Earth's surface temperature is about 14°C (58°F) on average—more than 30°C warmer than it would be if there were no natural greenhouse effect.
- Human interference with the natural greenhouse effect produces the **anthropogenic greenhouse effect**.

2.2 Earth's Energy Budget

To understand the human influence on Earth's climate, it is important to first understand how the natural climate system works. The natural climate system is fundamentally a product of energy from the Sun. Sunlight entering Earth's atmosphere provides energy that is trapped, reflected, and released to space. These processes are described in Earth's Energy Budget.

2.2.1 Characteristics of Radiation

Energy from the Sun warms our world and stimulates a number of essential natural processes that make life possible. Photosynthesis, evaporation, wind, clouds, ocean currents, and the seasons all originate with energy that enters

our atmosphere after its 8-minute and 20-second journey from the Sun.

All matter in the universe with a temperature above absolute zero[1] contains heat and therefore radiates (releases) some amount of energy. Energy can be measured in terms of its wavelength and frequency, the two characteristics that define the electromagnetic spectrum (EM).[2] The characteristics of the radiation depend on the temperature of the radiating body.

With a temperature of about 5,500 degrees Celsius (10,000 degrees Fahrenheit), the surface of the Sun radiates energy in the *visible* and *near infrared* portion of the electromagnetic spectrum (**Figure 2.1**). The hotter something is, the shorter its peak wavelength of radiated energy is. The hottest objects in the universe radiate mostly gamma rays and X-rays. Cooler objects emit mostly longer-wavelength radiation, including visible light, thermal infrared, radio, and microwaves.

2.2.2 Sunlight

At an average distance of 150 million kilometers (93 million miles) from the Sun, the average energy of sunlight arriving at the top of Earth's atmosphere is known as the

---

[1] Absolute zero is the total absence of heat—a condition thought to be impossible to reach.

[2] Most of us think of light as something our eyes can see. But our eyes capture only a small portion of the total amount of light that surrounds us. The electromagnetic spectrum is a description of the entire range of light that exists in the universe, most of which is invisible to us. Light is a wave of alternating electric and magnetic fields. A wave is described by its frequency, measured in Hertz, which counts the number of waves that pass by a point in 1 second. A wave is also measured by its wavelength: the distance from the peak of one wave to the peak of the next. The larger the wave frequency, the smaller the wavelength is and vice versa.

total solar irradiance (also called the *solar constant*), about 1,361 watts[3] per square meter (10.8 square feet). However, this sunlight only hits the side of Earth facing the Sun. Basic geometry limits the amount of solar energy actually intercepted by Earth to half the total solar irradiance.

It is halved again by the fact that total solar irradiance is only achieved when sunlight hits a surface that is perpendicular to the path of incoming sunlight. Because of Earth's curved surface, only areas near the equator at midday come close to being perpendicular to the arriving energy. Everywhere else, the light arrives at an angle, reducing the amount of solar irradiance reaching Earth's surface (**Figure 2.2**). Hence, averaged over the entire planet, the amount of sunlight received at the top of the atmosphere is

only one-fourth of the total solar irradiance, approximately 340.25 W/m².

2.2.3 Heat Engine

The annual amount of incoming solar energy varies considerably from the tropics to the poles. At middle and high latitudes, it also varies considerably from season to season. One reason for this is that Earth's axis of rotation is tilted approximately 23.5 degrees from the vertical. This means that over the course of one year as Earth orbits the Sun, one hemisphere will be tilted toward the Sun receiving more sunlight and experiencing longer days. The other hemisphere meanwhile is tilted away from the Sun and receives less sunlight and experiences shorter days (**Figure 2.3**).

If Earth's axis of rotation were vertical, the size of the heating imbalance between equator and the poles would be the same year-round, and the seasons would not occur. However, because Earth's axis is tilted, the combination of more direct sunlight and longer days means that the pole tilted toward the Sun receives more incoming sunlight than the tropics. This creates summer. Simultaneously, at the opposite pole, it gets none, making it winter. Six months later the reverse will occur.

Even though illumination increases at the poles in the summer, bright white snow and sea ice reflect a significant portion of the incoming light, reducing the potential solar heating. Such differences in reflectiveness are known as Earth's albedo. The differences in albedo, and most importantly in the sunlight angle and the length of day at different latitudes, all lead to heating imbalances throughout the Earth system.

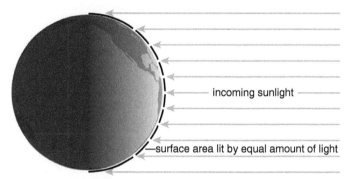

FIGURE 2.2 The curvature of Earth's surface, and the fact that only half of the planet can face the Sun at one time, reduces the amount of sunlight received into the atmosphere to one-fourth of the total solar irradiance, approximately 340.25 W/m² (10.8 ft²). From the equator to the poles, incoming sunlight intercepts Earth at smaller and smaller angles, and the light gets spread over larger and larger surface areas (line segments).

Source: NASA

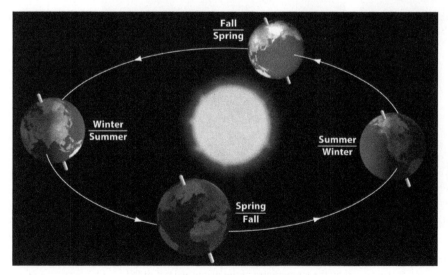

FIGURE 2.3 Earth's seasons are the result of a 23.5° tilt in the planet's axis with respect to the plane of the solar system (known as the "ecliptic"). Because of this tilt, different parts of the globe are oriented toward the Sun at different times of the year. Summer is warmer than winter (in each hemisphere) because the Sun's rays hit Earth at a more-direct angle than during winter and because days are much longer than nights. During winter, the Sun's rays hit Earth at a less-direct angle, and days are very short.

[3] A watt measures the amount of energy that something uses, or generates over time. A light bulb uses between 40 and 100 W, a microwave oven uses about 1,000 W. The average amount of sunlight at the top of the atmosphere on the side facing the sun is 1,361 W/m² (10.8 ft²).

At any place on Earth, the net heating is the difference between the amount of incoming sunlight and the amount of heat radiated by Earth back to space. In the tropics, there is a net energy surplus because the amount of sunlight absorbed is larger than the amount of heat radiated. In the polar regions, however, there is an annual energy deficit because the amount of heat radiated to space is larger than the amount of absorbed sunlight (owing to the albedo).

This net heating imbalance between the equator and poles drives the heat engine consisting of atmospheric and oceanic circulation. Evaporation, convection, rainfall, winds, and ocean currents are all part of Earth's heat engine. Earth's heat engine not only works to circulate solar energy from the equator toward the poles but also from the surface and lower atmosphere back to space.

If heat is not radiated to space at the same rate it is received from the Sun, the temperature becomes out of equilibrium and continuously increases or decreases. However, when this system is in equilibrium, the temperature doesn't rise or fall because the surface and the atmosphere are simultaneously radiating heat to space. This net flow of energy into and out of the Earth system is termed the energy budget.[4]

2.2.4 Radiative Equilibrium

Earth's energy system will always seek equilibrium. This means that with constant solar radiation and unchanging conditions on Earth (no changes in albedo or atmospheric chemistry), Earth's planetary temperature will on its own

shift toward a state called the radiative equilibrium. The basic principle underlying this is the Stefan–Boltzmann Law (described more on the next page), which dictates that the hotter an object is, the more radiation it emits. . . powerfully leading to an equilibrium state.

The Stefan–Boltzmann Law says that the radiation (heat) emitted from a surface is proportional to the fourth power of the temperature of that surface: if the surface temperature doubles, the radiated energy increases by a factor of 16 (2 to the fourth power). Thus, as the energy absorbed by Earth's surface causes the temperature to rise, a rapidly increasing amount of heat is radiated to space, counteracting the surface warming. The amount of heat lost to space is large compared to the smaller increase in surface temperature, a process referred to as radiative cooling. This is the primary mechanism that prevents runaway heating on Earth.

This simple concept is fundamental to understanding Earth's climate. Radiative cooling provides a powerful negative feedback, preventing runaway temperature changes and producing radiative equilibrium to the climate system. Global temperature will be relatively stable under these conditions (**Figure 2.4**).

2.2.5 Energy Entering the Climate System

As we learned, averaged over the entire planet, the amount of sunlight received at the top of the atmosphere is approximately 340.25 W/m². About 29 percent of this energy is

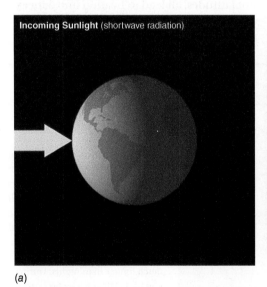

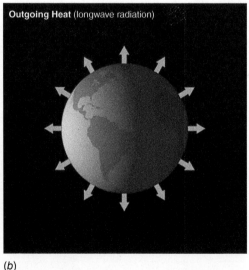

(a)   (b)

FIGURE 2.4 Radiative equilibrium occurs when (a) incoming energy from the Sun equals (b) the outgoing energy emitted by Earth. If equilibrium is not achieved, then the temperature of the planet will rise or fall accordingly.

Source: NASA

---

[4] VIDEO: NASA Aqua CERES - Tracking Earth's Heat Balance, https://www .youtube.com/watch?v=uVkfh89iyeU

reflected back to space by the tops of clouds, atmosphere particles, and bright ground surfaces such as snowfields, sea ice, and glaciers. Reflected energy does not influence the Earth's climate system. Albedo is measured on a scale from zero to one (or as a percentage). An albedo of one (or 100 percent) describes a white surface that does not absorb any incoming energy (and therefore is not warmed by sunlight). An albedo of zero describes a black surface that completely absorbs sunlight and, as a result, becomes warm. Earth has a complex surface environment with albedo values that range across this entire spectrum.

Because Earth's surface has both high albedo regions such as glaciers and sea ice and low albedo regions such as oceans and other dark surfaces, the amount of reflected and absorbed sunlight varies from place to place. This leads to uneven heating of Earth's surface and drives air movement that creates atmospheric circulation—part of the heat engine. We describe the system of winds, and the weather patterns they create, in **Chapter 3**.

Another 23 percent of incoming solar energy is absorbed by water vapor, dark particles, and ozone in the atmosphere. The atoms and molecules that make up the atmosphere absorb different types of radiation to varying degrees. Oxygen of two types, O_2 and O_3 (called **ozone**), provide the important service of absorbing high-energy radiation such as ultraviolet light that is damaging to living tissue. The sunburn you may get when exposed to strong sunlight reveals that some of this dangerous radiation still makes it

to Earth's surface. Other types of atmospheric gases also absorb sunlight at various frequencies. Water vapor, for instance, is very effective at absorbing sunlight. Incoming solar energy that is absorbed is converted into heat energy, which also causes the absorbing substance to emit its own radiation. This warms the atmosphere.

The remaining 48 percent of incoming sunlight passes through the atmosphere and is absorbed by dark surface environments such as the ocean and other water bodies, rock, and soil. Thus, between the atmosphere and the surface, roughly 71 percent of incoming solar radiation is absorbed somewhere among Earth's environments (**Figure 2.5**).

2.2.6 Surface Temperature

Everything that has a temperature above absolute zero emits radiation. And the hotter an object is, the shorter the wavelength of the emitted energy. Therefore, the Sun, being very hot, emits shortwave radiation, which enters Earth's atmosphere and hits the surface. This energy is absorbed and warms atoms and molecules in the atmosphere and on the surface. Once warmed, and being less hot than the Sun, these materials give off long-wave radiation in the form of **thermal infrared radiation** (or heat). Thus, gases, liquids, and solids in the atmosphere and on Earth's surface that are heated by the Sun radiate long-wave infrared radiation.

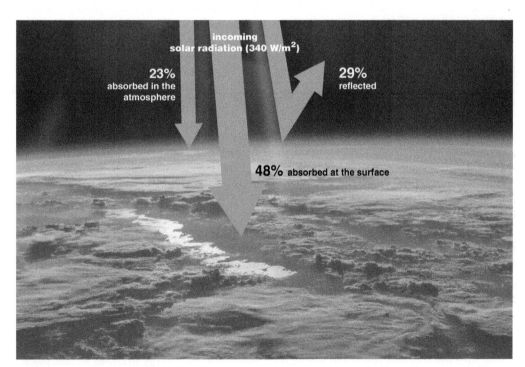

FIGURE 2.5 On average, 340.25 W/m² of solar energy enters the atmosphere. Of this 29% is reflected back into space by clouds, bright particles in the air, and reflective surfaces such as snowfields, sea ice, and glaciers. Certain gases, dark particles, and ozone absorb approximately 23% of incoming energy. The remaining 48% is absorbed at the surface.

Source: N. Hulbirt

As we discussed earlier, the Stefan–Boltzmann Law describes how the energy radiating from a mass or a surface is proportional to the fourth power of the absolute temperature of that surface. Thus, as Earth's surface absorbs energy and it's temperature rises, the atmosphere must radiate an increasing amount of heat to space in order to maintain radiative equilibrium. The amount of heat lost to space is large compared to the smaller increase in temperature (radiative cooling). This mechanism is the primary reason that runaway heating does not occur on Earth.

To maintain radiative equilibrium, Earth must radiate out to space as much energy as it receives. We know it absorbs about 71 percent of the incoming solar irradiance—340.25 W/m².

$$0.71 \times 340.25\,\text{W/m}^2 = 241.57\,\text{W/m}^2$$

Thus, Earth must radiate out to space 241.57 W/m² of energy in order to keep the climate stable.

Using the Stefan–Boltzmann Law we can calculate Earth's radiative equilibrium temperature, the temperature necessary to emit 241.57 W/m². The law is expressed as the relationship $I = \sigma T^4$, which equates the radiant energy (I) to the fourth power of the temperature of a radiating body (T^4) times a constant of proportionality (σ) equal to 5.67×10^{-8}. This can be reorganized as $T = (I/\sigma)^{0.25}$ and solved as

$$T = \left(241.57/5.67 \times 10^{-8}\right)^{0.25}$$

$$T = 255.484\,\text{K, or} -17.65°\text{C}$$

$$(0.23°\text{F}).$$

Our calculation shows that if Earth's surface temperature is in equilibrium with incoming solar radiation it must be a chilly –17.65 degrees Celsius. Note that here we use the symbol K. This is a temperature scale called the Kelvin Scale (K), which is used by climate scientists. It is designed so that zero degrees K is defined as absolute zero (at absolute zero, a hypothetical temperature, all molecular movement stops—all actual temperatures are above absolute zero). The size of one unit Kelvin is the same as the size of one degree Celsius. Units of Kelvin do not use the degree (°) symbol.

However, a global array of thermometers and ocean surface measurements tells us that Earth's mean surface temperature over the period 1951 to 1980 was actually 287 Kelvin, or 14 degrees Celsius (57.2 degrees Fahrenheit). Where did the extra 31 degrees Celsius come from? Why is Earth warmer than its radiative equilibrium temperature? Recent years have been much warmer; the 2017 mean was 14.74 degrees Celsius, and the warmest year 2016 was 14.84 degrees Celsius.

Under steady state conditions, the measured long-wave radiation leaving Earth into space would be the same as that determined by the radiative equilibrium temperature (using the Stefan–Boltzmann Law). However, the real surface temperature is much warmer than the radiative equilibrium temperature. This is because the temperature of Earth's surface is influenced by trapped radiation associated with certain gases in the atmosphere that absorb and then reradiate long-wave infrared energy. As we learned in **Chapter 1**, this is the greenhouse effect and these gases are called greenhouse gases.

The difference between Earth's radiative equilibrium temperature (–17.65 degrees Celsius [0.23 degrees Fahrenheit]) and the actual surface temperature (about 14 degrees Celsius [57.2 degrees Fahrenheit]) is a good indicator of the strength of the greenhouse effect. If the planet had no atmosphere (or an atmosphere with no greenhouse gases), it's average surface temperature would be equal to the radiative equilibrium temperature. This would be true, for example, for the moon.

Without a greenhouse effect, Earth's mean surface temperature would be well below freezing, and life on Earth would be very different—if there was life at all.

2.2.7 Natural Greenhouse Effect

The natural greenhouse effect exists because of the capacity of certain gases in the atmosphere to capture heat from the Sun, and heat coming off Earth's surface that would otherwise escape to space. As these gases gain warmth, they radiate energy in all directions and in so doing raise the temperature of the atmosphere above the temperature achieved by the Sun's energy alone.

The greenhouse effect is a major influence on climate. For instance, water vapor, the most powerful greenhouse gas, plays an important role in the difference between nighttime and daytime temperatures. You get much larger daily temperature shifts in regions that are very dry (e.g., Arizona, where the daily range between maximum and minimum temperatures can be 27 degrees Celsius [50 degrees Fahrenheit] or more) than in regions near the ocean (e.g., Hawaii where the daily range is only about 8 degrees Celsius [15 degrees Fahrenheit]). This is because the atmosphere in areas with high humidity continues to radiate heat throughout the night, whereas in dry areas this is not the case.

This is true on other planets as well. The energy arriving at a planet surface comes from two sources: the Sun and the atmosphere. The energy from the Sun varies over the day, with large values during daytime and zero at night. The energy from the atmosphere (the greenhouse effect), on the other hand, is constant throughout the day and night. Take the planet Mercury, which has no atmosphere. Energy arriving at the planet surface comes solely from the Sun and is thousands of Watts per square meter during the day. At night, with no atmosphere, there is zero energy directed at the planet surface. This leads to huge variations in surface temperature between day and night: daytime temperatures are 430 degrees Celsius (800 degrees Fahrenheit), while at night they are −170 degrees Celsius (−280 degrees Fahrenheit). The temperature on Mercury changes over 1,000 degrees Fahrenheit between a single night and day!

For Venus, on the other hand, energy arriving from the atmosphere is a constant 17,000 W/m², while energy from the Sun is about 400 W/m² during the day and zero at night. Thus, energy arriving at the surface of Venus is approximately 17,400 W/m² during the day and 17,000 W/m² during the night. This very small change in energy over a day means that there is little change in temperature between day and night. Venus, in fact, has a "runaway" greenhouse effect stemming from its geologic history.

Early in the history of the Solar System, Venus may have had a global ocean. But as the brightness of the early Sun increased, the amount of water vapor in the atmosphere increased, raising the temperature and consequently increasing the evaporation of the ocean, leading eventually to the situation in which the oceans boiled and all of the water vapor entered the atmosphere. Today, the Venusian atmosphere is mostly carbon dioxide (CO_2), another powerful greenhouse gas, which developed when the surface of Venus got so hot that carbon trapped in rocks sublimated into the atmosphere and reacted with oxygen to form carbon dioxide. The carbon dioxide atmosphere on Venus is 92 times denser than Earth's atmosphere at the surface. The greenhouse effect gives Venus the hottest surface in the Solar System.

2.2.8 Surface Energy Budget

On Earth, a balanced energy budget is achieved through radiation. That's because the only form of energy exchange between planet Earth and the rest of the universe is via radiation. However, within Earth's atmosphere and at the surface, other forms of energy transfer also occur.

Earth's surface absorbs 48 percent of the solar radiation that enters the atmosphere. For the energy budget at Earth's surface to balance, processes on the ground must get rid of the 48 percent of incoming solar energy that the ocean and land surfaces absorb. This occurs by energy leaving the surface through evaporation, convection/conduction, and emission of thermal infrared energy (**Figure 2.6**).

Evaporation occurs when liquid water molecules absorb solar radiation and begin to intensely vibrate, freeing them from the liquid phase to enter the gas (or vapor) phase. As long as they maintain this energy, they will remain as water vapor. But cool air, at elevations of a few thousand feet in the atmosphere, remove this energy, causing condensation to take place. Condensation turns water vapor back into liquid water particles, most commonly in the form of clouds in the tropics. This conversion from gas to liquid releases **latent heat** into the atmosphere.

Latent heat is energy released or absorbed in the atmosphere related to changes in phase between liquids, gases, and solids. Evaporation from tropical oceans, and the subsequent release of latent heat to form precipitation (because water vapor has condensed into water particles), are the

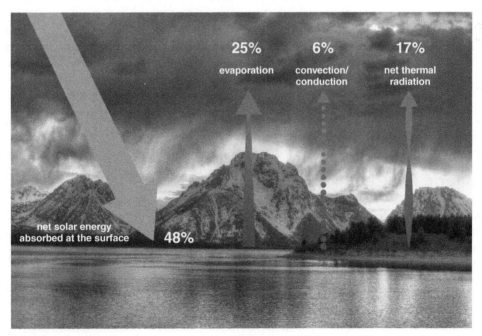

FIGURE 2.6 Processes that radiate energy from Earth's surface include evaporation, convection/conduction, and infrared radiation (net thermal radiation).

Source: N. Hulbirt

primary drivers of the atmospheric heat engine. Evaporation accounts for 25 percent of the energy movement from Earth's surface to the atmosphere.

Six percent of the heat leaving Earth's surface is carried by convection/conduction of **sensible heat**. While latent heat is related to changes in phase between liquids, gases, and solids. Sensible heat is related to changes in temperature of a gas or object with no change in phase. In this case, convection is a rising current of warm air that forms because it is heated by coming in direct contact with the warm surface. Warm air rises, carrying heat away from Earth's surface. Conduction consists of energy transfer directly atom to atom and represents the flow of energy along a temperature gradient from Earth's surface to the atmosphere.

Lastly, as Earth's surface is heated by the Sun, it reradiates that warmth in the form of thermal infrared radiation. The surface releases this long-wave radiation at the same time that downward radiation is arriving from the atmosphere. The atmosphere releases radiation equally in both directions. Because it is absorbing $241.57 \, W/m^2$ from space, it must also be releasing $241.57 \, W/m^2$ to the surface, if energy balance is to be maintained. Thus, proportionally, 100 percent of the arriving heat flows downward to the ground, and 117 percent flows upward; a net of 17 percent leaves the surface in the form of thermal infrared radiation. The additional 17 percent radiation leaving the ground is a product of radiative cooling, the process described by the Stefan–Boltzmann Law where emitted radiation is proportional to the fourth power of the absolute temperature of the radiating surface.

In summary, of the 48 percent of solar radiation that is absorbed by Earth's surface, 25 percent is moved back into the atmosphere by evaporation, 6 percent as sensible heat moved by convection/conduction, and 17 percent by thermal infrared radiation. The atmosphere has already absorbed 23 percent of incoming energy and, with the contribution from the surface, now accounts for approximately 71 percent of incoming solar radiation. Because 29 percent of the Sun's radiation was initially reflected back to space, all (71% + 29% = 100%) of the $340.25 \, W/m^2$ ($10.8 \, ft^2$) of incoming solar radiation is accounted for.

2.2.9 Atmospheric Energy Budget

Clouds, dark particles, water vapor, and ozone absorb 23 percent of incoming solar radiation to the atmosphere. As we saw in the surface energy budget, evaporation and convection/conduction transfer 25 percent and 6 percent (resp.) of incoming solar energy from the surface to the atmosphere. Thus, the equivalent of 54 percent of solar radiation resides in the atmosphere. However, satellite measurements indicate that the atmosphere radiates thermal infrared energy to space equivalent to 59 percent of the incoming solar radiation. Where does the roughly 5 percent difference come from? It comes from the natural greenhouse effect.

Earth's atmosphere is mostly composed of nitrogen (78 percent) and oxygen (21 percent). The gas argon composes 0.9 percent on average and gases such as carbon dioxide, nitrous oxides, methane, and ozone are trace gases that account for about one tenth of one percent of the atmosphere. Just as nitrogen, oxygen, and argon are transparent to solar energy in the visible wavelengths, they are also transparent to outgoing long-wave infrared radiation.

However, water vapor, carbon dioxide, methane, and other greenhouse gases are opaque to infrared energy. This means they absorb infrared energy coming from Earth's surface as part of the surface energy budget. Recall that Earth's surface radiates in the infrared wavelengths a net 17 percent of the incoming solar radiation. However, the amount of thermal infrared energy that directly escapes to space is only about 12 percent. Greenhouse gas molecules in the atmosphere absorb the difference, roughly 5 percent.

When greenhouse gas molecules absorb thermal infrared energy, their temperature rises, and they radiate thermal infrared energy in all directions. Heat radiated upward continues to encounter greenhouse gas molecules; those molecules absorb the heat, their temperature rises, and the amount of heat they radiate increases. Heat continues to pass upward until at an altitude of roughly 5 to 6 km (3.1 to 3.7 mi), the concentration of greenhouse gases in the overlying atmosphere is so small that heat can radiate freely to space.

Heat radiated by greenhouse gas molecules also spreads downward and ultimately comes back into contact with Earth's surface, where it is absorbed. The temperature of the surface becomes warmer than it would be if it were heated only by direct solar radiation. This additional heating of Earth's surface by the atmosphere is the natural greenhouse effect. The more greenhouse gases there are in the atmosphere, the more Earth's surface temperature will increase.[5] Because of this additional warming, Earth's mean surface temperature is a comfortable 14 degrees Celsius (57.2 degrees Fahrenheit) under normal conditions. However, if greenhouse gases are added or removed, the temperature will rise or fall in response.

Recall that the difference between Earth's radiative equilibrium temperature and the actual surface temperature is a good indicator of the strength of the natural greenhouse effect. The temperature difference comes from greenhouse gases that radiate heat from the atmosphere downward to Earth's surface.

The heat radiated from greenhouse gases to the surface is called **back radiation**. As we explained earlier, back radiation is equivalent to 100 percent of the incoming solar energy. Radiative cooling prevents Earth's surface from overheating because it releases 117 percent to the atmosphere for a net upward heat flow of 17 percent (**Figure 2.7**).

---

[5] Hansen, J., et al. (2010) Global surface temperature change, *Reviews of Geophysics*, 48, RG4004, doi:10.1029/2010RG000345. See http://www.giss.nasa.gov/research/news/20110113

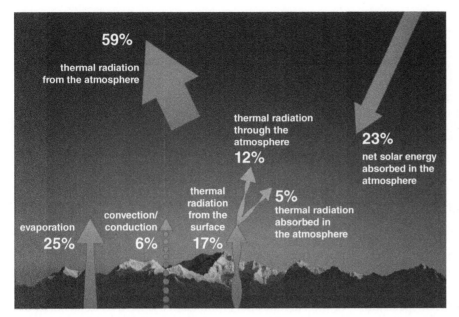

FIGURE 2.7 The atmosphere radiates 59% of the incoming solar radiation back to space. This energy comes from a combination of sources: 23% comes from absorbed sunlight that is reradiated to space, 25% comes from surface heat delivered by evaporation, 6% from surface heat delivered by convection/conduction, and 5% from infrared radiation emitted by greenhouse gases. The remaining thermal infrared from the surface (12%) passes through the atmosphere and escapes out to space, and so it is not included in balancing the quantities (i.e. loss to space (59%) = gain from various sources (23% + 25% + 6% + 5%)).

Source: N. Hulbirt

Some of this heat escapes directly to space, and the rest is transferred among greenhouse gases to higher and higher levels of the atmosphere.

Eventually, the energy leaving the top of the atmosphere matches the amount of incoming solar energy. The natural greenhouse effect does not cause runaway heating in most cases because solar radiation is remarkably constant and in a natural setting the amount of greenhouse gas in the atmosphere is relatively stable. If either of these were to change, so would the surface temperature. In fact, today the surface temperature is not stable because the natural greenhouse effect is being augmented by the release of additional greenhouse gases by various types of human activities (discussed in detail later); this creates the anthropogenic greenhouse effect.

2.2.10 Radiative Forcing

Radiative forcing describes processes in Earth's climate system that cause an imbalance between the amount of sunlight entering the atmosphere and the amount of energy radiating to space. As we learned earlier, when there is a difference between these two, temperatures will either rise or fall. These destabilizing influences are also called climate forcings.

There are a variety of natural physical and chemical climate forcings that can affect the global energy balance: changes in the Sun's brightness, variations in the shape of Earth's orbit and its axis of rotation that occur over thousands of years (called Milankovitch cycles, discussed in detail in **Chapter 4**), large volcanic eruptions that inject light-reflecting particles (ash and sulfate aerosols) into the stratosphere, and others.

Anthropogenic climate forcings include dark particles that absorb sunlight causing heating and bright particles that reflect sunlight causing cooling (both a form of human pollution in the atmosphere), deforestation that changes how the surface reflects and absorbs sunlight and the storage or release of carbon dioxide, the rising concentration of atmospheric carbon dioxide and other greenhouse gases that decrease heat radiated to space, concentrated animal feeding operations that produce large amounts of methane (a powerful greenhouse gas), and others.

A climate forcing can trigger feedbacks that intensify (positive feedback) or weaken (negative feedback) the original forcing. An example of a feedback is when global warming causes the loss of ice at the poles, which reduces Earth's albedo, causing less reflection of incident solar radiation, leading to more absorption by the atmosphere and surface, causing further warming that leads to more ice loss.

As we have learned already, greenhouses gases cause important climate forcings by absorbing thermal infrared radiation leaving Earth's surface. Infrared energy radiates in a range of wavelengths and depending on the wavelength, some, none, or all of the radiation may be absorbed by greenhouse gases. For instance, water vapor is a powerful absorber of infrared energy; but at wavelengths between about 8 and 15 μm (the μm is a unit of a micrometer, or one-millionth of 1 meter) less than half of the radiation is absorbed. It's as if a water vapor window has been left open to allow for radiative cooling of Earth's surface.

However, carbon dioxide is a strong absorber of thermal infrared energy in the range of 12 to 13 μm. This means that CO_2 absorbs radiation in a part of the window that is left open by water vapor (**Figure 2.8**). That is, some wavelengths of outgoing thermal infrared energy that water vapor would have let escape to space are instead absorbed by carbon dioxide. A similar water vapor window between 3 and 5 μm is also partially closed by carbon dioxide at about 4 to 4.5 μm.

As we have learned, once solar radiation warms Earth's surface, the surface emits long-wave radiation in the

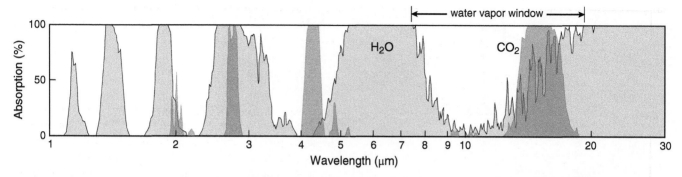

FIGURE 2.8 Greenhouse gases have unique wavelength ranges of long-wave thermal infrared energy absorption: they absorb energy at some wavelengths but are transparent at others. The absorption pattern of water vapor (blue, light grey) and carbon dioxide (pink, dark grey) overlap in some cases, and not in others. Carbon dioxide is not as strong a greenhouse gas as water vapor, but it absorbs energy in wavelengths (12 to 13 and 4 to 4.5 μm) that water vapor does not, partially closing "windows" through which heat radiated by the surface would normally escape to space.
Source: N. Hulbirt

thermal infrared portion of the electromagnetic spectrum. Various greenhouse gases absorb this infrared energy. In Figure 2.8, we see how carbon dioxide and water vapor overlap to increase the amount of long-wave radiation that is absorbed. **Figure 2.9** expands on this idea and shows the absorption characteristics of other types of gases (shaded areas represent the percent of radiation absorbed by each gas).

As a general characteristic, an effective absorber of infrared radiation will have a broader absorption profile, which means that it can absorb a wider spectrum (a *band*) of wavelengths. In Figure 2.9, the top panel shows the absorption characteristics of the greenhouse gas nitrous oxide (N_2O), which is very effective in two narrow infrared regions centered on about 4.5 and 7.9 μm. Methane (CH_4) strongly absorbs thermal infrared in two narrow bands at about 4 and 7.6 μm.

In the third panel, oxygen (O_2) and ozone (O_3) are weak absorbers at 15 μm, but ozone is a strong absorber in a moderately wide band centered on 9 to 10 μm. Notice also that both molecules of oxygen are highly effective absorbers in the ultraviolet wavelengths (as mentioned earlier). Water vapor and carbon dioxide, both powerful absorbers of thermal infrared radiation, were discussed earlier. The bottom panel shows the percent of radiation absorbed by all of the atmospheric gases.

The rising concentration of greenhouse gases in the atmosphere means that although Earth still absorbs about 71 percent of the incoming solar energy, an equivalent amount of heat is no longer leaving. This is called **Earth's energy imbalance**. The energy imbalance is the difference between the amount of solar energy absorbed by Earth and the amount of energy the planet radiates to space as heat. If the imbalance is positive, more energy coming in than going out, we can expect Earth to become warmer in the future—but cooler if the imbalance is negative.

Earth's energy imbalance is thus the single most crucial measure of the status of Earth's climate, and it defines expectations for future climate change. The exact amount of the current energy imbalance is hard to determine, but it appears to be approximately +0.58 ± 0.15 W/m². Measuring the energy gained by the ocean, the land, melting ice around the world, and satellite measurements allowed calculation of this number.[6]

Rising greenhouse gas concentrations cause an energy imbalance immediately. But it may take decades for the full impact of a forcing to be felt. This lag between when an imbalance occurs and when the impact on surface temperature becomes fully apparent is mostly because of the immense capacity of the global ocean to absorb heat. The heat capacity of the oceans gives the climate a thermal inertia that can make surface warming or cooling more gradual, but it can't stop a change from occurring.

Recall the radiative equilibrium concept. The existence of radiative forcing (i.e., anthropogenic greenhouse gases) has pushed Earth's energy system out of balance and it is shifting to a new equilibrium temperature. That alone is a source of concern but it is an even greater source of worry that the new temperature is additionally shifting upward as we continue to emit greenhouses gases and as various positive feedbacks (e.g., albedo decrease) further raise the temperature.

Changes in global climate observed so far are only part of the full response we can expect from the current energy imbalance. Thus far, global average surface temperature is approximately 1.0 degrees Celsius (1.8 degrees Fahrenheit) hotter than the "preindustrial" climate of the 1800s. Surface temperature will likely rise another 0.6 degrees Celsius (1.0 degrees Fahrenheit) or so in response to the existing energy imbalance.[7] If the concentration of greenhouse gases stabilizes, then Earth's climate will once again come into equilibrium, although with the

---

[6] Hansen, J., et al. (2011) Earth's energy imbalance and implications, *Atmospheric Chemistry and Physics*, 11, 13421–13449, doi:10.5194/acp-11-13421-2011

[7] NASA https://earthobservatory.nasa.gov/Features/EnergyBalance/page7.php

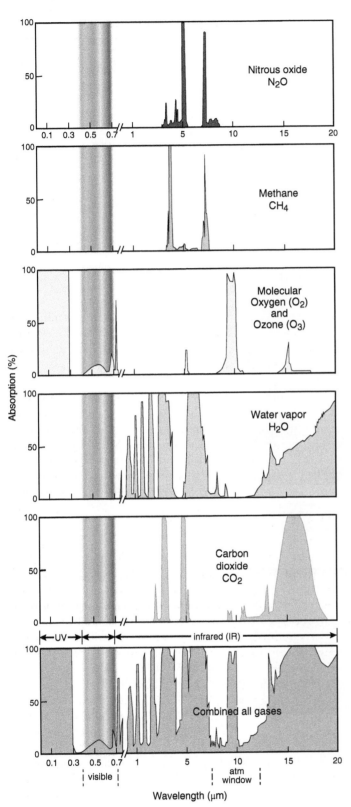

FIGURE 2.9 Absorption of various wavelengths of radiation by various types of gases in the atmosphere. The shaded area represents the percent of radiation absorbed by each gas. Nitrous oxide, methane, ozone, water vapor, and carbon dioxide show strong absorption characteristics in the infrared portion of the electromagnetic spectrum. Oxygen (O_2) and ozone (O_3) show strong absorption in the ultraviolet portion of the spectrum, as well as some absorption at infrared wavelengths.

Source: N. Hulbirt

"thermostat"—global average surface temperature—set at a higher temperature than it was before humans began to change the atmosphere.

2.2.11 Sunspots

Increases in solar output are associated with times of high sunspot activity.[8] Sunspots are dark patches that appear on the Sun's surface. These sunspots look small compared to the Sun's massive size, but in fact many individual sunspots are larger than the entire size of planet Earth. The number of sunspots on the Sun's surface has been observed, and recorded, for the past 400 years.

Over the entire time of monitoring, the amount of solar irradiance has not changed. However, over shorter periods of time, solar energy does change. This is an important area of research and has led to the creation of programs[9] that monitor the Sun in order to improve understanding of why it fluctuates. If changes in the amount of sunlight entering Earth's atmosphere are large enough, it has the potential to change the climate; however, this has not been the case in at least several centuries (**Figure 2.10**).[10]

In Figure 2.10, notice the relatively low level of total solar irradiance during the **Maunder Minimum** and again during the **Dalton Minimum**. These occurred in the second half of a period of cooling mostly located in the North Atlantic region known as the **Little Ice Age** (LIA).[11] The LIA is thought to have occurred in two phases between 1300 and 1870. The coldest period occurred when winter temperatures in Europe ranged from 1 to 1.5 degrees Celsius (1.8 to 2.7 degrees Fahrenheit) colder than average.

Researchers still debate the degree to which these solar minima were responsible for the LIA. Other causes have

[8] VIDEO: NASA Solar Cycle, https://www.youtube.com/watch?v=sASbVkK-p0w

[9] Why NASA keeps an eye on the sun's irradiance. https://www.nasa.gov/topics/solarsystem/features/sun-brightness.html

[10] Spotty sunspot record gets a makeover. http://www.nature.com/news/spotty-sunspot-record-gets-a-makeover-1.18145. See also Clette, F. and Lefèvre, L. (2016) The new sunspot number: assembling all corrections, *Solar Physics*, 291, 2629, doi:10.1007/s11207-016-1014-y. Svalgaard, L. and Schatten K.H. (2016) Reconstruction of the sunspot group number: the backbone method, *Solar Physics*, 291, 2653, doi:10.1007/s11207-015-0815-8

[11] The Little Ice Age (LIA) is a period between about 1300 and 1870 during which Europe and North America experienced colder winters than during the 20th century. The period can be divided into two phases, the first beginning around 1300 and continuing until the late 1400s. The Intergovernmental Panel on Climate Change Third Assessment Report concluded that the LIA was most likely a time of regional climate change and not a globally synchronous period of global cooling. Researchers working on the LIA no longer expect records to agree on either the start or end dates of the period or the global extent of the event. This is because of the apparent influence of local conditions. Many scientists view the LIA as a largely northern hemisphere or North Atlantic phenomenon. Several causes for the LIA have been proposed. These include decreases in solar radiation, increased volcanic activity, changes in ocean circulation, decreases in human population, and climate variability that is not well understood.

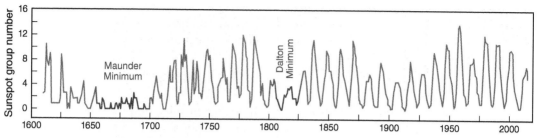

FIGURE 2.10 This graph shows the sunspot *group number* as measured over the past 400 years.[12] The group number is a standardized method for counting the presence of sunspots.[13] The Maunder Minimum is a time of low group number between 1645 and 1715, when sunspots were scarce and the winters harsh. Another period of low solar activity was in the early 1800s called the Dalton Minimum. Satellite measurements, begun in 1978, reveal that the Sun's output fluctuates by about 1 W/m² (or one-tenth of one percent) on approximately 11-year cycles.

Source: N. Hulbirt

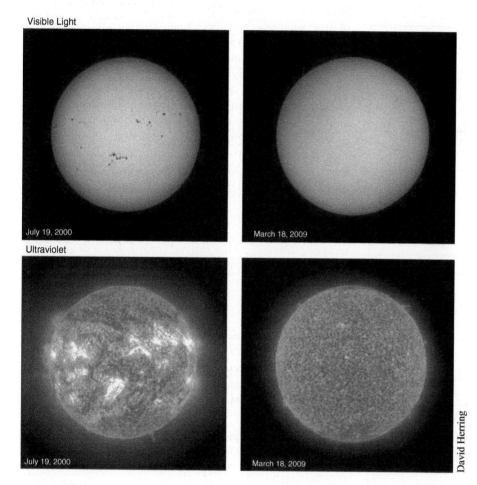

FIGURE 2.11 Instruments on solar monitoring satellites allow researchers to make images of the Sun using different wavelengths of light. In the top pair of images, the Sun's surface is viewed in visible wavelengths. In the bottom pair of images (from the same date as the top pair), the hotter atmosphere above the Sun's surface is shown in ultraviolet light. Here, the temperature is about ten times hotter than the surface. Sunspots can be seen in the upper-left image during a period of high solar activity. The upper-right image shows a period of low solar activity with few sunspots. In the lower-left image, very bright patches are associated with surface activity called *plage*. They're typically seen above *faculae* and occur during periods of high solar output. The image in the lower right shows the solar atmosphere during a period of low activity.

Source: NOAA

been proposed including volcanic eruption[14] (eruptions produce aerosols that reflect sunlight and cause temporary global cooling), changes in ocean circulation, decreased human populations, and natural climate variability. Astronomers at the time recorded only 50 sunspot groups over a 30-year period in the 1600s—indicting a very quiet Sun.

The LIA is not thought to be a globally synchronous event and thus does not represent cooling on a global scale. Rather, it may represent largely independent regional climate changes whose origin is still debated.[15]

Sunspots look like freckles on the face of the Sun (**Figure 2.11**). They are cool regions (hence, they are

[12] Corrected sunspot history suggests climate change not due to natural solar trends, https://phys.org/news/2015-08-sunspot-history-climate-due-natural.html
[13] Clette and Lefèvre (2016); Svalgaard and Schatten, (2016)

[14] Miller, G., et al. (2012) Abrupt onset of the little ice age triggered by volcanism and sustained by sea-ice-ocean feedbacks, *Geophysical Research Letters*, 39, L02708, doi: 10.1029/2011GL050168
[15] Mann, M.E., et al. (2009) Global signatures and dynamical origins of the little ice age and medieval climate anomaly, *Science*, 326, 1256–1260.

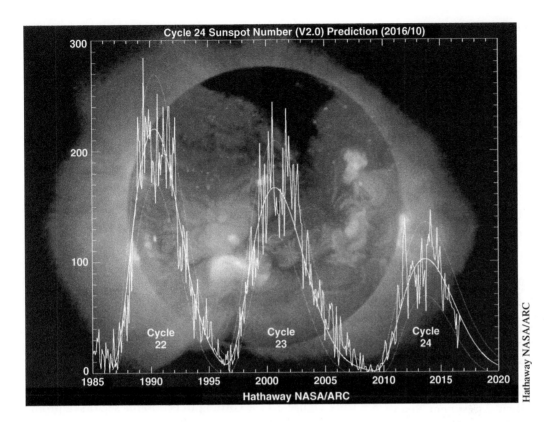

FIGURE 2.12 A prolonged but temporary lull in the Sun's activity between 2005 and 2010 did not prevent Earth's atmosphere from absorbing more solar energy than it radiated back into space. Despite the decrease in sunlight, Earth nonetheless accumulated 0.58 W of excess heat per square meter more than it released back to space. Because more energy was coming in than leaving, this imbalance drove global warming, a result of heat trapping by anthropogenic greenhouse gases.

Source: NASA

dark), but bright regions called **faculae** surround them. Faculae are a few percent brighter than the average surface of the Sun and they more than compensate for the cooling associated with sunspots. Hence, periods of sunspot formation are times of greater solar irradiance (by approximately 1 W/m²). Observations of sunspot activity reveal that they fluctuate on an 11-year cycle—commonly known as "the sunspot cycle." Figure 2.10 shows this regular 11-year cycle of solar brightening and dimming.

The full range of solar variability is still poorly understood. It is important that scientists continue monitoring total solar irradiance in order to advance scientific understanding of the Sun and its role in Earth's climate system.

Climate scientists have studied the heat-trapping properties of the atmosphere for over a century. See the box *Spotlight on Climate Change*, "The Earliest Climate Scientists."

2.3 Anthropogenic Forcing

Radiative forcing allows scientists to identify imbalances in Earth's energy budget and thereby learn more about the component parts. As part of a recent sunspot cycle, a solar lull from 2005 to 2010 offered just such an opportunity (**Figure 2.12**). The solar constant typically declines by about a 0.1 percent during low periods in a sunspot cycle. These solar minimums occur about every 11 years and last a year or so. However, the most recent minimum persisted more than two years longer than normal, making it the longest minimum recorded during the satellite era; NASA scientists

saw that this was an opportunity to assess the impact of the Sun's energy on the temperature of Earth's surface.

NASA scientists studied[16] and calculated that despite the decrease in sunlight, Earth nonetheless accumulated 0.58 W of excess heat per m² more than it released back to space. This extra heat was more than twice as much as the reduction in solar energy between the maximum and minimum points of the sunspot cycle (0.25 W/m²). Lead researcher on the study, Dr. James Hansen, stated "The fact that we still see a positive imbalance despite the prolonged solar minimum isn't a surprise given what we've learned about the climate system, but it's worth noting because this provides unequivocal evidence that the Sun is not the dominant driver of global warming."

2.3.1 Global Warming Potential

As we have learned, molecule for molecule, some greenhouse gases are stronger than others. Each differs in its ability to absorb heat and in the length of time it resides in the atmosphere.[17] The ability to absorb heat and warm the atmosphere is expressed by its **global warming**

---

[16] Hansen et al. (2011)
[17] Schmidt, G., et al. (2010) The Attribution of the present-day total greenhouse effect, *Journal of Geophysical Research*, 115, D20106, doi:10.1029/2010JD014287. See also Lacis, A., et al. (2010) Atmospheric CO₂: Principal control knob governing Earth's temperature, *Science*, 330, 356–359, doi:10.1126/science.1190653

potential (GWP), usually compared to CO_2 over some given time period.

Methane traps 21 times more heat per molecule than carbon dioxide. Nitrous oxide absorbs 270 times more heat per molecule than CO_2. Fluorocarbons are the most heat absorbent, with GWPs that are up to 30,000 times those of CO_2. The GWPs of various gases are very useful for understanding the impact of human emissions and determining which specific changes in emissions can accomplish the most positive effect in mitigating global warming; they also allow the attribution of warming to various types of human activities.[18]

As greenhouse gases accumulate in the atmosphere, the amount of heat they trap also increases. Some factors offset this process, including the reflection of sunlight by aerosols and increases in the reflectivity of land cover and perhaps clouds (though it is becoming more apparent[19] that clouds provide a positive feedback to warming). Calculating the amount that they change the overall radiative forcing in W/m^2 allows researchers to assess the overall impact of compounds that alter the balance between radiation entering and exiting the atmosphere.

This was done by the Intergovernmental Panel on Climate Change in their Assessment Report 4 (AR4) published in 2007[20] and again in their Assessment Report 5 (AR5) published in 2013.[21] **Figure 2.13** provides the results and allows us to compare the changes in Earth's energy budget that resulted from the increase in greenhouse gases over the intervening years between the two publications. The total net anthropogenic forcing increased from 1.66 to about 2.3 W/m^2 as a result of greenhouse gas emissions over the intervening 5 years.

Figure 2.13 shows how anthropogenic gases accumulated between the publication of AR4 and AR5. By 2015, the increase in radiative forcing since 1990 had increased 37 percent.[22] Of the greenhouse gases analyzed by the IPCC, carbon dioxide accounts for the largest share of radiative forcing since 1990, and its contribution continues to grow

at a steady rate. In fact, carbon dioxide alone accounts for 30 percent of the increase in radiative forcing since 1990.

As we see again and again, greenhouse gases produced by human activities have caused an overall warming influence on Earth's climate since 1750, and the largest contributor is carbon dioxide, followed by methane and black carbon. Although aerosol pollution and certain other activities have caused temporary and localized cooling, the net result is that human activities on the whole are responsible for warming the atmosphere. Throughout this text, we explore the impacts of this warming in detail.

2.3.2 Assigning Radiation Values to Human Behavior

Another approach to understanding the factors driving Earth's radiation imbalance is to assign radiation values to specific kinds of human behavior. Each segment of the economy, such as operating automobiles, doing agricultural work, generating power, or burning dung to boil water (e.g., common in India), emits a specific combination of gases and aerosols that influence the greenhouse effect in different ways and at different times. By reorganizing Figure 2.13 according to economic sectors, a profile emerges of how humans are affecting climate.

In this version[23] of radiative forcing, the impacts of various human activities can be calculated for the near term (2020) and the end of the century (2100). For the next decade or so, cars, trucks, and buses emerge as the greatest contributor to atmospheric warming (**Figure 2.14**). Motor vehicles release greenhouse gases that promote warming, and they emit few aerosols that counteract it. The next most important contributor is burning household biofuels, primarily wood and animal dung for heating and cooking. Third in line is the methane produced by livestock, particularly methane-producing cattle (whose numbers have grown enormously above natural levels due to industrial agriculture operations and whose methane emissions are amplified by a diet dedicated to rapid growth).

But the picture changes somewhat by the end of this century. Assuming that greenhouse gas emissions today remain relatively constant in the near-term future, electric power generation will overtake road transportation as the biggest promoter of warming, and the industrial sector will shift from the smallest contribution in 2020 to the third largest by 2100. These changes will occur because the aerosols produced by household biofuels have short lifetimes in the atmosphere and eventually rain out. But power generation and industrialization generate long-lived CO_2, and their impacts will accumulate and intensify over time.

[18] Unger, N., et al. (2010) Attribution of Climate forcing to economic sectors, *Proceedings of the National Academy of Science*, 107(8), 3382–3387, www.pnas.org/cgi/doi/10.1073/pnas.0906548107

[19] Lauer, A., et al. (2010) The impact of global warming on marine boundary layer clouds over the Eastern Pacific—A regional model study, *Journal of Climate*, 23(21), 5844, doi: 10.1175/2010JCLI3666.1. See Cloud study predicts more global warming, *ScienceDaily*, http://www.sciencedaily.com/releases/2010/11/101122172010.htm

[20] IPCC (2007) Climate Change 2007: Synthesis Report. Contribution of Working Groups I, II and III to the Fourth Assessment Report of the Intergovernmental Panel on Climate Change [Core Writing Team, Pachauri, R.K and Reisinger, A. (eds.)]. IPCC, Geneva, Switzerland, 104 pp.

[21] IPCC (2013) Summary for Policymakers. In: *Climate Change 2013: The Physical Science Basis. Contribution of Working Group I to the Fifth Assessment Report of the Intergovernmental Panel on Climate Change* [Stocker, T.F., et al. (eds.)]. Cambridge University Press, Cambridge, United Kingdom and New York, NY, USA.

[22] U.S. Environmental Protection Agency on Climate change indicators, https://www.epa.gov/climate-indicators/climate-change-indicators-climate-forcing#ref2.

[23] Unger et al. (2010)

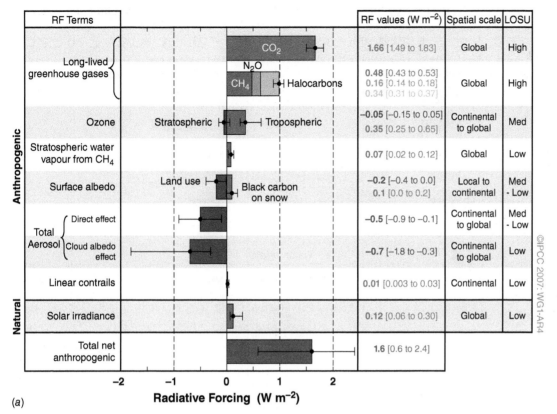

(a)

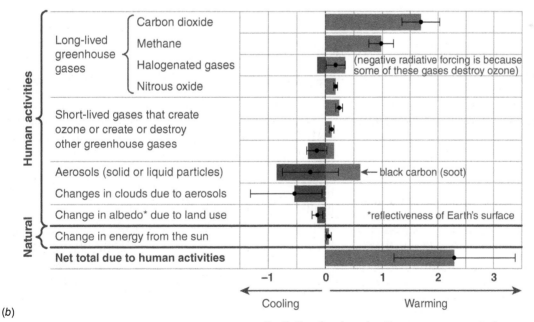

(b)

FIGURE 2.13 Estimates of global average radiative forcing (RF) made by the IPCC, AR4 (a) and AR5 (b). Radiative forcing is the net effect of various factors that cool *(bars to the left)* or warm *(bars to the right)* the atmosphere. RF is measured in W/m² *(bottom axis)*. In both (a) and (b) the *top three quarters* of the box reports on human-induced RF, and the lower portion reports on long-term changes in the Sun, the only persistent natural factor (volcanic and ENSO effects are short-lived). In both panels, the total net effect of human activities is strong warming *(*Net Total, *bottom)*, and CO_2 is the most important human factor. Total anthropogenic RF in 2006 (a, 1.66 W/m²) increased to about 2.3 W/m² (b) in 2011 between the publication of the two reports because of the continued accumulation of anthropogenic greenhouse gases.

Source: Intergovernmental Panel on Climate Change, Assessment Report 4

Source: Intergovernmental Panel on Climate Change, Assessment Report 5

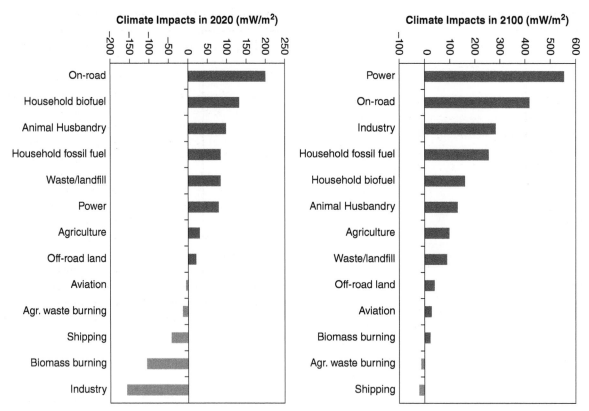

FIGURE 2.14 By 2020 *(left)*, transportation, household biofuels, and animal husbandry have the greatest warming impact on the climate, and the shipping, biomass burning, and industrial sectors have a cooling impact. Data plotted as yearly mean radiative forcing (mW/m²). By 2100 *(right)*, the power and industrial sectors become strong contributors to global warming as the impacts of long-lived carbon dioxide accumulate.

Source: NASA

2.3.3 Radiative Forcing That Promotes Cooling

Industrialization releases a high number of sulfates[24] and other aerosols (fine liquid or solid particles), leading to a significant amount of cooling. Biomass burning (such as tropical forest fires, deforestation, and savannah and shrub fires) produces black soot and greenhouse gases, but it also emits particles that block solar radiation. Poor air quality can produce health problems, however, and many developed countries have been reducing aerosol emissions through technology improvements driven by policies promoting public health (e.g., the Clean Air Act in the United States passed in 1963 and amendments in 1970, 1977,

and 1990 eliminated many cooling aerosols in the United States). By reducing air pollution, such efforts also decrease the cooling effect of aerosol production, likely leading to accelerated global warming.

These results indicate that to reduce radiative forcing caused by human activities, policy makers can focus on decreasing emissions from transportation, household biofuel, and animal husbandry. Targeting the transportation sector may be particularly effective, because it would yield both short-term and longer-term climate benefits. Public health research indicates that traffic-related particulate matter is more toxic than particulates from the power sector,[25] and by reducing industrial particles there are benefits for human health. To protect Earth's climate in the longer-term[26] and tackle concerns about climate change toward the end of this century, emphasis can be placed on reducing emissions from the power and

---

[24] The sulfate anion (SO_4) consists of a central sulfur atom surrounded by four oxygen atoms in a tetrahedral arrangement. Sulfates occur as microscopic particles in the atmosphere (aerosols) that result from the combustion of fossil fuels and biomass. They increase the acidity of the atmosphere and form acid rain. The main effect of sulfates on the climate involves scattering of sunlight. This increases Earth's albedo. This effect leads to a cooling of about 0.4 W/m² relative to preindustrial values, partially offsetting the larger (about 2.4 W/m²) warming effect of greenhouse gases. The effect is strongly geographically controlled and is largest downstream of large industrial areas.

---

[25] Grahame, T. and Schlesinger, R. (2007) Health effects of airborne particulate matter: Do we know enough to consider regulating specific particle types or sources?, *Inhalation Toxicology*, 19, 457–481.

[26] VIDEO: Over 10 years ago NASA Scientist James Hansen spoke about the Urgency of the climate crisis. https://www.youtube.com/watch?v=f0hHlxaYNb0

industry sectors,[27] a conclusion that is consistent with findings of other research.[28]

2.3.4 Greenhouse Gases

Earth's atmosphere is composed mostly (99 percent) of oxygen and nitrogen, but neither of these gases absorbs infrared energy, so they do not play a role in warming Earth. There are six principal greenhouse gases in Earth's atmosphere that absorb long-wave, thermal infrared radiation and keep Earth warm:

- Carbon dioxide (CO_2)
- Methane (CH_4)
- Ozone (O_3)
- Nitrous oxide (N_2O)
- Chlorofluorocarbons (CFCs)
- Water vapor (H_2O)

Combined, these gases make up less than 1 percent of the atmosphere, but their heat-trapping ability is strong. Because greenhouse gases are efficient at trapping long-wave radiation from Earth's surface (**Figure 2.15**), even their small percentage is enough to keep temperatures in the ideal range for liquid water (and life) to exist on Earth. If the abundance of these gases increases, more heat is trapped. If their abundance decreases, less heat is trapped. Theoretically, as the concentration of greenhouse gases

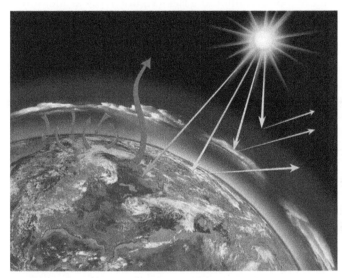

FIGURE 2.15 Energy from the Sun powers the climate system. About 29% of incoming solar radiation is reflected off high-albedo regions, the atmosphere absorbs about 23%, and 48% is absorbed by Earth's surface. Heat that is emitted from Earth's surface can be absorbed and reemitted in all directions by greenhouse gases. As the concentration of greenhouse gases grows in the atmosphere, more heat accumulates.

grows and trap greater amounts of heat that would otherwise be escaping to space, sensors in space should detect a cooling Earth, while on the surface it should be getting warmer. Indeed, this is exactly what satellites observe.[29]

Each greenhouse gas contributes differently to warming the atmosphere. Its role is affected by both the characteristics of the gas and its abundance. For example, methane is about eight times stronger at trapping heat than carbon dioxide, but it is not as abundant, so its total contribution is smaller. However, as its abundance increases, its role in global warming increases. Also, different gases have different residence times in the atmosphere: Water recycles within a few hours to a few days, methane resides only a decade or so, and carbon dioxide may stay in the atmosphere a few decades to over 1,000 years.

These gases can be described by their net contribution to the greenhouse effect[30] : Water vapor contributes about 50 percent, clouds contribute approximately 25 percent, carbon dioxide contributes about 20 percent,[31] and the other gases such as methane and ozone contribute minor amounts.[32]

Water Vapor as a Feedback Although this accounting identifies water vapor as the dominant greenhouse gas, in reality individual water molecules only reside in the atmosphere for a few days. But as the temperature of the atmosphere rises, more water is evaporated and the total concentration of water vapor increases. Hence, it accumulates in the atmosphere as a *positive feedback* to the warming caused by other gases, principally carbon dioxide. As a result, the most powerful greenhouse gas is carbon dioxide because once it is in the atmosphere it is only removed in any significant abundance by geologic processes (such as the formation of limestone, $CaCO_3$), requiring long time scales to complete. Additionally, its ability to trap heat is magnified because it drives the positive feedback of other gases, such as water vapor.

Clouds Clouds are an important though still poorly understood contributor to the greenhouse effect. This is because they absorb and emit infrared radiation similar to greenhouse gases, but they also reflect sunlight, a cooling effect. Different types of clouds do more of one than the other depending on whether they are high or low altitude,

[27] Unger et al. (2010)

[28] Jacobson, M. (2004) The short-term cooling but long-term global warming due to biomass burning, *Journal of Climate*, 17, 2909–2926.

[29] Harries, J. (2001) Increases in Greenhouse forcing inferred from the outgoing longwave radiation spectra of the Earth in 1970 and 1997, *Nature*, 410, 355–357.

[30] Schmidt et al. (2010)

[31] Lacis et al. (2010) A discussion is available at NASA, Carbon dioxide control's Earth's temperature, http://www.nasa.gov/topics/earth/features/co2-temperature.html

[32] Trenberth, K.E., et al. (2009) Earth's global energy budget, *Bulletin of the American Meteorological Society*, 90 (3), 311–324, doi: 10.1175/2008BAMS2634.1. See also Trenberth, K.E.(2009) An imperative for climate change planning: Tracking Earth's global energy, *Current Opinion in Environmental Sustainability*, 1(1), 19–27, http://www.sciencedirect.com/science/article/pii/S1877343509000025

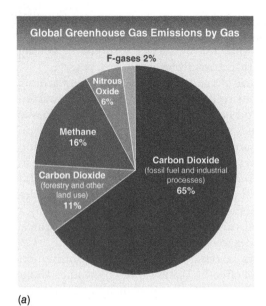

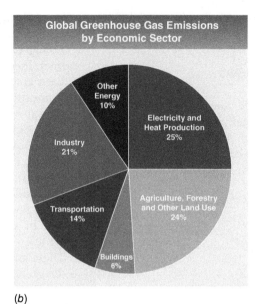

FIGURE 2.16 (a) The primary greenhouse gases emitted globally include carbon dioxide from fossil fuels, industrial processes, forestry, and other land uses; methane; nitrous oxide; and the fluorinated gases. (b) The primary economic activities that lead to greenhouse gas production include electricity and heat production (25%), industrial activities (21%), agriculture, forestry, and other land uses (24%), transportation (14%), energy generation in buildings (6%), and other forms of energy uses (10%) such as fuel extraction and processing.

Source: U.S. EPA

more or less abundant, thick or thin, and depending on how these factors change in a warming world. Generally speaking, as the planet warms, clouds have a cooling effect if there are more low-level clouds or fewer high-level clouds. If the opposite were true, clouds would cause more warming. To work out the overall effect, scientists need to know which types of clouds are increasing or decreasing.[33]

The net effect of clouds on climate change depends on many variables and cloud research is a growing and major area of study. For example,[34] as climate warms, more water evaporates; thus it should be cloudier, with thicker and denser clouds. However, warmer air requires more water molecules to reach saturation and condense into clouds, thus limiting cloud development. Similarly, although summer is warmer and more humid than winter, the sky is not noticeably cloudier. Despite the complexities of these and other factors, several studies indicate that clouds are, in fact, amplifying global warming.[35, 36] Clouds are discussed in-depth in **Chapter 6**.

Emissions The National Oceanic and Atmospheric Administration (NOAA) created the **Annual Greenhouse**

Gas Index (AGGI).[37] The AGGI is a yearly report on the combined influence of long-lived atmospheric greenhouse gases on Earth's surface temperature. The index compares the warming influence of these gases each year to their influence in 1990, the year that countries who signed the U.N. Kyoto Protocol agreed to use as a benchmark for their efforts to reduce emissions. By the end of 2015, the warming influence of greenhouse gases (**Figure 2.16**) as measured by the AGGI had risen 37 percent above the 1990 baseline.

Carbon Dioxide (CO_2) Fossil fuel use is the primary source of CO_2 (**Figure 2.17**). The way in which people use land is also an important source of CO_2, especially when it involves deforestation, land clearing for agriculture, and degradation of soils. However, proper land use can also remove CO_2 from the atmosphere through reforestation, improvement of soils, and other activities.

Carbon dioxide and the other greenhouse gases have increased in abundance because of human activity, but the reasons differ for each gas. The amount of carbon dioxide in the atmosphere has varied significantly during Earth's history, and it began doing so long before modern humans inhabited the planet. There are four ways human activity

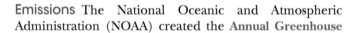

[33] VIDEO: How climate models treat clouds, https://www.youtube.com/watch?v=_YZxqRM97eo

[34] NASA, Clouds and climate change: The thick and thin of it, http://www.giss.nasa.gov/research/briefs/delgenio_03/

[35] Clement, A.C., et al. (2009) Observational and model evidence for positive low-level cloud feedback, *Science*, 325(5939), 460–464. See also Lauer et al. (2010); see also Dessler, A.E., et al. (2013) Stratospheric water vapor feedback, *Proceedings of the National Academy of Sciences*, 110, doi: 10.1073/pnas.1310344110, 18,087-18,091

[36] VIDEO: The role of clouds in climate change. https://www.youtube.com/watch?v=TW33e9J3fRc

[37] NOAA's Annual Greenhouse Gas Index https://esrl.noaa.gov/gmd/aggi/. The AGGI is a measure of the warming influence of long-lived trace gases and how their influence is changing each year. The index was designed to enhance the connection between scientists and society by providing a normalized standard that can be easily understood and followed. The warming influence of long-lived greenhouse gases is well understood by scientists and has been reported by NOAA through a range of national and international assessments. Nevertheless, the language of scientists often eludes policy makers, educators, and the general public. This index is designed to help bridge that gap. The AGGI provides a way for this warming influence to be presented as a simple index.

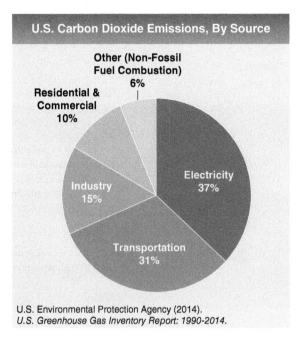

U.S. Carbon Dioxide Emissions, By Source

Other (Non-Fossil Fuel Combustion) 6%

Residential & Commercial 10%

Industry 15%

Electricity 37%

Transportation 31%

U.S. Environmental Protection Agency (2014).
U.S. Greenhouse Gas Inventory Report: 1990-2014.

FIGURE 2.17 In the United States, the main human activity that emits CO_2 is the combustion of fossil fuels (coal, natural gas, and oil) for energy, transportation, and industrial activities. Heating and cooling residential and commercial properties also generates carbon dioxide.

Source: U.S. EPA

is known to have caused the observed rapid increase in atmospheric CO_2 over the last few centuries.

1. Various national statistics are collected that account for fossil fuel consumption, combined with knowledge of how much atmospheric CO_2 is produced per unit of fossil fuel (e.g. gallons of gasoline). These are tracked by various entities and provide reproducible datasets.

2. By examining the ratio of various carbon isotopes in the atmosphere. The burning of long-buried fossil fuels releases CO_2 containing carbon of different isotopic ratios to those of living plants, enabling distinction between natural and human-caused contributions to CO_2 concentration.

3. Higher atmospheric CO_2 concentrations in the northern hemisphere, where most of the world's population lives (and emissions originate from), compared to the southern hemisphere. This difference has increased as anthropogenic emissions have increased.

4. Atmospheric O_2 levels are decreasing in Earth's atmosphere as it reacts with the carbon in fossil fuels to form CO_2.

Natural sources of carbon dioxide include volcanic outgassing (volcanoes exhaust CO_2 released from molten rock during and between eruptions), animal respiration, and decay of organic matter (decaying tissue is made of carbon, C, which combines with O_2 in the atmosphere to make CO_2).

The air we inhale is roughly 78 percent by volume nitrogen (N_2), 21 percent oxygen (O_2), 0.96 percent argon (Ar), and 0.04 percent carbon dioxide (CO_2), helium (He), water (H_2O), and other gases. The permanent gases we exhale are roughly 4 to 5 percent more carbon dioxide and 4 to 5 percent less oxygen than was inhaled (the difference in O_2 is used to fuel our metabolism, and CO_2 is a waste product that we expel in our breath).

Scientists can measure the concentration of past atmospheric carbon dioxide and other gases by analyzing air bubbles trapped in ice cores (**Figure 2.18**) and the chemistry of ancient sediments[38] and by employing other techniques using **geologic proxies** of climate such as the chemical composition of fossils (e.g., corals) and various plankton from freshwater and marine ecosystems. These methods have helped scientists understand long-term trends in carbon dioxide variability and global climate change caused by natural factors. In **Chapter 4**, we expand on the field of paleoclimatology and what scientists have learned about past climate changes.

Excess carbon dioxide is added to the atmosphere by human activities, in particular, the burning of fossil fuels (e.g., oil, natural gas, and coal), the burning of solid waste for fuel (e.g., dung, peat, wood, mostly in Asia), deforestation (e.g., logging and clearing land for farming and development), industrial agricultural (a large CO_2 source and a source of other greenhouse gases such as methane), and cement production (cement plants account for five percent of global emissions of CO_2[39]).

These anthropogenic emissions are at the center of research on global warming. The sources and heat-trapping properties of greenhouse gases are undisputed, but there is uncertainty about the details of how Earth's climate will respond to increasing concentrations of the various gases.

There is wide consensus among scientists around the world (including leading climate researchers in the United States, like those at the National Climate Data Center[40] in Asheville, North Carolina, and at the National Center for Atmospheric Research[41] in Boulder, Colorado) that if anthropogenic emissions of carbon dioxide continue to rise, by the end of this century average global temperatures will increase, perhaps by as much as 2.4 to 6.4 degrees Celsius (4.3 to 11.5 degrees Fahrenheit).[42] Scientists also agree that industrial emissions have been the dominant influence

[38] Tripati, A.K., et al. (2009) Coupling of CO_2 and ice sheet stability over major climate transitions of the last 20 million years, *Science*, 326(5958), 1394–1397, http://www.sciencemag.org/cgi/content/abstract/1178296

[39] Rosenthal, E. (2007) Cement industry is at center of climate change debate, *New York Times*, October 26, http://www.nytimes.com/2007/10/26/business/worldbusiness/26cement.html

[40] NOAA Satellite and Information Service, NCDC frequently asked questions, http://www.ncdc.noaa.gov/faqs/index.html

[41] Climate of the future at NCAR, Learn more about climate, http://ncar.ucar.edu/learn-more-about/climate

[42] This is the IPCC-AR4 best estimate for a "high scenario." IPCC (2007)

Antarctic time series for CO₂, CH₄, and temperature

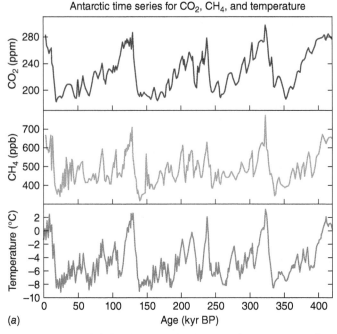

Karim Agabi/Photo Researchers

FIGURE 2.18 (a) Global carbon dioxide content (CO_2 in parts per million [ppm]), methane content (CH_4 in parts per billion [ppb]), and temperature (°C) over the past 400,000 years have been measured using fossil air trapped in ice in Antarctica. (b) Scientists drill ice cores on mountain glaciers, as well as in Greenland and Antarctica, to obtain evidence of past atmospheric composition.

on climate change for the past 50 years, overwhelming natural causes.[43]

In 2002, the National Aeronautical and Space Agency launched a new satellite into space. NASA's Atmospheric Infrared Sounder, or AIRS,[44] significantly increased weather forecasting accuracy within a couple of years by providing extraordinary three-dimensional maps of clouds, air temperature, water vapor, and other gases throughout the atmosphere's weather-making layer (the troposphere).

Fifteen years later, AIRS[45] continues to be a valuable asset for forecasters worldwide, sending 7 billion observations streaming into forecasting centers every day. AIRS is able to make so many detailed observations because it records 2,378 wavelengths of heat radiation in the air below the satellite. Having more wavelengths allows researchers to see a very highly resolved three-dimensional structure of the atmosphere that provides a much sharper picture than was previously available.

Among the observations that are made by AIRS is the measurement of carbon dioxide, making it possible to measure the global distribution of carbon dioxide in the troposphere every day (**Figure 2.19**).[46] One of the natural processes that this data has allowed scientists to better understand is the short-term movement of carbon dioxide

between plants, the ocean, human emissions, watersheds and soil, and other places where it is stored and released. About half of the carbon released by human activity enters the atmosphere where it contributes to global warming. Oceans, watersheds, and plants absorb the rest as part of the natural movement of carbon through Earth environments—a process known as the **carbon cycle**, explored more in **Chapter 4**.

In the geologic past, climate changes occurred naturally. In the past half million years, carbon dioxide concentration remained between about 180 ppm (parts per million) during ice ages (also called glacial periods), and 280 ppm during warm periods (also called interglacial periods), such as today. But carbon dioxide content has been much greater in other periods of Earth's history.

Estimates of carbon dioxide content during geologic history are based on the chemistry of fossilized soils, fossil plants, and fossil shells of plankton. These indicate that concentrations as high as 1,000 to 4,000 ppm[47] may have occurred for sustained periods, and even reached twice this level. The cause of such high levels is controversial among researchers who study this history: Episodes of extreme global volcanism, changes in land surface area as a result of plate tectonics, reorganization of ocean circulation, absence of polar ice, mountain building, and other

[43] Lacis et al. (2010)
[44] VIDEO: AIRS https://www.jpl.nasa.gov/news/news.php?feature=6836
[45] AIRS 15th birthday: https://www.jpl.nasa.gov/news/news.php?feature=6836
[46] NASA AIRS, https://airs.jpl.nasa.gov/data/carbon_dioxide

[47] Berner, R.A. (1997) The rise of plants and their effect on weathering and atmospheric CO_2, *Science*, 276, 544–546. See also Pagani, M., et al. (1999) Miocene evolution of atmospheric carbon dioxide, *Paleoceanography*, 14, 273–292. See Carbon dioxide in Earth's atmosphere, *Wikipedia*, http://en.wikipedia.org/wiki/Carbon_dioxide_in_Earth's_atmosphere

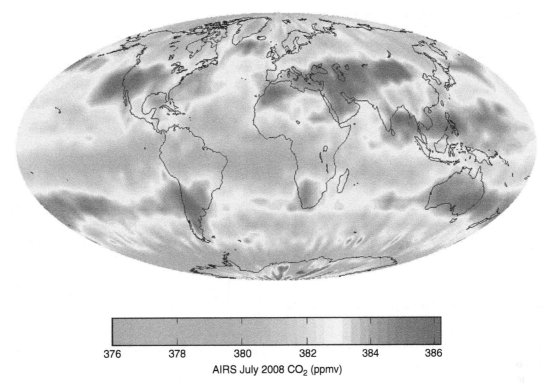

FIGURE 2.19 The level of carbon dioxide in Earth's atmosphere has been on the rise since the late 19th century. Originally collected in flasks by hand, and then by instrument, CO_2 is now measured by a global network of monitoring stations and satellites. One of these, the Atmospheric Infrared Sounder (AIRS),[48] has been able to pinpoint the influence of specific carbon dioxide sources. For instance, it identified a large amount of carbon dioxide cycling around 40°S to 50°S latitude— the Roaring 40s—fed by two huge anthropogenic sources: a coal liquefaction plant in South Africa that is the largest single source of carbon dioxide on Earth, and a cluster of power generation plants in eastern Australia.[49]

Source: NASA

mechanisms have all been suggested. However, it is clear that the level of only 180 ppm during glaciations is not far from the lowest that has ever occurred since the rise of macroscopic life on Earth in the past half billion years.

In the 200 years since the Industrial Revolution in the early 1800s, humans have altered Earth's environment through agricultural and industrial practices. The growth of the human population and activities such as deforestation and burning of fossil fuels have affected the mixture of gases in the atmosphere. We know from ice cores that the amount of carbon dioxide in the atmosphere prior to the Industrial Revolution was about 280 ppm. Today (2018) the global average concentration of carbon dioxide is about 411 ppm and rising, higher than at any other time in the past 15 million years.[50]

Although carbon dioxide is not the most effective absorber of heat compared to other greenhouse gases, it is one of the most abundant, and once in the atmosphere

it can stay there for a very long time. Carbon dioxide can reside in the atmosphere for more than 1,000 years, resulting in essentially irreversible climate change.[51]

Monthly records of atmospheric CO_2 concentration collected at the Mauna Loa Observatory in Hawaii[52] show seasonal oscillations superimposed on a long-term increase in CO_2 in the atmosphere (**Figure 2.20**). This increase is attributed to the human activities associated with the rise of modern industry, primarily the burning of fossil fuels and deforestation. The most recent monthly average (May 2018) CO_2 concentration measured at Mauna Loa observatory was 411.25 ppm.[53]

Of all the greenhouse gases released by human activities, carbon dioxide is the largest individual contributor to the anthropogenic greenhouse effect, accounting for about 60 percent of the excess heating compared to the other

[48] NASA Jet Propulsion Lab, AIRS and carbon dioxide: From measurement to science, http://airs.jpl.nasa.gov/story_archive/Measuring_CO2_from_Space/Measurement_to_Sci

[49] VIDEO: NASA A year in the life of Earth's CO_2, https://www.youtube.com/watch?v=x1SgmFa0r04

[50] Tripati et al. (2009)

[51] Solomon, S., et al. (2009) Irreversible climate change due to carbon dioxide emissions, *Proceedings of the National Academy of Science*, 106, 1704–1709, doi: 10.1073/pnas.-9128211-6

[52] The NOAA Earth System Research Laboratory on Mauna Loa, Hawaii, measures carbon dioxide daily: http://www.esrl.noaa.gov/gmd/ccgg/trends

[53] Recent monthly averages, https://www.esrl.noaa.gov/gmd/ccgg/trends/index.html

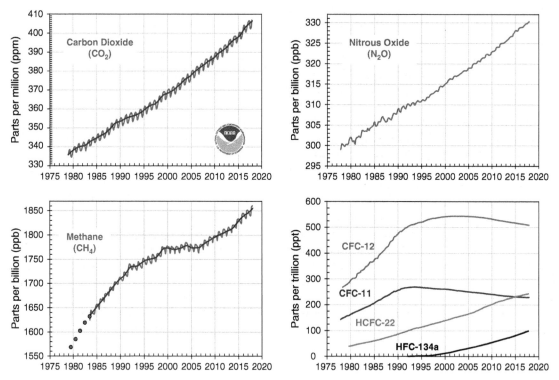

FIGURE 2.20 Concentration of the most important greenhouse gases in Earth's atmosphere. Clockwise from upper left: Levels of carbon dioxide and nitrous oxide continue to climb. Levels of CFCs have declined since the Montreal Protocol (see text for discussion) was implemented in 1987. The concentration of methane stabilized early in this century as a result of droughts and a temporary decline in industrial emissions, but it has since returned to its previous pattern of steady increases.

Source: NOAA

human sources. After a rapid increase in global emissions of around 3 percent per year between 2000 and 2013, emissions only grew by 0.4 percent per year between 2013 and 2016. However, in 2017, global energy-related carbon dioxide emissions increased by 1.4 percent to 32.5 billion tons, a record high.

In 2014, CO_2 accounted for about 80.9 percent of U.S. anthropogenic greenhouse gas emissions. The main human activity that emits CO_2 is the combustion of fossil fuels (coal, natural gas, and oil) for energy and transportation, although certain industrial processes and land use changes also emit CO_2. The main sources of CO_2 emissions in the United States include electricity generation (37 percent), transportation (31 percent), industrial activities that combust fossil fuels (15 percent), residential and commercial energy uses (10 percent), and other sources of carbon dioxide such as land use (6 percent).

How did the global recession of 2008 affect carbon dioxide emissions? After a 1 percent decline in 2009, global CO_2 emissions increased[54] by more than 6 percent in 2010, producing the highest annual net increase in carbon pollution

ever measured. The world pumped about 564 million more tons of carbon into the air in 2010 than it did in 2009, a rate that exceeded the worst-case scenario of the Intergovernmental Panel on Climate Change Assessment Report 3 in 2007. Carbon dioxide production went up in most of the major economies, led by increases in China (10 percent increased emissions) and India (9 percent increased emissions).[55] The average annual growth rate in CO_2 emissions over the three years of the global recession, including a 1 percent increase in 2008 when the first impacts became visible, was 1.7 percent, almost equal to the long-term annual average of 1.9 percent for the preceding two decades back to 1990.

However, by 2014 emissions of CO_2 had lessened to just 0.7 percent growth, they held nearly steady (0.1 percent growth) in 2015, and in 2016 growth was only 0.2 percent. This was a notable slowdown in emission growth, compared to an average rate of 3.5 percent in the 2000's and 1.8 percent over the most recent decade, 2006 to 2015.

In 2017, global energy demand increased by 2.1 percent, compared with 0.9 percent the previous year and

[54] Bornstein, S. (2011) Biggest jump ever seen in global warming gases, Associated Press, November 4, http://news.yahoo.com/biggest-jump-ever-seen-global-warming-gases-183955211.html

[55] PBL Netherlands Environmental Assessment Agency (2011) Long-term trend in global CO_2 emissions, 2011 report, September 21, http://www.pbl.nl/en/publications/2011/long-term-trend-in-global-co2-emissions-2011-report

0.9 percent on average over the previous five years. More than 40 percent of the growth in 2017 was driven by China and India; 72 percent of the rise was met by fossil fuels, a quarter by renewables and the remainder by nuclear.

Global energy-related CO_2 emissions grew by 1.4 percent in 2017, reaching a historic high of 32.5 billion tons (Gt), a resumption of growth after three years of global emissions remaining flat. The increase in CO_2 emissions, however, was not universal. While most major economies saw a rise, some others experienced declines, including the United States, United Kingdom, Mexico, and Japan. The biggest drop came from the United States, mainly because of higher deployment of renewables.[56]

However, while emissions declined, the accumulation of CO_2 in the atmosphere accelerated. Data from the National Oceanic and Atmospheric Administration Earth System Research Laboratory[57] showed that the growth rate of atmospheric CO_2 in 2015 reached an all-time high of 3.02 ppm/year. In 2016, measurements showed an accumulation of 2.98 ppm/year, the second highest on record, and in 2017 carbon dioxide grew by 1.95 ppm/year. It is not unusual for there to be a mismatch between emissions of anthropogenic carbon dioxide and the observed growth rate in the atmosphere. Carbon dioxide participates in the global carbon cycle, a complex system of carbon sources and storage sites (both short term and long term) that mean one year's releases will not be reflected immediately in the annual monitoring data. The good news is that human reliance on carbon as an energy source appears to be decreasing and, if continued, should eventually be reflected in the atmospheric monitoring data.

Methane (CH_4) Natural sources of methane include the activity of microbes and insects in wetlands, seawater, and soils; wildfires; and the release of gases stored in ocean sediments. Methane is also emitted during the production and transport of coal, natural gas, and oil. Methane emissions additionally result from livestock[58] and other agricultural practices and by the decay of organic waste in municipal solid waste landfills (**Figure 2.21**). About 60 percent of annual methane emissions come from anthropogenic sources.[59]

Domestic livestock such as cattle, buffalo, sheep, goats, and camels produce large amounts of CH_4 as part of their normal digestive process. Also, when animals' manure is stored or managed in lagoons or holding tanks, CH_4 is produced. Because humans raise these animals for food,

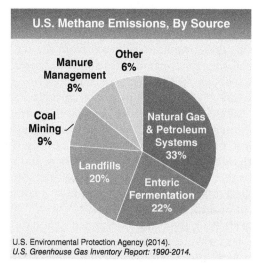

FIGURE 2.21 U.S. emissions of methane (CH_4). Methane is the second most prevalent greenhouse gas emitted by human activities. In 2014, CH_4 accounted for about 11% of all U.S. greenhouse gas emissions from human activities.

Source: U.S. EPA

the emissions are considered human-related. Methane accounts for about 11 percent of all U.S. greenhouse gas emissions from human activities.

The present global atmospheric concentration of methane is more than 1,800 parts per billion (ppb), more than double what it was before the Industrial Revolution. Methane levels increased steadily in the 1980s, but the rate of increase slowed in the 1990s and was close to zero from 2000 to 2007.

Researchers attribute this lull to a temporary decrease in emissions during the 1990s related to the decline of industry and farming when the former Soviet Union collapsed, along with a slowdown in wetland emissions during prolonged droughts. Now, with methane emissions accelerating again, scientists warn that a more typical rate of increase will have a significant impact on climate.[60]

Methane is 25 times more potent as a greenhouse gas than carbon dioxide, but there is far less of it in the atmosphere and it is measured in parts per billion. When related climate effects are taken into account, methane's overall climate impact is less than half that of carbon dioxide; thus, methane is second only to carbon dioxide as a cause of global warming.

Methane is also trapped in ice, glaciers, frozen seafloor sediment, and the permafrost in tundra and under the rapidly disappearing sea ice of the Arctic Ocean[61]; as

[56] International Energy Agency (2018) *Global energy & CO2 status report*, March. http://www.iea.org/publications/freepublications/publication/GECO2017.pdf

[57] NOAA ESRL https://www.esrl.noaa.gov/gmd/ccgg/trends/gr.html

[58] Reisinger, A. and Clark, H. (2018) How much do direct livestock emissions actually contribute to global warming? *Global Change Biology*, v. 24.4, April, p. 1749-1761, doi:10.1111/gcb.13975

[59] U.S. greenhouse gas inventory report: 1990–2014, https://www.epa.gov/ghgemissions/us-greenhouse-gas-inventory-report-1990-2014

[60] National Science Foundation, Methane releases from Arctic Continental Shelf, http://www.nsf.gov/news/news_summ.jsp?cntn_id=116532

[61] Kort, E., et al. (2012) Atmospheric observations of Arctic ocean methane emissions up to 82° north, *Nature Geoscience*, doi: 10.1038/ngeo1452

melting of all these frozen sources occurs, the gas is released to the atmosphere. There is fear that as frozen regions thaw, methane released from the ice will add to atmospheric concentrations and constitute a positive feedback.[62, 63]

In the case of methane, climate change melts permafrost (**Figure 2.22**), releasing more methane, causing more warming, melting more permafrost, releasing more methane, and so on. Unlike CO_2, methane is destroyed by reactions with other chemicals in the atmosphere and soil, so its atmospheric lifetime is about a decade. But if it is released rapidly and in large quantities, it could drive a potent positive feedback process.[64]

In March 2010, the National Science Foundation (NSF) issued a remarkable press release warning that methane escaping from the Arctic continental shelf of Siberia has been observed to be much larger and faster than anticipated.[66] Researchers identified a section of the Arctic Ocean seafloor with vast stores of frozen methane showing signs of instability and widespread venting of the gas.

FIGURE 2.22 Permafrost bluffs from Barter Island in northeastern Alaska. As permafrost thaws, quantities of methane are liberated; as the soil warms, microbes digest vegetation contained in the frozen ground, releasing methane as a by-product. Potentially, this process could set a feedback cycle into motion, amplifying atmospheric warming, increasing permafrost thaw, and promoting the release of more methane.[65]

Source: Photograph by Ben Jones of the USGS Alaska Science Center

A paper published in *Science*[67] showed that permafrost under the East Siberian Arctic Shelf, long thought to be an impermeable barrier sealing in methane, is instead perforated and leaking gas into the atmosphere. The amount of methane being released from just this one area is comparable to the amount coming out of all of the world's oceans combined.

High volumes of methane release by decaying permafrost in the oceans and on the land have been identified by some researchers as a potential climate tipping point. A **tipping point**[68] occurs when a positive feedback cannot be recovered, implying that human efforts to decrease greenhouse gas emissions might not be sufficient to stop the warming.

A paper[69] published in 2013 put a threshold on the question of when frozen methane in permafrost will reach a tipping point—1.5 degrees Celsius (2.7 degrees Fahrenheit) above preindustrial greenhouse gas levels. Evidence from Siberian caves suggests that a global temperature rise of 1.5 degrees Celsius could see permanently frozen ground thaw over a large area of Siberia, threatening release of carbon from soils, and damage to natural and human environments. Such a thaw could release over 1,000 billion tons of carbon dioxide and methane into the atmosphere, strongly enhancing global warming.

Another important source of methane is industrial agriculture. Livestock now use 30 percent of Earth's entire land surface. This mostly includes permanent pasture but also counts 33 percent of the global arable land that is used for producing feed for livestock. Agriculture is also a major driver of deforestation as forests are cleared to create new pastures. This is especially true in Latin America where, for example, some 70 percent of former forests in the Amazon have been turned over to grazing.

For these reasons, the *Food and Agriculture Organization of the United Nations* (FAO) has declared that livestock are a major threat to the environment.[70] Agriculture is responsible for 18 percent of the total release of greenhouse gases worldwide (this is more than the whole transportation sector), and cattle-breeding is a major factor in producing these greenhouse gas emissions.

The fear of methane release from permafrost extends to the seafloor as well. See discussion of the "Clathrate Gun Hypothesis" in the *Spotlight on Climate Change*.

[62] McGuire, A., et al. (2009) Sensitivity of the carbon cycle in the Arctic to climate change, *Ecological Monographs*, 79(4), 523–555, doi: 10.1890/08-2025.1. See http://www.esajournals.org/doi/abs/10.1890/08-2025.1

[63] VIDEO: Thawing permafrost: Changing planet, http://www.youtube.com/watch?v=yN4OdKPy9rM

[64] VIDEO: Permafrost: The ticking time bomb, https://www.youtube.com/watch?v=FLCgybStZ4g

[65] NOAA, Hot on methane's trail, http://researchmatters.noaa.gov/news/Pages/NOAAHotonMethane%27sTrail.aspx

[66] National Science Foundation, http://www.nsf.gov/news/news_summ.jsp?cntn_id=116532&org=NSF&from=news

[67] Shakova, N., et al. (2010) Extensive methane venting to the atmosphere from sediments of the East Siberian Arctic Shelf, *Science*, 327(5970), 1246–1250.

[68] http://www.sciencedaily.com/releases/2007/05/070531073748.htm

[69] Vaks, A., et al. (2013) Speleothems reveal 500,000-year history of siberian permafrost, *Science*, 340(6129), 183–186, doi: 10.1126/science.1228729

[70] http://www.fao.org/Newsroom/en/news/2006/1000448/index.html

Box 2.1 | Spotlight on Climate Change

The Clathrate Gun Hypothesis

Although now discounted as unlikely,[71] the **Clathrate Gun Hypothesis**, when first proposed,[72] captured the attention of the scientific world.

The hypothesis described the possibility of warming ocean water (or falling sea level) resulting in the release of massive quantities of methane by the melting of icy methane hydrate. Methane hydrate is a naturally forming substance in which a large amount of methane is trapped within an icy crystalline latticework called a clathrate (**Figure 2.23**).

Originally thought to occur only in the outer regions of the solar system where temperatures are low and water ice is common, large deposits of methane clathrate have been found within the shallow sediments of the seafloor.

These deposits are thought to form where methane gas in the crust migrates along fault planes toward the surface and precipitates or crystallizes on contact with cold seawater. In 2008, research on Antarctic Vostok and EPICA Dome C ice cores revealed that methane clathrates were also present in deep Antarctic ice cores and record a history of atmospheric methane concentrations, dating to 800,000 years ago.

The Clathrate Gun Hypothesis supposes that climate change can trigger the sudden release of methane from clathrate compounds (essentially seabed permafrost) buried on the seafloor. Because methane itself is a powerful greenhouse gas, a positive feedback effect would lead to further temperature rise and further methane clathrate destabilization—in effect initiating a runaway process as irreversible, once started, akin to firing a gun.

In its original form,[73] the hypothesis proposed that the "clathrate gun" could cause abrupt runaway warming

FIGURE 2.23 Sample of methane clathrate taken from the seafloor off Oregon.

Source: https://commons.wikimedia.org/wiki/File:Gashydrat_mit_Struktur.jpg

on a timescale less than a human lifetime. It was thought to be responsible for warming events in and at the end of the last glacial maximum; however, this is now thought to be unlikely.

There is, however, evidence that a runaway methane clathrate breakdown may have caused significant alteration of the ocean environment (acidification and/or stratification) and of the atmosphere on a number of occasions in geologic history. Over timescales of tens of thousands of years, the Paleocene-Eocene Thermal Maximum (56 million years ago) and the Permian-Triassic extinction event (252 million years ago) have been proposed as candidates.[74]

Tropospheric Ozone (O_3) The role of ozone[75] is complicated. There is "good ozone" in the upper atmosphere that blocks the Sun's harmful UV radiation and does not play a role in climate change. There is "bad ozone" at ground level that damages people's lungs and contributes to smog. There is also mid-altitude ozone that acts as a greenhouse gas.

Greenhouse ozone, also known as **tropospheric ozone**, is on the rise because cars and coal-fired power plants release air pollutants that react with oxygen to produce more tropospheric ozone. Natural sources of ozone include chemical reactions that occur among carbon monoxide (CO), hydrocarbons, and nitrous oxides (N_2O), as well as lightning and wildfires.

[71] Sowers, T., (2006) Late Quaternary atmospheric CH_4 isotope record suggests marine clathrates are stable, *Science*, 311(5762), 838–840.

[72] Kennett, J.P., et al. (2000) Carbon isotopic evidence for methane hydrate instability during Quaternary interstadials, *Science*, 288(5463), 128–133.

[73] Kennett, J.P., et al. (2003) *Methane Hydrates in Quaternary Climate Change: The Clathrate Gun Hypothesis.* American Geophysical Union, Washington D.C. ISBN 0-87590-296-0.

[74] The day the earth nearly died, Horizon, December 5, 2002. BBC.

[75] NASA page on ozone, http://www.nasa.gov/missions/earth/f-ozone.html; the U.S. Environmental Protection Agency also has an ozone report that explains the role of ozone in climate; see http://www.epa.gov/oar/oaqps/gooduphigh/

Human activities increase ozone concentrations indirectly by emitting pollutants that are precursors of ozone. These include CO, N_2O, sulfur dioxide (SO_2), and hydrocarbons that result from the burning of biomass and fossil fuels. Ozone is a strong absorber of heat, but it does not stay in the atmosphere for long, only a few weeks to a few months. Nonetheless, its concentration is increasing at a rapid rate. Concentrations of ozone have risen by around 30 percent since the preindustrial era, and ozone is now considered by the IPCC to be the third most important greenhouse gas after carbon dioxide and methane.[76]

Nitrous Oxide (N_2O) In 2014, nitrous oxide (N_2O) accounted for about 6 percent of all U.S. greenhouse gas emissions from human activities (**Figure 2.24**). The compound is emitted when people add nitrogen to soil through the use of synthetic fertilizers. Agricultural soil management is the largest source of N_2O emissions in the United States, accounting for about 79 percent of total U.S. emissions. Nitrous oxide is also emitted during the breakdown of nitrogen in livestock manure and urine, which contributes 4 percent of emissions.

Nitrous oxide is also emitted when transportation fuels are burned. Motor vehicles, including passenger cars and trucks, are the primary source of emissions from transportation. The amount of N_2O emitted from transportation depends on the type of fuel and vehicle technology, maintenance, and operating practices.

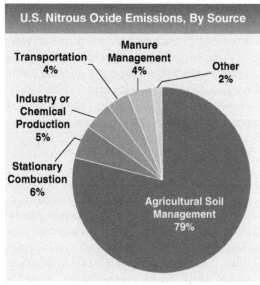

FIGURE 2.24 Globally, about 40% of total N_2O emissions come from human activities. Nitrous oxide is emitted from agriculture, transportation, and industry activities.

Source: U.S. EPA

Nitrous oxide is naturally present in the atmosphere as part of Earth's nitrogen cycle and has a variety of natural sources. However, human activities such as agriculture, fossil fuel combustion, wastewater management, and industrial processes are increasing the amount of N_2O in the atmosphere.

Nitrous oxide molecules stay in the atmosphere for an average of 114 years before being removed by a natural process or destroyed through chemical reactions. The impact of 1 pound of N_2O on warming the atmosphere is almost 300 times that of 1 pound of carbon dioxide.

Nitrous oxide[77] is a clear, colorless gas with powerful greenhouse properties. Because it has a long atmospheric lifetime and is hundreds of times more powerful than carbon dioxide at trapping heat, it is important to track where N_2O comes from and where it is stored in nature. The main natural source of nitrous oxide is the activity of microbes in swamps, soil, rainforests, and the ocean surface. Primary human sources of this greenhouse gas include fertilizers, industrial production of nylon and nitric acid, the burning of fossil fuels, and solid waste. The present atmospheric concentration of N_2O is about 330 ppb. Human activities have caused it to increase by over 16 percent since the beginning of the Industrial Revolution.

Nitrous oxide is also produced by permafrost thaw.[78] About 25 percent of the land surface in the Northern Hemisphere is underlain by permafrost. As global warming thaws these soils, it does not initially stimulate nitrous oxide production. However, as meltwater from the frozen soils flows back into the thawed sediment, it stimulates increased nitrous oxide production more than 20 times greater than it forms otherwise. Nearly one-third of the nitrous oxide produced in this process escapes into the atmosphere, adding to the positive feedback aspects of permafrost thaw.

Fluorinated Gases A number of very powerful heat-absorbing greenhouse gases in the atmosphere do not occur naturally. These include hydrofluorocarbons (HFCs), perfluorocarbons (PFCs), sulfur hexafluoride (SF_6), and nitrogen trifluoride (NF_3), all of which are produced by industrial processes (**Figure 2.25**).

CFCs are used as coolants in air conditioning (Freon is a CFC), aerosol sprays, and the manufacture of plastics and polystyrene. CFCs did not exist on Earth before humans created them in the 1920s. They are very stable compounds, have long atmospheric lifetimes, and are now abundant enough to cause global changes in air chemistry and climate.

Fluorinated gases are emitted through a variety of industrial processes such as aluminum and semiconductor manufacturing. Many fluorinated gases have very high GWPs relative to other greenhouse gases, so small atmospheric

[76] National Climatic Data Center, http://www.ncdc.noaa.gov/oa/climate/gases.html#introduction

[77] Environmental Protection Agency, Nitrous oxide: Sources and emissions, http://www.epa.gov/nitrousoxide/sources.html

[78] Elberling, B., et al. (2010) High nitrous oxide production from thawing permafrost, *Nature Geoscience*, 3, 332–335, doi:10.1038/ngeo803

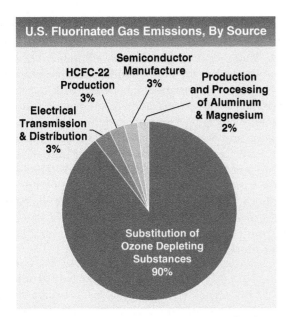

FIGURE 2.25 There are four main categories of fluorinated gases—hydrofluorocarbons (HFCs), perfluorocarbons (PFCs), sulfur hexafluoride (SF_6), and nitrogen trifluoride (NF_3). Overall, fluorinated gas emissions in the United States have increased by about 77% between 1990 and 2014. This increase has been driven by a 258% increase in emissions of hydrofluorocarbons (HFCs) since 1990, as they have been widely used as a substitute for ozone-depleting substances.

Source: U.S. EPA

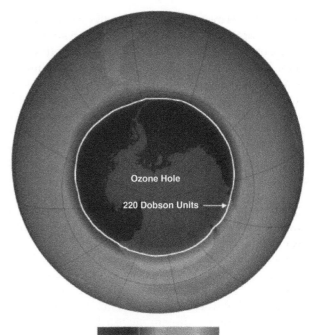

FIGURE 2.26 The ozone hole over Antarctica (October 2004). The dark region in the center has the least ozone; the surrounding lighter region has more ozone. Ozone abundance is measured in Dobson units; one Dobson unit is the number of molecules of ozone that would be required to create a layer of pure ozone 0.01 mm thick at a temperature of 0 degrees Celsius and a pressure of 1 atmosphere (the air pressure at sea level).

Source: NASA

concentrations can have large effects on global temperatures. They can also have long atmospheric lifetimes—in some cases, lasting thousands of years. Like other long-lived greenhouse gases, fluorinated gases are well mixed in the atmosphere, quickly spreading around the world after they are emitted. Fluorinated gases are removed from the atmosphere only when they are destroyed by sunlight in the upper atmosphere. In general, fluorinated gases are the most potent and longest lasting type of greenhouse gases emitted by human activities.

Fluorocarbons contribute to warming by enhancing the greenhouse effect in the lower atmosphere. CFCs also chemically react with and destroy ozone (O_3) in the upper atmosphere, creating the "ozone hole" over the Southern Hemisphere (**Figure 2.26**). Depending on where ozone resides, it can protect or harm life on Earth. Most ozone resides in the stratosphere, where it shields Earth's surface from harmful ultraviolet (UV) radiation emitted by the Sun. However, because chlorofluorocarbons destroy ozone, stratospheric ozone had been declining at a rate of about 4 percent per decade until the 1980's when they became outlawed by international treaty. At the same time, a much stronger, but seasonal decrease in ozone over Earth's poles

has opened an "ozone hole" over the Antarctic. Without ozone, humans are more likely to develop skin cancer,[79] cataracts, and impaired immune systems. Closer to Earth in the troposphere, ozone is a harmful pollutant that causes damage to lung tissue and plants.

The good news is that the effects of many CFCs are reversible. Thanks to the Montreal Protocol, signed by 27 nations in 1987,[80] CFCs were recognized as dangerous pollutants and their production and use was significantly reduced. The United States, one of the signers of the Protocol, banned the use of CFCs in aerosols and ceased their production by 1995. CFCs already in the atmosphere have lifetimes of 75 to 150 years, so ozone depletion could continue for decades. However, the first signs that the ozone hole in the Southern Hemisphere is beginning to heal have surfaced,[81] and scientists are hopeful that the trend of ozone depletion in the stratosphere over the Antarctic may be reversing.

Unfortunately, an ozone hole has opened in the Arctic that is nearly as large as the hole over the Antarctic. Depletion

[79] Overexposure to UV radiation is believed to be contributing to the increase in melanoma, the most fatal of all skin cancers. Since 1990, the risk of developing melanoma has more than doubled. http://www.epa.gov/oar/oaqps/gooduphigh/good.html#1

[80] See the Wiki entry on Montreal Protocol, *Wikipedia*, http://en.wikipedia.org/wiki/Montreal_Protocol. The same style of treaty is being considered by governments today as a way to reach agreement on reducing greenhouse gas production.

[81] Salby, M., et al. (2011) Rebound of Antarctic ozone, *Geophysical Research Letters*, 38, L09702, doi:10.1029/2011GL047266

of Arctic ozone is mainly due to unusually cold temperatures in the stratosphere that drive reactions involving CFCs and that destroy ozone. Researchers[82] have calculated that over the past 30 years the stratosphere in cold Arctic winters has cooled down by about 1 degree Celsius (1.8 degrees Fahrenheit) per decade. What is driving the cooling trend? The likely culprit is that when heat is trapped by greenhouse gases in the troposphere it produces cooling in the overlying stratosphere. If this trend to colder stratospheric temperatures continues, the Arctic ozone hole can be expected to persist and widen. Further decrease in temperature by just another 1 degree Celsius (1.8 degree Fahrenheit) would be sufficient to cause a nearly complete destruction of the Arctic ozone layer in certain areas including densely populated areas in northern Russia, Greenland, and Norway.

Water Vapor (H_2O) Earth's climate is able to support life because of the natural greenhouse effect and the availability of water. Water vapor (a gas) is a key component of both of these processes. It is the most abundant and powerful greenhouse gas and an important link between Earth's surface and its atmosphere. The concentration of water in the atmosphere is constantly changing, controlled by the balance between evaporation and precipitation (rain and snowfall). In fact, the average water molecule spends only about nine days in the air before precipitating back to Earth's surface (**Figure 2.27**).

Water vapor constitutes as much as 2 percent of the atmosphere and accounts for the largest percentage of the natural greenhouse effect. The abundance of water vapor varies from one spot to another based on natural evaporation and precipitation.

FIGURE 2.27 Nearly 577,000 km³ (138,500 mi³) of water circulates through the water cycle every year. The cycle consists of five major processes: condensation (cloud formation), precipitation (rain and snowfall), infiltration (water soaking into the ground), runoff (water draining off the land in streams), and evapotranspiration (evaporation plus transpiration; transpiration is a process wherein plants take water in through the roots and release it through the leaves). These processes keep water continuously moving through Earth's environments.[83]

---

[82] Sinnhuber, B.-M, et al. (2011) Arctic winter 2010/2011 at the brink of an ozone hole, *Geophysical Research Letters*, 38, doi: 10.1029/2011GL049784

[83] VIDEO: Water Cycle Animation, http://www.youtube.com/watch?v=Az2xdNu0ZRk

Normally, human activity does not significantly affect water vapor concentrations except in local circumstances (such as irrigating fields or building reservoirs in arid areas). However, as global warming increases the average temperature of the troposphere, the rate of evaporation increases; hence, the amount of water vapor increases in a warmer atmosphere—a powerful positive feedback effect.[84] Increases in other heat-trapping gases, such as carbon dioxide, lead to more heating and thus more water vapor (increased water vapor in the atmosphere has already been observed[85]).

This increase in atmospheric water vapor, in turn, produces increased heating, more water vapor, and so on. Like methane being released from melting permafrost, water vapor can drive a positive feedback in the global warming system.

Basic theory, observations, and climate models all show the increase in water vapor is around 6 to 7.5 percent per degree Celsius (or per 1.8 degrees Fahrenheit) warming of the lower atmosphere. Notably, a study by NASA[86] confirmed that the heat-amplifying effect of water vapor is potent enough to double the climate warming caused by increased levels of carbon dioxide in the atmosphere. So if there is a 1 degree Celsius (1.8 degrees Fahrenheit) change caused by CO_2, the resulting water vapor increase will cause the temperature to go up another 1 degree Celsius (1.8 degrees Fahrenheit).

A general rule has developed among climate scientists who study the water cycle: "In a warmer world wet places will get wetter and dry places will get drier." This is based on the simple observation that places that already experience abundant rainfall will see more moisture as air temperature rises and humidity increases, and dry places, such as around 30°N and 30°S latitudes, where many of the world's great deserts lie, will continue to see the same dry air.

This was unambiguously confirmed by a study[87] of the 50-year salinity history (1950 to 2000) of the ocean surface. Researchers tracked the changing salinity of ocean water using shipboard data and the 3,500-float armada of the Argo array.[88] Gauging the oceans' changing salinity reveals the movement of water between the atmosphere and ocean; salinity drops where there is more rain and salinity rises where there is more evaporation.

Researchers discovered that areas of high rainfall, such as the high-latitude and equatorial parts of the oceans, became even less salty during the period of study. In the middle latitudes, where evaporation dominates, ocean salinity increased. The scientists calculated that over 50 years the water cycle had sped up roughly 4 percent while the surface warmed 0.5 degree Celsius (0.9 degree Fahrenheit), an 8 percent increase per degree Celsius of warming. Because the water cycle over land behaves the same way as over the oceans, and because much of the rain over land comes from the ocean, these results likely apply to rainfall on the continents as well.

If (as predicted) the world warms 2 to 3 degrees Celsius (3.6 to 5.4 degrees Fahrenheit) by mid-century, the water cycle could accelerate 16 to 24 percent. This would be an ominous development for several reasons: Evaporation carries heat from the surface to the atmosphere that can fuel violent storms, from tornadoes to tropical cyclones, and increased evaporation would enhance this relationship. Increasing rainfall in wet places can lead to more-severe and more-frequent flooding. Decreasing rainfall in arid and semiarid regions would mean longer and more-intense droughts.

Aerosols Burning fossil fuels not only produces heat-trapping gases but also produces aerosols, fine solid particles, or liquid droplets suspended in the atmosphere that scatter (reflect) or absorb sunlight. Scattering behavior increases Earth's albedo, the tendency to reflect sunlight, and thus has a cooling effect.

On the other hand, heat absorption, such as by particles of black soot produced by biomass burning, has a warming effect. Most anthropogenic aerosols are sulfates (SO_4) that are released with the pollution from burning coal, wood, dung, and petroleum. So much aerosol production accompanied industrial growth in the middle of the 20th century that it may have caused global cooling occurred in the decades of the 1950s to 1970s. Today we track atmospheric particles with sensors aboard NASA's Terra satellite (**Figure 2.28**).

Volcanic eruptions can have the same effect. They blast huge clouds of particles and gases (including sulfur dioxide, SO_2) into the atmosphere. Most of these particles stay in the troposphere and fall out within a few days to weeks. But if a volcanic eruption is especially large, particles may penetrate into the stratosphere and can remain in the air for years.

In the upper atmosphere, sulfur dioxide converts into tiny, persistent sulfuric acid (called *sulfate*) particles that reflect sunlight. Especially massive eruptions can produce global cooling. For example, Mount Pinatubo in the Philippines erupted in June 1991 and cooled the planet nearly 1 degree Celsius (1.8 degree Fahrenheit), temporarily

[84] Water vapor, *Wikipedia*, http://en.wikipedia.org/wiki/Water_vapor

[85] Santer, B.D., et al. (2007) Identification of human-induced changes in atmospheric moisture content, *Proceedings of the National Academy of Sciences*, 104(39), 15248–15253, http://www.pnas.org/content/104/39/15248.full.pdf

[86] NASA/Goddard Space Flight Center (2008) Water vapor confirmed as major player in climate change, *ScienceDaily*, November 18, http://www.sciencedaily.com/releases/2008/11/081117193013.htm

[87] Durack, P., et al. (2012) Ocean salinities reveal strong global water cycle intensification during 1950 to 2000, *Science*, 336(6080), 455–458, doi: 10.1126/science.1212222

[88] Argo is an observation system for the oceans that provides real-time data for use in climate, weather, oceanographic, and fisheries research. Argo consists of a large collection of small drifting oceanic robotic probes deployed worldwide. The probes float as deep as 2 km. Once every 10 days, the probes surface, measuring conductivity and temperature profiles to the surface. From these, salinity and density can be calculated. The data are transmitted to scientists onshore via satellite.

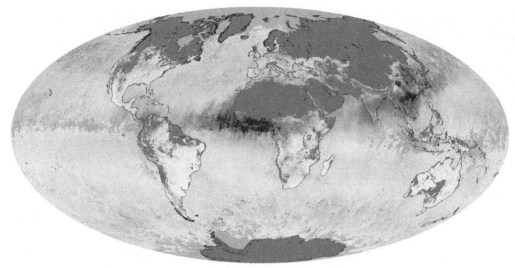

FIGURE 2.28 This image from NASA's Terra satellite shows the concentration of aerosols in March 2010. A dark plume extends west from Africa, where thick dust blew over the Atlantic Ocean. Dark patches also cover parts of China and Southeast Asia where aerosols cloud the sky. Dust contributed to the aerosols in northern Asia, but smoke is the likely culprit for high aerosols in southern Asia. Fires burned extensively in Southeast Asia through March, veiling the region in a pall of smoke. See the NASA Earth Observatory website (http://earthobservatory.nasa.gov) for more satellite imagery.

Source: NASA

offsetting the greenhouse effect for more than one year. Other major recent volcanic eruptions that produced temporary global cooling were from Mt. Agung (1963, Indonesia) and El Chichon (1982, Mexico).

Scientists continue to investigate the role that stratospheric aerosols play in the climate system. One study[89] concluded that global climate models used to project future patterns in Earth's climate miss an important cooling factor if they do not account for the influence of stratospheric aerosol or do not include recent changes in stratospheric aerosol levels.

Researchers found that a previously unmeasured increase in the abundance of particles high in the atmosphere has offset about a third of the warming influence of carbon dioxide change during the first decade of the 21st century. Since 2000, stratospheric aerosols have caused a slower rate of climate warming than would have occurred without them. The reasons for this increase are not known, but because there were no large-scale volcanic eruptions over the period, the particles could have come from several sources: smaller volcanic eruptions, sulfur compounds from Earth's surface such as biomass burning and industrial emissions, and even meteoric dust arriving from space.

When we burn coal, animal dung, diesel fuel, wood, vegetable oil, and other fuels made of biomass, part of the exhaust is black soot. Soot consists of microscopic particles of carbon that are carried into the atmosphere and

that contribute to global warming. Soot has been found to cause climate changes in areas of higher latitude where ice and snow are more common.[90]

Typically, ice and snow reflect sunlight rather than absorb it, owing to their white background. When black soot collects on the snow, the particles absorb heat, accelerating the melting of snow and ice, and replace part of the reflective white surface with heat-absorbing black particles. As the snow and ice disappear, the water and barren earth that are revealed also absorb heat; hence, the formerly reflective surface is replaced by heat-absorbing water and rock.

This process has been linked to accelerated melting of the Greenland ice sheet,[91] and to ice around the world.[92] Using satellite data from 1981 to 2012, researchers[93] have studied changes in Greenland's albedo related to black soot as well as the growth of dark microbes (including algae) on the surface of the ice.[94] At first, data showed

[89] Solomon, S., et al. (2011) The persistently variable "background" stratospheric aerosol layer and global climate change, *Science*, 333(6044), 866–870, doi: 10.1126/science.1206027

[90] Xu, B., et al. (2009) Black soot and the survival of Tibetan glaciers, *Proceedings of the National Academies of Sciences*, 106(52), 22114–22118, http://www.pnas.org/content/early/2009/12/07/0910444106.

[91] Thomas, J.L., et al. (2017) Quantifying black carbon deposition over the Greenland ice sheet from forest fires in Canada, *Geophysical Research Letters*, doi: 10.1002/2017GL073701

[92] Alia, L., et al. (2017) Dissolved black carbon in the global cryosphere: Concentrations and chemical signatures, *Geophysical Research Letters*, doi: 10.1002/2017GL073485

[93] Tedesco, M., et al. (2016) The darkening of the Greenland ice sheet: Trends, drivers, and projections (1981–2100), *The Cryosphere*, 10, 477–496, https://doi.org/10.5194/tc-10-477-2016

[94] VIDEO: Alive and Well, Microbes add to melting of Greenland ice sheet, https://www.yaleclimateconnections.org/2017/03/often-overlooked-melting-influence-of-dark-snow

little change, but starting around 1996, Greenland ice began absorbing about 2 percent more solar radiation per decade as a result of darkening. At the same time, summer temperatures in Greenland increased at a rate of about 0.74 degrees Celsius (1.3 degrees Fahrenheit) per decade, allowing more snow to melt. This accelerated melting fueled a positive feedback loop, which revealed more dark soot, leading to more melting.

The strengthening of summer melting in 1996 is related to a shift in atmospheric circulation tied to the North Atlantic Oscillation, a natural weather cycle that favors more incoming solar radiation and warmer, moist air from the south. Later records show those conditions shifted again in 2013 to 2014 to favor less melting, but the damage was already done—the ice sheet had become more sensitive. In 2015, melting spiked again and extended its reach to cover more than half of the Greenland ice sheet.

According to computer simulations, soot may be responsible for 25 percent of observed global warming over the past century. One study[95] found that soot may be contributing to the trend of early spring in the Northern Hemisphere. Earlier springs are a contributing factor to the thinning of Arctic sea ice and the melting of glaciers and permafrost.

In summary, aerosols can have a cooling effect (e.g., sulfate particles from volcanic eruptions and industrialization) or a warming effect (e.g., black soot from biomass burning) on climate.

2.3.5 Emissions by Country

Each of the world's countries contributes different amounts of heat-trapping gases to the atmosphere. The top carbon dioxide emitters, in order, are China, the United States, the European Union, India, the Russian Federation, Japan, and Canada. These rankings include CO_2 emissions from fossil fuel combustion, as well as cement manufacturing and gas flaring. Together, these sources represent a large proportion of total global CO_2 emissions (**Figure 2.29**).

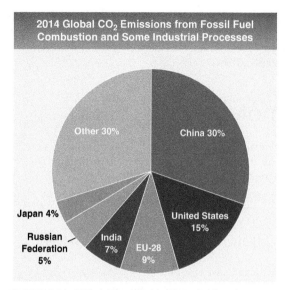

FIGURE 2.29 Global CO_2 emissions from fossil fuel combustion and other industrial processes.
Source: U.S. EPA

The picture that emerges from these figures is one where—in general—developed countries and major emerging economies lead in total carbon dioxide emissions. Developed nations typically have high carbon dioxide emissions per capita, while some developing countries lead in the growth rate of carbon dioxide emissions. Obviously, these uneven contributions to the climate problem are at the core of the challenges the world community faces in finding effective and equitable solutions.

Emissions and sinks related to changes in land use are not included in these estimates. However, changes in land use can be important: estimates indicate that net global greenhouse gas emissions from agriculture, forestry, and other land use were over 8 billion metric tons of CO_2 equivalent, or about 24 percent of total global greenhouse gas emissions. In areas such as the United States and Europe, changes in land use associated with human activities have the net effect of absorbing CO_2, partially offsetting the emissions from deforestation in other regions.

Animations and Videos

1. NASA Aqua CERES: Tracking Earth's Heat Balance: https://www.youtube.com/watch?v=uVkfh89iyeU

2. NASA Solar Cycle: https://www.youtube.com/watch?v=sASbVkK-p0w

3. NASA The Greenhouse Effect: https://www.youtube.com/watch?v=ZzCA60WnoMk

4. NASA Scientist James Hansen Talks about the Urgency of the Climate Crisis: https://www.youtube.com/watch?v=f0hHlxaYNb0

5. How do climate models treat clouds? https://www.youtube.com/watch?v=_YZxqRM97eo

6. The effect of clouds on global warming. https://www.youtube.com/watch?v=TW33e9J3fRc

7. AIRS https://www.jpl.nasa.gov/news/news.php?feature=6836

8. NASA A Year in the Life of Earth's CO2 https://www.youtube.com/watch?v=x1SgmFa0r04 https://www.youtube.com/watch?v=x1SgmFa0r04

9. Thawing Permafrost: Changing Planet http://www.youtube.com/watch?v=yN4OdKPy9rM

---

[95] Gutro, R. (2005) NASA study finds soot may be changing the Arctic environment, *NASA News Archive*, March.

10. Permafrost: The Ticking Time Bomb: https://www.youtube .com/watch?v=FLCgybStZ4g

11. Water Cycle Animation http://www.youtube.com/watch?v= Az2xdNu0ZRk

12. Alive and Well: Microbes Add to Melting of Greenland Ice Sheet: https://www.yaleclimateconnections.org/2017/ 03/often-overlooked-melting-influence-of-dark-snow

Climate Change Essay

1. Write a short essay that summarizes the main points made by Bill Nye in this video. Climate Change 101 with Bill Nye: National Geographic https://www.youtube.com/watch?v=EtW2rrLHs08

Comprehension Questions

1. What is radiative equilibrium?

2. How does Earth's heat machine move excess heat away from the equator where solar radiation is greatest?

3. What is the total solar irradiance and why is only 25% of it intercepted by Earth?

4. What are sunspots and how are they relevant to climate change?

5. Describe the surface energy budget.

6. Describe the atmospheric energy budget.

7. What is radiative forcing?

8. Explain the energy budget depicted in a two-layer model of the atmosphere.

9. List the greenhouse gases in order of GWP.

10. According to the paper by Unger et al. (2010), which five human activities have the greatest impacts on radiative forcing in the near future?

11. For each greenhouse gas, describe at least one process that increases the amount of the gas in the atmosphere.

12. Why is carbon dioxide considered the most important greenhouse gas?

Thinking Critically

1. What could you and your friends do to decrease the impacts of global warming?

2. Describe some ways Earth's surface is becoming less reflective. What is decreased reflection's impact on climate change?

3. Describe the role of ozone and nitrous oxide as greenhouse gases. What steps can policy makers take to reduce their future impact?

4. You are a politician running for President of the United States. Describe what steps you would take to address climate change at the end of the century.

5. Why have carbon dioxide emissions increased so dramatically over the past 150 years?

6. How is the water cycle likely to change, as the atmosphere gets warmer?

7. Study the human impacts on radiative forcing by the end of the century as modeled by Unger et al. (2010). Identify three activities that concern you; what can the United States do to mitigate their impacts?

8. Is climate change "dangerous," or is that too strong a word? Why?

Activities

1. Visit the *Climate Literacy* website http://www.globalchange.gov/ resources/educators/climate-literacy and discover answers to the following questions.
   (a) Why is it important for everyone to become informed on climate science?
   (b) What are the essential principles of climate science?
   (c) Describe the primary ways to improve understanding of the climate system.
   (d) Describe how climate varies over space and time.

2. Explore this article at ABC News: http://abcnews.go.com/ Technology/GlobalWarming/global-warming-common-misconceptions/story?id=9159877

   (a) What are the seven common misconceptions about global warming?

3. Explore the NASA Vital Signs website: https://climate.nasa .gov/news
   (a) Pick two news articles and describe them.
   (b) Go to the "Solutions" link. What did you learn about mitigation and adaptation?

4. Figures 2.15 and 2.27 are not labeled.
   (a) Label Figure 2.15 with the components of the Greenhouse Effect (natural or anthropogenic).
   (b) Label Figure 2.27 with the components of the water cycle.

Key Terms

Aerosols

Albedo

Annual Greenhouse Gas Index

Anthropogenic green-
    house effect

Back radiation

Carbon Cycle

Clathrate Gun Hypothesis

Climate forcings

Conduction

Convection

Dalton Minimum

Earth's energy imbalance

Electromagnetic spectrum

Energy budget

Evaporation

Faculae

Geologic proxies

Global warming potential

Heat engine

Latent heat

Little Ice Age

Maunder Minimum

Milankovitch cycles

Natural greenhouse effect

Negative feedback

Net heating

Ozone

Positive feedback

Radiative cooling

Radiative equilibrium

Radiative forcing

Sensible heat

Stefan-Boltzmann Law

Thermal infrared radiation

Tipping point

Total solar irradiance

Tropospheric ozone

CHAPTER 3

The Climate System

LEARNING OBJECTIVE

Circulation in the atmosphere and oceans carries heat away from the equator and toward the poles. This circulation combines with the amount of sunlight and other local characteristics to determine the local **climate**. Earth is characterized by a number of climate types, each reflecting the influence of the Sun, the ocean, the atmosphere, and local topography. Understanding how global warming influences these local characteristics is an important aspect of studying climate change.

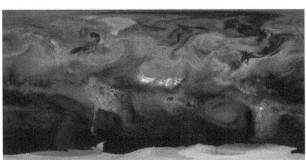

NASA's Goddard Space Flight Center/B. Putman

An ultra-high-resolution NASA computer model has given scientists a stunning new look at how carbon dioxide in the atmosphere travels around the globe. Plumes of carbon dioxide swirl and shift as winds disperse the greenhouse gas away from its sources. This simulation also illustrates differences in carbon dioxide levels in the northern and southern hemispheres and distinct swings in global carbon dioxide concentrations as the growth cycle of plants and trees changes with the seasons.[1]

Source: NASA

3.1 Chapter Summary

Earth's surface can be categorized into a number of climate zones that are created by the amount of sunlight they receive and the heat that is distributed by the general circulation of the atmosphere and oceans.

In this chapter, you will learn that:

- The **general circulation of the atmosphere** is a system of winds that transport heat from the equator, where solar heating is greatest, toward the cooler poles. This pattern gives rise to Earth's climate zones.
- The basic components of global atmospheric circulation, known as the "three-cell model," are the **Hadley Cell**, the **Ferrel Cell**, and the **Polar Cell**. There is one of each cell type in the Northern Hemisphere and one of each in the Southern Hemisphere.
- Global atmospheric circulation is a primary factor determining variations in temperature, precipitation, surface winds, storminess, and, hence, the **weather** and climate.

- In areas where relatively warm moist air is rising, such as near the equator and around 60° latitude, there is ample precipitation in all seasons.
- At the poles and around 30° latitude there are few clouds and little precipitation. This forms regions of arid climate. Some of the world's largest deserts are clustered around 30° N and 30° S latitudes.
- It is possible to describe a number of global climate zones on the basis of annual and monthly averages of precipitation and temperature. The Köppen–Trewartha climate classification system is based on six major types: tropical, dry, subtropical, temperate and continental, boreal, and polar. Each of these is further divided into one or more subcategories.
- The **Intertropical Convergence Zone** (ITCZ) is a belt of low atmospheric pressure (also called the

---

[1] VIDEO: NASA, A year in the life of Earth's CO_2, https://www.youtube.com/watch?v=x1SgmFa0r04. To read about the model, see the article here: https://www.nasa.gov/press/goddard/2014/november/nasa-computer-model-provides-a-new-portrait-of- carbon-dioxide.

"Doldrums"), high humidity, and warm temperature created by ascending air located at or near the equator.

- Weather patterns are often related to variations in the jet stream, a river of winds that are found in four locations (two in the Northern and two in the Southern Hemisphere).
- Because the Arctic is heating faster than the rest of the world, the thermal gradient driving the jet stream has weakened. The jet stream has responded by losing energy and developing large, slow-moving meanders. These meanders produce strong frontal systems, prolonged heat waves, record-setting cold, and intensified storminess.
- The oceans influence the weather and climate. Ocean water moderates air temperatures by absorbing heat from the atmosphere and the Sun and transporting that heat toward the poles as well as down toward the seafloor.
- The overall outlook for the global ocean is not healthy. Warming, acidification, and anoxia have been identified as the "deadly trio" that threatens mass extinctions in the marine ecosystem.

3.2 Weather and Climate

Weather[2] is the short-term state of the atmosphere at a given location. It affects the well-being of humans, plants, and animals and the quality of our food and water supply. Weather is somewhat predictable because of our understanding of Earth's global climate patterns. For instance, in certain seasons we can expect precipitation events of either rain or snow, and these can be predicted a few days in advance by a combination of computer modeling and the modern technology of satellites and radar.[3]

Climate is the long-term average weather pattern in a particular region and is the result of interactions among land, ocean, atmosphere, water in all of its forms, and living organisms. Climate[4] is described by many weather elements, such as temperature, precipitation, humidity, sunshine, wind, and more. Both climate and weather result from processes that accumulate and move heat within and between the atmosphere and the oceans, and from the tropics to the poles.

To understand both natural and human influences on global climate, we must explore the physical processes that

govern heat movement in the atmosphere. Later in this chapter, we examine heat movement in the oceans by systems of currents.

3.3 The Atmosphere

It is worth remembering that global warming and the greenhouse effect are not exactly the same thing. Global warming is, essentially, the greenhouse effect intensified by human activities—the discharge of carbon dioxide, and other greenhouse gases, into the atmosphere. The atmosphere can act like a thin versus a thick blanket, depending on the concentration of greenhouse gases.

Because greenhouse gas emissions, and the global warming they cause, are dramatically changing Earth's climate, it is worth familiarizing ourselves with how the atmosphere works and the distribution and characteristics of Earth's major climate zones.

3.3.1 Layers of the Atmosphere

The atmosphere[5] is the envelope of gases that surround Earth, extending from its surface to an altitude of about 145 km (90 mi). Around the world, the composition of the atmosphere is similar, but when looked at in cross section, the atmosphere is not a uniform blanket of air. It can best be described as having four layers, each with distinct properties, such as temperature and chemical composition. The red line in **Figure 3.1** shows how atmospheric temperature changes with altitude.

- **Thermosphere.** The highest layer of the atmosphere, the thermosphere (also called the *ionosphere*), gradually merges with space. Temperatures increase with altitude in the thermosphere because it is heated by cosmic radiation from space.
- **Mesosphere.** Below the thermosphere is the mesosphere, which extends to an altitude of about 80 km (50 mi). This layer grows cooler with increasing altitude.
- **Stratosphere.** Below the mesosphere is the stratosphere, where the protective ozone layer absorbs much of the Sun's harmful ultraviolet radiation. This layer extends to an altitude of about 50 km (31 mi). It becomes hotter with increasing altitude, and it is vital to the survival of plants and animals on Earth because it blocks intense ultraviolet radiation from the Sun that damages living tissue.
- **Troposphere.** In the layer nearest Earth, the troposphere (or *weather zone*), the air becomes colder with increasing altitude; you might have noticed this if you

---

[2] Weather, *Wikipedia*: http://en.wikipedia.org/wiki/Weather.

[3] Radar (the use of pulses of radio waves to remotely measure objects) is used in weather forecasting to identify various types of precipitation (rain, snow, hail, etc.). Weather radars can detect the motion of rain droplets in addition to the intensity of precipitation. This is used to characterize storms and their potential to cause severe weather.

[4] Climate, *Wikipedia*: http://en.wikipedia.org/wiki/Climate

[5] Earth's Atmosphere, *Wikipedia*: http://en.wikipedia.org/wiki/Earth%27s_atmosphere

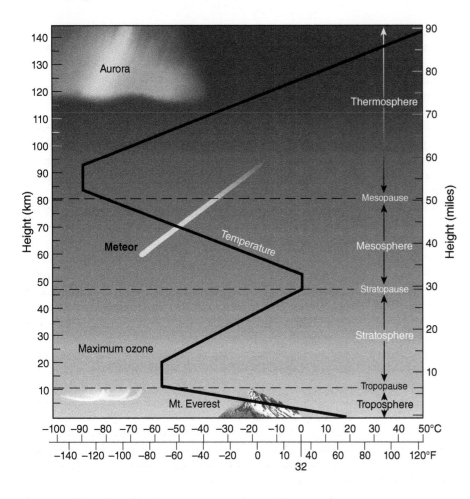

FIGURE 3.1 Around the world, the composition of the atmosphere is similar, but when looked at in cross section, the atmosphere is not a uniform blanket of air. It has several layers, each with distinct properties, such as temperature and chemical composition. The red (black) line shows how atmospheric temperature typically changes with altitude.

have ever hiked in the mountains. This layer extends to an altitude of about 8 km (5 mi) in the polar regions and up to nearly 17 km (10.5 mi) above the equator. It is also the layer where greenhouse gases are trapped and global warming occurs.

The boundaries between these layers are called *pauses*. For example, the tropopause is the boundary between the troposphere and the stratosphere.

There is little vertical mixing of gases between layers of the atmosphere, and one layer can be warming while another is cooling. As we saw in **Chapter 1** (**Figure 1.11**), global warming in the troposphere, the layer closest to Earth's surface, causes cooling in the stratosphere[6] because as more heat is trapped in the lower atmosphere, less heat reaches the upper atmosphere.[7] To an observer in space, Earth would appear to be cooling, but that is only true of the upper atmosphere.

3.3.2 General Circulation of the Atmosphere

An essential component of climate, the atmosphere is the most rapidly changing and dynamic of Earth's physical systems, and it constantly interacts with Earth's other systems: the **hydrosphere** (water in all its forms), **biosphere** (living organisms), and **lithosphere** (rock, soil, and Earth's geology). In the troposphere, global winds circulate the air and interact with the ocean surface, mixing water vapor and heat.[8] Close to Earth's surface, atmospheric circulation is so vigorous that air can travel around the world in less than a month.

Global circulation (**Figure 3.2**) is essentially driven by heat from the Sun and by the rotation of Earth. The worldwide system of winds that transport warm air from the equator (where there is the greatest solar heating) toward the cooler high latitudes is called the general circulation of the atmosphere. This pattern gives rise to Earth's climate zones.[9]

Global atmospheric circulation is a primary factor determining variations in temperature, precipitation, surface

[6] Fu, Q., et al. (2004) Contribution of Stratospheric cooling to satellite-inferred tropospheric temperature trends, *Nature*, 429, 55–58.

[7] Laštovička, J., et al. (2006) Global change in the upper atmosphere, *Science*, 314 (5803), 1253–1254, doi: 10.1126/science.1135134; see also Santer, B.D., et al. (2011) The reproducibility of observational estimates of surface and atmospheric temperature change, *Science*, 334(6060), 1232–1233.

[8] VIDEO: Global Circulation of the Atmosphere, https://www.youtube.com/watch?v=DHrapzHPCSA

[9] VIDEO: Science on a Sphere Movie about Global Circulation, https://www.youtube.com/watch?v=0j6oi1fdo5E

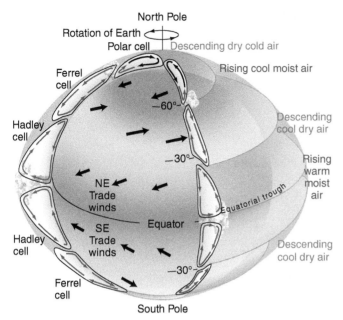

FIGURE 3.2 The general circulation of the atmosphere (known as the "three-cell model") is driven by heat from the Sun and rotation of the planet.

winds, storminess, and, hence, the weather and climate. The basic components of global atmospheric circulation, known as the "three-cell model", are the Hadley Cell, the Ferrel Cell, and the Polar Cell. There is one of each cell type in the Northern Hemisphere and one of each in the Southern Hemisphere.

Atmospheric circulation starts with the basic principle that hot air rises and cool air sinks. Therefore, air heated by the Sun rises at the equator, where solar heating is greatest. As the air moves toward the poles, it cools and eventually sinks. Rising air causes **low air pressure** (at the equator and 60 degrees latitude), and sinking air causes **high air pressure** (at 30 degrees latitude, and the poles).

If Earth were perfectly still and smooth, we might have a single cell in each hemisphere where hot air rises at the equator, moves north or south toward the poles, and then sinks to ground level as it cools at the poles. This air would then flow back to the equator along the ground surface. We would see this pattern expressed in the Northern Hemisphere as a constant north wind and in the Southern Hemisphere as a constant south wind.

Fortunately, however, Earth is neither still nor smooth. Earth spins on its axis, causing dramatic changes in heating related to the day and night, and large mountain ranges deflect the direction of surface winds. Life on Earth is much more interesting this way.

The Coriolis Force In 1856, **William Ferrel** (1817–1891) demonstrated that owing to the rotation of Earth on its axis, air and water currents moving distances of tens to hundreds of kilometers tend to be deflected to the right in the Northern Hemisphere and to the left in the Southern

Hemisphere. This phenomenon is known as the **Coriolis force**, named after the French scientist **Gaspard-Gustave Coriolis** (1792–1843), who described the transfer of energy in rotating systems.

The Hadley Cell By the time an air mass has risen at the equator and traveled to about 30° latitude, it has precipitated most of its moisture and cooled sufficiently to sink back to Earth's surface (forming an area of high pressure, i.e., arid). This air must flow away when it reaches the surface, and so it splits to move back either toward the equator or toward the pole. The air that flows back to the equator is reheated and rises again to repeat the process. This completes the Hadley Cell.

The Polar Cell and the Ferrel Cell At the poles, cold, dry, dense air descends; forming an area of high pressure that is arid. Air flows away from the high pressure and toward the equator. By the time this air nears 60° latitude, it begins to meet air flowing poleward from the south. When these two air masses meet, they have nowhere to go but up. As they rise, they cool and lose moisture, causing high precipitation. Once high in the atmosphere, they must head poleward, where they cool and sink again, or toward the equator, where they meet the Hadley Cell flow heading poleward from the equator and sink. The circulatory cell sinking at the poles and rising at 60° latitude is the Polar Cell, and the cell sinking at 30° latitude and rising at 60° latitude, is the Ferrel Cell.

Because surface winds in a Hadley Cell are moving south (in the Northern Hemisphere) when they are deflected to the right, they turn westward and are called the **northeast trade winds**. In the Southern Hemisphere, they turn left to become the **southeast trade winds**. The surface winds in the Northern Hemisphere's Ferrel Cell are moving north, and when deflected right they become the **mid-latitude westerlies**. The surface winds in the northern Polar Cell are heading south, and when deflected right they become the **polar easterlies**. Check a globe to convince yourself of these patterns and figure out what part of the global atmospheric circulation system you live in.

3.3.3 How Global Circulation Affects Climate

As air rises, it cools and expands. This is due to the increased distance from the warming effects of Earth's surface and the lower air pressure found at higher altitudes. As a rising air mass cools and expands, so does the water vapor contained in it. As the water vapor cools and expands, more water condenses than evaporates, causing water droplets and then clouds to form. Continued condensation produces precipitation, which falls as rain or snow. Therefore, in areas where relatively warm moist air is rising, such as

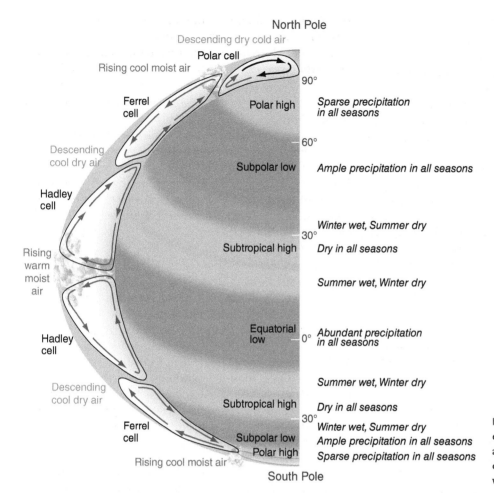

FIGURE 3.3 Global climate zones are governed by atmospheric circulation; rising and sinking air associated with circulation cells control the movement of surface winds, water vapor, and aspects of the temperature.

near the equator and around 60° latitude, there is ample precipitation in all seasons.

The opposite is also true: Air warms and compresses as it sinks closer to Earth's surface. This causes evaporation to exceed condensation. No clouds form in locations with lots of sinking air. These areas, such as at the poles and around 30° latitude, have few clouds and little precipitation, thus forming a great belt of arid climate (and deserts) that girdle the globe. Many of the world's deserts are clustered around 30° N and 30° S latitudes for this reason (**Figure 3.3**).

3.4 Climate Zones

It is possible to describe certain regular patterns to the world's climate on the basis of annual and monthly averages of precipitation and temperature. **Wladimir Köppen** (1846–1940) devised the most widely accepted climate zone classification system.[10] His system is based on a subdivision of the world's climate zones into five major types

(tropical, dry, temperate, continental, and polar), each of which is further divided into one or more subcategories.

Glenn Trewartha (1896–1984), an American geographer, reworked the Köppen classification system in 1968[11] (updated again in 1980[12]) and reclassified the temperate category into three types (subtropical, temperate and continental, and boreal) to produce what is considered a more true or "real-world" reflection of the global climate.[13] Otherwise, the tropical, dry, and polar climates remain the same as the original Köppen climate classification.

We will use the combined Köppen–Trewartha system in this chapter.[14] The zones we will cover are as follows: Tropical, Dry, Subtropical, Temperate and Continental, Boreal, and Polar. We also address the special category of "high-altitude climate," found in the upper reaches of the world's mountain systems.

---

[10] The Köppen climate classification scheme divides climates into five main climate groups: A (tropical), B (dry), C (temperate), D (continental), and E (polar). A second letter is added to create subcategories that indicate the seasonal precipitation type, while a third letter indicates the level of heat.

[11] Trewartha, G.T. (1968) *An Introduction to Climate.* McGraw-Hill, New York, NY.

[12] Trewartha, G.T. and Horn, L.H. (1980) *Introduction to Climate*, 5th ed. McGraw-Hill, New York, NY.

[13] Wikipedia page on the Trewartha system, https://en.wikipedia.org/wiki/Trewartha_climate_classification

[14] Belda, M., et al. (2014) Climate classification revisited: From Koppen to Trewartha, *Climate Research*, 59, 1–13, doi:10.3354/cr01204

3.4.1 Climate Phenomena

Several factors influence the development of these climate regions. A primary control, of course, is the amount of sunlight that a climate zone receives and this depends on the latitude of the location. As discussed in **Chapter 2**, because of the Earth's axial tilt, during the course of the year the Northern and Southern hemispheres each receive changing amounts of solar radiation.

The seasonal shift influences the location of the Intertropical Convergence Zone (ITCZ), a belt of low atmospheric pressure (also called the "Doldrums"), high humidity and warm temperature created by ascending air located at or near the equator. The trade winds from the Southern and Northern Hemispheres converge at the ITCZ to create rising air currents that shift location depending on the season of the year.

The **subtropical high-pressure zones** are areas of atmospheric high pressure located at about 30 degrees north and south latitude (also referred to as the "Horse Latitudes"). These are formed by vertically descending air currents from the Hadley Cell that creates arid conditions and some of the world's greatest deserts.

Two other processes (**Figure 3.4**) related to atmospheric circulation that control climate are the polar front and the polar vortex. The **polar front** is located at the boundary of the Ferrel and Polar circulation cells. The polar front is an area of rising air, cloud development, and precipitation that forms at a transitional zone separating arctic air masses from tropical air masses. As the polar front shifts to the north and south with the seasons, it influences storminess, precipitation, and temperature.

The **polar vortex** is a large area of low pressure and cold air surrounding both of Earth's poles that weakens in summer and strengthens in winter. The term *vortex* refers to an organized persistent westerly flow of air that characterizes the system.

During the winter in the Northern Hemisphere, the polar vortex may expand and send cold air southward in association with large meanders in the jet stream. This occurs fairly regularly during wintertime and is often associated with large outbreaks of Arctic air in the United States. Because the jet stream is developing unusually large meanders as a result of global warming, these outbreaks of the polar vortex have been responsible for record-setting low temperatures and snow storms in temperate latitudes in recent decades. Meandering of the jet stream is discussed further in a later part of this chapter.

3.4.2 Köppen–Trewartha Climate Zones

We will briefly review six Köppen–Trewartha climate zones: Tropical, Dry, Subtropical, Temperate and Continental, Boreal, and Polar. Each of these is discussed on the following pages.

Tropical Climate The global region known as the **tropics** (**Figure 3.5**) falls between the northern latitude of the

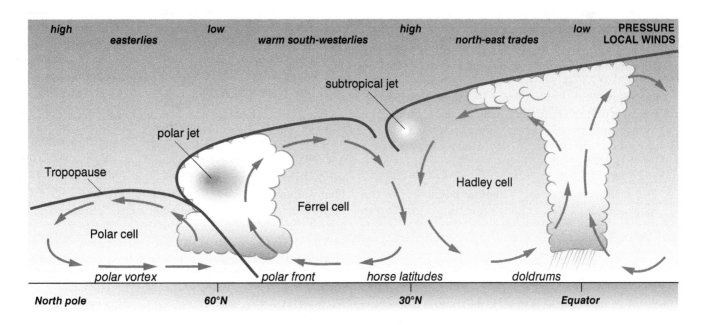

FIGURE 3.4 The major climate processes that control precipitation and air temperature are based on the three-cell model of atmospheric circulation (shown in Figures 3.2 and 3.3).

Source: N. Hulbirt

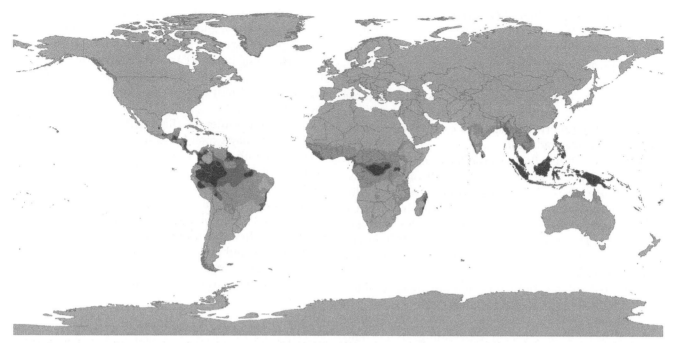

FIGURE 3.5 The tropics are the region of Earth surrounding the equator. They lie between two lines of latitude: the Tropic of Cancer in the Northern Hemisphere and the Tropic of Capricorn in the Southern Hemisphere. Three subcategories are shown: tropical wet climate (dark), tropical monsoon climate (medium), and the tropical wet and dry climate (light).

Source: Wikipedia, Public Domain Source: https://commons.wikimedia.org/wiki/File:Koppen_World_Map_Af_Am_Aw.png

Tropic of Cancer and the southern latitude of the **Tropic of Capricorn**.[15] The general pattern of tropical climate is warm temperatures, however, depending on atmospheric circulation patterns, humidity can be variable. There are three general tropical climate subcategories: *Tropical Wet* (or Rainforest) *Climate, Tropical Monsoon Climate,* and *Tropical Wet and Dry* (or Savanna) *Climate.*

- *Tropical Wet* (or Rainforest) *Climate:* Regions that experience large quantities of precipitation (all 12 months experience at least 60 mm [2.4 in] of rainfall) throughout the year. Monthly temperature variations in this climate are less than 3°C (5.4°F). Because of intense surface heating and high humidity, pronounced cloud formation occurs in the early afternoon almost every day. Daily highs average 32°C (89.6°F), while nighttime temperatures average 22°C (71.6°F).
- *Tropical Monsoon Climate:* Regions dominated by the **monsoon winds** (that change direction according to

the seasons and bring a turn in the weather). This climate pattern is based on seasonal shifts in rain. Annual rainfall is equal to or greater than the tropical wet climate, but most of the precipitation falls in the 7 to 9 hottest months. During the dry season, very little rainfall occurs.

- *Tropical Wet and Dry* (or Savanna) *Climate:* Regions that have an extended dry season in winter. Precipitation during the wet season is usually less than 1,000 mm (39.4 in), and only during the summer season. By comparison, New York City's annual average precipitation is about 46 in.

The most important climate control in regard to the tropical climate types relates to the position of the ITCZ. The ITCZ is an area of low pressure, high rainfall, and warm temperature that marks the point where the trade winds from the Southern Hemisphere and the Northern Hemisphere converge. These characteristics make it an important ingredient in atmospheric circulation and give it a critical role in the formation of the Hadley Cell.

The ITCZ's location varies throughout the year (**Figure 3.6**), and while it remains near the equator, the ITCZ over land drifts farther north or south than the ITCZ over oceans. This is due to greater variations in land temperatures compared to the ocean. The location of the ITCZ can range across 40 to 45 degrees of latitude centered on the equator depending on the pattern of land and ocean. Despite these variations, the ITCZ relates closely to the altitude of the Sun and marks the point where the Sun is highest in the

---

[15] The Tropic of Cancer is the circle marking the latitude 23.5°N, where the Sun is directly overhead at noon on June 21, the beginning of summer in the Northern Hemisphere. The Tropic of Capricorn is the circle marking the latitude 23.5°S where the Sun is directly overhead at noon on December 21, the beginning of winter in the Northern Hemisphere. When the lines were named 2,000 years ago, the Sun was in the constellation of Capricorn during the winter solstice and Cancer during the summer solstice (hence the names). Now, due to the precession of the equinoxes, the Sun is no longer in these constellations during these times, but the names remain.

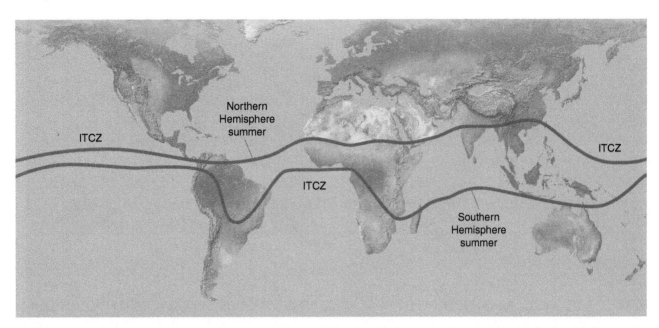

FIGURE 3.6 The position of the Intertropical Convergence Zone (ITCZ) varies with the seasons. Over land, it moves back and forth across the equator following the path of the Sun. Over the oceans, where the convergence zone is better defined, the seasonal cycle is less pronounced as the convection that characterizes the ITCZ is limited by the distribution of ocean temperatures. Sometimes, a double ITCZ forms, with one located north and another south of the equator, one of which is usually stronger than the other. When this occurs, a narrow ridge of high pressure forms between the two convergence zones.

Source: N. Hulbirt

sky. In temperate latitudes, migration of the Sun between the Tropic of Cancer and Tropic of Capricorn is responsible for creating the seasons. But in tropical latitudes, migration of the Sun is responsible for shifts in the ITCZ low-pressure belt, and thus it produces seasonal tropical rains as the ITCZ migrates north and south during the year.

Dry Climate Dry climate zones (**Figure 3.7**) extend from 20 to 35 degrees latitude north and south of the equator. They are also found in large continental regions of the mid-latitudes often surrounded by mountains. The cause for dry climate may originate from several sources, but in all cases a general rule is that **evapotranspiration** exceeds precipitation. There are two types of dry climate: *Dry Arid Climate*, and *Dry Semiarid Climate*.

- *Dry Arid*, a true desert climate. This climate type covers 12% of the global land surface and is dominated by vegetation that needs very little water.
- *Dry Semiarid*, a grassland (steppe) climate. This climate type covers 14% of the global land surface. It receives more precipitation than the Dry Arid zone, either from seasonal movement of the ITCZ or from **mid-latitude cyclones** (storms developed by troughs in the Polar Jetstream—discussed later in this chapter).

Several factors in addition to atmospheric circulation contribute to dry climate conditions. The first is called the **Orographic effect**, which causes an arid area to develop on the upwind side of a mountain system. It occurs when a mountain range forces the prevailing winds to rise (called "orographic uplift") into cooler air. The rising air cools and expands, and the rate of condensation exceeds the rate of evaporation, causing clouds to form and rain to fall.

As a result, the windward side of these mountains can be extremely wet. In contrast, on the leeward side of the mountains, the sinking air warms and compresses—behaving just as it did when sinking as part of the Hadley Cell. The resulting increase in evaporation limits cloud formation and rainfall, leading to arid and semiarid conditions (**Figure 3.8**).

The dry area downwind from the mountains can be very well defined; it is called a **rain shadow**, reflecting the lack of rain. In the western United States, much of the desert in Nevada lies in the rain shadow of the Sierra Nevada Mountains of eastern California. This range traps moisture coming off the Pacific Ocean in winds that blow from west to east, the mid-latitude westerlies.

A second factor contributing to dry local climate conditions has to do with the distance moisture is transported in the atmosphere. Since moisture gets into the atmosphere principally by evaporation from the ocean, a location that is a long distance from the ocean may experience low rainfall.

The vast mountainous central Asian deserts of Kazakhstan, Afghanistan, Mongolia, and the Tarim Basin and Gobi Desert of northern China are far downwind from any oceans (**Figure 3.9**). Their extreme interior location ensures that by the time air masses reach them most of the moisture has been extracted from the air. Adding to the

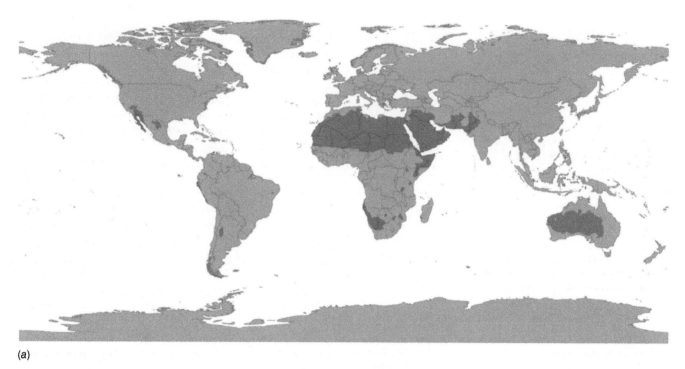

(a)

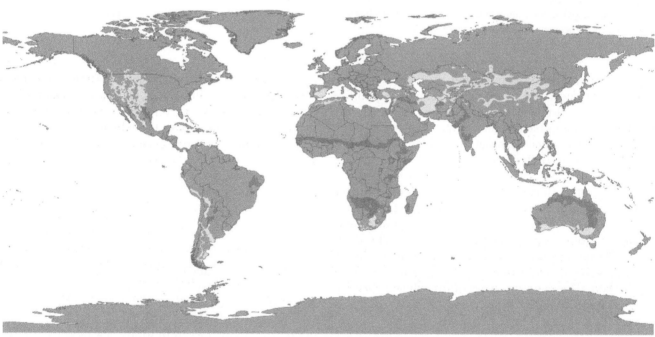

(b)

FIGURE 3.7 The Dry Climate Zone includes (a) the Dry Arid (true desert) regions, which consist of hot desert (red, dark) and cold desert (pink, light) types, and (b) the Dry Semiarid (steppe climate) regions, which consist of hot semiarid (gold, faint dark) and cold semiarid (yellow, light).

Source: Wikipedia, Public Domain Source: https://commons.wikimedia.org/wiki/File:Koppen_World_Map_BSh_BSk.png

extreme dryness of this location, the Himalayan Mountain system represents the best example of an orographic effect on the planet. Located in the lee of these lofty mountains, central Asian deserts experience a double effect of distance from the ocean and a major orographic barrier.

At the land–sea boundary where a cold ocean current flows next to a tropical coast, cold air will move onshore over hot land and quickly and dramatically heat up and expand. This process causes high rates of evaporation that produces few clouds and little rain. The driest regions on large regions of Mexico, the southern United States, North Africa, the Middle East, and Southeast Asia. Subtropical climates are characterized by warm to hot summers and cool to mild winters with infrequent frost. There are two subcategories in the subtropical zone: *Dry Summer Climate* (or Mediterranean, southern Europe where seasonal rainfall is concentrated in the cooler months) and *Humid Subtropical Climate* (or monsoon, where rainfall is often concentrated in the warmest months).

Related to the dry, sinking air of the Hadley Cell, many of the world's deserts are located within the subtropics. In this climate, areas bordering warm oceans are prone to locally heavy rainfall from tropical cyclones (e.g., eastern Florida), which can contribute a significant percentage of the annual rainfall. The humid or monsoon subtropical climate is often located on the western side of the subtropical high (sinking arm of the Hadley Cell). Here, unstable tropical air masses in summer bring convective overturning in the form of rising warm air that leads to the development of clouds followed by frequent tropical downpours in the hot season. In the winter (dry season), the monsoon retreats, and the drier trade winds bring more stable air and dry weather.

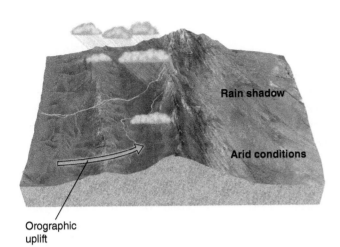

FIGURE 3.8 Humid air is forced upward by orographic uplift in the presence of mountain ranges. An arid rain shadow is produced on the downwind side of the mountains.

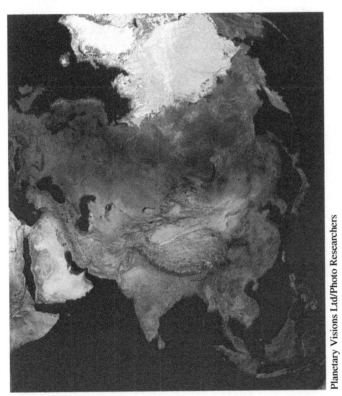

Planetary Visions Ltd/Photo Researchers

FIGURE 3.9 The great deserts of central Asia are far from the ocean, which means that by the time air has reached them they have lost most of their moisture.

Earth, the deserts along the west coasts of South America (Peru and Chile) and Africa (Namibia and Angola) are particularly arid, in part due to the presence of a cold coastal current located just offshore (refer to Figure 3.7).

Subtropical Climate The subtropical climate zone (**Figure 3.12**) includes locations that experience 8 or more months with a mean temperature of 10 degrees Celsius (50 degrees Fahrenheit) or higher. This zone includes

Temperate and Continental Climate In the temperate and continental climate zone (**Figure 3.13**), the average temperature is around 10 degrees Celsius (50 degrees Fahrenheit) for 4 to 8 months. This type of climate is found in the middle latitudes between the subtropical and boreal climates. There are two types of climate that are typically included in the temperate group: *Temperate Marine Climate* and *Temperate Continental Climate.*

- *Temperate marine climate* has mild winters. In this climate zone, the average temperature for all 12 months is 0°C (32°F) or above. The maritime climate is strongly influenced by the oceans, which maintain fairly steady temperatures across the seasons. Since the prevailing winds are westerly in the temperate zones, the western edge of continents in these areas experiences the maritime climate most commonly. Such regions include western Europe, in particular the United Kingdom, and western North America at latitudes between 40° and 60° North. This is a relatively humid climate with ample precipitation in all seasons. It is found on the western windward side of the continents in the temperate zone.

- *Temperate continental climate* is found in the interiors of the middle-latitude continents. Severe winters and summers characterize this climate zone because it lies far from the moderating effect of the ocean. Annual temperature ranges are high with extreme weather fairly common. This includes cold snaps, heat waves, blizzards, and heavy downpours possibly occurring throughout the year. Precipitation occurs throughout the year with maximum concentration during summers.

BOX 3.1 | Spotlight on Climate Change

Desertification Drives Tides of Humanity

Desertification means, simply, expanding deserts. It decreases the availability of natural resources, especially water and food. It can lead to political unrest and outright conflict[16] as one group of people, in order to maintain its own supply level, appropriates resources from another group or migrates into the resource-rich lands of another country.

Thirty-eight percent of the world's land surface is in danger of desertification as unsustainable land-use practices, often related to deforestation and frequent soil tillage for food production, combine with growing aridity related to climate change. Vulnerable regions include coastal areas, the Prairies, the Mediterranean region, the savannah, the temperate Steppes, the temperate deserts, tropical and subtropical Steppes, and the tropical and subtropical deserts.[17]

Poor land management practices combined with the effects of climate change have forced 50 million people to leave their homes, according to the United Nations. Each year nearly 30 million acres of land are lost. That's land where 20 million tons of grain could have been grown (**Figure 3.10**).

Around the world, the loss of arable land is driving individuals and entire families to nearby cities that are swelling with the growing population. But many of these cities lack sufficient resources and are unable to provide basic services such as clean water, food, and shelter for these displaced populations.

As a result, environmental refugees are looking to the developed world for help, to take them in, and to give them new lives. In the next few years, as desertification spreads, this migration will become a tide of humanity looking for new homes and by the middle of the century the world could see anywhere from 25 million to 1 billion climate-related migrants looking for new homes.[18]

In fact, drawing one of the strongest links yet between global warming and human conflict, researchers have shown[19] that an extreme drought in Syria between 2006 and 2009 was most likely due to climate change,

and that the drought was a factor in the violent uprising that began there in 2011. Rather than natural climate variability, a century-long trend toward warmer and drier conditions in the Eastern Mediterranean was shown to be the primary cause of the drought.

Studies reveal that the extreme dryness, combined with other factors, including misguided agricultural and water-use policies of the Syrian government, caused crop failures that led to the migration of as many as 1.5 million people from rural to urban areas. This, in turn, added to social stresses that eventually resulted in the uprising against President Bashar al-Assad in March 2011.[20]

Desertification is worldwide. North Africa, Brazil, China, Australia, the Middle East, Latin America, and the Southwest United States are all experiencing the loss of formerly productive land because of poor land management and spreading drought[21] related to climate change.

Extreme temperatures in Brazil are causing desertification and drought in certain parts of the country,

FIGURE 3.10 Characteristics of desertification include destruction of native vegetation, unusually high rates of soil erosion, declines in surface water supplies, rising levels of water and topsoil saltiness, and widespread lowering of groundwater tables.

[16] O'Loughlin, J., et al. (2014) Effects of temperature and precipitation variability on the risk of violence in Sub-Saharan Africa, 1980–2012, *PNAS*, doi: 10.1073/pnas.1411899111

[17] Núñez, M., et al. (2010) Assessing potential desertification environmental impact in life cycle assessment part 1: Methodological aspects, *The International Journal of Life Cycle Assessment*, 15(1), 67, doi: 10.1007/s11367-009-0126-0

[18] Desertification: http://www.climatechangenews.com/2012/04/27/climate-change-desertification-and-migration-connecting-the-dots

[19] Kelley, C.P., et al. (2014) Climate change in the fertile crescent and implications of the recent syrian drought, *PNAS*, 112(11), 3241–3246, doi: 10.1073/pnas.1421533112

[20] Conflict and Drought: https://www.nytimes.com/2015/03/03/science/earth/study-links-syria-conflict-to-drought-caused-by-climate-change.html

[21] Drought can have different causes depending on location and other natural factors. However studies link more intense droughts to climate change. As more greenhouse gas emissions are released into the air, causing air temperatures to increase, more moisture evaporates from land and lakes, rivers, and other water bodies. Warmer temperatures also increase evaporation in plant soils, which affects plant life and can reduce rainfall even more. When rainfall does come to drought-stricken areas, the drier soils it hits are less able to absorb the water, increasing the likelihood of flooding.

threatening the lives of over 35 million people.[22] In China, more than a quarter of the entire country is now degraded or turning to desert, thanks to overgrazing by livestock, overcultivation, excessive water use, and expanding drought.[23] In Australia, where three-quarters of the land is arid or semiarid, the effects of land degradation cause soil erosion, soil degradation, altered stream flow regimes, increased soil salinity, and loss of biodiversity.[24]

Desertification is a phenomenon that ranks among the greatest environmental challenges of our time. Degraded soils are vulnerable to being washed away by weather events fueled by global warming. Deforestation, which removes trees that help knit landscapes together, is detrimental to soil health. The world has lost a third of its arable land due to erosion or pollution in the past 40 years, with potentially disastrous consequences as global demand for food soars.[25] Soil erosion is related to continual disturbance by crop planting and harvesting. If soil is repeatedly turned over, it is exposed to oxygen and its carbon is released into the atmosphere (as carbon dioxide), causing it to fail to bind as effectively. This loss of integrity impacts soil's ability to store water, which neutralizes its role as a buffer to floods and a fruitful base for plants.

Desertification threatens the livelihoods of some of the poorest and most vulnerable populations on the planet. In the end, desertification, accelerated by global warming, may end up being responsible for changing the population of the entire planet as people looking for food and clean water search for new homes.

The **Sahel** region of North Africa lies along the border between the shifting sands of the Sahara in the north and the moist tropical rain forest regions of central Africa to the south. This swath stretches from the Atlantic to the Indian Ocean and includes parts or all of the countries of Senegal, Mauritania, Mali, Burkina Faso, Niger, Nigeria, Chad, Sudan, South Sudan Ethiopia, Eritrea, Djibouti, and Somalia. More than 300 million people live in this region of dry and unstable weather, many of them poor farmers or nomadic herders. They rely on local rains or migrate to follow the rains.

Unfortunately, the number of people in the Sahel exceeds the carrying capacity of the land, and their huge livestock herds denude vegetation that is already stressed by low rainfall. During the periodic multiyear droughts, the remaining native vegetation withers, crops fail, and millions of livestock animals die. Ultimately, in the wake of livestock losses, crop failures, and low water supplies, millions of people are exposed to famine, sickness, and often violence—all brought on by the desertification of their lands (**Figure 3.11**).

Many of the worst famines and humanitarian crises of the last 50 years have occurred in the Sahel during

FIGURE 3.11 The advance of desertification in the Sahel threatens food and water supplies, producing a population of environmental refugees. Displaced populations settle on the outskirts of existing towns, as here in El Fasher, northern Sudan, where the distinguishing feature is white plastic sheeting. These new arrivals add to an already heavy burden on the surrounding desert environment.

United Nations Environment Programme

[22] Desertification in Brazil: http://www.telesurtv.net/english/news/Desertification-Threatens-35-Million-Brazilians-20160313-0016.html
[23] Desertification in China: http://www.businessinsider.com/china-is-turning-into-a-desert-and-its-causing-problems-across-asia-2016-5
[24] Desertification: http://theconversation.com/on-dangerous-ground-land-degradation-is-turning-soils-into-deserts-94100
[25] Food security: https://www.theguardian.com/environment/2015/dec/02/arable-land-soil-food-security-shortage

droughts. These crises have been compounded by political strife, started in no small part by individuals under stress stemming from desertification.

In an effort to combat desertification, and thereby diffuse ensuing political unrest, the United Nations adopted the **Convention to Combat Desertification**,[26] which over 110 countries have joined. Its purpose is to ensure the long-term productivity of inhabited dry areas. Recognizing that human activities contribute to the process of desertification, the Convention promotes behaviors that reduce those contributions. It also recognizes that desertification adds to the lethal nature of political unrest and the likelihood of civil war. Clearly, starving people will find it more difficult to get along with their neighbors, and desertification is one of the basic causes of their starvation.

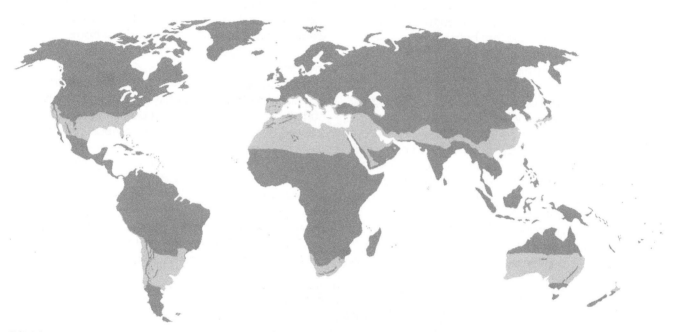

FIGURE 3.12 The subtropics (yellow, light) are a climate zone located roughly between the tropic circle of latitude at 23.5° (Tropic of Cancer and Tropic of Capricorn) and the temperature latitudes (locations higher than 40° latitude).

Source: Wikipedia, Public Domain https://commons.wikimedia.org/wiki/File:Subtropical.png

Boreal Climate The boreal (also called subarctic) climate (**Figure 3.14**) is characterized by long, very cold winters, and short, cool to mild summers. It is found on continental landmasses where the moderating influence of the ocean is absent. The boreal zone, found 50 degrees to 70 degrees north latitude poleward of the humid continental climate, has some of the most extreme seasonal temperature variations found on the planet: in winter, temperatures can drop to –40 degrees Celsius (–40 degrees Fahrenheit) and in summer they may exceed 30 degrees Celsius (86 degrees Fahrenheit). Summers are short (no more than 3 months) and winters are long.

With 5 to 7 consecutive months where the average temperature is below freezing, all moisture in the soil and subsoil freezes solidly to depths of many feet. Summer warmth is insufficient to thaw more than a few feet below the surface, so **permafrost** is found under most areas of this climate zone. Summer temperatures may thaw the permafrost to depths of 0.6 to 4.3 meters (2 to 14 feet).

Most boreal climates have very little precipitation, typically no more than 380 millimeters (15 inches) over an entire year. Away from the coasts, precipitation occurs mostly in the warmer months; while in coastal areas with subarctic climates the heaviest precipitation is usually during the autumn months when the relative temperature contrast between the ocean and the land is greatest.

Polar Climate Globally, polar climates cover more than 20 percent of Earth's surface. Much of the polar region is considered a desert because the level of precipitation is low. As we discussed earlier, air in the descending arm of the polar circulation cell warms and contracts as it sinks closer to Earth's surface. This causes evaporation to exceed condensation. No clouds form in locations with lots of sinking air. These areas have few clouds and little precipitation.

A lack of warm summers characterizes the polar climate zone. On average, each month has a temperature

---

[26] United Nations site on the Convention to Combat Desertification: http://www2.unccd.int

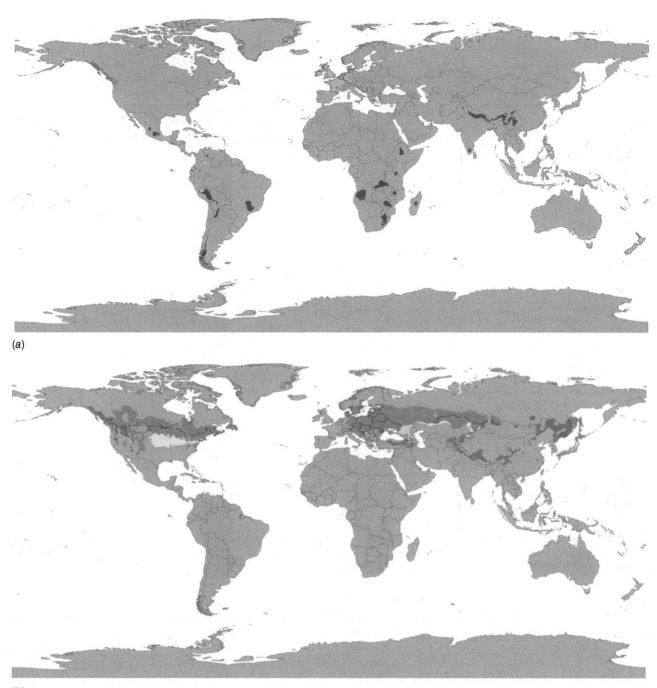

(a)

(b)

FIGURE 3.13 The temperate marine and temperate continental climates. (a) Throughout the year precipitation is both ample and reliable in temperate marine climates and the temperature tends to be cool with infrequent extremes. Temperate marine subcategories include subtropical and subpolar marine temperate climate (dark) and an intermediate (light) subcategory. (b) Temperate continental climates are typified by large seasonal temperature differences, with warm to hot (and often humid) summers and cold (sometimes severely cold) winters. Precipitation is usually evenly distributed through the year. Subcategories include dry summers (light) with warmer and cooler varieties; dry winter (dark) with warmer and cooler varieties; and more evenly distributed annual precipitation with warmer and cooler varieties.

Source: Wikipedia, Public Domain https://commons.wikimedia.org/wiki/File:Koppen_World_Map_Cfb_Cfc_Cwb_Cwc.png and https://commons.wikimedia.org/wiki/File:Koppen_World_Map_Dfa_Dwa_Dsa_Dfb_Dwb_Dsb.png

of less than 10 degrees Celsius (50 degrees Fahrenheit). In the summer, the Sun shines for long hours, and in the winter, sunlit daytime is very short, or absent altogether. A polar climate has cool summers and very cold winters, which results in treeless tundra, glaciers, or a permanent or semipermanent layer of ice. Two types of polar climate have been defined: *Tundra Climate* and *Ice Cap Climate*.

- *Tundra Climate* (**Figure 3.15**) is characterized by specialized vegetation adapted to long freezing winter

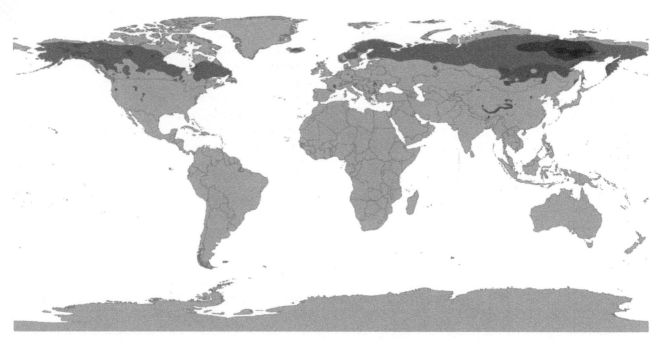

FIGURE 3.14 The Boreal (or subarctic) climate zone is characterized by long, cold winters and short, mild summers. Subcategories include: dry summers, with cold summer and very cold winter varieties; dry winter (medium) with cold summer and very cold winter varieties; and regions lacking a dry season (light) with cold summer and very cold winter (dark) varieties.

Source: Wikipedia, Public Domain https://commons.wikimedia.org/wiki/File:Koppen_World_Map_Dfc_Dwc_Dsc_Dfd_Dwd_Dsd.png

FIGURE 3.15 Tundra Climate is characterized by at least one month with an average temperature above freezing (high enough to melt snow), but no month with an average temperature that exceeds 10°C (50°F).

Source: Wikipedia, Public Domain https://commons.wikimedia.org/wiki/File:800px-Map-Tundra.png

conditions and short cool summers. Typically, there is at least one month whose average temperature is above freezing during which a layer of meltwater forms atop the permafrost creating a tundra bog. Tundra is treeless and other plants are highly specialized.

- *Ice Cap Climate* is characterized by year-round freezing temperatures where no single month has an average temperature above 0°C (32°F). No plants grow here as

the ground is perennially snow-covered and there is no soil. Accumulating snow historically builds glacial conditions. However, due to global warming, many ice cap climates are in a state of meltdown and are retracting at the edges. Many high-altitude locations have climates that meet the Ice Cap criteria, but this is a result of high elevation and forms a different zone known as the *High-Altitude Climate* zone.

High-Altitude Climate It is worth noting a special climate zone related to high-elevation mountain systems. High-Altitude (or Alpine) Climate forms as a result of high elevation. High elevations typically have cooler temperatures than lower elevations. This is because hot air, formed at ground level by infrared radiation from the surface, expands, and in growing less dense, rises into the atmosphere transferring heat upward. The rising parcel of air will come to equilibrium when it has the same density as the air at a given altitude.

This process has a representative pressure–temperature curve with the quality that as pressure gets lower, the temperature decreases at a rate of approximately 9.8 degrees Celsius per kilometer of altitude (5.4 degrees Fahrenheit per 1,000 feet). The presence of water vapor can complicate this process through the process of condensation (forming clouds and precipitation). The combined effect of this, and other complications of atmospheric circulation, is that a rise in elevation of 100 meters (330 feet) on a mountain is roughly equivalent to a shift of 80 kilometers (45 miles or 0.75 degrees of latitude) toward the poles. Thus, a rise in elevation on a mountain slope will see a shift from rain to snow precipitation, a change in vegetation, and an increase in windiness (proximity to the ocean, and other environmental factors, can change this relationship significantly).

BOX 3.2 | Spotlight on Climate Change

Arctic Meltdown and Dangerous Weather

Weather disasters cost the United States a record $306 billion in 2017. There is a component of these disasters that is related to the loss of Arctic sea ice because of Arctic Amplification.

Uncapping the Arctic Ocean

Arctic sea ice is in a state of collapse with records being set for summer retreat, winter retreat, decreasing surface area extent, and decreasing total ice volume. The pace of melting is rapid and records are being broken every few years in all these categories.

Primarily driven by warming air and ocean temperatures, natural variability, and increases in extreme weather systems that have in the past been confined to lower latitudes, studies have concluded that anthropogenic climate change is the underlying cause of these trends.[27] (**Figure 3.16**)

Arctic sea ice has undergone a stunning 50% drop in summer surface area, and a 65% reduction in ice volume since the period of satellite monitoring began in the mid-1970s.[28] Only a few decades ago, the North Pole was covered in ice 3 to 3.6 m (10 to 12 ft) thick. Subsurface ice ridges in some areas were known to extend into water depths as much as 45 m (150 ft).

Sep 13, 2017

NASA's Scientific Visualization Studio/Helen-Nicole Kostis

FIGURE 3.16 Arctic sea ice minimum 2017 versus 30 yr average minimum sea ice extent (line around periphery).

Source: NASA

[27] Ding, Q., et al. (2017) Influence of high-latitude atmospheric circulation changes on summertime Arctic Sea ice, *Nature Climate Change*, doi: 10.1038/nclimate3241

[28] Lindsay, R. and Schweiger, A. (2015) Arctic Sea ice thickness loss determined using subsurface, aircraft, and satellite observations, *The Cryosphere*, 9, 269–283, doi:10.5194/tc-9-269-2015

Now, most of that thick ice is gone and the total volume of Arctic sea ice in late summer has declined by 75% in only a few decades.[29] Data reveal that arctic winter warming exceeds summer warming by a factor of 4 or more. This is strongly linked to winter sea ice retreat through the release of ocean heat gained during the summer because of summer sea ice retreat.[30]

Feedbacks

Scientists are afraid that sea ice loss will trigger other climate feedbacks.

The most basic feedback is related to significant decreases in Earth's albedo (albedo is how much light is reflected back to space [discussed in **Chapter 2**]) as white sea ice converts to dark ocean water. Newly exposed Arctic Ocean water absorbs sunlight that it once reflected back to space. Consequently, water temperatures have climbed by as much as 3.9°C (7°F) above the long-term average. This heat adds to the excess heat being trapped by global greenhouse gases, by perhaps as much as 25%.

As Arctic seawater heats up, in turn warming the air above, rising temperatures spread and increase the amount of snow that melts on adjacent lands. Today, in midsummer, Arctic land area covered by snow has decreased by several million square kilometers compared to five decades ago.[31] Now these dark lands absorb more heat and further warm the Arctic—and the planet.

Another worrisome feedback is the rise in atmospheric **water vapor**. With air temperatures rising in many parts of the Arctic by several degrees in recent decades, there has been a simultaneous rise in water vapor. This is because warmer air holds more moisture. A powerful greenhouse gas, water vapor concentration has gone up by more than 20%, adding to Arctic, and global, warming.

Melting Greenland

Until the 1980s, the Greenland ice sheet was not subject to extensive summer melting. Today it is losing an average of 281 billion tons of ice each year,[32] more than a staggering 1 trillion tons of ice every 4 years.[33] For several decades, this made Greenland the single largest contributor to global sea-level rise (**Figure 3.17**). Today, however, the biggest contributer is Antarctica.

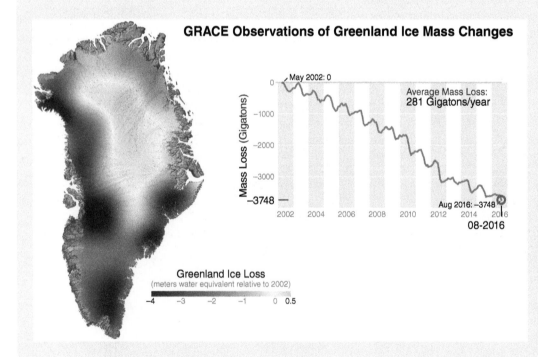

GRACE Observations of Greenland Ice Mass Changes

Greenland Ice Loss
(meters water equivalent relative to 2002)

FIGURE 3.17 The mass of the Greenland ice sheet has rapidly declined in the 21st century due to surface melting and iceberg calving. The NASA/German Aerospace Center Gravity Recovery and Climate Experiment (GRACE) satellite indicates that between 2002 and 2016, Greenland shed approximately 280 gigatons of ice per year, causing global sea level to rise by 0.8 mm (0.03 in) per year.

Source: NASA

[29] Laxon, W.S., et al. (2013) CryoSat-2 estimates of Arctic Sea ice thickness and volume, *Geophysical Research Letters*, doi:10.1002/grl.5019

[30] Bintanja, R. and van der Linden, E.C. (2013) The changing seasonal climate in the Arctic, *Scientific Reports*, 3, article number: 1556, doi:10.1038/srep01556

[31] NASA: https://earthobservatory.nasa.gov/IOTD/view.php?id=80102

[32] NASA Climate Vital Signs: https://climate.nasa.gov/vital-signs/land-ice

[33] Greenland ice loss: https://www.washingtonpost.com/news/energy-environment/wp/2016/07/19/greenland-lost-a-trillion-tons-of-ice-in-just-four-years/?utm_term=.f02419336989

At first, Greenland melting was confined to low-lying areas near the coast. But over the past two decades the boundary between annually melting and annually accumulating snow has migrated to higher elevations on the ice at over 40 m (130 ft) per year. In recent years, the melting has spread to the whole ice sheet and in 2012, 97% of the Greenland ice cap experienced continuous surface melting for nearly 2 weeks.[34] One group of researchers estimate there is a 50% probability that by the year 2025 this level of melting will happen every year.[35]

Jet Stream

As the Arctic warms faster than the rest of the planet, the temperature difference between the temperate regions and the Arctic weakens. It is this temperature difference that propels the fast moving west-to-east river of air known as the polar jet stream. This atmospheric feature separates cold air to the north from warm air to the south.

The jet stream follows a wavy path as it wraps around the northern hemisphere between 30° and 60°N of the equator. However, now that the Arctic region is warming so rapidly, the jet stream is slowing and the meanders are growing more pronounced.[36] Upper-level winds around the Northern Hemisphere have slowed during the closing months of the year, exactly when sea ice loss exerts its strongest effect on the north–south temperature gradient. Over the North American continent and the North Atlantic Ocean, winds have slowed dramatically, by approximately 14% since 1980.[37] **(Figure 3.18)**

A large southerly directed loop in the jet stream increases the probability of driving cold Arctic air far to the south, such as into the southern states, and lingering for many days to a few weeks. This was the condition that led parts of the central and eastern United States to experience record cold conditions in late December 2017 through early January 2018. Similarly, a large loop to the north will pull tropical air into northern regions, increasing the probability of a deadly heat wave that lingers for many weeks in the summer.

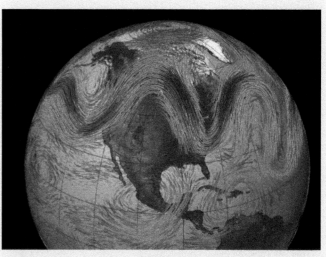

FIGURE 3.18 As the temperature difference decreases between the Arctic and the temperate regions to the south, the jet stream slows and large meanders develop that transport pockets of cold air to the south, and warm air to the north. These are responsible for episodes of slow-moving extreme weather.[38]

Source: NASA

These and other forms of extreme weather events are the indirect result of global warming.

Dangerous Climate

The United States has sustained 219 weather and climate disasters since 1980 where overall damages/costs reached or exceeded $1 billion.[39] The total cost of these 219 events exceeds $1.5 trillion. In 2017, there were 16 weather and climate disaster events with losses exceeding $1 billion each across the United States. These events included 1 drought event, 2 flooding events, 1 freeze event, 8 severe storm events, 3 tropical cyclone events, and 1 wildfire event. Overall, these events resulted in the deaths of over 4000 people and had significant economic effects on the areas impacted. The 1980 to 2017 annual average is 5.8 events; the annual average for the most recent 5 years (2013 to 2017) is 11.6 events. **(Figure 3.19)**.[40]

[34] NASA: https://www.nasa.gov/topics/earth/features/greenland-melt.html
[35] NASA http://www.climatecentral.org/news/widespread-melting-in-greenland-a-sign-of-things-to-come-16018
[36] Arctic warming: http://e360.yale.edu/features/unusually_warm_arctic_climate_turmoil_jennifer_francis
[37] Weird weather: http://e360.yale.edu/features/linking_weird_weather_to_rapid_warming_of_the_arctic
[38] Data from NASA Aerial Superhighway https://svs.gsfc.nasa.gov/10902
[39] This includes consumer price index adjustments.
[40] NOAA https://www.climate.gov/news-features/blogs/beyond-data/2016-historic-year-billion-dollar-weather-and-climate-disasters-us

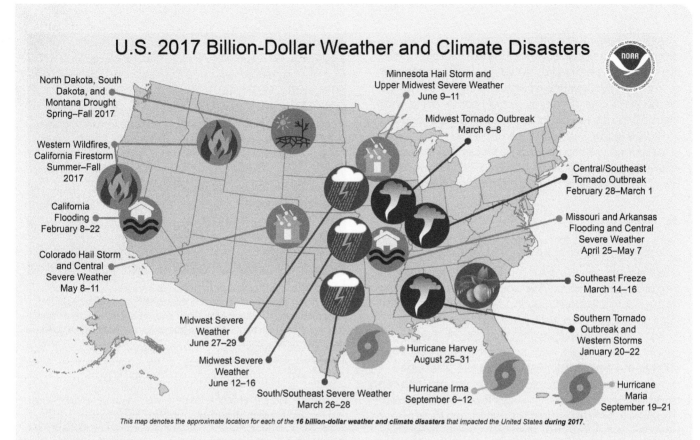

FIGURE 3.19 Location and type of the 16 weather and climate disasters in 2017 with losses exceeding $1 billion. The majority of events occurred in the middle of the country from the Central Plains to Texas and Louisiana.[41]

Source: NOAA

Arctic Amplification is far more than a global warming curiosity. It is a destructive and potentially catastrophic phenomenon that contributes strongly to global sea-level rise, heat waves, cold snaps, and the rise of global temperatures.

3.5 Jet Stream

A jet stream is a high-elevation current of air that is moving faster than the surrounding air. It is located anywhere from 9 to 16 kilometers (5 to 10 miles) above Earth's surface and can be more than 1 kilometer thick. Jet streams typically run for several thousand kilometers around the globe. They consist of very strong winds reaching speeds of over 300 kilometers/hour (200 miles/hour) that move weather systems around the planet. Jet streams are found at the top of the troposphere, just below the tropopause.

The Northern and Southern Hemispheres each have two jet streams, a polar jet stream and a subtropical jet stream

(**Figure 3.20**). In the Northern Hemisphere, the polar jet runs across the continental United States, Europe, and Siberia, while the subtropical jet runs north of the equator through Central America, the Caribbean Sea, North Africa, the Middle East, and Southeast Asia.

3.5.1 Jet Meanders and Extreme Weather

In the Northern Hemisphere, the polar jet tends to separate a region of cold air to the north from a region of warm air to the south. The polar jet can develop meanders, which exert a lot of control on local weather patterns. A meander that extends to the south brings cold, dry air down from the Arctic, forming a low-pressure center. Meanders that extend to the north draw warm, dry air from the south and form a high-pressure ridge.

---

[41] Map from NOAA 2016: A historic year for billion-dollar weather and climate disasters in U.S. https://www.climate.gov/news-features/blogs/beyond-data/2016-historic-year-billion-dollar-weather-and-climate-disasters-us

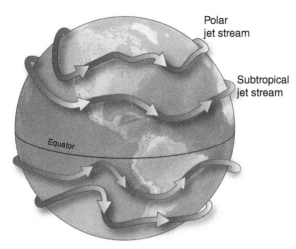

FIGURE 3.20 Jet streams are powerful currents of air in the upper troposphere that affect the weather and climate. Both the Southern and Northern Hemispheres have two jet streams, a polar jet and a subtropical jet. All four of these rapidly moving air currents flow to the east.

Source: N. Hulbirt

The Arctic is warming more than twice as fast as the rest of the planet. A report[42] by the Arctic Monitoring and Assessment Program finds that the region was warmer between 2011 and 2014 than at any time since records began in 1900. The rapid warming is accelerating the melting of glaciers and sea ice, and boosting sea-level rise. The extent of snow cover across the Arctic regions of North America and Eurasia each June has decreased by half compared with observations before 2000. The Arctic Ocean could be largely free of sea ice in summer as early as the late 2030s, less than two decades from now. Near-surface permafrost in the high Arctic and other very cold areas has warmed by more than 0.5 degree Celsius since 2007 to 2009, and the layer of the ground that thaws in summer has deepened in most areas where permafrost is monitored.

Global warming is affecting the jet stream in important ways.[43] Because the Arctic is warming faster than other parts of the planet (Arctic Amplification[44]), the thermal gradient that moves heat from the tropics to the poles is weakening. This slows the jet stream and it develops large, slow-moving meanders that can produce extreme weather.[45] Extreme weather is growing across the globe and it is costing communities a great deal of money to manage the consequences.[46]

Meanders in the Northern Hemisphere polar jet move to the east, and because they represent the borderline between two bodies of air that have different temperatures and humidity, **weather fronts** tend to develop along the boundary. In a meander that draws down Arctic air, the leading boundary becomes a low-pressure **cold front** that brings cold air, and bands of rainy, snowy conditions (**Figure 3.21**). Sometimes these can develop into a **cut-off low**, which is a quasi-stable region of stormy weather. Similarly, a **warm front** may develop if the warm body of air is more powerful. A warm front usually produces a long cloudy period of light and drizzly rain.

According to a report in Nature Climate Change,[47] 74 percent of the world's population will be exposed to deadly heat waves by 2100 if carbon dioxide emissions continue to rise at current rates. Even if emissions are aggressively reduced, it is expected that 48 percent of the world's human population will be affected. Unprecedented summer warmth and flooding, forest fires, heat waves, drought and torrential rain, and extreme weather events are occurring more often,[48] and many scientists view changes in the jet stream as one reason why.[49]

Research[50] is improving our understanding of how the path of a jet stream is changed by global warming. Data indicate that the jet streams are slowly rising in altitude and moving toward the poles, but are dipping into the deep south during winters causing severe winter weather. The movement toward the poles means that the dry area typically located under the sinking air of the Hadley Cell is migrating toward the poles. This shifts the location of storms, drought, record-setting temperature events (warm and cold) and contributes to the spread of desertification and changing water supplies.

It is common to assume that global warming leads to more extreme weather because warmer temperatures mean more heat waves, hotter summers mean worse drought, the warmer atmosphere holds more moisture so when it rains or snows we tend to see greater amounts of precipitation, and other direct influences of warmer air on the weather.

[42] Snow Water Ice and Permafrost, Summary for Policymakers, Arctic Monitoring and Assessment Program (AMAP) (2017) http://www.amap.no/documents/doc/Snow-Water-Ice-and-Permafrost.-Summary-for-Policymakers/1532

[43] Weird weather: http://e360.yale.edu/features/linking_weird_weather_to_rapid_warming_of_the_arctic

[44] Arctic Amplification is the name given to the observation that the warming trend in the Arctic has been almost twice as large as the global average in recent decades. Scientists think this is because the loss of Arctic sea ice and snow cover reduces the albedo and leads to rapid warming in an area that was previously highly reflective. The dark land and dark seawater are now absorbing sunlight that would have previously reflected back to space. Greenhouse gases trap the heat radiating from these surface environments as they warm.

[45] Francis, J.A. and Vavrus, S.J. (2012) Evidence linking Arctic Amplification to extreme weather in mid-latitudes, *Geophysics Research Letters*, 39, L0681.

[46] VIDEO: Climate.gov https://www.youtube.com/watch?v=ihnfzGcqVx8&t=36s

[47] Mora, C., et al. (2017) Global risk of deadly heat, *Nature Climate Change*, doi: 10.1038/NCLIMATE3322

[48] Horton, R.M., et al. (2016) A review of recent advances in research on extreme heat events, *Current Climate Change Reports*, doi: 10.1007/s40641-016-0042-x

[49] Coumou, D., et al. (2015) The weakening summer circulation in the Northern Hemisphere mid-latitudes, *Science*, 348, 324–327.

[50] Mann, M.E., et al. (2017) Influence of Anthropogenic climate change on planetary wave resonance and extreme weather events, *Scientific Reports*, 7, 45242, doi: 10.1038/srep45242

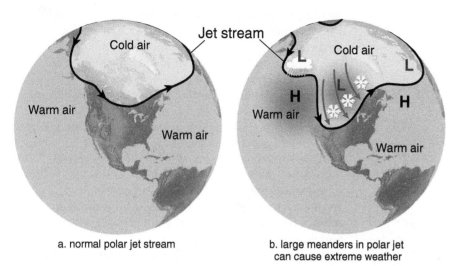

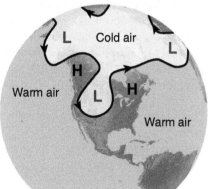

a. normal polar jet stream

b. large meanders in polar jet can cause extreme weather

c. cold low pressure system

d. cut-off low, persistent stormy weather

FIGURE 3.21 Meanders in the polar jet draw cold, dry air out of the Arctic and sometimes set record low temperatures and high amounts of snowfall in temperate continental regions. Northerly meanders draw warm, dry air out of the tropics and may create heat waves and drought. Meanders in the polar jet stream create regions of low and high pressure. Meanders move to the east, taking these weather patterns with them. The boundary between a low and high can develop a cold front that is characterized by cold, rainy, or snowy conditions. At times a *cut-off low* can develop, which results in persistent stormy weather.

Source: N. Hulbirt

However, with the new understanding of how the polar jet is changing in the Northern Hemisphere, researchers now realize that an increase in extreme weather is more complex than previously believed. It is also the result of climate change influencing the behavior of the jet stream in a way that favors more extreme and persistent weather irregularities.[51]

3.6 Ocean Currents Carry Heat

Atmospheric processes distribute heat, water vapor, and wind across the face of the planet, which, in turn, determines the level of precipitation, the character of the seasons, how cold or warm it is at various times of the year—in short, the climate. Oceans, because they carry heat from the tropics toward the poles, also play a significant role in regulating the climate.

Climate change is the product of changes in the accumulation and movement of heat in the atmosphere and oceans. The oceans[52] influence many other natural systems on the planet, and they especially affect the weather and climate. Ocean water moderates surface temperatures by absorbing heat from the Sun and transporting that heat toward the poles as well as downward toward the seafloor. Restless ocean currents[53] distribute this heat around the globe, warming the land and air during winter and cooling it in summer. The fact that cold water is denser and so is heavier than warm water also plays into the way ocean water circulates.

3.6.1 Ocean Circulation

There are basically two types of large-scale oceanic circulation: surface circulation, which is stimulated by winds and

---

[51] Extreme weather: https://www.csmonitor.com/Science/2017/0327/How-climate-change-may-be-driving-extreme-weather

[52] There are five oceans: the Atlantic, Pacific, Indian, Arctic, and Southern oceans. Smaller bodies, known as "seas," include the Mediterranean and China seas.

[53] Explore the science of oceanography at NASA, NASA Science Earth, NASA Oceanography http://nasascience.nasa.gov/earth-science/oceanography

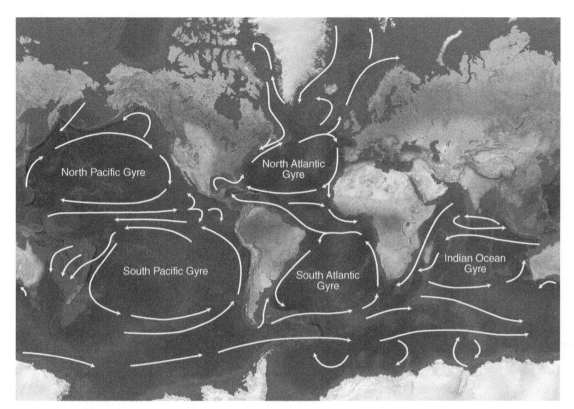

FIGURE 3.22 There are five major basin-wide gyres, each controlled by the interaction of winds and the Coriolis force. Currents within the gyres carry heat from the equator toward the poles and thus strongly influence climate.

the Coriolis force, and deep circulation, which is the result of cool salty water at the poles sinking and moving through the lower ocean. Both are driven by the exchange of heat.

The general pattern of circulation consists of surface currents carrying warm water away from the tropics toward the poles and, in the process, releasing heat to the atmosphere. Conditions at the poles cool this surface water. It is also salty as it has experienced evaporation during the course of its journey, increasing the concentration of dissolved salt. Since it is denser than surrounding ocean water, it sinks to the deep ocean, creating cold, salty currents along the seafloor and at mid depths in the ocean.

This process is especially pronounced in the North Atlantic and in the Southern Ocean along the coastal waters of Antarctica, where cooling is the strongest. Deep ocean water gradually returns to the surface nearly everywhere in the ocean. Once at the surface, it is carried back to the tropics by surface currents, where it is warmed again and the cycle begins anew. The more efficient the cycle, the more heat is transferred from the tropics to the poles, and the more this heat warms the climate.[54]

Surface Currents Owing to Earth's rotation (Coriolis force), ocean currents are deflected to the right in the Northern Hemisphere and to the left in the Southern Hemisphere. In surface circulation, this process creates

large-scale circulation systems called gyres that sweep the major ocean basins.

There are five major basin-wide gyres (**Figure 3.22**) including the North Atlantic, South Atlantic, North Pacific, South Pacific, and Indian Ocean gyres. Each gyre is composed of a strong and narrow western boundary current and a weak and broad eastern boundary current. Each of the five major gyres in the oceans has parallel systems of currents, and these currents each carry heat and govern climate where they flow. The surface circulation of the North Pacific Gyre is a typical example of how winds and the Coriolis force combine to create surface circulation.

In the North Pacific atmosphere, the sinking column of the Hadley Cell produces a current of air that blows toward the equator along the surface but is deflected to the west (right) by the Coriolis force. This southwest-flowing wind is known as the Pacific trade wind. The Pacific trade wind drives the North Equatorial Current to the west just north of the equator at about 15°N latitude. This current is deflected north (again to the right) near the Philippines to create the warm western boundary current known as the Japan or Kuroshio Current. The Kuroshio Current carries warm water away from the tropics until it turns to the east at approximately 45°N latitude and becomes the North Pacific Current, which moves across the basin toward North America.

As it approaches the North American continent, the North Pacific Current splits, sending one arm north to circulate through the Gulf of Alaska and the Bering Sea as the Alaska Current. The southern arm becomes the cool,

---

[54] VIDEO: The ocean - A driving force for weather and climate, https://www.youtube.com/watch?v=6vgvTeuoDWY

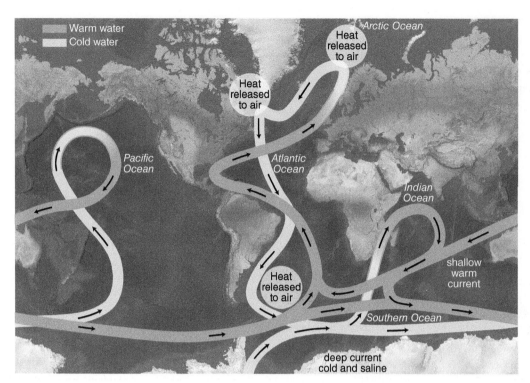

FIGURE 3.23 The thermohaline circulation is a global pattern of currents that carries heat, dissolved gas, and other compounds on a round trip that can take over 1,000 years to complete.

slow-moving eastern boundary current called the California Current. The California Current moves from about 60°N to 15°N latitude and merges with the North Equatorial Current. From there it once again travels thousands of miles westward across the basin to Asia. Each of the five major gyres in the oceans has similar systems of currents, and these each carry heat and strongly influence climate where they flow.

3.6.2 Deep Circulation

In the North Atlantic basin, the western boundary current of the North Atlantic Gyre is known as the Gulf Stream. The Gulf Stream carries warm tropical water from the Caribbean to the cold waters of the North Atlantic. As it moves, the Gulf Stream cools and evaporates, thus greatly increasing its density. By the time it arrives in the North Atlantic as a cold, salty body of water, it can no longer stay afloat and begins a long descent toward the seafloor of 2 to 4 kilometers (1 to 2.5 miles), where it becomes a deep current known as the North Atlantic Deep Water.

The North Atlantic Deep Water travels south through the Atlantic and eventually joins similar deep water that is forming in the Southern Ocean. These waters then become the Circumpolar Deep Water, which journeys throughout the Southern Ocean. An arm of the Circumpolar Deep Water migrates into the North Pacific and there, after a voyage of approximately 35,000 kilometers (22,000 miles), water that originated in the North Atlantic Gulf Stream eventually surfaces into the sunshine.

It has been estimated that up to 1,300 years can pass before the cycle is completed and water returns to its place of origin.[55] This **thermohaline circulation** (**Figure 3.23**), also called the **oceanic conveyor belt**,[56] travels through the entire world's oceans.[57] The process connects all seawater in a truly global system that transports both energy (heat) and matter (solids, dissolved compounds, and gases), and in doing so influences global climate.

3.7 Global Warming Is Changing the Ocean

The ocean covers 70 percent of Earth and is the largest single component of the planet's surface. It exerts a vast influence on the climate because it contains a huge amount of heat energy. The top 2 meters (6.5 feet) of ocean water carries as much heat as the entire overlying atmosphere.[58] When ocean currents carry this heat around the globe, the warmth provides the driving force behind the weather and climate.

[55] Lozier, S. (2010) Deconstructing the conveyor belt, *Science*, 328(5985), 1507–1511, doi: 10.1126/science.1189250.

[56] Thermohaline Circulation, *Wikipedia*, http://en.wikipedia.org/wiki/Thermohaline_circulation

[57] Mann, A. (2010) Ocean-conveyor belt model stirred up, *Nature* News, September 12, http://www.nature.com/news/2010/100912/full/news.2010.461.html, doi:10.1038/news.2010.461

[58] NOAA "Modeling Sea Surface Temperature," *ClimateWatch*, http://www.climatewatch.noaa.gov/image/2009/modeling-sea-surface-temperature

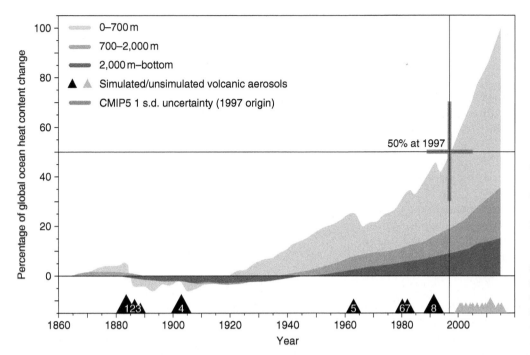

FIGURE 3.24 Studies show that the upper 700 m (2,297 ft) of the oceans is absorbing heat owing to global warming. The graph shows how the heat content of the oceans has changed as a percentage of the total heat uptake over the period 1865 to 2015. Lightest region shows heating in the upper 700 m, medium shows heating from 700 to 2,000 m (2,297 to 6,562 ft), and darkest region shows heating from a depth of 2,000 m to the sea floor. Black and gray triangles mark aerosol release related to volcanic eruptions.

Source: Glecker, et al. (2016)

The ocean influences the atmosphere, but the atmosphere also influences the ocean, in that excess heat from global warming is largely stored in the waters of the ocean. This heat drives up the sea surface temperature (and the temperature of the deep ocean), which further influences climate, and also influences critical aspects of the marine environment, such as the biology, chemistry, and physical attributes of the sea.

3.7.1 Warming Oceans

Global warming has raised Earth's average surface temperature by about 1 degree Celsius (1.8 degrees Fahrenheit)[59] over the past century. In the oceans, the sea surface temperature has increased only half that amount.[60] However, because the ocean holds additional heat in its lower reaches, the sea surface temperature tells only part of the story.

Greenhouse gases have trapped excess heat in the troposphere leading to global warming. The ocean has absorbed more than 90 percent of this extra warmth[61] and nearly 30 percent of the carbon dioxide generated by human consumption of fossil fuels. As a result, the world's oceans are warming at a quickening rate, with the past 20 years accounting for half of the increase in ocean heat content that has occurred since preindustrial times.[62] Scientists discovered that much of the extra heat in the ocean is buried

deep underwater, with 35 percent of the additional warmth found at depths below 700 m (2,300 ft). This means far more heat is present in the deepest reaches of the ocean than 20 years ago, when it contained just 20 percent of the extra heat produced from the release of greenhouse gases since the industrial revolution (**Figure 3.24**).

The repercussions of a warmer ocean are far-reaching. Ecosystems from polar to tropical regions are already feeling the effects, driving entire groups of species such as plankton, jellyfish, turtles, commercial fisheries,[63] and seabirds up to 10 degrees of latitude toward the poles. This is causing the loss of breeding grounds for turtles and seabirds, and affecting the breeding success of marine mammals. Warming oceans are affecting fish stocks in some areas by damaging fish habitats and causing fish species to move to cooler waters. This is expected to lead to reduced fisheries catches in regions across the globe.

The direct impacts are already being felt by human communities including reductions in fish stocks and crop yields, more extreme weather events, and increased risk from water-borne diseases.[64] Warmer water also causes stress to corals,[65] plankton, pelagic fish and other marine plants and animals.[66]

[59] NASA discussion: http://earthobservatory.nasa.gov/IOTD/view.php?id=89469

[60] NOAA discussion: https://www.ncdc.noaa.gov/data-access/marine-ocean-data/extended-reconstructed-sea-surface-temperature-ersst-v4

[61] IPCC: http://www.climatechange2013.org/images/report/WG1AR5_Chapter03_FINAL.pdf

[62] Glecker, P.J., et al. (2016) Industrial era global ocean heat uptake doubles in recent decades, *Nature Climate Change*, doi:10.1038/nclimate2915

[63] Kleisner, K.M., et al. (2017) Marine species distribution shifts on the U.S. northeast continental shelf under continued ocean warming, *Progress in Oceanography*, 153(24), doi: 10.1016/j.pocean.2017.04.001

[64] International Union for Conservation of Nature 2016 report *Explaining Ocean Warming: Causes, Scale, Effects and Consequences*, https://portals.iucn.org/library/node/46254

[65] Hughes, T.P., et al. (2017) Global warming and recurrent mass bleaching of corals, *Nature*, 543(7645), 373, doi: 10.1038/nature21707. See also, Heron, S.F., et al. (2016) Warming trends and bleaching stress of the world's coral reefs 1985–2012, *Scientific Reports*, 6(38402), doi: 10.1038/srep38402.

[66] Ocean warming: https://www.theguardian.com/environment/2017/jan/08/fish-ocean-warming-migration-sea

Water expands as it warms, this is one cause of sea-level rise.[67] Sea-level rise, a phenomenon that we study in detail in **Chapter 5**, threatens many of the world's major cities, the global economy, coastal and marine ecosystems, and the livelihoods of thousands of communities and tens of millions of individuals. Because of sea-level rise, many communities in the United States and abroad will see waves of displaced populations moving to new areas looking for new homes and livelihoods.[68]

All living organisms are the product of evolution, a process that selects populations of species on the basis of their ability to successfully reproduce under given environmental conditions. When those conditions change, the natural framework that gave rise to a species is threatened. Marine ecosystems are faced with extinction[69] if they are not able to migrate away from warming waters and move into cooler regions. Marine ecosystems adapted to polar waters are left with no options whatsoever.

One study[70] found that even though warming ocean water threatens many reef species, there may be a refuge adjacent to equatorial islands where deep, cool, nutrient-rich water rises to the surface. The Pacific nation of Kiribati was found to have 33 atoll islets that are bathed in cooling waters that rise from below along the westward flanks of each atoll. These conditions result from currents 100 to 200 meters (330 to 660 feet) deep that run counter (known as the Equatorial Undercurrent) to the surface flow driven by the (east-to-west blowing) trade winds. When the (west-to-east flowing) undercurrent encounters the submerged slope of an atoll, it is forced to the surface and envelops the reef there with cool water. Conditions such as these may be right for protecting coral species in the future as oceans continue to store heat.

The ocean is warming because it absorbs most of the extra heat being added to the climate system from the buildup of greenhouse gases in the atmosphere.[71] The warmer atmosphere leads to a warmer ocean, and ocean circulation carries the warm water across Earth's surface as well as into the depths[72] of the sea,[73] although most of the

heat is accumulating in the ocean's near-surface layers. In fact, according to the United Nations Intergovernmental Panel on Climate Change (IPCC) assessment report number 4 (AR4),[74] during the past 50 years the ocean has stored more than 90 percent of the increase in heat content of the Earth system.[75] A popular climate change website, Skeptical Science,[76] has provided a graphic that dramatically illustrates this point (**Figure 3.25**).

3.7.2 Phytoplankton

Marine life across the planet depends on tiny ocean plants called **phytoplankton** (**Figure 3.26**). They contain chlorophyll and require sunlight in order to live and grow. Most

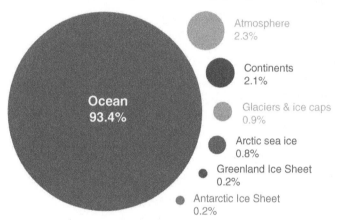

FIGURE 3.25 Warming of the ocean accounts for about 93% of the increase in global heat between 1971 and 2010, with warming of the upper ocean accounting for about 64% of the total.[77]

Source: Skeptical Science Graphics

[67] Current Sea Level Rise, *Wikipedia*, http://en.wikipedia.org/wiki/Current_sea_level_rise

[68] https://phys.org/news/2016-05-tide-migration-accompanies-sea.html

[69] In 2011, an international panel of marine experts warned that the world's ocean is at high risk of entering a phase of extinction of marine species unprecedented in human history. See the report at International Program on the State of the Ocean, home page, http://www.stateoftheocean.org; see also "Multiple Ocean Stresses Threaten 'Globally Significant' Marine Extinction, Experts Warn," *ScienceDaily,* http://www.sciencedaily.com/releases/2011/06/110621101453.htm

[70] Karnauskas, K. and Cohen, A. (2012) Equatorial refuge amid tropical warming, *Nature Climate Change,* 2, 530–534, doi:10.1038/nclimate1499

[71] https://phys.org/news/2017-03-ocean.html

[72] Volkov, D.L., et al. (2017) Decade-long deep-ocean warming detected in the subtropical South Pacific, *Geophysical Research Letters,* 44, 927–936, doi:10.1002/2016GL071661

[73] Song, Y.T. and Colberg, F. (2011) Deep ocean warming assessed from altimeters, gravity recovery and climate experiment, in situ measurements, and a non-Boussinesq ocean general circulation model, *Journal of Geophysical Research,* 116, C02020, doi:10.1029/2010JC006601

[74] Under the joint auspices of the United Nations Environmental Program (UNEP) and the World Meteorological Organization (WMO), the Intergovernmental Panel on Climate Change (IPCC) produces global assessments of climate change every 5 to 7 years, representing the state of understanding. Past IPCC reports have been published in 1990, 1995, 2001, 2007, and 2013; the next report is scheduled for sometime around 2020. The IPCC does not carry out original research, nor does it do the work of monitoring climate or related phenomena itself. Its primary role is publishing special assessments on topics relevant to the implementation of the United Nations Framework Convention on Climate Change, which is an international treaty that acknowledges the possibility of harmful climate change. For example, see IPCC, "Special Report on Managing the Risks of Extreme Events and Disasters to Advance Climate Change Adaptation," http://www.ipcc.ch

[75] Rhein, M., et al. (2013) Observations: Ocean. In: *Climate Change 2013: The Physical Science Basis. Contribution of Working Group I to the Fifth Assessment Report of the Intergovernmental Panel on Climate Change* [Stocker, T.F., et al. (eds.)]. Cambridge University Press, Cambridge, United Kingdom and New York, NY.

[76] SkepticalScience.com "Explaining Climate Change Science and Rebutting Global Warming Misinformation," http://www.skepticalscience.com

[77] Data for this figure is from Rhien et al., 2013.

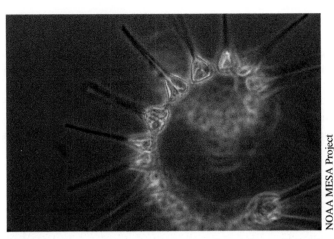

NOAA MESA Project

FIGURE 3.26 Phytoplankton form the base of several marine food webs. In a balanced ecosystem, they provide food for a wide range of sea creatures including whales, shrimp, snails, and jellyfish.

Source: NOAA

phytoplankton are buoyant and float in the upper part of the ocean where sunlight penetrates the water. Phytoplankton requires inorganic nutrients such as nitrates, phosphates, and sulfur, which they convert into proteins, fats, and carbohydrates.

In a balanced ecosystem, phytoplankton provide food for a wide range of sea creatures including whales, shrimp, snails, and jellyfish. They form the fundamental basis of the marine food chain and most life in the oceans depend on a healthy phytoplankton population.

When too many nutrients are available, phytoplankton may grow out of control and form harmful **algal blooms**. These blooms can produce extremely toxic compounds that have harmful effects on fish, shellfish, mammals, birds, and even people. They are commonly known as "red tides." One of the best-known red tides occurs nearly every summer along Florida's Gulf Coast. Microscopic algae that produce toxins that kill fish and make shellfish dangerous to eat cause this bloom. The toxins may also make the surrounding air difficult to breathe. As the name suggests, the algae often turn the water red.

Studies show that many algal species flourish when wind and water currents are favorable. In other cases, harmful algal blooms may be linked to an excess of nutrients in the water. This occurs when phosphorus and nitrogen from sources such as fertilizer and sewage flow into bays, rivers, and the sea and build up at a rate that overnourishes the natural algae in the environment. Some algal blooms appear in the aftermath of natural phenomena like sluggish water circulation, unusually high water temperatures, and extreme weather events such as hurricanes, floods, and drought.

Large algal blooms can create **dead zones**, regions of anoxic waters that suffocate other forms of life in the area. Excess nutrients that run off land or are piped as wastewater into rivers and coasts can stimulate an overgrowth of algae, which then sinks and decomposes in the water. Bacteria

involved in the decomposition process consume oxygen and deplete the natural supply available to healthy marine life. Dead zones occur in many areas of the United States, particularly along the East Coast, the Gulf of Mexico, and the Great Lakes, but there is no part of the country or the world that is immune. The second largest dead zone in the world is located in the United States, in the northern Gulf of Mexico. It has been estimated[78] that there are over 400 dead zones around the world where polluted waters lead to algal blooms.

The surface temperature of the ocean influences where and when phytoplankton grow. Because they cannot survive in extremely warm water, global warming is having a negative impact on them, which, in turn, is affecting the entire web of organisms in the ocean.

Despite the occurrence of harmful algal blooms, phytoplankton is essential to marine food chains. They underpin most of the life in the oceans. According to researchers,[79] phytoplankton account for half of all the production of organic matter on Earth, a colossal characteristic of a microscopic organism; however, worldwide phytoplankton levels are down 40 percent since the 1950s. In fact, their numbers have been declining in 8 out of 10 ocean regions at a global rate of about 1 percent per year. Scientists identify rising sea surface temperatures as the cause of this decline, because warmer water makes it hard for phytoplankton to metabolize vital nutrients. As these tiny components of the food chain decline, the entire marine ecosystem is being affected.

There are more than 7,000 different species of algae. Most live in the oceans, but they also live in fresh water and even on land. A critically important aspect of marine phytoplankton is that they produce about 330 billion tons of oxygen each year. It is estimated these marine plants produce between 60 and 80 percent of the oxygen in the atmosphere. This means that all life on Earth depends on phytoplankton for the oxygen in the air.

If ocean warming continues to high levels, such as 6 degrees Celsius (10.8 degrees Fahrenheit) above preindustrial temperatures, it could disrupt oxygen production by impeding the process of photosynthesis.[80] Roughly two-thirds of the planet's total atmospheric oxygen is produced by ocean phytoplankton—and therefore disruption or decrease of this process would result in the depletion of atmospheric oxygen on a global scale.

At the same time that global warming is threatening to disrupt the oxygen produced by phytoplankton, researchers have revealed that warming ocean water is leading to a decrease in the dissolved oxygen content of the world's oceans, with greatest impacts in cool polar waters.

[78] Dead Zone: https://en.wikipedia.org/wiki/Dead_zone_(ecology)
[79] Boyce, S. et al. (2010) Global phytoplankton decline over the past century, *Nature*, 466, 591–596, doi:10.1038/nature09268. See Borenstein, B., Climate change: Plankton in big decline, foundation of ocean's food web, *Huffington Post*, http://www.huffingtonpost.com/2010/07/29/climate-change-plankton-i_n_663488.html
[80] Sekerci, Y. and Petrovskii, S. (2015) Mathematical modeling of plankton–oxygen dynamics under the climate change, *Bulletin of Mathematical Biology*, doi: 10.1007/s11538-015-0126-0

3.7.3 Declining Oxygen Content

Seawater contains an abundance of dissolved oxygen that all marine animals breathe to stay alive. It has long been established in physics that cold water holds more dissolved oxygen than warm water does—this is one reason that cold polar seas are teeming with life while tropical oceans are blue, clear, and relatively poorly populated with living creatures. Thus, as global warming raises the temperature of marine waters, it is self-evident that the amount of dissolved oxygen will decrease. This is a worrisome and potentially disastrous consequence if allowed to continue to an ecosystem-threatening level.

Now scientists have analyzed data indicating that the amount of dissolved oxygen in the oceans has been declining for more than a half century (**Figure 3.27**).[81] The data

a. 100m

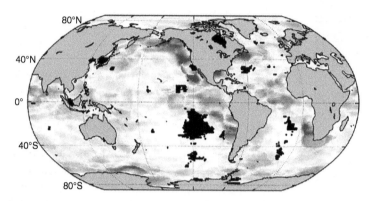

b. 400m

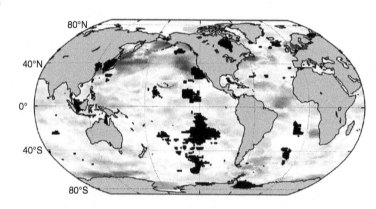

c. 700m

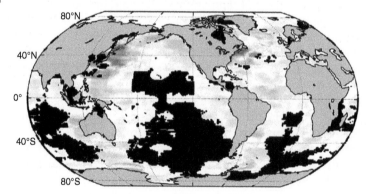

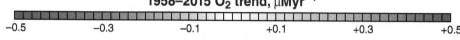

1958–2015 O$_2$ trend, µMyr^{-1}

−0.5   −0.3   −0.1   +0.1   +0.3   +0.5

FIGURE 3.27 Global map of the linear trend of dissolved oxygen at depths of 100, 400, and 700 m (328, 1,312, and 2,297 ft). The amount of oxygen in the world's oceans is measured in micromolar (µM).[82] This map shows the trend of changes in µM per year (black: no data; red [faint grey]: increasing; blue [dark grey]: decreasing). Researchers have documented that globally a 2% decrease in dissolved oxygen has occurred in the oceans over the past half century, with greatest losses in cool waters of the North Pacific.

Source: Taka Ito, Georgia Institute of Technology

[81] Ito, T., et al. (2017) Upper ocean O$_2$ trends: 1958–2015, *Geophysical Research Letters*, doi: 10.1002/2017GL073613

[82] One molar (M) is defined as 6.02×10^{23} atoms per 1 L of solution (1µM = 0.000001 M).

show that the ocean oxygen level has been falling more rapidly than the corresponding rise in water temperature. Falling oxygen levels in water have the potential to impact the habitat of marine organisms worldwide and in recent years has led to more frequent anoxic events that killed or displaced populations of fish, crabs, and many other organisms.

Researchers calculated that the rate of oxygen decline is about two to three times faster than what was predicted from the decrease of solubility associated with simple warming of the water. The total decline amounts to more than 2 percent over the past 50 years,[83] but only about 15 percent of this loss is attributable to the fact that warm seawater has trouble holding oxygen.

The majority of the oxygen in the ocean is absorbed from the atmosphere at the surface or created by photosynthesizing phytoplankton. Ocean currents then mix that more highly oxygenated water with subsurface water. But rising ocean water temperatures near the surface have made it more buoyant and harder for the warmer surface waters to mix downward with the cooler subsurface waters. In addition, in polar seas, melting sea ice and glaciers on land has added more freshwater to the ocean surface. Freshwater is more buoyant than saltwater and thus floats on the ocean surface. This also hampers the natural mixing and leads to increased ocean layering (termed *stratification*).

When the ocean is stratified, oxygen in the surface waters does not mix down into deeper waters, leaving them oxygen deprived. This means there is less oxygen getting to the ocean. Because large fishes in particular avoid or do not survive in areas with low oxygen content, these changes can have far-reaching biological consequences.

Another aspect of this phenomenon in the North Pacific and Arctic oceans is that the reduction of sea ice has led to more plankton growth. With more plankton growth comes more plankton decomposition. Decomposition decreases oxygen levels even further and dead zones are starting to develop along the shoreline. Fish can't thrive under these conditions, and this is a dangerous threat to both the ecosystem and the economy as commercial fishing is a main component of the economy in northern communities.

But that's not the only problem. These areas are pumping out the powerful greenhouse gas nitrous oxide, which is a by-product of the chemical reactions that take place in dead zones.[84] Nitrous oxide is potent. It lasts in the atmosphere for over a century and contributes strongly to global warming. This means that the effects of climate change on the world's oceans cause more global warming, a powerful, and not previously predicted, positive feedback effect.

3.7.4 Acidifying Oceans

In addition to absorbing heat, the oceans have absorbed about half of the carbon dioxide[85] emitted by humans over the past two centuries. This is a great environmental service[86] that has slowed warming of the atmosphere; unfortunately, the chemistry of the ocean is changing as a result.

Increasing ocean acidity, brought on by dissolved carbon dioxide (CO_2) that mixes with seawater (H_2O) to form carbonic acid (H_2CO_3), makes it difficult for calcifying organisms (corals, mollusks, and many types of plankton) to secrete the calcium carbonate ($CaCO_3$) they need for their skeletal components. Calcium carbonate—the stuff of which shells, corals, and many types of plankton are made—is impeded from forming in low levels of carbonic acid and openly dissolves in the presence of high levels of carbonic acid. This ocean acidification[87] is one of the consequences of carbon dioxide buildup that could have a serious impact on the world's ocean ecology, which depends on the secretion of calcium carbonate by thousands of different species.

For instance, one study[88] predicts that rising carbon emissions and acidification of seawater might kill off the ocean's coral reefs by 2050. That forecast from 2007 was validated by research[89] a decade later that makes essentially the same prediction, though largely on the basis of coral bleaching caused by warmer water.

Coral bleaching is a stress response that happens when an increase in water temperature causes the expulsion of algae that grow inside coral, turning the reefs white and eliminating their main energy source (**Figure 3.28**). Whether ocean acidification or ocean warming causes it, coral bleaching is directly linked to global warming. Bleaching doesn't kill coral right away—if temperatures drop, the algae has the chance to recolonize. But if temperatures remain high, soon the coral will die, removing the natural habitat for many species of marine life. But the reality may be more complicated than this.

---

[83] Schmidtko, S. et al. (2017) Decline in global oceanic oxygen content during the past five decades, *Nature*, 542(7641), 335–339.

[84] N$_2$O: http://www.umces.edu/hpl/release/2010/mar/03/aquatic-"dead-zones"-contributing-climate-change

[85] Sabine, C.L., et al. (2004) The oceanic sink for anthropogenic CO$_2$, *Science*, 305(5682), 367–371, doi: 10.1126/science.1097403

[86] Pickrell, J. (2004) Oceans found to absorb half of all man-made carbon dioxide, *National Geographic News*, July 15, http://news.nationalgeographic.com/news/2004/07/0715_040715_oceancarbon.html

[87] Zeebe, R.E., et al. (2008) Carbon emissions and acidification, *Science*, 321(5885), 51–52, doi: 10.1126/science.1159124

[88] Hoegh-Guldberg, O., et al. (2007) Coral reefs under rapid climate change and ocean acidification, *Science*, 318(5857), 1737–1742.

[89] Heron, et al. (2016)

NOAA

FIGURE 3.28 In 2015 NOAA declared that a global coral bleaching event was under way.[90] Coral bleaching occurs when coral polyps expel algae that live inside their tissues. Normally, coral polyps live in a symbiotic relationship with the algae and that relationship is crucial for the coral and hence for the health of the whole reef. Bleached corals continue to live for a period of time. But if the water temperature does not return to normal so that the algae recolonize the polyp, the coral will die.

Source: NOAA

An analysis[91] of coral along the entire length of Australia's Great Barrier Reef found that as ocean temperatures and acidity rise, some species of corals are likely to succeed, some species might not, and the mix of species making up any single reef will change. The loss of healthy coral reefs affects all the species that dwell there (such as turtles, seals, mollusks, crabs, and fish), as well as the animals that depend on reef habitats as a food source (including seabirds, mammals, and humans). One quarter of all sea animals spends time in coral reef environments during their life cycle. Acidification has already been seen to damage the ability of oyster larvae on the Oregon coast to successfully develop their shells and grow at a pace that would allow them to be commercially harvested.[92]

Ocean acidification is measured using the pH scale. pH is a number scale[93] that ranges from 1 (acidic) to 14 (basic); a pH of 7 is considered neutral. To calibrate coral response to acidification, scientists have used field studies[94] of locations where carbon dioxide seeps out of the ocean floor on the submerged slopes of a volcano.[95] It was found that as pH declines from 8.1 to 7.8, equivalent to the seawater change expected if atmospheric carbon dioxide concentration increases from 410 ppm[96] (today's atmospheric concentration of CO_2) to 750 ppm (possible by the end of this century), reefs show a reduction in coral diversity, recruitment (new populations of coral on barren substrate), and abundances of reef-building corals. Reef development ceased altogether below a pH of 7.7. Researchers concluded that these responses are consistent with previous model results and that together with temperature stress of warming seawater will probably lead to severely reduced resiliency of Indo-Pacific coral reefs this century.

There are economic impacts as well. Tourism tied to coral reefs and commercial fisheries generate billions of dollars in revenue annually. Biodiversity, food supplies, and economics thus could all be affected by the impacts of warming and acidification. Reef loss is a complex issue, however. Reefs can suffer from coastal pollution,[97] overfishing,[98] and other types of human stresses as well as at the hands of warming and acidification. Exactly what roles warming temperatures, ocean acidification, and other anthropogenic impacts play in global marine health have

[90] NOAA News: http://www.noaanews.noaa.gov/stories2015/100815-noaa-declares-third-ever-global-coral-bleaching-event.html

[91] Hughes, T.P., et al. (2010) Assembly rules of reef corals are flexible along a steep climatic gradient, *Current Biology*, 22(8), 736–741.

[92] Barton, A., et al. (2012) The Pacific oyster, *Crassostrea gigas,* shows negative correlation to naturally elevated carbon dioxide levels: Implications for near-term ocean acidification effects, *Limnology and Oceanography,* 57(3), 698, doi: 10.4319/lo.2012.57.3.0698

[93] Acidic and basic are two extremes that describe chemicals, just as hot and cold are two extremes that describe temperature. Mixing acids and bases can cancel out their extreme effects, much as mixing hot and cold water can even out the water temperature. A substance that is neither acidic nor basic is neutral. See EPA, "What Is pH?" http://www.epa.gov/acidrain/measure/ph.html

[94] Fabricius, K., et al. (2011) Losers and winners in coral reefs acclimatized to elevated carbon dioxide concentrations, *Nature Climate Change*, 1, 165–169, http://www.reefrelieffounders.com/science/2011/06/07/nature-com-climate-change-losers-and-winners-in-coral-reefs-acclimatized-to-elevated-carbon-dioxide-concentrations-by-katharina-e-fabricius-et-al

[95] Carbon dioxide dissolved in the ocean reacts with seawater to form carbonic acid. Carbonic acid produces positively charged hydrogen ions (that lower pH) and negatively charged bicarbonate ions. Bicarbonate ions may lose a hydrogen ion (further lowering pH) to produce the carbonate ion (used by marine organisms to build calcium carbonate exoskeletons). Ocean acidification decreases the bicarbonate to carbonate reaction and leads to some carbonate recombining with hydrogen to form bicarbonate. The result is that acidification reduces carbonate available for corals, some plankton, oysters, clams and other organisms.

[96] Ppm means "parts per million." It is a measurement of abundance (or concentration) the same way that "per cent" means parts per hundred. In this case, ppm means molecules of CO_2 per million molecules of air.

[97] "Mass extinctions and 'Rise of Slime' Predicted for Oceans," *Science Daily*, http://www.sciencedaily.com/releases/2008/08/080813144405.htm

[98] Climate Change Refuge for Corals Discovered: https://www.sciencedaily.com/releases/2017/05/170517120556.htm

yet to be fully defined by researchers, but they are all negative factors.[99]

So large that it can be seen from space, the Great Barrier Reef of Australia is the largest single coral ecosystem on the planet and has been designated a World Heritage Site, meaning that it is recognized by the United Nations as being of outstanding international importance and therefore as deserving special protection. The world watched in shock then as in 1998, 2002, and again in 2014–2017 the Great Barrier Reef was hit with severe bleaching events that killed wide swaths of corals. In 2016, huge sections of the reef, stretching across hundreds of kilometers of its most pristine northern sector, were found to be dead, killed by overheated seawater. More southerly sections around the middle of the reef that barely escaped in 2016 experienced bleaching in 2017.

As the coral mortality unfolded intermittently for nearly two decades, the reef has endured compounding effects that have greatly reduced any remaining coral refuges. In a paper[100] published in 2017, scientists showed that the cumulative footprint of these multiple bleaching events has expanded to encompass virtually all of the Great Barrier Reef.

The 2014–2017 bleaching event proved the most severe, affecting 91 percent of individual reefs along the entire 2,300 km (1,429 mi) system. Notably, scientists studying the reef have in the past concluded that protecting reefs from fishing, and improving water quality will likely help bleached reefs recover in the longer term; however, those measures made no difference to the amount of bleaching that occurred on the Great Barrier Reef (**Figure 3.29**).

With the Great Barrier Reef having been hit by a string of deadly bleaching events over the past two decades, and more recently 3 years in a row, scientific experts are telling the Australian government that, because of climate change, the national plan to protect the reef is no longer viable.[101]

Back-to-back bleaching events have killed almost half the reef and actions such as strengthening laws against poaching, limiting pollutants, and enhancing environmental protection are insufficient steps when bleaching from global warming is causing such massive damage. The federal and Queensland government's *Reef 2050 Long Term Sustainability Plan*[102] was released in 2015, with its central vision to "ensure the Great Barrier Reef continues to improve on its outstanding universal values". The plan is part of an effort to satisfy the United Nations Environmental, Scientific, and Cultural Organization (UNESCO) World Heritage Centre that the reef is not in danger and does not need to be added to their list of world heritage sites in danger. However, since the bleaching of 2014–2017, reef experts are saying that because of climate change, the 2050 plan goals to conserve the reef are not achievable. Thus, even in a location where

FIGURE 3.29 Severe bleaching occurred on the Great Barrier Reef in 1998, 2002, 2014–2017. In 2016, only 8.9% of 1,156 surveyed reefs escaped with no bleaching, compared to 42.4% of 631 reefs in 2002 and 44.7% of 638 in 1998. Scientists now fear that much of the ecosystem will fail to recover given that ocean warming is highly likely to continue.

Source: Dr. Gergely Torda, ARC Centre of Excellence for Coral Reef Studies

[99] Wiese, E., Scientists: Global warming could kill coral reefs by 2050, *USA Today* http://www.usatoday.com/weather/climate/globalwarming/2007-12-13-coral-reefs_N.htm

[100] Hughes et al. (2017)

[101] GBR plan: https://www.theguardian.com/environment/2017/may/25/great-barrier-reef-2050-plan-no-longer-achievable-due-to-climate-change-experts-say

[102] GBR plan: http://www.environment.gov.au/marine/gbr/long-term-sustainability-plan

human communities are working to minimize negative impacts on local reefs, the problems of ocean warming and ocean acidification are now developing with such intensity that local conservation steps may be completely ineffective.

3.7.5 Deadly Trio

The ocean, like the atmosphere, has always been considered too big to fail. From every shore we discharge pollutants in massive quantities. Industrialized fleets harvest fish with unforgiving efficiency using huge swaths of steel-cable netting that strip-mine the seafloor and water column wiping out entire ecosystems. And now, because we have treated the atmosphere with equal contempt, the ocean is turning hotter, anoxic, acidic, polluted, and, in many locations, devoid of life. The ocean is absorbing carbon dioxide and heat from the air as a massive environmental service that greatly benefits life on land but, if continued, effectively guaranteeing the ocean a slow and painful death.

Anoxia As we studied earlier, global warming is reducing marine oxygen content by warming (making more buoyant) surface water that resists mixing with deeper water. This stratification of the ocean seals the lower water column from mixing with the atmosphere and slows the distribution of dissolved oxygen below the surface. Combined with the fact that warm water holds less dissolved gas, the dissolved oxygen content of the world's ocean has decreased 2 percent over the past half century.

Plastic The ocean is the final dumping ground for many human products, and chief among these is plastic. Plastic debris sweeps into the ocean every day from our coastal communities, as lost fishing tackle, through illegal dumping, as litter, and from purposeful dumping by unregulated nations. The waves and sun degrade plastic to small pieces that are eaten by over 90 percent of seabirds, fish, marine mammals, and even plankton. As they degrade, plastic releases dimethyl sulfide,[103] a compound that is also released by plankton and algae when they break down (usually when they are being eaten). Hence, plastic smells like food to many marine species.

Plastic ingested by marine animals causes many negative effects. The sharp edges of plastic pieces puncture internal organs. Cavity space intended for legitimate food is filled by plastic, leaving little room for real forms of nutrition. Baby seabirds are fed plastic by their parents, and as their stomachs fill up, they starve to death. Many animals are entangled in orphan nets and fishing line and eventually lose their strength and die.

Humans need to be concerned because plastic that has been ingested by marine creatures that we eat, such as fish, releases toxic compounds. Toxins in plastics have been linked to disruption of the endocrine system, various types of cancer, birth defects, immune system problems, and childhood developmental issues. Plastic pieces in the digestive systems of fish release these compounds and are incorporated into their muscle, the very part of the fish that is sold commercially for protein.

Warming and Acidification Ocean water has absorbed more than 90 percent of the excess heat and nearly 30 percent of the carbon dioxide generated by human consumption of fossil fuels. As a result, the world's oceans are warming at a quickening rate, with the past 20 years accounting for half of the increase in ocean heat content that has occurred since preindustrial times.[104] Most of the warming overall is occurring in the Southern Hemisphere and is contributing to the subsurface melting of Antarctic ice shelves. Since the 1990s the atmosphere in the polar regions has been warming at about twice the average rate of global warming.

Ocean warming is affecting humans in direct ways and the impacts are already being felt, including effects on fish stocks and crop yields, more extreme weather events, and increased risk from water-borne diseases. Ocean warming is already affecting ecosystems from polar to tropical regions, driving entire groups of species such as plankton, jellyfish, turtles, and seabirds up to 10 degrees of latitude toward the poles, causing the loss of breeding grounds for turtles and seabirds, and affecting the breeding success of marine mammals.[105]

By damaging fish habitats and causing fish species to move to cooler waters, warming oceans are affecting fish stocks in some areas and are expected to reduce catches in tropical regions. In East Africa and the Western Indian Ocean, for example, ocean warming has reduced the abundance of some fish species by killing parts of the coral reefs they depend on, adding to losses caused by overfishing and destructive fishing techniques. In Southeast Asia, harvests from marine fisheries are expected to fall by 10 to 30 percent by 2050 relative to 1970 to 2000, as the distributions of fish species shift.[106]

Ocean warming is also causing increased disease in plant and animal populations, and impacting human health as pathogens spread more easily in warmer waters, including cholera-bearing bacteria and harmful algal blooms that cause neurological diseases like ciguatera.[107]

Warming oceans are also affecting the weather, with a range of additional effects on humans. The number of severe hurricanes has increased at a rate of around 25 to 30 percent per degree of global warming. Ocean warming

---

[103] Marine plastic: http://www.sciencemag.org/news/2016/11/why-do-seabirds-eat-plastic-they-think-it-smells-tasty

[104] Glecker et al. (2016)

[105] Laffoley, D. and Baxter, J.M., eds (2016) *Explaining Ocean Warming: Causes, Scale, Effects and Consequences.* Full report. Gland, Switzerland: IUCN. p. 456.

[106] Laffoley and Baxter (2016).

[107] Laffoley and Baxter (2016).

has led to increased rainfall in mid-latitudes and monsoon areas, and less rain in various subtropical regions. These changes will have impacts on crop yields in important food-producing regions such as North America and India.[108]

3.8 Outlook

The overall picture for the global ocean is not healthy. The emission of greenhouse gases into the atmosphere is changing the marine environment along with the rest of the planet. The ocean has warmed by 0.7 degree Celsius (1.26 degrees Fahrenheit) since the 19th century, damaging corals and encouraging organisms to migrate toward the poles in search of cooler waters. Greater concentrations of carbon dioxide in the water are making it more acidic. That tends to harm creatures such as crabs and oysters, whose calcium carbonate shells suffer with changes in marine chemistry.

Warming and acidification, pollution, and anoxia[109] have been identified as the deadly trio that threatens mass extinctions in the marine ecosystem. Anoxia (also referred to as hypoxia) produces oceanic dead zones, where excess nutrients (particularly nitrogen and phosphorous) from fertilizers used in agriculture and human sewage collect in coastal waters. These nutrients fuel massive, short-lived blooms of phytoplankton. The algae produce oxygen during the day (through the process of photosynthesis), but at night they take oxygen out of the water column (through the process of respiration), and when they die, the decay process takes additional oxygen out of the water. The net result produces anoxia, regions where marine life cannot be supported owing to oxygen deficiency. Oceanographers first began noticing dead zones in the 1970s. In 2004, 146 dead zones[110] were reported, and by 2008 the number had increased to 405.[111] Today, the number of dead zones is estimated to be close to 500.[112]

Scientists have concluded[113] that the combination of stressors (warming, acidification, pollution, and anoxia) on the ocean today is creating the conditions associated with every previous major extinction of species in Earth's history.[114] The rate of degeneration in the ocean is faster than anyone has predicted. Many of the negative impacts we have already discussed are greater than the worst predictions, and although difficult to assess because of the unprecedented rate of change, the first steps to globally significant extinction may have already begun with a rise in the threat to marine species, such as reef-forming corals, open-ocean fishing stocks, and phytoplankton.

For example, experts[115] have determined that the rate at which carbon is being absorbed by the ocean is already far greater now than at the time of the last globally significant extinction of marine species: some 55 million years ago, when up to 50 percent of some groups of deep-sea animals were wiped out. Researchers point to these events as possible signals that extinction is under way:

- Persistent global coral bleaching, stretching from 2014 to 2017, has been the longest, most widespread, and most damaging on record.[116] This event came on the heels of damaging events in 1998[117] and 2010.[118] Prior to the strong El Niño of 1998, there was no such thing as a global bleaching event; since then we have had 4.
- Overfishing, which has reduced some commercial fish stocks and populations of by-catch species by more than 90%.[119]
- The widespread release of pollutants, including plastics, flame-retardant chemicals, and synthetic compounds found in detergents. The presence of these pollutants has even been traced to the polar seas, where they are being absorbed by tiny plastic particles in the ocean that are in turn ingested by marine creatures.[120]
- Plastic pollution[121] of the oceans has turned into an epidemic.[122] Hundreds of thousands of tons of plastic[123] are dumped into the oceans every year with devastating consequences for sea life of all types.[124] For instance, over 90% of all seabirds have plastic pieces in their stomachs that cause malnutrition and eventually death.[125]

---

[108] Laffoley and Baxter (2016).

[109] Biello, D. (2008) Oceanic dead zones continue to spread, *Scientific American*, http://www.scientificamerican.com/article.cfm?id=oceanic-dead-zones-spread

[110] Dead Zone, *Wikipedia*: http://en.wikipedia.org/wiki/Dead_zone_%28ecology%29#cite_note-sfgate-1

[111] Diaz, R.J. and Rosenberg, R. (2008) Spreading dead zones and consequences for marine ecosystems, *Science*, 321(5891), 926–929.

[112] UNESCO (2017) Marine Pollution, http://www.unesco.org/new/en/natural-sciences/ioc-oceans/focus-areas/rio-20-ocean/blueprint-for-the-future-we-want/marine-pollution/facts-and-figures-on-marine-pollution/

[113] Rogers, A.D. and Laffoley, D.A. (2011) International Earth system expert workshop on ocean stresses and impacts, Summary Report IPSO Oxford, 18 pp. http://www.stateoftheocean.org/ipso-2011-workshop-summary.cfm

[114] VIDEO: Ocean Threats, at the end of the chapter.

[115] Rogers and Laffoley (2011)

[116] NOAA: https://coralreefwatch.noaa.gov/satellite/analyses_guidance/global_coral_bleaching_2014-17_status.php

[117] Wilkinson, C. (1988) *The 1997-1998 Mass Bleaching Event Around the World*, http://www.oceandocs.net/bitstream/1834/545/1/BleachWilkin1998.pdf

[118] NOAA: http://www.noaanews.noaa.gov/stories2015/100815-noaa-declares-third-ever-global-coral-bleaching-event.html

[119] Rogers and Laffoley (2011)

[120] Multiple Ocean Stresses Threaten 'Globally Significant' Marine Extinction, Experts Warn: https://www.sciencedaily.com/releases/2011/06/110621101453.htm

[121] Marine plastic: https://www.plasticoceans.org/microplastics-bbc/

[122] NOAA: https://marinedebris.noaa.gov/info/plastic.html

[123] Plastic debris in the open ocean: http://www.pnas.org/content/111/28/10239.full

[124] Marine plastic: https://www.ecowatch.com/22-facts-about-plastic-pollution-and-10-things-we-can-do-about-it-1881885971.html

[125] Marine plastic: http://news.nationalgeographic.com/2015/09/15092-plastic-seabirds-albatross-australia

Are the oceans in trouble? The weight of scientific evidence indicates that the combined effects of warming, acidification, anoxia, and human pollution of various types are assembling a deadly framework for marine ecosystems.

Animations and Videos

1. Global Circulation of the Atmosphere, http://www.youtube.com/watch?v=DHrapzHPCSA
2. Ocean Currents, http://www.montereyinstitute.org/noaa/lesson08.html
3. Thawing Permafrost-Changing Planet, http://www.youtube.com/watch?v=yN4OdKPy9rM
4. Water Cycle Animation, http://www.youtube.com/watch?v=Az2xdNu0ZRk
5. NASA, "Vital Signs of the Planet," https://climate.nasa.gov//
6. NASA Scientist James Hansen talks about the urgency of the climate crisis, https://www.youtube.com/watch?v=y09kyx9YgUM

Ocean Threats

1. Ocean extinction, http://www.youtube.com/watch?v=vkNns0-w79Q&feature=player_embedded
2. Ocean pollution, http://www.youtube.com/watch?v=QxZs0B5yqqo&feature=player_embedded
3. The Coral Story, http://www.youtube.com/watch?v=3WH_6PgNIQI&feature=player_embedded#at=62
4. Vital Role of the Oceans, http://www.youtube.com/watch?v=ayb7zpXSs0g&feature=player_embedded
5. The Speed of Ocean Change, http://www.youtube.com/watch?v=Giua4EmwPgw&feature=player_embedded#at=12
6. An Overview of Threats, http://www.youtube.com/watch?feature=player_embedded&v=sup3XxHmBoo

Comprehension Questions

1. How is Earth's atmosphere organized?
2. Explain the global circulation of the atmosphere.
3. What characteristics form a climate zone?
4. Identify three ways that dry climate zones can form.
5. How does temperate marine climate differ from temperate continental climate?
6. What climate zone do you live in? Describe its distinguishing characteristics.
7. Pick two climate zones. Describe how they are likely to change in a warming world.
8. How do ocean currents transport heat and why is it important?
9. Describe the deadly trio and how they put the oceans at risk.
10. Explain how the dissolved oxygen content is decreasing in the oceans.

Thinking Critically

1. What could you and your friends do to decrease the impacts of global warming?
2. Describe some ways Earth's surface is becoming less reflective. What is the impact of decreased albedo on climate change?
3. How are global warming and desertification related?
4. Describe the negative impacts of Arctic amplification. What steps can policy makers take to reduce these impacts?
5. Describe the negative impacts of desertification. What steps can policy makers take to reduce these impacts?
6. You are a politician running for President of the United States. Describe what steps you would take to address climate change.
7. How does climate change affect the weather?
8. Why have carbon dioxide emissions increased so dramatically over the past 150 years?
9. How are changes in the water cycle likely to influence climate zones, as the atmosphere gets warmer?
10. The ocean is changing. How will anoxia, warming, and acidification combine to affect marine ecosystems? What role does marine pollution play as a catalyst in the impacts of these stressors?
11. Is climate change "dangerous," or is that too strong a word? Why?

Activities

1. Visit the *Climate Literacy* website http://www.globalchange. gov/resources/educators/climate-literacy and answer the following questions.
   (a) Why is it important for everyone to become informed about climate science?
   (b) What are the essential principles of climate science?
   (c) Describe the primary ways to improve understanding of the climate system.
   (d) Describe how climate varies over space and time.

2. Explore this article at ABC News: http://abcnews.go.com/ Technology/GlobalWarming/global-warming-common-misconceptions/story?id=9159877
   (a) What are the seven common misconceptions about global warming?

3. Open the report on ocean warming by the International Union for Conservation of Nature: https://www.iucn.org/ news/secretariat/201609/latest-ocean-warming-review-reveals-extent-impacts-nature-and-humans
   (a) How is ocean warming affecting marine ecosystems?
   (b) In what ways does ocean warming affect humans?
   (c) How does ocean warming affect coral reefs?

   (d) What is the relationship between ocean warming and El Niño events?
   (e) What are the impacts of ocean warming in the high-latitude regions of the planet?
   (f) What steps can policy makers and stakeholders take to reduce the impacts of ocean warming?

4. Study Figure 3.19 and read this article from NOAA about billion dollar weather disasters in the United States. https://www. climate.gov/news-features/blogs/beyond-data/2016-historic-year-billion-dollar-weather-and-climate-disasters-us
   (a) What is causing the rise in weather disasters worldwide and in the United States?
   (b) What are the predominate types of disasters striking the United States?
   (c) How are the predominate types of disasters related to global warming?
   (d) What other factors besides global warming contribute to the rise of weather disasters?
   (e) Why are the disasters in the United States clustered in the middle of the country?

Key Terms

Algal blooms
Biosphere
Climate
Cold front
Convention to Combat
  Desertification
Coral bleaching
Coriolis force
Coriolis, Gaspard-Gustave
  (1792–1843)
Cut-off low
Dead zones
Desertification
Evapotranspiration

Ferrel Cell
Ferrel, William (1817–1891)
General circulation of the
  atmosphere
Hadley Cell
High air pressure
Hydrosphere
Intertropical Convergence
  Zone
Köppen, Wladimir
  (1846–1940)
Lithosphere
Low air pressure
Mid-latitude cyclones

Mid-latitude westerlies
Monsoon winds
Northeast trade winds
Ocean acidification
Oceanic conveyor belt
Orographic effect
Permafrost
Phytoplankton
Polar Cell
Polar easterlies
Polar front
Polar vortex
Rain shadow
Sahel

Southeast trade winds
Subtropical high-
  pressure zones
Thermohaline circulation
Trewartha, Glenn (1896–1984)
Tropic of Cancer
Tropic of Capricorn
Tropics
Warm front
Water vapor
Weather
Weather fronts

CHAPTER 4

Are Humans the Cause?

LEARNING OBJECTIVE

Despite rigorous testing, scientists are unable to identify a natural process that is responsible for modern global warming. The accumulation of heat-trapping greenhouse gases in the atmosphere resulting from various human activities has been shown through modeling, theoretical calculations, empirical evidence, direct observations, and simple correlation to be the clear and undisputed cause of global warming.

A massive block of ice, about 1 to 2 km (1 mi) long, has broken off Antarctica's Pine Island Glacier and floated into the adjacent bay. Pine Island Glacier is estimated to deliver about 79 km³ (19 mi³) of ice to the bay each year. This is one of the principal means by which ice moves from the interior of the West Antarctic Ice Sheet to the ocean, where it melts and contributes to sea-level rise. Pine Island Glacier has been thinning at an accelerating rate so that now it is the largest single contributor to sea-level rise in Antarctica. Several studies[1] have concluded that Pine Island Glacier is in a state of irreversible retreat.[2]

Source: National Aeronautics and Space Administration

4.1 Chapter Summary

Climate change has been a natural process throughout geologic history. But modern global warming is not the product of the Sun, natural cycles, or bad data. Every imaginable test has been applied to the hypothesis that humans are causing global warming. The simplest, most objective explanation for the many independent lines of clear, factual evidence is that human land use and greenhouse gas emissions are the cause of climate change.

In this chapter, you will learn that:

- Burning coal, oil, and natural gas instantly releases carbon that took millions of years to accumulate in Earth's crust.

- Over 36 billion tons of carbon dioxide are released into the atmosphere annually as a result of industrialization and deforestation[3]—a rise of about 40% since the mid-1800s—and has resulted in a disruption of the carbon cycle.[4]

- Vigorous scientific testing has established that the best explanation for modern global warming is that humans are the cause.

[1] Favier, I, et al. (2014) Retreat of Pine Island Glacier controlled by Marine Ice-Sheet Instability, *Nature Climate Change* 4, 117–121, doi:10.1038/nclimate2094

[2] Pine Island Glacier: http://www.bbc.com/news/science-environment-25729750

[3] CO₂ Now Site that tracks carbon emissions http://co2now.org

[4] VIDEO: Carbon Cycle. https://www.khanacademy.org/science/biology/ecology/biogeochemical-cycles/v/carbon-cycle

- Global warming is not part of a natural climate cycle.
- Earth's climate has changed throughout geologic history. Over the past 500,000 years or so, the climate system has been characterized by swings in temperature, from glacial to interglacial and back again.
- Our knowledge of paleoclimate comes from "climate proxies": chemical and other types of clues stored in ice and sediment that identify past climate change.
- Variations in the intensity and timing of heat from the Sun due to changes in how Earth is exposed to sunlight are the most likely cause of glacial/interglacial cycles.
- The popular notion of natural "climate cycles" is overly simplistic. Actual paleoclimate is the product of complex interactions among solar and Earthly factors.
- Climate change is governed by positive and negative feedbacks that make the timing of climate changes irregular. These feedbacks can suppress or enhance temperature and other climate processes in unpredictable ways.
- Global warming is not caused by the Sun, it did not stop during the first decade of the 21st century, scientists do not disagree that climate is warming and that humans are the cause, and today's warming is not simply a repeat of the recent past or part of a natural cycle.
- Claims that global warming is based on unreliable data have been rigorously tested[5] and are simply not true.

4.2 Mitigating Global Warming Requires Managing Carbon

Many of the chemical compounds found on Earth's surface move between the air, the water cycle, Earth's crust, and living organisms. Along the way, they go through biologic, geologic, and chemical exchanges and reactions, as does everything they come in contact with. The worldwide movement of these chemical compounds as they pass through, interact with, change, and are changed by Earth's atmosphere, crust, water supply, and life forms is known as global biogeochemical cycling.[6,7]

The key elements required for life move through biogeochemical cycles; they include oxygen, carbon, phosphorus, sulfur, and nitrogen. The rates at which elements and compounds move between places where they are temporarily stored (reservoirs) and where they are exchanged (processes) can be measured directly and modeled using computer programs.

We will focus on carbon because it forms the greenhouse gas carbon dioxide (CO_2), the primary cause of global warming. Carbon is also the key element in methane (CH_4), a powerful greenhouse gas as well. Carbon is absolutely essential to life. Its presence or absence helps define whether a molecule is considered organic (formed by living organisms) or inorganic (not formed by living organisms). Every living creature on Earth needs carbon either for building tissue and a skeletal system, or for energy (or, in the case of plant and animal life, for both). If you don't count the water in our bodies, you are about half carbon. Carbon is also found in other forms such as the rocks **limestone** (made of the mineral calcite, $CaCO_3$) and shale, and in wood, plastic, diamonds, and graphite.

In order to manage carbon dioxide, it is necessary to understand how and where carbon is stored and exchanged in the Earth system of biogeochemical cycles. Hence, researchers have spent a lot of effort to define various aspects of the carbon cycle.

4.2.1 The Carbon Cycle

One of the most important cycles that affects global climate is that of the element carbon. The global carbon cycle, one of the major biogeochemical cycles, can be divided into two components: geological and biological. The geological carbon cycle operates on a timescale of millions of years, whereas the biological carbon cycle operates on a timescale of days to thousands of years.

Most carbon stored on Earth is in the form of geologic (long-term) reservoirs (e.g., coal, oil, limestone). However, for the past 200 years, humans have been moving carbon out of these reservoirs and into the atmosphere at greater rates than natural processes can move it back. This phenomenon is explained in the carbon cycle.

The global carbon cycle[8] describes the many forms that carbon takes in various reservoirs and processes. These include the following:

- **Rocks** in the crust, such as limestone ($CaCO_3$) and carbon-rich shale[9]
- Gases in the atmosphere, such as carbon dioxide (CO_2) and methane (CH_4)
- Carbon dioxide dissolved in water (oceans and fresh water)
- Organic material in ecosystems, such as the simple carbohydrate glucose ($C_6H_{12}O_6$), found in plants and animals

[5] Menne, M., et al. (2010) On the reliability of the U.S. surface temperature record, *Journal of Geophysical Research*, 115, D11108, doi: 10.1029/2009JD013094

[6] Biogeochemical Cycle, *Wikipedia*, http://en.wikipedia.org/wiki/Biogeochemical_cycle

[7] VIDEO: Biogeochemical cycles. https://www.khanacademy.org/science/biology/ecology/biogeochemical-cycles/v/biogeochemical-cycles

[8] The carbon cycle explained by scientists at NASA, http://earthobservatory.nasa.gov/Features/CarbonCycle

[9] Shale is a rock composed of mud (silt and clay particles) that has been compressed by burial in Earth's crust. Water in the mud is forced to migrate toward areas of lower pressure leaving behind grains of silt and clay. These grains are compacted into tight layers, forming the rock shale. Some forms of shale are rich with carbon from algae that are buried with the mud grains. During the burial process, under specific conditions of rising temperature and pressure, the buried algae may turn to oil and produce oil shale.

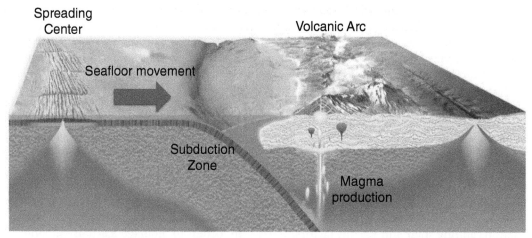

FIGURE 4.1 The theory of plate tectonics describes the formation of new oceanic crust at a spreading center. The crust moves, like a conveyor belt, toward a subduction zone where it is recycled into the mantle. Along the way it collects a thick layer of skeletal debris composed of fossilized silica and calcite plankton—these are recycled back into the mantle. Magma produced at a volcanic arc, located above the subduction zone, will release the silica as lava and the carbon (formerly in the calcite) as carbon dioxide gas.

Geologic Carbon Cycle The geological carbon cycle describes where carbon interacts with the **rock cycle**. The rock cycle consists of the natural processes of **weathering** (chemical and physical processes that destroy rock and minerals), **crystallization** (formation of new minerals), and **lithification** (formation of rocks out of minerals and broken pieces of other rocks).

In the atmosphere, carbon will join with two oxygen atoms to form carbon dioxide. Carbon dioxide will react with water to form **carbonic acid** (H_2CO_3). Carbonic acid condenses and precipitates to Earth as rain and reacts with the most common groups of minerals – the silicates—and turns them to clay. This reaction is one of the most important forms of weathering as it is responsible for changing many types of rocks and minerals into other forms (clay is itself a type of mineral). The process of silicate minerals reacting with carbonic acid is called **hydrolysis**.

Hydrolysis releases many types of dissolved ions and molecules that are carried into the ocean by running water in streams. In the ocean, some of these dissolved molecules are used by plankton and other forms of life to build tissues and skeletal components, the most common of which are calcite-producing plankton (foraminifera and coccolithophores) and silica-producing plankton (diatoms and radiolarians). When calcite-producing plankton die and fall to the seafloor, they lithify and make the rock limestone, and the skeletal components of silica-producing plankton form the rock chert.

But the ocean floor is not still. It moves like a treadmill from a **spreading center** (where new oceanic crust is produced from rock in Earth's mantle) to a **subduction zone** (where the seafloor is recycled back into the mantle). Along the way (a trip that usually takes more than 100 million

years), the oceanic crust collects thick layers of fossil plankton that have settled through the water column and accumulated on the seafloor.[10]

As the seafloor and its sedimentary load are recycled at the subduction zone, chemical reactions and high temperatures in the upper mantle produce magma containing carbon and silica (and a lot of other elements). Some of this magma is erupted by volcanoes at the subduction zone. The carbon is erupted as carbon dioxide and the silica is erupted in lava, and both are released back onto Earth's surface to engage in more of the rock cycle. This process is described by the theory of *plate tectonics* (**Figure 4.1**), which provides the fundamental concepts for understanding the field of geology.

Weathering, new seafloor production, plankton deposition, subduction, and volcanic eruptions—these all control atmospheric carbon dioxide concentrations over time periods of millions of years.

Biological Carbon Cycle As you have seen in the description of the rock cycle and plate tectonics, living organisms play an important role in the movement of carbon between land, ocean, and atmosphere. Fundamentally, it is the processes of **photosynthesis** and **respiration** that drive the movement of carbon in the biological carbon cycle. Almost all living organisms depend on the production of carbohydrates as a source of energy to drive metabolism. Carbohydrates are sugar, and they are produced from sunlight and carbon dioxide during photosynthesis. Those molecules are also broken down in the process of respiration, which

---

[10]VIDEO: Tectonic plates. https://www.youtube.com/watch?v=7nxITuot-ko

RECENT MONTHLY MEAN CO₂ AT MAUNA LOA

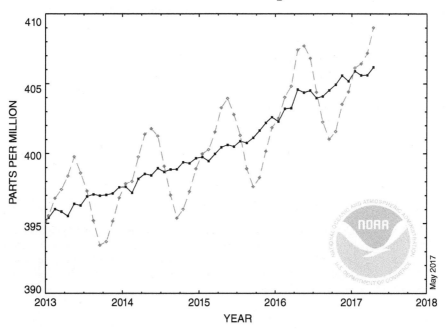

FIGURE 4.2 This graph shows 4 years of the Mauna Loa Observatory CO_2 record from 2013 to 2017. The red line, measured monthly, cycles up and down with the summer and winter seasons in the Northern Hemisphere. A downward phase in the annual cycle of CO_2 represents the season of photosynthesis (spring and summer) that pulls CO_2 out of the atmosphere, and an upward phase represents winter, the season of respiration that releases CO_2 back into the atmosphere. The rising slope of the black line (also monthly values, but with seasonal signal removed) is due to anthropogenic carbon dioxide accumulation in the atmosphere.[11]

Source: NOAA

uses the sugar to produce energy needed for movement, growth, repair of tissue, and reproduction.

Plants take in carbon dioxide from the air during photosynthesis and release it back to the atmosphere during respiration in the following chemical reactions:

Photosynthesis:

Energy (sunlight) + $6CO_2$ (from the atmosphere) + $H_2O \rightarrow C_6H_{12}O_6$ (carbohydrate) + $6O_2$

Respiration:

$C_6H_{12}O_6$ (carbohydrate) + $6O_2 \rightarrow 6CO_2$ (released back to the atmosphere) + $6H_2O$ + energy (sunlight)

Green plants use photosynthesis to turn atmospheric carbon dioxide into carbohydrates, an important source of energy. Respiration is the energy-producing step. In respiration, energy is released from the carbohydrates so that it

can be used in metabolism. In a sense, this changes the carbohydrate "fuel" back into carbon dioxide, which is then released into the atmosphere.

The amount of carbon taken up by photosynthesis and released back to the atmosphere by respiration each year is about 1,000 times greater than the amount of carbon that moves through the geological cycle on an annual basis. Thus, the biological carbon cycle operates much faster than the geological carbon cycle as it repeatedly cycles through roughly the same volume of CO_2 year after year.

The Northern Hemisphere growing season, 6 months of spring and summer, draws down the amount of CO_2 in the atmosphere because plant life use it for photosynthesis. In Northern Hemisphere winter, the amount of CO_2 climbs again as a result of respiration, which releases carbon dioxide back into the atmosphere. This effect can be seen in the red line in **Figure 4.2**. The Northern Hemisphere dominates this annual cycle because it is composed of 60 percent land (where the majority of plant life accumulate) and 40 percent water. The Southern Hemisphere is 20 percent land and 80 percent water.

The long-term storage of carbon in rocks and minerals has global implications. If that carbon is released into the

---

[11] Plot and data from NOAA ESRL, https://www.esrl.noaa.gov/gmd/ccgg/trends

atmosphere, it will drive up the amount of heat-trapping carbon dioxide (black line in Figure 4.2). This is exactly how global warming is happening. Over time (decades), humans are increasing the amount of carbon dioxide in the atmosphere by burning fossil fuels, which are composed of carbon that was previously buried and part of the geologic carbon cycle. While year after year, the biological carbon cycle temporarily increases and then decreases a much larger amount.

The Carbon Budget

Most of the carbon on Earth is contained in the rocks of Earth's crust; it has been deposited slowly over tens of millions of years mostly in the form of marine plankton that form limestone as they are buried on the seafloor, but algae that are buried on the floor of freshwater lakes also constitute an important source of carbon burial. Carbon is stored in the crust in two forms: *Oxidized carbon* is buried as **carbonate** (CO_3) such as limestone, which is composed of calcium carbonate, $CaCO_3$ (the mineral calcite), and *reduced carbon* is buried as organic matter (such as dead plant and animal tissue).

Figure 4.3 illustrates the movement (gigatons[12] per year) and storage (gigatons) of carbon on Earth. From year to year, several hundred gigatons of carbon are exchanged

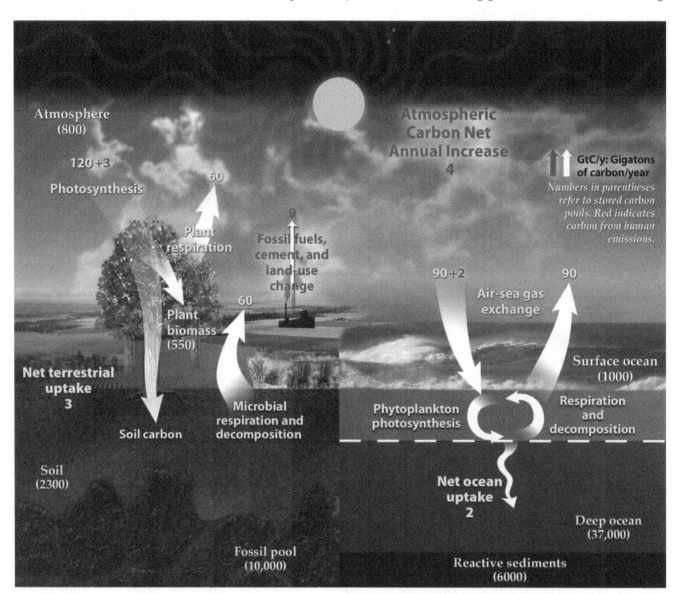

FIGURE 4.3 Carbon is naturally cycled through Earth's atmosphere, oceans, living organisms, and the crust (arrows, gigatons of carbon per year) or stored in global reservoirs (gigatons). Human contributions to the carbon cycle are shown (dark numbers; gigatons of carbon per year).[13]

Source: U.S. Department of Energy

---

[12] One gigaton = 1 billion tons.

[13] U.S. Department of Energy, https://public.ornl.gov/site/gallery/detail .cfm?id=445&topic=&citation=&general=carbon%20cycle&restsection =BERPublic

between the ocean, the atmosphere, and the biosphere. The ocean contains the largest active pool of carbon near Earth's surface. The natural movement of carbon between the atmosphere, ocean, terrestrial ecosystems, and sediments is fairly balanced from one year to the next, so that carbon levels in the reservoirs would be roughly stable without human influence.

However, by burning fossil fuels, deforestation, and other types of activities, humans now emit roughly 9 billion tons of carbon each year, of which about 3 billion are stored in the biosphere through photosynthesis and 2 billion are dissolved in the ocean by mixing between the atmosphere and surface waters. The remaining 4 billion tons stay in the atmosphere as greenhouse gases and cause an increase in the greenhouse effect. While this does not represent a large annual contribution to the carbon cycle compared to the hundreds of gigatons that move in the natural system, over time, year after year, the anthropogenic carbon builds up and constitutes a significant and growing component of the carbon cycle.

Carbon in the atmosphere exists mostly in the form of carbon dioxide and methane. This carbon leaves the atmosphere largely through photosynthesis and enters the land and ocean biospheres. Carbon dioxide also dissolves directly from the atmosphere into bodies of water (oceans, lakes, etc.), as well as dissolving in raindrops that fall through the atmosphere. When dissolved in water, carbon dioxide reacts with water molecules and forms carbonic acid, H_2CO_3 (as we discussed earlier) that contributes to ocean acidity.

On land, the terrestrial biosphere includes all forms of organic carbon in living and dead organisms as well as carbon that is stored in soils. About 550 billion tons of carbon is stored above ground in plants and other living organisms, while soil holds approximately 2,300 billion tons of carbon. About one-third of soil carbon is stored as calcium carbonate (the mineral calcite)—inorganic carbon.

Carbon can cycle out of the terrestrial biosphere in several ways: anthropogenic combustion of fossil fuels, deforestation, and plant respiration all release it rapidly into the atmosphere. Carbon on land can also be exported into the oceans through rivers or remain in soils. Carbon stored in soil can remain there for thousands of years before being washed into rivers by erosion or released into the atmosphere through soil respiration. Soil respiration is the same as plant respiration, but it is driven by organisms that live in the soil such as plant roots, microbes, fungus, and animals of various types (insects and worms).

Between 1989 and 2008 soil respiration increased by about 0.1 percent per year. In 2008, the global total of CO_2 released from the soil reached roughly 98 gigatons, about 10 times more carbon than humans are now putting into the atmosphere each year by burning fossil fuels. There are a few plausible explanations for this trend, but the most likely explanation is that increasing temperatures due to global warming have increased rates of decomposition of soil organic matter, which has increased the release of CO_2.

The ocean contains the largest single quantity of actively cycled carbon. Carbon enters the ocean mainly through the dissolution of atmospheric carbon dioxide, which is converted into dissolved inorganic carbon (DIC) in the form of carbonate (H_2CO_3). Surface waters hold a large amount of DIC (1,000 gigatons), which is exchanged rapidly with the atmosphere. The amount of DIC in deep water is higher than in surface waters, and is stored there for centuries. Deep water and surface water exchange carbon principally through thermohaline circulation (discussed in **Chapter 3**).

Carbon can also enter the ocean through rivers as dissolved organic carbon (DOC). DOC comes from decaying organisms, humic acid, and microparticles of organic material (that are not really dissolved at all). Organisms in the marine food chain scavenge some DOC that is suspended in the water, and some falls through the water column and collects on the seafloor in carbon-rich layers as dead soft tissue.

Because the ocean stores dissolved CO_2 that would otherwise be in the atmosphere, it has reduced the amount of anthropogenic global warming. However, the rate of dissolved CO_2 removal from the ocean is dependent on the weathering of rocks, and this process takes place slower than current rates of human greenhouse gas emissions. Thus, ocean CO_2 storage will decrease in the future. Also, CO_2 absorption makes water more acidic, which affects marine organisms. The rate of increasing oceanic acidity could slow the biological precipitation of calcium carbonate. This will ultimately decrease the ocean's capacity to absorb carbon dioxide.

Carbon moves through the carbon cycle via several processes:

1. **Limestone that forms under the ocean and surface waters traps carbon.**

   Most of Earth's carbon is contained in limestone, which provides effective long-term storage of carbon that has been taken from the atmosphere and transferred to the crust. The process of storing carbon in this way occurs in several steps:

   a. Carbon dioxide is constantly moving from the atmosphere to the ocean and other surface waters, where it dissolves into the bicarbonate ion (HCO_3^-). In fact, it has been calculated that the oceans have absorbed about half of the carbon dioxide released by human activities. This (simplified) chemical reaction describes the process:

   $$2CO_2 + 2H_2O \rightarrow 2HCO_3^- + 2H^+$$

   b. The bicarbonate ion combines with dissolved calcium (Ca^{2+}) in seawater to form calcium carbonate ($CaCO_3$, the mineral calcite), which is the primary component of limestone. The reaction is called **calcification**:

   $$2HCO_3^- + Ca^{2+} \rightarrow CaCO_3(\text{limestone}) + CO_2 + H_2O$$

In the first reaction, two molecules of CO_2 are taken from the atmosphere; in the second, one molecule of CO_2 is released. Thus, one molecule of carbon dioxide is stored in the limestone.

   c. As limestone is formed, some atmospheric CO_2 is trapped and buried in the most stable of forms, rock. Coral and other marine organisms, such as mollusks, some types of algae, and the plankton animal foraminifera, are also excellent calcifiers (makers of limestone).

2. **Limestone that forms in fresh water traps calcium.**
Most calcification occurs in the ocean, but some also occurs in fresh water. Have you ever seen stalagmites and stalactites in caves? These are made of limestone that was formed by the same chemical calcification reaction but without the help of plants and animals. Freshwater calcification often occurs by evaporation; wherein dissolved compounds precipitate (form a solid mineral) because the water they are dissolved in evaporates.

3. **The weathering of limestone consumes atmospheric CO_2, which contains carbon.**
The movement of carbon doesn't end with calcification. Once formed, weathering, a natural process involving chemical reactions between rocks and atmospheric gases, can eventually break down limestone. Weathering consumes atmospheric CO_2 in a chemical reaction that is essentially the reverse of calcification:

$$CaCO_3 + CO_2 + H_2O \rightarrow Ca^{2+} + 2HCO_3^-$$

4. **The weathering of silica rocks, mostly in Earth's crust, also uses CO_2.**
The weathering of types of rocks other than limestone also uses CO_2. These rocks are known as **silica rocks**, represented here as the mineral $CaSiO_3$, which account for most of the rocks in Earth's crust. In this case, a silica rock is changed by reaction with carbon dioxide and water into dissolved calcium, bicarbonate, and silica:

$$CaSiO_3 + 2CO_2 + H_2O \rightarrow Ca^{2+} + 2HCO_3^- + SiO_2$$

This is the process called hydrolysis that we discussed earlier in the chapter.

5. **Photosynthesis converts CO_2 into organic carbon.**
Plants and some forms of bacteria use inorganic CO_2 and convert it into organic carbon (carbohydrates), which is then consumed by all other forms of life, from zooplankton to humans, through the food chain. This is the process of photosynthesis that we discussed earlier.

6. **Carbon returns to the atmosphere as gaseous CO_2, a by-product of respiration and the decay of organic matter.**

Some organic carbon created by plants during photosynthesis is consumed by animals and transferred through the food chain to higher forms of life. Eventually, the organic matter decays or is used in respiration, and the carbon is returned to the atmosphere as gaseous CO_2.

The cycling of carbon through photosynthesis and respiration is so rapid and efficient that all of the CO_2 in the atmosphere is estimated to pass through the global ecosystem every 4 to 5 years.

4.2.2 The Imbalance of the Carbon Cycle and Its Impact on Climate Change

Understanding the above details of the carbon cycle, scientists are able to use computer software to measure, track, and model the movement of carbon dioxide and other forms of carbon throughout the carbon cycle. What they have learned is that many global events have changed the carbon cycle in the past: the coming and going of ice ages, changes in the land surface and ocean currents related to plate tectonics, increased volcanic activity also related to plate tectonics, and others.

What scientists have also found is that as a result of human activities, more carbon is being released into the air than at any time in recent geologic history, resulting in the presence of more methane and carbon dioxide in the atmosphere, in turn resulting in a severe perturbation of the carbon cycle. For example, in 2017 approximately 4.8 billion tons of carbon dioxide was released into the atmosphere by removing forests, which store carbon in tree trunks, leaves, and organic-rich soil, and replacing them with crops or grasslands that store significantly less carbon.[14]

Burning fossil fuels (coal, oil, and natural gas) releases carbon that took millions of years to accumulate. In 2017, emissions from fossil fuel use and industry totaled approximately 36.8 billion tons of carbon dioxide into the atmosphere. These disruptions to the carbon cycle have caused the amount of carbon dioxide in the atmosphere and the ocean to rise significantly above natural levels. Of the total 41.5 billion tons of carbon dioxide released in 2017 through burning fossil fuels (e.g., industrial power generation) and land-use changes (e.g., deforestation), approximately 46 percent (16.9 billion tons) remains in the atmosphere. Another 30 percent (12.45 billion tons) enters the photosynthesis portion of the carbon cycle (e.g., forest uptake), and 24 percent (9.96 billion tons) is absorbed by the ocean. About 6 percent or 2.49 billion tons of carbon dioxide is unaccounted for (known as the "budget imbalance").[15]

---

[14] Global Carbon Project provides updates on annual carbon emissions and sources: http://www.globalcarbonproject.org/carbonbudget/index.htm
[15] Global Carbon Project: http://www.globalcarbonproject.org/carbon-budget/index.htm

The release of extra carbon to the atmosphere is causing the planet surface to get dramatically warmer by enhancing the greenhouse effect (thereby threatening ecosystems worldwide[16]) and excess carbon dissolved in the oceans is causing the water to grow more acidic, putting marine life in danger.[17]

Since the preindustrial era, roughly 30 percent of human CO_2 emissions have been taken up by the ocean and about 30 percent by the land. The remaining 40 percent of emissions have led to the observed increase in the concentration of CO_2 in the atmosphere (**Figure 4.4**).

After a decade of rapid growth in global CO_2 emissions, which increased at an average annual rate of 4 percent, much smaller increases were registered in 2012 (0.8 percent), 2013 (1.5 percent), and 2014 (0.5 percent). In 2015 and 2016, emissions did not grow at all. Nevertheless, in 2017, a worldwide increase in economic activity pushed global CO_2 emissions to a new record estimated at 41.5 billion tons. Instead of a hoped-for downturn in carbon emissions, the 2017 increase is a sign that time is running out on our ability to keep warming well below 2 degrees Celsius let alone 1.5 degrees Celsius. For more discussion on recent trends in global emissions, read *Spotlight on Climate Change*, "Greenhouse Gas Emissions Forecast to Continue Growing."

4.2.3 What Temperature Has Carbon Emissions Committed Us To?

To model future global warming and its impacts, researchers base their calculations on greenhouse gas concentrations (called Representative Concentration Pathways, RCPs) that would force a range of temperature outcomes (measured in watts per square meter) by the year 2100. For instance, RCP8.5 is a greenhouse gas emission scenario that would result in 8.5 W/m² of increased heat by 2100. Four modeling scenarios are typically used: a stringent mitigation scenario (RCP2.6), two intermediate scenarios (RCP4.5 and RCP6.0), and one scenario with very high greenhouse gas emissions (RCP8.5). Scenarios in which humans fail to engage in efforts to limit greenhouse gas emissions lead to pathways ranging between RCP6.0 and RCP8.5. RCP2.6 represents a scenario that attempts to keep global warming below 2 degrees Celsius above preindustrial temperatures.

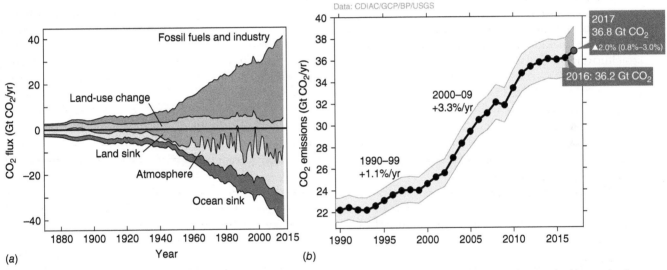

FIGURE 4.4 (a) Annual movement of CO_2 (billions of tons per year) in terms of sources (fossil fuels, land use) and reservoirs (oceans, land, atmosphere). Very roughly about 30% of the anthropogenic CO_2 emissions have been taken up by the ocean and about 30% by land. The remaining 40% of emissions have led to an increase in the concentration of CO_2 in the atmosphere. (b) Global emissions of carbon dioxide (gigatons) by fossil fuel use and industrial activities have been growing steadily since the onset of the Industrial Revolution. In 2010, global CO_2 emissions due to fossil fuel use and industrial activities grew by 5.9%, the largest annual increase on record. However, since 2012, global emissions have slowed considerably, and from 2014 to 2016 they were nearly stable. Nevertheless, in 2017 global emissions from fossil fuel use and industrial activities rose by approximately 2% indicating that the world must continue to wait for a long-hoped for downturn in carbon emissions in order to achieve the targets of the 2015 Paris Accord.[18]

Source: http://folk.uio.no/roberan/GCP2017.shtml

[16] Rosenzweig, C., et al. (2008) Attributing physical and biological impacts to anthropogenic climate change, *Nature*, 453(7193), 353–357, doi:10.1038/nature06937

[17] NOAA: https://www.pmel.noaa.gov/co2/story/What+is+Ocean+Acidification%3F

[18] Global Carbon Project: http://www.globalcarbonproject.org/carbonbudget/index.htm

BOX 4.1 | Spotlight on Climate Change

Greenhouse Gas Emissions Forecast to Continue Growing

The demand for energy around the world grew by 2% in 2017, compared with 0.9% the previous year and 0.9% on average over the previous 5 years. More than 40% of the growth was driven by China and India; 72% of the rise was met by fossil fuels, one-quarter by renewable sources of energy, and the remainder by nuclear power.

In an average year, the global demand for energy accounts for about two-thirds of total greenhouse gas emissions and 80% of carbon dioxide. Any effort to reduce emissions and mitigate climate change must involve the energy sector. In 2017, lifted by strong global economic growth, most of the increased demand for energy was met with oil, gas, and coal. However, renewable forms of energy made notable gains.

Here are the statistics[19]:

1. Globally, energy-related carbon dioxide emissions grew by about 2% in 2017.
2. World oil demand rose by 1.6% in 2017, a rate that was more than twice the annual average seen over the last decade. Increasing demand for oil from the transportation industry (e.g., sport-utility vehicles and light trucks), and the petrochemical industry (e.g., manufacture of plastics) boosted this growth.
3. Global natural gas demand grew by 3%. China alone accounted for almost 30% of this growth. Over the past decade, half of global gas demand growth came from the power sector; last year, however, over 80% of the rise came from industry and buildings.
4. Global coal demand rose by about 1% in 2017, reversing a declining trend seen over the last 2 years. Demand came mainly from Asia, driven by an increase in coal-fired electricity generation.
5. Renewables saw the highest growth rate of any energy source in 2017, meeting one-quarter of the increase in global energy demand. China and the United States led this unprecedented growth, contributing around 50% of the increase in renewables-based electricity generation, followed by the European Union, India and Japan. Wind power accounted for 36% of the growth in renewables-based power output.

6. World electricity demand increased by 3.1%, significantly higher than the overall increase in energy demand. China and India accounted for 70% of this growth. Output from nuclear plants rose, as a significant amount of new nuclear capacity saw its first full year of operation.
7. Improvements in global energy efficiency slowed down dramatically in 2017. This was the result of weaker policies as well as lower prices for traditional energy. Global energy intensity improved by only 1.7% in 2017, compared with an average of 2.3% over the last 3 years.

Global economic growth, concentrated in Asia (India and China), drove the increased demand for energy. Although renewables surged to meet about 25% of this demand growth, increased use of fossil fuels drove a 2% rise in carbon dioxide emissions, after 3 years of remaining flat.

Meeting the temperature objectives of the Paris Agreement requires urgently reducing greenhouse gas emissions. It has been shown[20] that carbon dioxide reductions equaling 50% each decade, beginning in the 2020s, can lead to net-zero emissions around mid-century. This pathway is necessary to limit warming to 2°C. The question is: Are projections of near-term energy use consistent with this?

According to the International Energy Agency,[21] major shifts in global energy use set the stage for the near-term future. These are:

1. Energy demand expands by 30% over the next two decades, the equivalent of adding another China and India to today's energy use. Driving this increase are a global economy growing at 3.4% per year, expanding population from 7.4 billion to more than 9 billion in 2040, and urbanization that adds a city the size of Shanghai to the world's urban population every four months.
2. Thirty percent of the growth in energy demand comes from India, whose share of global energy use rises to 11% by 2040. Developing countries in Asia account for two-thirds of global energy growth. The rest comes mainly from the Middle East, Africa, and Latin America.
3. Natural gas, renewables, and energy efficiency play major roles in meeting the growth in energy demand.

[19] IEA (2018) Global Energy & CO$_2$ Status Report 2017, International Energy Agency: http://www.iea.org/publications/freepublications/publication/GECO2017.pdf

[20] Rockström, J. et al. (2017) A roadmap for rapid decarbonization, *Science*, 355(6331), 1269, doi: 10.1126/science.aah3443

[21] IEA (2017) World Energy Outlook, International Energy Agency: https://www.iea.org/weo2017/#section-4-5

Renewable energy meets 40% of the increase. In India, the share of new demand met by coal drops from three-quarters in 2016 to less than half in 2040.

4. Oil demand continues to grow to 2040 but at a steadily decreasing pace. Natural gas use rises by 45% to 2040.

5. Renewable energy captures two-thirds of global investment in power plants by 2040. They become the most affordable source of new energy. Rapid expansion of solar power, led by China and India, helps it become the largest source of low-carbon capacity by 2040. By 2040, the share of all renewables in total power generation reaches 40%.

6. Oil demand continues to rise by 2040: oil use to produce petrochemicals is the largest source of growth, closely followed by rising consumption for trucks (fuel-efficiency policies cover 80% of global car sales today, but only 50% of global truck sales), for aviation and for shipping.

In their 2017 report, the U.S. Energy Information Administration[22] forecasts global energy use to the year 2040. They project the following patterns: coal sustaining a 20-year-long plateau, natural gas plentiful and growing, carbon-free wind and solar growing rapidly in percentage terms but not fast enough to bring emissions down in absolute terms, and petroleum holding its own as the main source of energy for transportation, despite the arrival of electric vehicles.

With populations growing and developing nations getting richer, total energy consumption will keep climbing despite gains in energy efficiency. Fossil fuels hold a 77% market share, and, as a result, greenhouse gas emissions will increase in parallel. The Energy Information Administration forecasts that worldwide emissions of carbon dioxide from the burning of fossil fuels will grow 16% by the year 2040 (compared to 2015). This is far away from the necessary 50% decrease per decade called for by Rockstrom et al. (2017) in order to meet the Paris Agreement target of 2°C.[23]

Little time remains for the world to get its emissions under control. The Paris agreement is predicated on a single global carbon budget that countries are collectively using up each year. The longer humanity waits to reduce emissions, the more aggressive future measures will need to be to keep the total carbon released to the atmosphere under budget. However, global energy consumption is still dominated by fossil fuels. A small fluctuation in coal use from one year to the next can wipe out a seemingly dramatic expansion in renewable energy (**Figure 4.5**).

Poised at a critical time, how we behave in the next decade will determine the nature of future human prosperity on Earth. The Paris Agreement translates into a finite planetary carbon budget. Rogeli et al. (2015) found that to achieve a 50% chance of limiting warming to 1.5°C by 2100, and a greater than 66% probability of meeting the 2°C target, global carbon dioxide emissions must peak no later than 2020. Gross emissions must decline thereafter from ~40 billion tons of carbon dioxide per year in 2020 to ~24 by 2030, ~14 by 2040, and ~5 by 2050 (Rogel et al., 2015). On the other hand, Raftery et al. (2017) calculated that the likely range of global temperature, increase is 2.0–4.9°C, with a median 3.2°C and a 5% (1%) chance that it will be less than 2°C (1.5°C).

Literally everything that we depend upon, from the air we breathe, freshwater, and food, to transportation, shelter, and security are all contingent on climate. Because of the fast pace at which climate change is happening, in many locations, these requirements for human life teeter on the edge of sustainability. They are naturally adapted, or humanly engineered, to the climate of yesterday.

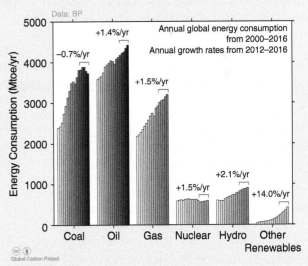

FIGURE 4.5 World energy consumption (in millions of tons of oil equivalent per year) is still strongly dominated by fossil fuels. A minor fluctuation in coal or oil use can wipe out a seemingly dramatic expansion in renewable energy.

Source: Global Carbon Project: http://folk.uio.no/roberan/img/GCP2017/PNG/s22_EnergyChanges.png

[22] EIA (2017) International Energy Outlook 2017, U.S. Energy Information Administration: https://www.eia.gov/outlooks/ieo/pdf/0484(2017).pdf

[23] VIDEO: The Carbon Law. http://www.stockholmresilience.org/research/research-news/2017-03-23-curbing-emissions-with-a-new-carbon-law.html

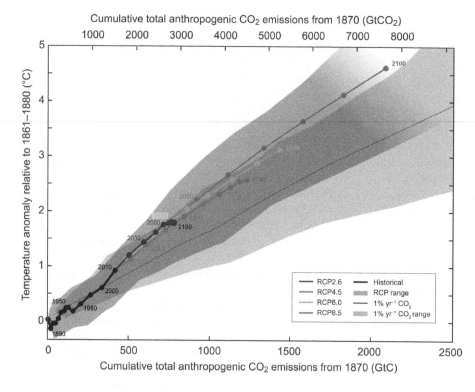

FIGURE 4.6 Global mean temperature change for a number of scenarios as a function of cumulative CO_2 emissions from preindustrial conditions, with time progressing along each individual line for each scenario. (Figure source DeAngelo et al. (2017), from IPCC (2013) Summary for policymakers. T.F. Stocker, et al. (eds.) Cambridge University Press, 1–30)

Climate models identify a nearly linear relationship between total CO_2 emissions and increases in global mean temperature (**Figure 4.6**). Thus, to limit warming to 1.5 degrees Celsius (or 2 degrees Celsius) per the Paris Agreement, the total amount of CO_2 added to the atmosphere must be stabilized at or below some critical amount. This can be conceptualized as a "carbon budget."

The 4th National Climate Assessment of the US Global Change Research Program[24] provides a status report on the global carbon budget (their Table 14.1). Between 1870 and 2015, burning fossil fuels and widespread deforestation led to net carbon dioxide emissions totaling 560 billion tons of carbon (GtC). To provide a two-thirds (66 percent) chance of avoiding 2 degrees Celsius of warming, total, cumulative emissions must not exceed 1,000 GtC. Including the effects of other types of GHGs lowers the amount of allowable carbon to 790 GtC. Thus, to stand a 66 percent chance of avoiding 2 degrees Celsius global warming, no more than 230 GtC of additional CO_2 can be emitted globally.

This threshold is exceeded by the year 2037 if a restricted emissions pathway is followed (RCP4.5) and by 2033 with higher emissions (RCP8.5). To limit warming to 1.5 degrees Celsius, the allowable budget is about 590 GtC. With annual emissions of about 40 GtC, the 1.5 degrees Celsius budget is either already or will soon be, exceeded.

4.3 Paleoclimate

We saw in **Chapter 1** that global warming poses serious threats to ecosystems and human communities[25]; thus, there has been public discussion of managing the impacts of warming by limiting the production of greenhouse gases and adapting to the inevitable consequences caused by gases that have already been released. Limiting greenhouse gas production and adaptation will be costly, however, and will require significant changes to human behavior.[26]

The public discussion on climate change includes personal worldviews and political ideologies; consequently, the theory that humans are responsible for global warming has been vigorously tested not only in science circles using the peer-review process, but also in the nonscientific court

[24] Climate Science Special Report, Chapter 14 Perspectives on Climate Change Mitigation: https://science2017.globalchange.gov/chapter/14/

[25] University Corporation for Atmospheric Research (UCAR) website "Understanding Climate Change," http://www2.ucar.edu/news/backgrounders/understanding-climate-change-global-warming

[26] Report on limiting greenhouse gas production, and congressional reaction: http://www.nytimes.com/2011/05/13/science/earth/13climate.html

of public opinion. Here we discuss some of the more widely cited tests.[27]

4.3.1 Is Global Warming Part of a Natural Climate Cycle?

This is an important question that deserves careful explanation.[28] If global warming is part of a natural climate cycle, then it suggests that changing human behavior to stop burning fossil fuels is essentially pointless; we might as well avoid the disruption to our lives that switching to noncarbon fuels will entail. If global warming is a natural phenomenon, then humanity must prepare for a future characterized by a warmer world over which we have little control.

If, however, global warming is *not* part of a natural process, if it is the result of releasing greenhouse gases into the atmosphere by human activities, then humans have an obligation to future generations to counteract the problem with significant and probably disruptive changes in our behavior. For our own good, and for the safety of our children and the global environment, we need to change our carbon-burning (e.g., oil, coal, natural gas) habits to prevent the most dangerous effects of climate change.

In either case, the answers involve making decisions that risk trillions of dollars of economic activity, global ecosystems, and the livelihoods of the entire human race. Thus, the field of past climate history, known as **paleoclimatology** , is a key point around which the entire discussion of climate change revolves. Knowing past climate history can improve our understanding of the question "Is today's global warming part of a natural climate cycle?"

To start, let's investigate paleoclimate in geologic history.

Earth's climate has changed throughout geologic history. In fact, it might surprise you to learn that the past half-million years or so mark one of the coolest phases in Earth's history.

One of the great challenges in studying past climate is that geologic materials that preserve a record of Earth's climate history are very difficult to find as you go farther back in time. Thus, our knowledge of how climate has changed over the course of many millions of years is based on rare evidence that has to be interpreted and tested by many researchers before it becomes widely accepted.

The past 542 million years of Earth history is known as the Phanerozoic Eon. Our understanding of Phanerozoic climate history comes from the interpretation of fossils and chemical clues in rocks and sediments preserved from this time. Although it is entirely likely that short-term annual and interannual climate processes have operated in various forms throughout geologic history, the evidence from hundreds of millions of years ago rarely preserves the details needed to define these processes. Instead, the ancient record of climate shows the effects of longer-term processes:

- Plate tectonics causing the clustering of continents at times in the high latitudes (promoting cooling) and at other times on the equator (promoting warming)
- Prolonged periods of volcanic outgassing (again the result of plate tectonics) that change climate because among the gases that volcanoes emit is carbon dioxide
- Chemical weathering of Earth's crust that involves hydrolysis, a chemical reaction that draws down carbon dioxide concentrations in the atmosphere
- Changes in ocean circulation owing to shifting continental positions (this is important because ocean currents carry heat toward the poles; if the currents change, the climate will change also)
- Release from the deep seafloor of large quantities of frozen methane (a powerful greenhouse gas) by warming ocean water, or from the tundra, where methane is stored in frozen soil
- Impacts by large meteorites that change climate
- Positive and negative feedbacks that amplify the effects of these processes

Throughout much of Earth's history, global temperature was largely warmer than today's. However, by approximately 55 million years ago global temperature started a long gradual cooling that led to a series of pronounced ice ages and warm periods that have characterized the past 500,000 years or so.

One explanation for this long 55-million-year trend of global cooling is that formation of the Himalayan Range and the Tibetan Plateau during the Paleocene (66 to 56 million years ago) and Eocene (56 to 34 million years ago) time periods caused an increase in chemical reactions (hydrolysis) between the atmosphere and newly exposed rock of the emerging mountain system. These reactions consumed carbon dioxide in the atmosphere, thus lowering the global CO_2 level, and thereby lowering the temperature (known as the "Uplift Weathering Hypothesis"[29]; see the *Spotlight on Climate Change*).

[27] VIDEO: Why reducing our carbon emissions matters. https://www.youtube.com/watch?v=rivf479bW8Q
[28] VIDEO: Sir David Attenborough. http://www.youtube.com/watch?v=S9ob9WdbXx0&feature=related

[29] Raymo, M., et al. (1988) Influence of late Cenozoic mountain building on ocean geochemical cycles, *Geology*, 16, 649–653. See also Raymo, M. and Ruddiman, W. (1992) Tectonic forcing of late Cenozoic climate, *Nature*, 359, 117–122.

BOX 4.2 | Spotlight on Climate Change

The Uplift Weathering Hypothesis

Scientists have discovered that starting about 55 million years ago, Earth's atmosphere began a long but steady cooling that persists today. It is most evident at the poles as it eventually led to the formation of the Antarctic and Greenland ice sheets.

What caused this cooling and why does it continue today? The answer may surprise you—it is caused by hydrolysis (introduced earlier in our discussion of the carbon cycle), a chemical weathering[30] process that relies on carbon dioxide pulled out of the atmosphere to attack rocks in Earth's crust. Rocks that are newly exposed to the air are most susceptible to hydrolysis, which is exactly why CO_2 began to decline 55 million years ago. That is when plate tectonics first began to raise the roof of the world—the Himalayan Range and the Tibetan Plateau. The slow uplift of the Himalayan region over the past 55 million years has exposed a huge volume of silicate rock to weathering (**Figure 4.7**). So much fresh rock was exposed that it literally

changed the chemistry of the entire atmosphere. The average CO_2 concentration dropped from approximately 1,500 parts per million (ppm) to an average 300 ppm—an amazing decrease of 80%![31] This explanation for 55 million years of global cooling is known as the **Uplift Weathering Hypothesis**.

You are already familiar with hydrolysis from our earlier discussion of the geologic carbon cycle. Hydrolysis is the most widespread chemical weathering process because it causes the decomposition of silicate minerals, the most common minerals in Earth's crust. Many types of minerals are decomposed at least partially by hydrolysis.

During hydrolysis, ions in a mineral react with hydrogen (H^+) and hydroxyl (OH^-) ions in water. The hydrogen ions replace some of the cations in the mineral, thereby changing the mineral's composition. In addition, other ions and compounds may be dissolved from the mineral and carried away in the water. Hydrolysis is not very effective in pure water, but when some acid (dissolved H^+) is added, it becomes a powerful

Cory Richards/National Geographic Creative

FIGURE 4.7 A massive amount of silicate rock is exposed to weathering in the Himalayan Range and the Tibetan Plateau. This weathering has reduced the amount of carbon dioxide in the atmosphere over the past 55 million years and is responsible for causing the long, drawn-out period of global cooling that characterizes recent geologic history.

Source: COREY RICHARDS/ National Geographic

---

[30] Weathering involves physical, chemical, and biological processes that degrade rocks (Earth's crust). Physical weathering involves mechanical breakdown of minerals and rocks. Chemical weathering is the chemical decomposition of minerals in rock resulting in the formation of new minerals, compounds that are dissolved in water, and gases that escape to the atmosphere. Biological weathering involves the chemical or physical activity of a living organism, ranging from bacteria to plants and animals. Chemical weathering is by far the most important weathering process because chemical reactions are constantly at work

on all rocks and minerals exposed to the atmosphere or hydrosphere. Weathering produces soil, which yields much of our food and the ores of critical metals, such as aluminum and iron. Almost everything humans need for a healthy lifestyle can be traced back to chemical reactions involved in weathering: the air we breathe, food we eat, and much more.

[31] Pearson, P.N. and Palmer, M.R. (2000) Atmospheric carbon dioxide concentrations over the past 60 million years, *Nature*, 406, 695–699, (17 August), doi:10.1038/35021000

agent of weathering. Acidic rainwater forms naturally when carbon dioxide dissolves in water, whether it is in the atmosphere or the ground, to produce carbonic acid. Carbonic acid is formed by the following reaction:

CO_2 (gas in the atmosphere or ground) $+ H_2O$ (water vapor in the atmosphere or ground) $= H_2CO_3$ (carbonic acid, usually precipitates as moisture)

Hydrolysis breaks down an abundant silicate mineral called feldspar, the most plentiful mineral in Earth's crust, and creates clay, the most plentiful sediment. The following equation is an example of a hydrolysis reaction. It involves feldspar that decomposes to produce a clay mineral. Note that by forming $2H_2CO_3$ (carbonic acid) this reaction pulls carbon dioxide out of the atmosphere and uses it to form clay (and other products). From the human timeframe, this constitutes a process by which CO_2 can be permanently removed from the atmosphere.

$$2KAlSi_3O_8 + 2H_2CO_3 + 9H_2O \rightarrow Al_2Si_2O_5(OH)_4 + 4H_4SiO_4 + 2K^+ + 2HCO_3^-$$

Feldspar + carbonic acid + water (yield) clay + dissolved silica + dissolved potassium and bicarbonate ions

Thus, the story of global cooling related to the tectonic rise of the Himalayan-Tibetan region is directly related to the geologic carbon cycle. Over geologic time, carbon dioxide enters Earth's atmosphere through volcanic outgassing (**Figure 4.8**), the activity of hot springs and geothermal vents, and through the oxidation of organic carbon in sediments. Volcanic outgassing alone is sufficient to contribute all the carbon dioxide in our atmosphere in only 4,000 years. This input must be balanced by withdrawal of carbon dioxide from the atmosphere; otherwise, the amount of carbon dioxide present would produce a runaway greenhouse effect.

As we learned in our carbon cycle discussion, one of the ways by which carbon dioxide is removed from the atmosphere is through photosynthesis by plants, which use CO_2 to make carbohydrates for energy. Another way CO_2 is removed is by chemical weathering of silicate minerals.

Carbon dioxide in the atmosphere easily combines with water during hydrolysis to form carbonic acid (H_2CO_3). The reaction of H_2CO_3 with continental rocks (represented by $CaSiO_3$ in Figure 4.8) produces $CaCO_3$ + SiO_2 and H_2O. Streams carry these products to the ocean, where the calcium carbonate ($CaCO_3$) and the silica (SiO_2) are dissolved in ocean water. Marine plankton use these dissolved compounds to build shells, which fall to the seafloor when they die—forming sediment called **biogenic ooze**.

Biogenic sediments composed of fossil plankton carry carbon into the rock cycle when the lithospheric

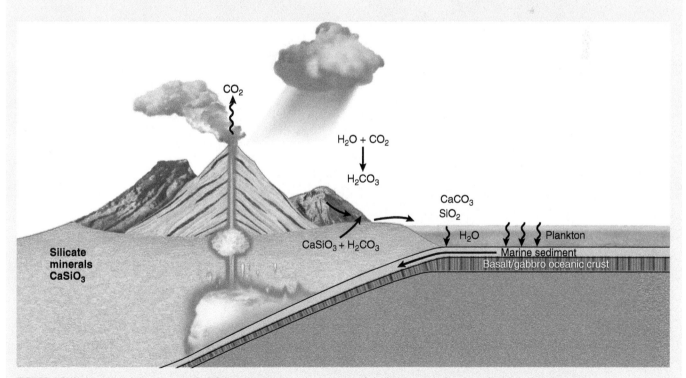

FIGURE 4.8 Weathering of silicate rock ($CaSiO_3$) and volcanic outgassing regulate the amount of carbon dioxide in the atmosphere.

plate they ride on is subducted and recycled. Limestone and other types of marine sedimentary rocks act as long-term storage sites for carbon. But ultimately these rocks and sediments are recycled into the mantle at a subduction zone. Plate subduction generates magma in the upper mantle that will migrate upward into the crust and lead to volcanic outgassing of carbon dioxide through a volcanic arc. Thus, the cycle is complete! Through this series of processes, the rock cycle regulates the amount of carbon dioxide in the atmosphere.

You may find it confusing to learn that global cooling is occurring when this entire book is about the problem of global warming. The explanation lies in the timescales of the two trends. Over millions of years, uplift of the Himalayas (**Figure 4.9**) has led to chemical weathering that has reduced the carbon dioxide in the atmosphere. This has cooled the air at the Earth's surface, playing a role in promoting several ice ages over the past million years.

More recently, pollution of the atmosphere with carbon dioxide and other greenhouse gases from anthropogenic burning of fossil fuels has caused the atmosphere to warm. Thus, the uplift weathering hypothesis describes a process that operates over tens of millions of years, while modern global warming is taking place over decades and centuries. These two very different processes, both involving carbon dioxide, influence the global climate on vastly different timescales.

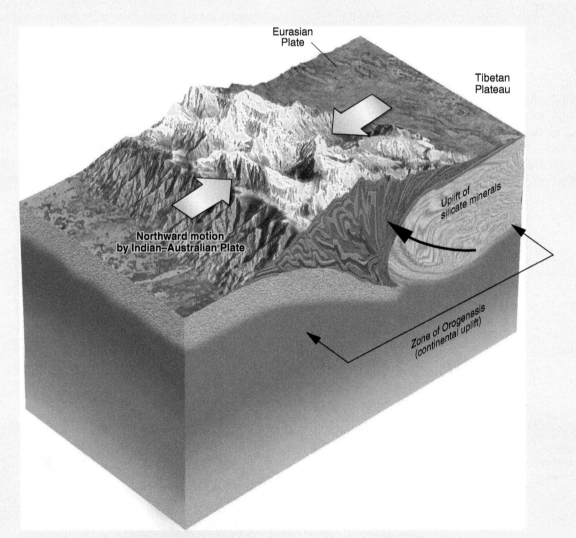

FIGURE 4.9 Tectonic uplift of the Tibetan–Himalayan region exposes silicate minerals to weathering by hydrolysis. Because of intense rains during the annual monsoon, hydrolysis occurs rapidly, on a large scale, and draws down the carbon dioxide content of the atmosphere. Geologists hypothesize that this process has been sufficient to produce slow, but persistent, global cooling over the past 55 million years.

As we have seen, Earth's climate is not steady.[32] This is especially true of the past 500,000 years or so, a time when the climate system has experienced great swings in temperature, from extreme states of cold (**glacials**) to dramatically warmer periods (*interglacials*). Glacials (simply known as *ice ages*[33]) are typified by the growth of massive continental ice sheets reaching across North America and Northern Europe. At their maximum, these glaciers were more than 4 kilometers (2.5 miles) thick in places; today, the ice sheets on Greenland and Antarctica are remnants of the most recent ice age. Accompanying the spread of ice sheets was dramatic growth in mountain glaciers, many of which expanded into ice caps that covered large areas of mountainous territory.

We currently live in the latest interglacial, known as the *Holocene Epoch*, which began about 10,000 years ago. The most recent ice age, occurring at the end of the **Pleistocene Epoch**, began approximately 75,000 years ago and peaked between 20,000 and 30,000 years ago. Formed over that period of about 50,000 years, the landscape of formerly glaciated areas is widespread and characterized by myriad glacial landforms that document this episode. For example, Long Island marks the location of the continental glacier formed during this ice age. This giant expanse of ice moved southward out of Canada and into what is today New York City. As climate warmed, melt water from the northward retreating ice deposited an enormous amount of sediment that now makes up Long Island, New Jersey, and Delaware, and barrier islands and beaches along the Mid-Atlantic region of the United States.

4.3.2 Climate Proxies

Layers of ice, fossilized plankton and coral, and other geologic materials can be analyzed to improve understanding of past climate changes. Improving our understanding of paleoclimate is one way to separate natural climate change from human-caused, or anthropogenic, climate change. We know the history of past climate changes because scientists can measure chemical telltales (known as **climate proxies**) of past climate in samples obtained by drilling in continental ice sheets and mountain glaciers,[34] as well as in sediment composed of fossilized plankton on the seafloor[35] (**Figure 4.10**).

In ice cores, frozen bubbles of carbon dioxide can be used to document the gas content of the atmosphere at times in the past. The longest ice cores (more than 3 kilometers in length) come from Antarctica and record up to 800,000 years of climate history. Greenland ice cores do not reach back as far into Earth history. They extend to about 120,000 years or so. Carbon dioxide is used as a climate proxy because it is directly related to the heat-trapping capacity of the atmosphere. But ice cores provide other measures of past climate as well. For instance, fossil snow also contains information about the temperature of the atmosphere and the amount of sunlight-blocking dust, and deep-sea cores can record changes in ocean chemistry that reveal the history of global ice volume.[36]

Deep-sea sediment is composed of the microscopic shells of fossil plankton. The chemistry of these shells—for instance, tiny plankton from the phylum Foraminifera (shown in Figure 4.10b)—provides chemical clues to the climate prevailing when they were formed. Deep-sea sediment cores offer a record of climate history extending hundreds of thousands to millions of years back through time.

Foraminifera use dissolved compounds and ions in seawater to precipitate microscopic shells of $CaCO_3$. Both calcite ($CaCO_3$) of a foraminifer's skeleton and a molecule of water (H_2O) in seawater contain oxygen (O). In nature, oxygen occurs most commonly as the isotope ^{16}O, but it is also found as ^{17}O and ^{18}O. (Isotopes[37] are atoms with a different mass number than other atoms of the same element.) Water (H_2O) molecules composed of the heavier isotope ^{18}O do not evaporate as readily as those composed of the lighter isotope ^{16}O. Likewise, in atmospheric water vapor, heavier water molecules with ^{18}O tend to precipitate (as rain and snow) more readily than those composed of lighter ^{16}O (**Figure 4.11**).

Both evaporation and precipitation of oxygen isotopes occur in relation to temperature. $H_2^{18}O$ tends to be left behind in greater abundance compared to $H_2^{16}O$ when water is evaporated, and it tends to be the first molecule to condense when rain and snow are forming. Hence, because most water vapor in the atmosphere is formed by evaporation in the tropical ocean, by the time it travels the long

---

[32] VIDEO: Ice Stories. http://www.youtube.com/watch?v=r81MtDimgSg

[33] Ice Age, *Wikipedia*, http://en.wikipedia.org/wiki/Ice_age

[34] National Science Foundation, Office of Polar Programs: http://www.nsf.gov/news/news_summ.jsp?cntn_id=115495&org=OPP&from=news

[35] Integrated Ocean Drilling Program: http://www.iodp.org

[36] The NOAA National Centers for Environmental Information archive serve paleoclimate data to the public. Their website provides access to descriptive information and explanatory notes, maps, searches, visualizations, and more. The data cover the globe, and while most span the last few millennia, some datasets extend back in time 100 million years. Most of the data are time series of geophysical or biological measurements and some include reconstructed climate variables such as temperature and precipitation. https://www.ncdc.noaa.gov/data-access/paleoclimatology-data/datasets

[37] Every atom in the known universe is a tiny structural unit consisting of electrons, protons, and (for many, but not all) neutrons. An atom's center, or nucleus, is composed of protons (large, heavy, and having a positive electrical charge, +) and neutrons (large, heavy, and having no electrical charge). The number of protons plus the number of neutrons makes the mass number. Some atoms of a given element can have a different mass number because they have a different number of neutrons. These are called *isotopes*. For example, carbon atoms normally contain 6 protons (the number of protons is called the *atomic number*, and it's what defines an element) and 6 neutrons; some, however, contain 7 or 8 neutrons. Hence, carbon always has an atomic number of 6, but its mass number may be 12, 13, or 14. These variations in mass number create isotopes of carbon that are written like this: ^{12}C, ^{13}C, and ^{14}C.

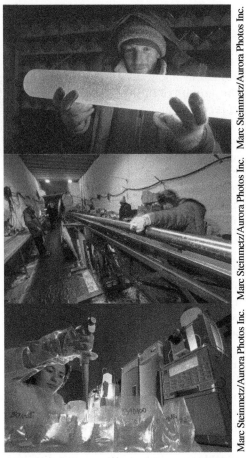

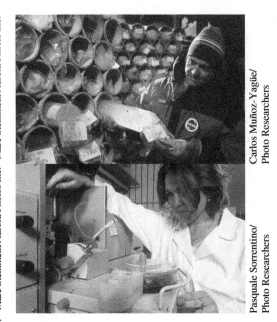

Marc Steinmetz/Aurora Photos Inc.   Marc Steinmetz/Aurora Photos Inc.   Marc Steinmetz/Aurora Photos Inc.

Carlos Muñoz-Yagüe/ Photo Researchers

Pasquale Sorrentino/ Photo Researchers

(a)

© JAMSTEC/IODP

© JAMSTEC

Thierry Berrod, Mona Lisa Production / Science Source

© JAMSTEC/IODP

FIGURE 4.10 Cores of ice and deep-sea sediments provide geologic samples that contain evidence of past climate. (a) Scientists from several nations have established collaborative drilling programs on the Greenland and Antarctic ice sheets, as well as on high-elevation ice caps in mountains. (b) The Integrated Ocean Drilling Program is funded by a consortium of nations interested in using sea-floor sediments to improve our understanding of Earth's history and natural processes.

(b)

distance to high latitudes and elevations where ice sheets and glaciers are located, it is relatively depleted of $H_2^{18}O$ and enriched in $H_2^{16}O$. This means that during an ice age vast amounts of $H_2^{16}O$ are locked up in global ice sheets for thousands of years. Because of this, the oceans are relatively enriched in $H_2^{18}O$ at the same time. Because the ratio of ^{18}O to ^{16}O in the shells of foraminifera mimics the ratio of these isotopes in seawater, the oxygen isotope content of these fossilized shells provides a record of changing global ice volume through time.

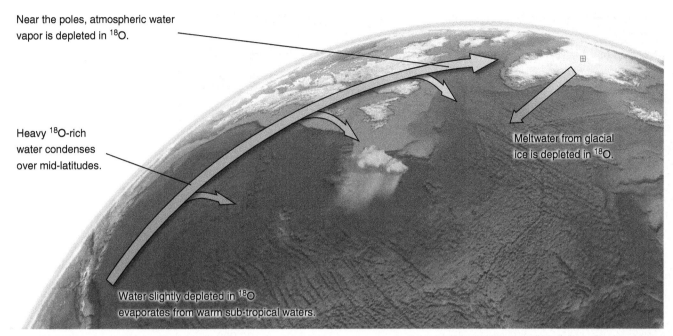

Near the poles, atmospheric water vapor is depleted in ^{18}O.

Heavy ^{18}O-rich water condenses over mid-latitudes.

Meltwater from glacial ice is depleted in ^{18}O.

Water slightly depleted in ^{18}O evaporates from warm sub-tropical waters.

FIGURE 4.11 The $H_2^{18}O$ water molecule does not evaporate as readily as the $H_2^{16}O$ molecule, and once in the atmosphere, water vapor composed of $H_2^{18}O$ tends to condense and precipitate more readily in cooling air than a molecule composed of $H_2^{16}O$. Because most water vapor originates from the tropical ocean, by the time it travels to high latitudes and high elevations where glaciers form, it is enriched in $H_2^{16}O$ relative to seawater. Hence, snow and ice are also relatively enriched in the $H_2^{16}O$ molecule.

Oxygen isotopes in fossil foraminifera provide a record of global ice volume, and in ice cores oxygen isotopes provide a record of changes in air temperature above the glacier. Because the atmosphere is so well mixed, the isotopic content of air above a glacier is easily related to the average temperature of the atmosphere; hence, the isotopic content of snow is useful as a proxy for global atmospheric temperature.

At the poles, as an air mass cools and water vapor condenses to snow, molecules of $H_2^{18}O$ condense more readily than do molecules of $H_2^{16}O$, depending on the temperature of the air. Typically, above a glacier, the condensation falls out of a cloud as snow. Thus, the oxygen isotopic content of snow (measured as the ratio of ^{18}O to ^{16}O) is a proxy for air temperature, and cores of glacial ice provide a record of variations in air temperature through time.

4.3.3 Paleoclimate Patterns

Because global ice volume and air temperature are related, the records of oxygen isotopes in foraminifera and in glacial ice show similar patterns.[38] These records provide researchers with two independent proxies for the history of global climate. Many researchers[39] have tested and verified the history of global ice volume preserved in deep-sea cores and the history of temperature preserved in glacial cores from every corner of the planet. Past episodes of cooler temperature reflected in ice cores strongly correlate to periods of increased global ice volume in marine sediments. Likewise, past episodes of warmer climate correlate well to periods of decreased ice volume (Table 4.1).

An example of this record is shown in **Figure 4.12**. Through time, the ratio of ^{18}O to ^{16}O in ice (a proxy for atmospheric temperature) and in marine foraminifera (a proxy for global ice volume) do indeed display strong agreement and provide researchers with a global guide for interpreting past climate patterns and events.

These natural archives show that global climate change is characterized by alternating warm periods and ice ages occurring approximately every 100,000 years. This history of cooling and warming has several important features:

- Major glacial and interglacial periods are repeated approximately every 100,000 years.
- Numerous minor episodes of cooling (called **stadials**) and warming (called **interstadials**) are spaced throughout the entire record.
- Global ice volume during the peak of the last interglacial (a period of time known as the **Eemian**[40]), approximately 125,000 years ago, was lower than at present, and global sea level was significantly higher.

---

[38] Discussion based on Fletcher, C.H. (2017) *Physical Geology: The Science of Earth 3rd ed.* Wiley, Hoboken, NJ.

[39] See various types of paleoclimate indicators used to reconstruct past climate at NOAA's Paleoclimatology Program website: http://www.ncdc.noaa.gov/paleo/recons.html

[40] Eemian, *Wikipedia*, http://en.wikipedia.org/wiki/Eemian_Stage

---

TABLE 4.1

Proxies for Air Temperature and Ice Volume in Geologic History

Measure	Sample Analyzed	Target of Analysis	Phenomenon for which Evidence Is a Proxy	Finding
Air temperature	Trapped air and the chemistry of samples from ice cores	Ratio of oxygen isotopes in ice and other types of chemical clues	Air temperature	The history of changing air temperature reflected in ice cores strongly correlates to the history of changing global ice volume in marine sediments.
Global ice volume	Foraminifera in deep-sea cores	Ratio of oxygen isotopes in foraminifera	Global ice volume	History of global ice volume correlates well with history of air temperature in ice cores.

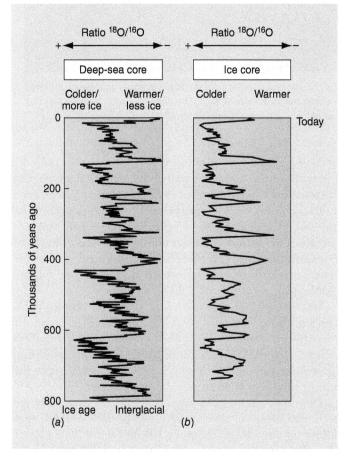

FIGURE 4.12 (a) The ratio of ^{18}O to ^{16}O in deep-sea cores of fossil foraminifera provides a proxy for global ice volume. (b) The ratio of ^{18}O to ^{16}O in cores of glacial ice documents changes in atmospheric temperature, confirming that decreased ice volume in deep-sea cores correlates to times of warmer atmosphere, whereas increased ice volume recorded in deep-sea cores correlates to times of cooler atmosphere. Notice that the ratio of $^{18}O/^{16}O$ increases with cooler climate in the deep-sea core record. In the ice-core record, the ratio decreases.

• The present interglacial, the Holocene Epoch, has lasted approximately 10,000 years.

The marine oxygen isotope record suggests that over the past 500,000 years, each glacial–interglacial period has lasted about 100,000 years; hence, there have been approximately five glaciations in this period.

During the length of a typical 100,000-year interglacial cycle, climate gradually cools and ice slowly expands until it reaches a peak. At the peak, glaciers cover Iceland, Scandinavia, the British Isles, and most of Canada southward to the Great Lakes. In the Southern Hemisphere, part of Chile is covered, and sea ice of Antarctica covers part of what is now the Southern Ocean. In mountainous regions, the snowline lowers by about 1,000 meters (3,280 feet) in altitude from the warmest to the coldest portions of a period.

Once ice cover reaches a maximum during a glacial episode, within a couple of thousand years global temperature rises again and the glaciers retreat to their minimum extent and volume (**Figure 4.13**). The last ice age culminated about 20,000 to 30,000 years ago, and by approximately 5,000 years ago most of the ice had melted (except for remnants on Greenland and Antarctica). Since then, glaciers have generally retreated to their smallest extent, with the exception of short-term climate fluctuations, such as the Little Ice Age, a cool period that lasted from about the 16th to the 19th century (discussed earlier in **Chapter 2**, and again later in this chapter).

Understanding of paleoclimate history extends beyond the Holocene and Pleistocene epochs. Using the properties of fossil soils, fossil leaf stomata density, boron isotope storage in seawater, and phytoplankton carbon isotope composition, researchers[41] have reconstructed the atmospheric composition of carbon dioxide since 65 million years ago (**Figure 4.14**). In the time interval between 65 and 23 million years, all proxy estimates of CO_2 concentration span a range of 300 to 1,500 ppm. Since 23 million years

• Following the last interglacial, global climate deteriorated in a long, drawn-out cooling phase, culminating approximately 20,000 to 30,000 years ago with a major glaciation.

---

[41] Masson-Delmotte, V. et al. (2013) Information from paleoclimate archives. In: *Climate Change 2013: The Physical Science Basis. Contribution of Working Group I to the Fifth Assessment Report of the Intergovernmental Panel on Climate Change* [Stocker, T.F., et al., eds. (2013)]. Cambridge University Press, Cambridge, United Kingdom and New York, NY.

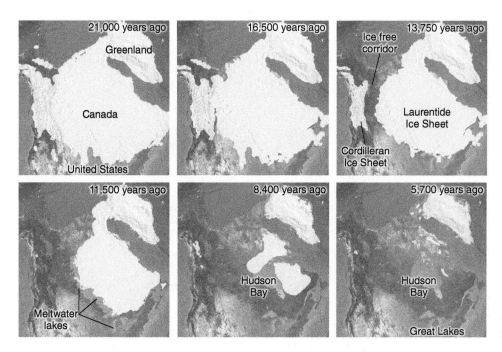

FIGURE 4.13 The history of retreating ice in North America as the last ice age ended.

ago, estimates of carbon dioxide are largely at preindustrial levels with the exception of the Middle Miocene Climatic Optimum (17 to 15 million years) and the Pliocene (5.3 to 2.6 million years), which have higher concentrations.

Assessing past changes in mineral dust accumulation using cores and other records (Figure 4.14) is important for understanding climate sensitivity and for knowing the history of nutrients supplied to the marine system (for instance, for experiments in iron fertilization[42]). Mineral dust accumulation is controlled by climate, with high rates of deposition interpreted as representing a shift to dry, cold, windy glacial conditions with abundant sediment supply and little vegetation cover. Depending on location, important dust sources include Australia, Asia, and northern South America (Southern Hemisphere) and Asian desert areas (Northern Hemisphere). Other regions (North Africa, Arabia, Central Asia) show complex, poorly understood behavior as mineral dust sources.

A 4 million year record of dust accumulation from the Southern Ocean shows reduced accumulation during the Pliocene compared to the Holocene followed by a significant rise by 2.7 million years ago when Northern Hemisphere ice volume increased. Central Antarctic ice-core records show that dust deposition is about 20 times higher during glacial conditions compared to interglacial conditions. This is consistent with ice cores from Greenland, which also show dust concentrations higher by a factor of 100 during glacial periods.

Global sea-level history can be estimated using proxies of global ice volume such as oxygen isotope ratios in marine fossils (foraminifera) and direct measurement of shoreline deposits. However, shoreline deposits are subject to vertical displacement wherein movement of the crust introduces uncertainties to interpreting them as a proxy for past sea level. This is the case with deposits from the east coast of the United States dating from the **mid-Pliocene Warm Period** (MPWP). Sedimentary rocks from Virginia through Florida provide a record of marine flooding during the MPWP.[43] These have been interpreted as a proxy for higher-than-present sea levels; however, they have been subject to uplift and contain large uncertainties, potentially eliminating their value as a sea-level proxy.[44]

Various proxies also record sea-level position during the last interglacial warm interval 125,000 years ago (the *Eemian*). Paleo-shoreline deposits and foraminifera proxies indicate that sea levels were 6 to 9 meters above the present levels.[45] However, as mentioned above, when interpreting paleo-shoreline evidence, it is important to account for vertical land motion resulting from tectonic factors and glacioisostatic adjustments (flexing of the crust with the addition and removal of ice).[46] A number of studies define both the elevation and chronology of last interglacial sea level with published dates ranging between 130,000 and 116,000 years ago. As a result, there is very high confidence

---

[42] Iron fertilization is the intentional introduction of iron dust into iron-poor areas of the ocean surface to stimulate phytoplankton growth. This is intended to enhance biological productivity in order to draw-down carbon dioxide from the atmosphere. See https://en.wikipedia.org/wiki/Iron_fertilization

---

[43] Dowsett, H.J. and Cronin, T. (1990) High eustatic sea level during the middle Pliocene: Evidence from the Southeastern U.S. Atlantic Coastal Plain, *Geology*, 18, 435, doi:10.1130/0091-7613(1990)018<0435:HESLDT>2.3.CO;2
[44] Rowley, D.B., et al. (2013) Dynamic topography change of the Eastern United States since 3 million years ago, *Science Express*, doi:10.1126/science. 1229180
[45] Dutton, A., et al. (2015) Sea-level rise due to polar ice-sheet mass loss during past warm periods, *Science*, 349(6244), doi: 10.1126/science.aaa4019
[46] Moucha, R., et al. (2008) Dynamic topography and long-term sea-level variations: there is no such thing as a stable continental platform, *Earth and Planetary Science Letters*, 271(101), doi:10.1016/j.epsl.2008.03.056

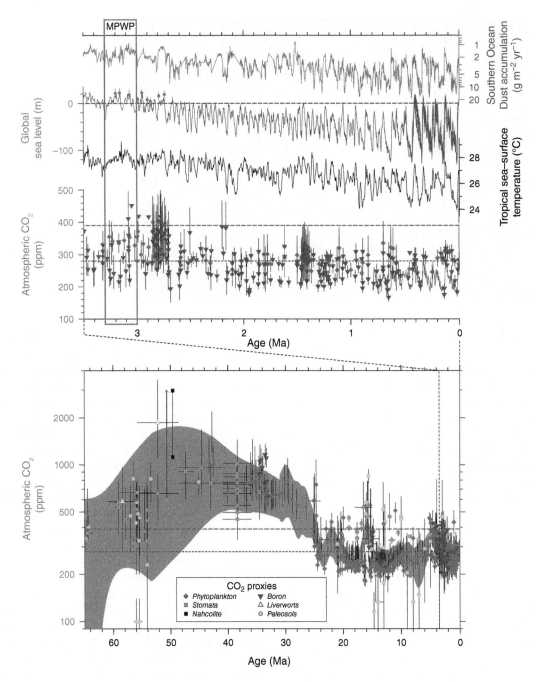

FIGURE 4.14 Paleoclimate research indicates that throughout much of Earth history, global temperature was significantly warmer than present until about 50 million years ago, when temperature began a long and slow cooling trend. *Lower panel:* Estimates of atmospheric carbon dioxide concentration using proxies from marine and terrestrial environments (see legend). CO_2 levels fell from about 1,500 ppm to less than 300 ppm over the period. Lower dashed line indicates recent preindustrial CO_2 (280 ppm). Upper dashed line indicates CO_2 level in 2011. Dark shading—one standard deviation uncertainty. *Upper panel:* Paleoclimate patterns in the past 3.5 million years. Top—dust mass accumulation (g/m²/year) from a Southern Ocean core. Second—global ice volume calibrated to sea level interpreted from oxygen isotope content of fossil marine foraminifera. The last 500,000 years of the record is based on oxygen isotope records of global ice volume calibrated with dated coral shoreline positions. Dots with bars—weighted mean sea-level estimates of Pliocene interglacials (see discussion of errors[47]). Black dashed line—modern sea level. Black—sea surface temperature using fossil plankton proxies of ocean temperature. *Lower graph:* CO_2 concentration from Antarctic ice cores, marine foraminifera (triangles), and marine phytoplankton (diamonds), plotted with two standard deviations uncertainty. Short dash and long dash lines—preindustrial and 2011 CO_2 concentrations (resp.). MPWP = mid-Pliocene Warm Period.[48]

Source: https://www.ipcc.ch/report/graphics/index.php?t=Assessment%20Reports&r=AR5%20-%20WG1

---

[47] Rowley et al. (2013)

[48] Source: FAQ 5.1, Figure 5.2 from Masson-Delmotte, V., et al. (2013)

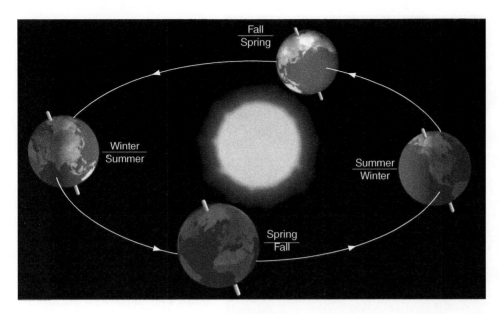

FIGURE 4.15 Earth's seasons are the result of a 23.5° tilt in the planet's axis. Because of this tilt, different parts of the globe are oriented toward the Sun at different times of the year. Summer is warmer than winter (in each hemisphere) because the Sun's rays hit Earth at a more-direct angle than during winter and because days are much longer than nights. During winter, the Sun's rays hit Earth at a less-direct angle, and days are very short.

that the maximum global mean sea level at the time was at least 6 meters higher than present (for several thousand years) and did not exceed 10 meters.

The field of paleoclimate studies encompasses the vast 4.6 billion year history of planet Earth. Consequently, the peer-reviewed literature is abundant. Our discussion has focused largely on recent geologic history because it is this period that has set the backdrop for anthropogenic global warming. Identifying past patterns of climate change is critical to understanding how the natural world has set the stage upon which humans have come to manipulate the global climate.

4.3.4 Orbital Parameters

Scientists are still uncertain about all the factors that drive variations in paleoclimate. There is, however, agreement that regular and predictable differences in Earth's exposure to solar radiation over the past half-million years must play an important role. This is because variations in the distribution of sunlight received on Earth's surface (known as **insolation**) can be mathematically calculated and they closely match the timing of ice ages and interglacials recorded by ice cores and deep-sea sediments. Three processes cause these variations in insolation: 1) changes in the geometry of Earth's orbit around the Sun, 2) tilting, and 3) wobbling of Earth's axis of rotation.

The Serbian mathematician **Milutin Milankovitch** (1879–1958) neatly described these three variables and calculated their timing.[49] These variables are known collectively as **orbital parameters**. They are eccentricity, changes in the geometry of Earth's orbit around the Sun; obliquity, changes in the angle at which Earth's axis of rotation tilts; and precession, a wobble in Earth's axis of rotation. These are each described next.

49 Milankovitch Cycles, *Wikipedia,* http://en.wikipedia.org/wiki/Milankovitch_cycles

Insolation To understand these orbital parameters, it is important to appreciate the effect of Earth's tilted axis on insolation, the amount of solar radiation received at Earth's surface through the year. Earth's axis is tilted an average of 23.5° from the vertical (**Figure 4.15**). As Earth orbits the Sun, this tilt means that during one portion of the year (summer) the Northern Hemisphere is tilted toward the Sun and receives greater insolation, whereas 6 months later (winter) it is tilted away from the Sun and receives less insolation. The reverse applies to the Southern Hemisphere: when it is summer in the Northern Hemisphere, it is winter in the Southern Hemisphere, and when it is winter in the Northern Hemisphere, it is summer in the Southern Hemisphere. These annual extremes in insolation create the seasons and their 6-month offset between the Northern and Southern Hemispheres.

Earth orbits the Sun on a flat plane called the ecliptic (aligned with the Sun's equator); however, three aspects of the geometry of this orbit change in a regular pattern under the influence of the combined gravity of Earth, the Moon, the Sun, and the other planets. These orbital parameters dictate the insolation reaching Earth's surface over time, which, in turn, regulates climate (**Figure 4.16**).

First Orbital Parameter: Eccentricity The shape of Earth's orbit changes from a nearly perfect circle to more elliptical and back again in a 100,000-year cycle and a 400,000-year cycle. The change in the shape of Earth's orbit is known as **eccentricity**. Eccentricity affects the amount of insolation received at the point in its orbit at which Earth is farthest from the Sun (aphelion) and at the point in its orbit at which Earth is closest to the Sun (perihelion). Eccentricity shifts the seasonal contrast in the Northern and Southern Hemispheres. For example, when Earth's orbit is more elliptical (less circular), one hemisphere has hot summers and cold winters, while the other has moderate summers and moderate winters. When Earth's orbit is

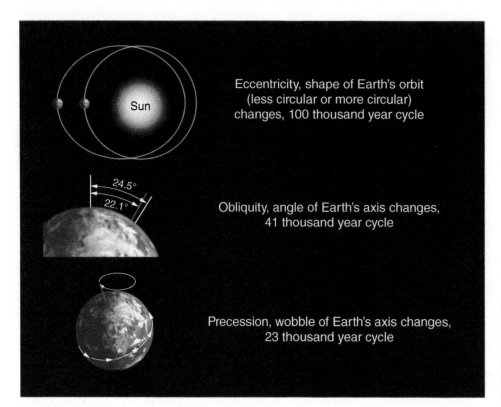

Eccentricity, shape of Earth's orbit (less circular or more circular) changes, 100 thousand year cycle

24.5°
22.1°

Obliquity, angle of Earth's axis changes, 41 thousand year cycle

Precession, wobble of Earth's axis changes, 23 thousand year cycle

FIGURE 4.16 The primary orbital parameters driving climate changes over the past half-million years are eccentricity, obliquity, and precession. These parameters regulate the intensity of insolation reaching Earth's surface, triggering changes in atmospheric temperature.

more circular, both hemispheres have similar contrasts in seasonal temperature.

Second Orbital Parameter: Obliquity The angle of Earth's axis of spin changes its tilt between 22.1° and 24.5° on a 41,000-year cycle. **Obliquity** describes the changing tilt of Earth's axis. Changes in obliquity cause large changes in the seasonal distribution of sunlight at high latitudes and in the length of the winter dark period at the poles. Changes in obliquity have little effect on low latitudes.

Third Orbital Parameter: Precession Finally, Earth's axis of spin slowly wobbles. Like a spinning top running out of energy, the axis wobbles toward and away from the Sun over the span of approximately 19,000 to 23,000 years. This wobbling or changing of Earth's tilt as it spins is known as **precession**. Precession affects the timing of aphelion and perihelion, and this has important implications for climate because it affects the seasonal balance of radiation. For example, when perihelion falls in January, winters in the Northern Hemisphere and summers in the Southern Hemisphere are slightly warmer than the corresponding seasons in the opposite hemispheres. The effects of precession on the amount of radiation reaching Earth are closely linked to the effects of obliquity (changes in tilt). The combined variation in precession and obliquity causes insolation changes of up to 15 percent at high latitudes, greatly influencing the growth and melting of ice sheets.

The predictable timing of the three orbital parameters has led to the idea that Milankovitch discovered a fundamental rhythm to terrestrial climate. These are known among paleoclimatologists as *Milankovitch cycles*. This is partially true, but also true is that the climate system is far more complex than the orbital parameters alone would suggest. Numerous relationships among a myriad of environmental factors produce a climate history with inexact timing, unpredictable rates, and ambiguous characteristics.

4.3.5 Short, Cool Summers

The variations in Earth–Sun geometry related to eccentricity, obliquity, and precession change how much sunlight each hemisphere receives during Earth's yearlong journey around the Sun. They also determine the time of year at which the seasons occur and the intensity of seasonal changes.

Milankovitch theorized that ice ages occur when orbital variations cause lands in the region of 65°N latitude (the approximate latitude of central Canada and northern Europe) to receive *less sunshine in the summer*. Why the Northern Hemisphere? Because the most continental landmass at the high latitudes is located in the Northern Hemisphere and glaciers form on land, not water.

When orbital parameters combine to create short, cool summers in the Northern Hemisphere, some snow from one year is likely to last into the following winter. This leads to the expansion of snow-covered areas from one year to the next and an increase in Earth's albedo, the tendency to reflect sunlight back to space. This, in turn, produces global cooling. Short, cool summers, increased

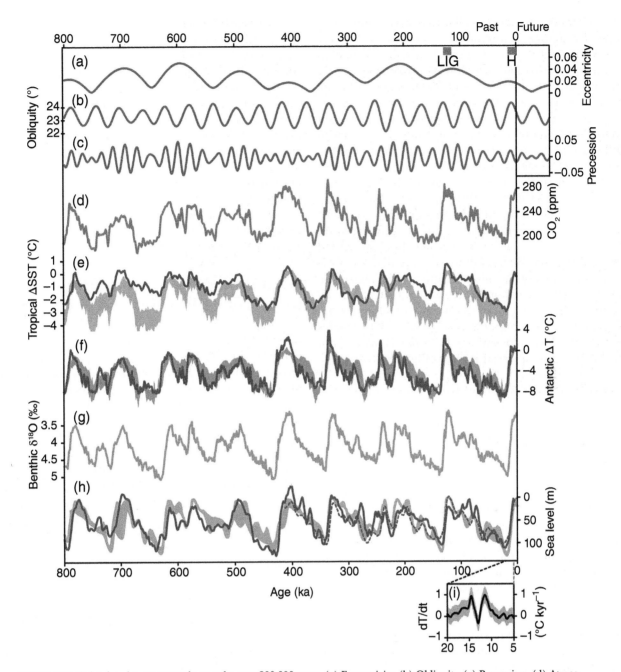

FIGURE 4.17 Orbital and proxy records over the past 800,000 years. (a) Eccentricity. (b) Obliquity. (c) Precession. (d) Atmospheric carbon dioxide from Antarctic ice cores. (e) Tropical sea surface temperature. (f) Antarctic temperature. (g) Oxygen isotope record of benthic fossils, a proxy for global ice volume and deep-ocean temperature. (h) Global sea level from various authors (dashed, solid, and shaded lines). There is high confidence that orbital parameters are the primary external driver of glacial cycles. However, atmospheric carbon dioxide content plays an important internal feedback role. Orbital scale variability in CO_2 concentrations over the last several hundred thousand years fluctuates with other proxy records of tropical sea surface temperature, Antarctic temperature, deep ocean temperature, and global sea level.[50] LIG = Last Interglacial, H = Holocene Epoch.

Source: https://www.ipcc.ch/report/graphics/index.php?t=Assessment%20Reports&r=AR5%20-%20WG1

albedo, and global cooling all generate a mutually reinforcing effect resulting in the formation of massive continental glaciers in northern Europe and Canada. Based on this reasoning and his calculations, Milankovitch predicted that the ice ages would peak every 100,000 and 41,000 years, with additional significant variations every 19,000 to 23,000 years.

Figure 4.17 plots variations in all three orbital parameters for the past one million years as well as a number of climate proxies (atmospheric CO_2, tropical sea surface temperature, Antarctic temperature, deep ocean temperature, and global ice volume or sea level). Indeed, according to

[50] Source: Figure 5.3 from Masson-Delmotte, V., et al. (2013)

any one of these proxies, Milankovitch predictions were accurate. Ice ages and interglacials occur roughly every 100,000 years, and the timing of stadials and interstadials varies on the more-rapid schedule approximated by his predictions of 41,000, 19,000 and 23,00 years, although that exact timing is not preserved in the sediment or ice-core records. The fact that geologic proxies do not record the exact timing of obliquity and precession has to do with the influence of Earth's surface environments on how the climate system changes over time.

If you compare the total solar forcing in Figure 4.17 (combined effect of the top three lines) to the paleoclimate record of any of the plotted proxies (colored lines), you notice that the timing and magnitude of the two do not exactly match. For instance, the solar forcing that led to the Eemian (LIG) is considerably greater than the forcing that is producing the modern interglacial. Also, the drop in insolation following the Eemian is greater than the drop at the culmination of the last ice age 20,000 to 30,000 years ago, yet the last ice age was much colder. What creates these disparities? The answer is that Earth's climate is not driven solely by insolation. It is also influenced by **climate feedbacks**.

Understanding the history of Earth's past climate, and thus addressing the question of whether modern global warming is a natural process, requires a familiarity with orbital parameters *and* climate feedbacks.

4.4 The Important Role of Climate Feedbacks

When scientists first tried to build computer models to simulate paleoclimate, they could not get them to reproduce past climate change unless they added fluctuations in carbon dioxide levels to accompany the changes in insolation caused by orbital parameters. This was an indication that it takes more than insolation alone to predict changes in the climate system. Although scientists are still improving the understanding of what causes natural changes in carbon dioxide levels, most believe that past episodes of climate warming were initiated by orbital forcing and then enhanced and extended by the rise of greenhouse gases. In other words, warming led to CO_2 rise, not the other way around. And the warming was initiated by the orbital parameters.

Because deep-sea cores reveal that carbon dioxide levels are much higher today than at any time in the past 15 million years,[51] pinning down the cause-and-effect relationship between carbon dioxide and climate change continues to be a focal point of modern research. In the case of paleoclimate, the cause-and-effect was related to climate feedbacks.

A climate feedback is a process taking place on Earth that amplifies (a *positive feedback*) or minimizes (a *negative feedback*) the effects of changes in insolation. Earth's environmental system generates positive and negative climate feedbacks that can enhance or suppress the timing and intensity of the Earth–Sun geometry. Climate feedbacks are responsible for the difference between the influence of orbital parameters and Earth's actual climate. That is, the climate is not solely controlled by sunlight; feedbacks are an equally powerful process, and it is the combined influence of feedbacks and orbital parameters that determines the climate.

Let's examine two case studies of how feedbacks influenced Earth's climate: the end of the last ice age and the **Younger Dryas** cold spell.

4.4.1 Why Warming Preceded CO₂ Increase at the End of the Ice Age

Ice-cores record past greenhouse gas levels as well as temperature; hence, they allow researchers to compare the history of the two. In the past, when the climate warmed, the change was accompanied by an increase in greenhouse gases, particularly carbon dioxide. However, increases in temperature *preceded* increases in carbon dioxide. This pattern is opposite to the present pattern, in which industrial greenhouse gases are *causing* increases in temperature. The difference is related to climate feedbacks.

One idea for understanding the role of feedbacks in paleoclimate was developed by scientists analyzing a core of marine sediments from the ocean floor near the Philippines.[52] That area of the Pacific contains foraminifera that live in tropical surface water. When they die and settle to the bottom, they preserve a record of changing tropical air temperatures.

But a different type of foraminifera lives on the deep sea-floor at the same location and is bathed in bottom waters (water that travels along the seafloor, not at the surface) fed from the Southern Ocean near Antarctica. These "benthic" foraminifera record the temperature of those cold southern waters. The fossils of both types of foraminifera (those that live in tropical surface water and those living in bottom waters) are deposited together on the seafloor. Upon radiocarbon dating[53] both types of foraminifera, scientists found that water from the Antarctic region warmed before waters in the topics—as much as 1,000 to 1,300 years earlier. The explanation for this difference, they believe, is

---

[51] Tripati, A., et al. (2009) Coupling of CO_2 and ice sheet stability over major climate transitions of the last 20 million years, *Science*, 326(5958), 1394–1397, http://www.sciencemag.org/cgi/content/abstract/1178296

[52] Stott, L., et al. (2007) Southern Hemisphere and deep sea warming led deglacial atmospheric CO_2 rise and tropical warming, *Science*, 318(5849), 435–438, doi: 10.1126/science.1143791

[53] Radiocarbon dating is a laboratory technique for determining how old a geologic sample is by using the constant rate of radioactive decay of the isotope carbon-14 in organic materials. The technique is typically only applicable to samples younger than 50,000 years old. See "Radiocarbon Dating," *Wikipedia*, http://en.wikipedia.org/wiki/Radiocarbon_dating

a positive climate feedback that enhanced warming initiated by orbital parameters.

First, predictable variations in Earth's eccentricity and obliquity increased the amount of sunlight hitting high southern latitudes during spring in the Southern Hemisphere. That increase warmed the Southern Ocean. As a result, sea ice shrank back toward Antarctica, uncovering and warming ocean waters that had been isolated from the atmosphere for millennia. As the Southern Ocean warmed, great quantities of CO_2 escaped into the atmosphere (warm water holds less dissolved gas than cold water). The released gas enhanced the greenhouse effect and caused warming in the global climate system. This process was responsible for driving climate out of its glacial state and into an interglacial state at the end of the last ice age. It explains how small temperature changes caused by orbital parameters led to larger temperature changes through a positive feedback in global carbon dioxide that warmed the world.

4.4.2 Rapid Climate Change: The Younger Dryas

When scientists first analyzed paleoclimate evidence in marine and glacial oxygen isotope records, they discovered that the Milankovitch theory predicted the occurrence of ice ages and interglacials with remarkable accuracy. But they also found something that required additional explanation: climate changes that appeared to have occurred very rapidly and that were not predicted by orbital parameters.

Because the Milankovitch theory tied climate change to slow and regular variations in Earth's orbit, it was assumed that climate variations would also be slow and regular. The discovery of rapid changes was a surprise. Here again, the answer lay in the climate feedback system.[54]

Cores[55] show that although it took thousands of years for Earth to totally emerge from the last ice age and warm to today's balmy climate, fully one third to one half of the warming—about 10 degrees Celsius (18 degrees Fahrenheit) at Greenland—occurred within mere decades, at least according to ice records in Greenland (**Figure 4.18**). At approximately 12,800 years ago, following the last ice age, temperatures in most of the Northern Hemisphere rapidly returned to near-glacial conditions and stayed there during a climate event called the "Younger Dryas" (named after the alpine flower *Dryas octopetala*). The cool episode lasted about 1,300 years, and by 11,500 years ago temperatures had warmed again. Ice-core records show that the recovery to warm conditions occurred with startling rapidity, less than a human generation. Changes of this magnitude would have a huge impact on modern human societies, and there is an urgent need to understand and predict such abrupt climate events so that we can better understand if one might occur again.

A look at marine sediments confirms that this pattern is present and may be a global characteristic of climate change. Hence, scientists conclude that although the general timing and pace of climate change are set by the orbital parameters, some feedback process must play an important role in its precise timing and magnitude. What might that process be? Global *thermohaline circulation* (discussed in **Chapter 3**) is thought to play a key role in the case of the Younger Dryas.

4.4.3 The Conveyor Belt Hypothesis

As we discussed in relation to Figure 3.23, today warm water near the equator in the Atlantic Ocean is carried to the north on the Gulf Stream, which flows from southwest to northeast in the western Atlantic (and carries more water

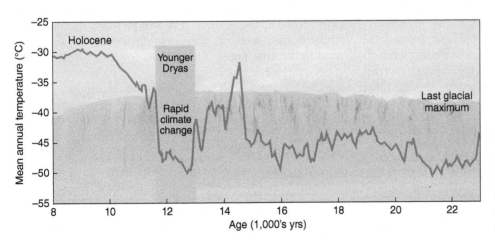

FIGURE 4.18 The Younger Dryas climate event was a dramatic cooling that lasted approximately 1,300 years during the transition between glacial and interglacial states. The return to warm conditions was equally rapid, occurring within the span of a single human lifetime.

[54] VIDEO: Richard Alley's Global Warming. http://www.youtube.com/watch?v=T4GThA35s1s&feature=relmfu

[55] Fairbanks, R.G. (1990) The age and origin of the 'Younger Dryas Climate Event' in Greenland ice cores, *Paleoceanography*, 5(6), 937–948, doi: 10.1029/PA005i006p00937

than all the world's rivers combined[56]). As this water moves to the north, it releases heat into the atmosphere through evaporation, and the heat, in turn, moves across England and Northern Europe and moderates the climate. These locations are much warmer than they would otherwise be for their latitude, because currents in the North Atlantic deliver heat to their region.

The Gulf Stream is part of a larger circulation system known as the Atlantic Meridional[57] Overturning Circulation, or **AMOC**. AMOC is the northern component of the thermohaline circulation characterized by a northward flow of warm, salty water in the upper layers of the Atlantic and a southward flow of colder water in the deep Atlantic. This ocean circulation system transports a substantial amount of heat from the tropics and Southern Hemisphere toward the North Atlantic, where the heat is transferred to the atmosphere. Changes in this circulation have a profound impact on the global climate system, as indicated by paleoclimate records. These include, for example, changes in African and Indian monsoon rainfall, atmospheric circulation of relevance to hurricanes, movement of heat through and out of the South Atlantic and Tropical Atlantic regions, and climate over North America and Western Europe.[58]

As it approaches Iceland and Greenland, North Atlantic surface water becomes cool and salty because of the high amount of evaporation it has experienced on its journey. This cool, salty water is very dense; as a result, it sinks thousands of meters into deep portions of the North Atlantic basin (making a deep current called the "North Atlantic Deep Water") before it can freeze. From there it is pulled southward toward the equator. Warm surface water from the tropics and South Atlantic continuously replaces the sinking water, releasing heat and setting up a global oceanic conveyor belt. The North Atlantic Deep Water moves into the South Atlantic and eventually the Southern Ocean where it is joined by similar cold, salty water coming off the Antarctic shelf. From there it continues its movement as a deep current through the Southern, Indian, and Pacific oceans.

Thermohaline circulation transports heat around the planet and hence plays an important role in global climatology.[59] Acting as a conveyor belt carrying heat from the equator into the North Atlantic, AMOC raises Arctic temperatures, discouraging the expansion of ice sheets. However, influxes of fresh water from melting ice on the lands that surround the North Atlantic (such as Greenland) can slow or shut down AMOC by preventing the formation of deep water. This process can ultimately lead to regional cooling in the North Atlantic, thereby regulating snowfall in the crucial region where ice sheets shrink and grow (65°N). Hence, a shutdown of AMOC could play a role in a negative climate feedback, beginning with ice melting (warming) that leads to glaciation (cooling).

The key to keeping the conveyer belt moving is the saltiness of the water, which increases the water's density and causes it to sink. Some scientists[60] believe that if too much fresh water enters the North Atlantic—for example, from melting Arctic glaciers and sea ice—the surface water would freeze before it could become dense enough to sink toward the bottom. This possibility was examined in a series of model experiments.[61] In one scenario, CO_2 emissions peak around 2040 and then decline. Under these conditions, AMOC does not collapse entirely, but eventually weakens by about 18 percent at the end of the century. In a second scenario assuming that anthropogenic greenhouse gas emissions do not decrease (the so-called "business as usual" scenario), AMOC loses about 37 percent of its strength by 2100. By 2290 to 2300, AMOC has diminished by approximately 75 percent and has a 44 percent chance of collapsing entirely.[62]

Slowing of AMOC A region of the North Atlantic south of Greenland and between Canada and Britain is becoming colder—an indicator of less northward heat transport (**Figure 4.19**). AMOC has already slowed 20 percent,[63] and despite record-setting global temperatures, observations show that in 2015 and 2016 the North Atlantic region south of Greenland cooled to the lowest recorded levels since 1880.[64] Research[65] has revealed that the overturning circulation started to weaken in the 1970s—possibly triggered by an unusual amount of sea ice traveling out of the Arctic ocean, melting, and causing freshening of North Atlantic surface water. The circulation started to recover

[56] NOAA http://oceanservice.noaa.gov/facts/gulfstreamspeed.html

[57] Meridional means "of or related to the meridian." Lines of meridian are used in global navigation. They trend north–south. Lines of latitude trend east–west (parallel to the equator). Oceanographers use the term "zonal" to refer to east–west (latitude) trends and "meridional" to refer to north–south trends. Thus, the AMOC refers to north–south circulation in the North Atlantic.

[58] Delworth, T.L., et al. (2008) The potential for abrupt change in the Atlantic Meridional overturning circulation. In: *Abrupt Climate Change*. A report by the U.S. Climate Change Science Program and the Subcommittee on Global Change Research. U.S. Geological Survey, Reston, VA, pp. 117–162.

[59] VIDEO: Thermohaline circulation. https://pmm.nasa.gov/education/videos/thermohaline-circulation-great-ocean-conveyor-belt

[60] Broecker, W.S. (2006) Was the Younger Dryas triggered by a flood? *Science*, 312(5777), 1146–1148, doi: 10.1126/science.1123253

[61] Bakker, P., et al. (2016) Fate of the Atlantic Meridional overturning circulation: Strong decline under continued warming and Greenland melting, *Geophysical Research Letters*, 44, 12,252–12,260, doi:10.1002/2016GL070457

[62] Geophysical Research Letters journal highlights: http://agupubs.onlinelibrary.wiley.com/hub/article/10.1002/2016GL070457/editor-highlight

[63] Send, U., et al. (2011) Observation of decadal change in the Atlantic Meridional overturning circulation using 10 years of continuous transport data, *Geophysical Research Letters*, 38, L24606, doi: 10.1029/2011GL049801

[64] Real Climate discussion: http://www.realclimate.org/index.php/archives/2016/10/q-a-about-the-gulf-stream-system-slowdown-and-the-atlantic-cold-blob

[65] Rahmstorf, S., et al. (2015) Exceptional twentieth-century slowdown in Atlantic Ocean overturning circulation, *Nature Climate Change*, doi: 10.1038/nclimate2554

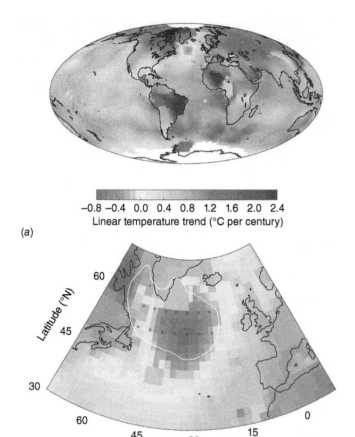

(a)

(b)

FIGURE 4.19 Scientists attribute record cooling in the North Atlantic to a slowdown of AMOC.[66] (a) Temperature trend for 1901 to 2013. White boxes indicate insufficient data. The cooling patch in central Africa is in a region of poor data coverage and may be an artifact of data irregularities. (b) Same analysis for the North Atlantic sector 1901 to 2000. Climate models show that adding fresh water to the North Atlantic cause strong reductions in AMOC. This produces cooling shown by the white line marking the 2°C (3.6°F) cooling contour. The geographic extent of the model-predicted cooling coincides well with observed cooling.[67]

Source: https://www.nature.com/nclimate/journal/v5/n5/fig_tab/nclimate2554_ft.html

in the 1990s, but this was only temporary, and now it has weakened further.

Studies continue to improve understanding of the conveyor belt, and, as a whole, are moving toward a consensus that the circulation is more fragile than originally thought. One study[68] found that if atmospheric carbon dioxide were to reach a concentration twice the natural level (approximately 500 to 600 ppm, entirely possible this century),

AMOC could collapse. Without the usual transport of warm water into the North Atlantic, scientists predict significant cooling over the North Atlantic, including in northwestern Europe, as well as in the Arctic.

A pair of studies in 2018 provides evidence that AMOC hasn't been running at peak strength since the mid-1800s, has weakened rapidly due to global warming, and is currently at its weakest point in the past 1,600 years.[69] If the system continues to weaken, it could disrupt weather patterns from the United States and Europe to the African Sahel, and cause more rapid increase in sea level on the U.S. East Coast. Researchers looked at climate model data and past sea-surface temperatures to reveal that AMOC has been weakening more rapidly since 1950 in response to recent global warming.[70] Together, the two new studies provide complementary evidence that the present-day AMOC is exceptionally weak, offering both a longer-term perspective and detailed insight into recent decadal changes.

The thermohaline circulation system conveys warm and cold currents to many parts of the world. Without cold deep water moving out of the North Atlantic back to the South Atlantic, a stronger warming pattern develops south of the equator causing more rain for northeastern Brazil and less rain for Central America. Also likely is a reduction in sea ice around the Antarctic. Exactly how this affects the rest of the global climate system is uncertain, and how fast a total collapse of AMOC could occur is debated among researchers, but if the situation mimics the Younger Dryas, the last time we think the conveyor collapsed, climate around the world may develop some unusual characteristics.

Disturbing the circulation will likely have a negative effect on the ocean ecosystem, and therefore fisheries and the associated livelihoods of many people in coastal areas. A slowdown also adds to the regional sea-level rise affecting cities like New York and Boston. Temperature changes can also influence weather systems on both sides of the Atlantic, in North America as well as Europe.

The conveyor belt circulation has long been considered a possible tipping point in the climate system. This would mean a relatively rapid and hard-to-reverse change. The 2013 report by the Intergovernmental Panel on Climate Change estimates there to be an up to one-in-ten chance that this could happen as early as within this century. However, expert surveys indicate that many researchers assess the risk to be higher.[71,72]

If the water in the north did not sink, the Gulf Stream would eventually stop moving warm water northward,

[66] Real Climate: http://www.realclimate.org/index.php/archives/2015/03/whats-going-on-in-the-north-atlantic

[67] Figure from Rahmstorf et al. (2015)

[68] Liu, W., et al. (2017) Overlooked possibility of a collapsed Atlantic Meridional overturning circulation in warming climate, *Science Advances*, doi: 10.1126/sciadv.1601666

[69] Thornalley, D.J.R., et al. (2018) Anomalously weak Labrador Sea convection and Atlantic overturning during the past 150 years, *Nature*, 556(7700), 227, doi: 10.1038/s41586-018-0007-4

[70] Caesar, L., et al. (2018) Observed fingerprint of a weakening Atlantic Ocean overturning circulation, *Nature*, 556(7700), 191, doi: 10.1038/s41586-018-0006-5

[71] AMOC: https://www.sciencedaily.com/releases/2015/03/150323132746.htm

[72] VIDEO: AMOC. https://www.youtube.com/watch?v=IQ8j6jFKSLQ

leaving Northern Europe cold and dry. Modeling[73] suggests that a low rate of meltwater addition to the North Atlantic would not significantly alter the circulation. However, a moderate or high rate of Greenland melting could make the thermohaline circulation weaken further. This further weakening would not necessarily make the global climate in the next two centuries cooler than in the late 20th century, but it would instead lessen the warming, especially in the northern high latitudes.

Hansen et al. A widely discussed[74] study published[75] in 2016 uses numerical climate models, paleoclimate data, and modern observations to simulate the effect of growing ice melt from Antarctica and Greenland. The study authors indicate that they chose a global warming scenario of 2 degrees Celsius (3.6 degrees Fahrenheit) because it replicates global changes that occurred during the Eemian and that led to sea-level rise of 6 to 9 meters above present (studies, in fact, assign the Eemian a warmth equivalent to the modern value of 1°C above pre-industrial). Two degrees Celsius is also the temperature that is identified by the 2015 Paris Accord as the target for stopping global warming.

Hansen et al.[76] find that 2 degrees Celsius warming produces sufficient meltwater from both Greenland and Antarctica to shut down the thermohaline circulation in the North and South Atlantic. If such a shutdown were to happen, it would trigger a number of globally disruptive reactions. Initially, simultaneous cooling of the waters near Greenland and Antarctica would lead to the accumulation of heat in the tropics and mid-latitudes. As the oceans would no longer be able to transport heat toward the poles effectively, the atmosphere would take over the job—and big storms are the best way of doing that. This would spawn frequent super-storms on the order of Hurricane Sandy (North Atlantic, 2015) or worse.

At the same time, the surface layer of cool fresh water in the high latitudes of the Southern Ocean (near Antarctica) produces an inversion in ocean temperatures, trapping mid-depth warm, salty water and limiting the loss of heat to the atmosphere. This type of layering in the ocean is called stratification. Stratification is observed today in the Amundsen Sea sector of West Antarctica, and it is responsible for significant melting at the base of the Pine Island ice shelf and others nearby. In the Hansen et al. model, stratification allows warm subsurface water to penetrate under the floating ice shelves of Antarctica and accelerate their decay.

Grounded ice shelves are known to hold back adjacent glaciers located on land and their loss would allow large portions of West Antarctic ice to move faster and faster into the sea, rapidly raising sea level.

Hansen et al. find that these feedbacks make ice sheets in contact with the ocean vulnerable to accelerating disintegration. Mass loss from the most vulnerable ice, sufficient to raise global sea-level several meters, is best approximated as exponential than by a more linear response. Doubling times of 10, 20, or 40 years yield multi-meter sea-level rise in about 50, 100, or 200 years.

The study concludes with a series of alarming points:

- Most of global CO_2 growth has occurred in the past several decades, and three-quarters of the 1°C (1.8°F) global warming since 1850 has occurred since 1975.
- Climate response to this CO_2 level, so far, is only partial.
- Full shutdown of the AMOC would be likely within the next several decades in such a climate-forcing scenario.
- If greenhouse gas emissions continue to grow, multi-meter sea-level rise would become practically unavoidable, probably within 50 to 150 years.
- Social disruption and economic consequences of such large sea-level rise, and the attendant increases in storms and climate extremes, could be devastating.
- It is not difficult to imagine that conflicts arising from forced migrations and economic collapse might make the planet ungovernable, threatening the fabric of civilization.
- Their model paints a very different picture than IPCC AR5 (2013).
- The basic features of their model scenario are already beginning to evolve in the real world including the following:
  - Accelerating mass losses from the Greenland and Antarctic ice sheets
  - Cooling of high-latitude oceans
  - Slowdown of conveyor belt circulation
  - Accelerating global mean sea-level rise
  - Ice shelf melting in West Antarctica by stratified mid-depth water in the Southern Ocean
  - Irreversible retreat of West Antarctic glaciers
- Their model underestimates sensitivity to freshwater forcing and the ocean stratification feedback, and the surface climate effects are likely to emerge sooner than their model suggests, if greenhouse gas climate forcing continues to grow.
- A fundamentally different climate phase, a "Hyper-Anthropocene," began in the latter half of the 18th century as improvements of the steam engine ushered in the industrial revolution and exponential growth of fossil fuel use.
- There is a possibility, a real danger, that we will hand young people and future generations a climate that is practically out of their control.

[73] Hu, A., et al. (2011) Effect of the potential melting of the Greenland ice sheet on the Meridional overturning circulation and global climate in the future, *Deep Sea Research* II, 58, 1914–1926.

[74] James Hansen: https://www.theguardian.com/science/2016/mar/22/sea-level-rise-james-hansen-climate-change-scientist

[75] Hansen, J., et al. (2016) Ice melt, sea level rise and superstorms: Evidence from Paleoclimate data, climate modeling, and modern observations that 2°C global warming could be dangerous, *Atmospheric Chemistry and Physics*, 16, 3761–3812, doi: 10.5194/acp-16-3761-2016

[76] Find the Hansen et al. paper here: http://www.atmos-chem-phys.net/16/3761/2016/acp-16-3761-2016.pdf

- 2°C warming, as negotiated by the Paris Agreement, is not "safe."
- We have a global emergency.

Hansen et al. state that predictions from their modeling are shown vividly in their Figure 16, which shows simulated climate four decades in the future (2056; **Figure 4.20**). Some major features include pronounced warming in the Arctic, cool sea surface temperatures at major overturning sites in the North Atlantic and Southern Ocean marking marine stratification, limited warming in Europe and the Southern Ocean, and warming of 1.5 to 2 degrees Celsius (2.7 to 3.6 degrees Fahrenheit) in the low latitudes.

Jim Hansen[77] and research colleague Makiko Sato[78] maintain on-line profiles where their writings, analysis, and latest graphics are available.

Feedback Examples The hypothesis of rapid climate change resulting from a shutdown in the thermohaline circulation system is called the conveyor belt hypothesis, and the paleoclimate record found in marine sediment and ice cores offers strong evidence that it is an important and common process in recent geologic history. Paleoclimate studies show that during the most recent interglacial cycle, when heat circulation in the North Atlantic Ocean slowed, Northern Europe's climate changed. Although the last ice age peaked about 20,000 to 30,000 years ago, the warming trend that followed it was interrupted by cold spells at 17,500 years ago and again at 12,800 years ago (the Younger Dryas).

These cold spells happened just after melting ice had diluted salty North Atlantic water, slowing AMOC. It is this idea that led to the movie *The Day After Tomorrow*. In the movie, global warming results in fresh water from melting ice stopping the thermohaline circulation, which, in turn, produces deadly cooling in the North Atlantic—a likely process that may be playing out today, but with Hollywood-style consequences in the movie that are unlikely to be so severe in reality.

Hence, we have seen two types of climate feedback:

- A positive climate feedback in the Antarctic that ended the last ice age. Predictable variations in Earth's tilt and orbit caused the initial warming that began the deglaciation process, which triggered the withdrawal of sea ice in the Southern Ocean. This led to additional warming of ocean water, reducing its ability to hold dissolved carbon dioxide. The carbon dioxide poured into the atmosphere and warmed the planet beyond the temperatures that would have been achieved by orbital parameters alone.
- A negative climate feedback late in the transitional phase between the last ice age and the modern interglacial. Warming at approximately 12,800 years ago produced abundant fresh water in the North Atlantic that diluted the salty Gulf Stream. This put a temporary end to AMOC and triggered rapid cooling in the Northern Hemisphere. Later, after a period of cooling lasting approximately 1,300 years, the Younger Dryas, thermohaline circulation once again became a source of heat transport throughout the world's oceans. This renewed circulation triggered global warming, apparently very rapidly, that would not have been predicted by orbital parameters alone.

These climate feedbacks, and others that are still being discovered (such as through modeling as exemplified by

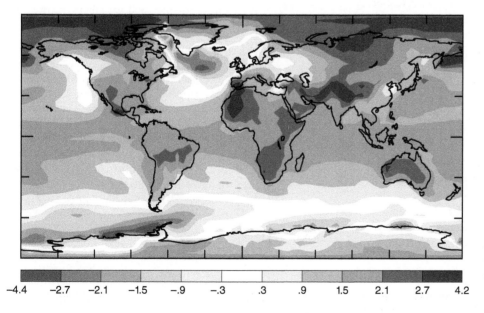

-4.4　-2.7　-2.1　-1.5　-.9　-.3　.3　.9　1.5　2.1　2.7　4.2

FIGURE 4.20 Surface air temperature (°C) change relative to 1880 to 1920 in 2055 to 2060 for continued carbon dioxide emissions.[79]

Source: Hansen et al. (2016)

[77] James Hansen website: http://www.columbia.edu/~jeh1
[78] Makiko Sato and James Hansen website: http://www.columbia.edu/~mhs119

[79] Hansen et al. (2016)

Hansen et al.), work in parallel with the orbital parameters to determine the final nature of Earth's climate. Of keen interest is the fact that 8,000 years ago marked when the orbital parameters were aligned for maximum warmth. The alignment since then has promoted global cooling, and Holocene temperature should theoretically have dropped roughly 1°C. One famous study[80] about Milankovitch cycles concluded, "this model predicts that the long-term cooling trend which began some 6,000 years ago will continue for the next 23,000 years."

Another study[81] showed that the influence of Milankovitch cycles predicts that a new continental ice sheet should be forming today in northeast Canada. The fact that Earth has not cooled over this interglacial as predicted has led to the **anthropogenic hypothesis**, which proposes that human agriculture (involving deforestation, rice wetland production, and animal husbandry) has controlled global climate for several thousand years.[82] It is clear that Milankovitch cycles, the major natural paleoclimate process, did not anticipate the global warming problem we face today.

4.4.4 Dansgaard–Oeschger Events

In some cases, global climate follows quasi-cyclic patterns of warming and cooling that are more frequent than Milankovitch cycles. As we have seen already, the popular notion of "natural cycles" is overly simplistic, and actual climate is the product of complex interactions among solar and terrestrial processes. Feedback processes that make cycle timing irregular and can suppress or enhance temperature and other climate processes in unpredictable ways in fact govern climate change.

An example of how feedback processes make climate cycle timing irregular and affect the surface temperature and other climate processes unpredictably, is the sequence of **Dansgaard–Oeschger** (DO) events that occurred during the last ice age in the North Atlantic (**Figure 4.21**). DO events are recorded in Greenland ice cores and in North Atlantic seafloor sediments. These rapid climate events have led some to pose an important scientific question: "Is present global warming part of a natural DO event?"[83]

DO events are rapid climate fluctuations, occurring approximately 25 times during the last glacial age, that are revealed in ice core and marine sediment records in the Northern Hemisphere. They take the form of rapid warming episodes, typically in a matter of decades, each followed by gradual cooling over a longer period. The pattern in the Southern Hemisphere is different, with slow warming and much smaller temperature fluctuations. However, orbital geometry does not predict these events.

Several explanations have been promoted to explain DO events, but their exact origin is still unclear. It has been

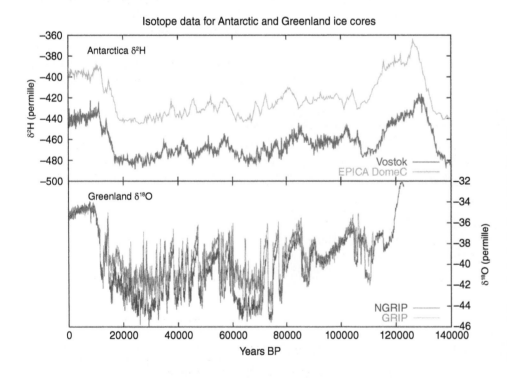

FIGURE 4.21 Climate change over 140,000 years (rising trends signify warming, falling trends signify cooling): Top panel (light and dark) are from Antarctic ice cores; Lower panel light and dark are from Greenland ice cores. Greenland ice cores use [18]O as a proxy for temperature, and Antarctic ice cores use an isotope of hydrogen, [2]H. Note the rapid climate changes in the Greenland ice core during the glacial age, between about 80,000 and 15,000 years ago, which barely register in the corresponding Antarctic record. These are Dansgaard–Oeschger events.

Source: https://en.wikipedia.org/wiki/Dansgaard–Oeschger_event

[80] Imbrie, J. and Imbrie, J.Z. (1980) Modeling the climatic response to orbital variations, *Science*, 207, 943–953, doi: 10.1126/science.207.4434.943

[81] Carlson, A.E., et al. (2008) Rapid early Holocene deglaciation of the Laurentide ice sheet, *Nature Geoscience*, 1, 620–624.

[82] Ruddiman, W. (2003) The anthropogenic greenhouse era began thousands of years ago, *Climatic Change*, 61, 261–293; Ruddiman, W. (2005) Cold climate during the closest stage 11 analog to recent millennia, *Quaternary Science Reviews*, 24, 1111–1121; Ruddiman, W. (2007) *Plows, Plagues, and Petroleum: How Humans Took Control of Climate.* Princeton University Press, Princeton.

[83] Bond Events, *Wikipedia*, http://en.wikipedia.org/wiki/Bond_event

hypothesized that they are the result of periodic collapses of thick glacier ice in Canada (ice buildup eventually collapses under its own weight) or changes in AMOC triggered by an influx of fresh water.[84] The question of whether DO events extend into the present interglacial is controversial, the last clear candidate for a DO event was 11,500 years ago (the Younger Dryas),[85] and it has been questioned if this event resulted in climate change that was truly global in extent.[86] In fact, some high-resolution records of climate extending 50,000 years into the past find no evidence of the Younger Dryas.[87]

4.4.5 Bipolar Seesaw

The fundamental problem with assigning modern global warming to cyclic-type DO events is the **bipolar seesaw**. Weak DO events are found in Antarctic ice cores, but their effect in the Southern Hemisphere is opposite in timing to the Northern Hemisphere. That is, cooling in the north is accompanied by warming in the south and vice versa.[88] Thermohaline circulation is thought to be related to this seesaw.[89]

The bipolar seesaw hypothesis goes like this: surface currents associated with thermohaline circulation carry heat from the Southern Hemisphere northward (joining AMOC) until increasing salinity and cooling cause it to sink at high northern latitudes, forming North Atlantic Deep Water, which returns southward as the North Atlantic Deep Water current. Strengthening of the system leads to more warm water carried into the North Atlantic. This warms the North Atlantic and cools the South Atlantic (that is, heat is rapidly exported from the southern region when the circulation is vigorous), with the converse occurring when the circulation weakens. Thus, the north and south Atlantic operate like a seesaw with one growing warmer while the other grows cooler: and vice versa.

The gradual cooling phase of a DO cycle may be associated with an influx of fresh water to the North Atlantic, which reduces the strength of the thermohaline circulation

(similar to what is happening today). As a result, less heat is exported from the tropical Atlantic and the South Atlantic to the North Atlantic. Warming is recorded in Antarctic ice-core proxies. Eventually, warming in Antarctica releases fresh water from melting glaciers in the region, which weakens Southern Hemisphere circulation. It is such an episode of enhanced meltwater production from Antarctica that Hansen et al.[90] propose will lead to marine stratification and ultimately to rapid sea-level rise, perhaps reaching several meters per century and could be under way today. Meanwhile, cooling in the North Atlantic has ended the freshwater pulse and the conveyor belt recovers its strength. The North Atlantic thermohaline circulation switches on (North Atlantic warming), and heat is exported from the South Atlantic (South Atlantic cooling). Thus, cooling/warming in the north is contemporaneous with warming/cooling in the south.[91]

Today, global warming is occurring simultaneously around the planet; hence, it cannot be tied to a modern-day DO event.[92] This was proved[93] by a detailed study of ice cores that capture a record of atmospheric temperature in both the Northern and Southern Hemispheres as well as by cores of ocean sediments[94]; no period of unequivocal simultaneous change in the north and south is seen over the past 20,000 years. This means that a fundamental characteristic of modern global warming has not existed since the last ice age: natural warming alternates between the hemispheres (the seesaw effect), but anthropogenic warming occurs across the entire planet simultaneously.

Here is another way to think of it: DO events lead to no net change in Earth's heat budget, due to offsetting trends in the Northern and Southern Hemispheres. Global warming is global and represents a significant net increase in the Earth's heat budget. In any case, DO events have not been shown to exist in the Holocene, nor have they been documented as global in extent.[95]

This rather lengthy treatment of paleoclimate was necessary to fully examine the hypothesis that global warming today is the result of a natural process. Despite testing the

[84] Maslin, M., et al. (2001) Synthesis of the nature and causes of rapid climate transitions during the Quaternary, *Geophysical Monograph*, 126, 9–52, http://www.essc.psu.edu/~dseidov/pdf_copies/maslin_seidov_levi_agu_book_2001.pdf

[85] Climate cycle: http://en.wikipedia.org/wiki/1500-year_climate_cycle

[86] Barrows, T., et al. (2007) Absence of cooling in New Zealand and the adjacent ocean during the Younger Dryas Chronozone, *Science*, 318 (5847), 86–89.

[87] Fleitmann, D., et al. (2009) Timing and climatic impact of Greenland Interstadials recorded in Stalagmites from Northern Turkey, *Geophysical Research Letters*, 36, L19707, doi: 10.1029/2009GL040050

[88] Barker, S., et al. (2009) Interhemispheric Atlantic seesaw response during the last deglaciation, *Nature*, 457(7233), 1097–1102, doi: 10.1038/nature07770. See discussion by Severinghaus, J.P. (2009) Climate change: Southern see-saw seen, *Nature*, 457(7233), 1093–1094.

[89] Stenni, B., et al. (2011) Expression of the bipolar seesaw in Antarctic climate records during the last deglaciation, *Nature Geoscience*, 4, 46–49, doi: 10.1038/ngeo1026

[90] Hansen et al. (2016)

[91] Bassis, J.N., et al. (2017) Heinrich Events triggered by ocean forcing and modulated by isostatic adjustment, *Nature*, 542(7641), 332, doi: 10.1038/nature21069

[92] Seidov, D. and Maslin, M. (2001) Atlantic Ocean heat piracy and the bipolar climate see-saw during Heinrich and Dansgaard-Oeschger events, *Journal of Quaternary Science*, 16(4), 321–328.

[93] Björck, S. (2011) Current global warming appears anomalous in relation to the climate of the last 20,000 years, *Climate Research*, 48(1), 5 doi: 10.3354/cr00873

[94] Hessler, I., et al. (2011) Impact of abrupt climate change in the tropical Southeast Atlantic during marine isotope stage (MIS) 3, *Paleoceanography*, 26, PA4209, doi: 10.1029/2011PA002118

[95] Abrupt climate: http://www.realclimate.org/index.php/archives/2006/11/revealed-secrets-of-abrupt-climate-shifts. Also see Blunier, T. and Brook, E.J. (2001) Timing of millennial-scale climate change in Antarctica and Greenland during the last glacial period, *Science*, 291, 109–112; Blunier, T., et al. (1998) Asynchrony of Antarctic and Greenland climate change during the last glacial period, *Nature*, 394, 739–743.

full list of options of climate processes known to exist in the recent past, no climate cycle candidates emerge with the characteristics to account for modern warming. In a few pages we test the last persistent issue, that medieval time was warmer than present. But first let's ask, "Is global warming caused by the Sun?"

4.5 Is Global Warming Caused by the Sun?

Until recently, it was thought that over the past few centuries there had been a steady increase in solar radiation. However, as we showed in **Chapter 2** (Figure 2.10 and repeated here, **Figure 4.22**), the Sun has not been increasing its output.[96] The most prominent feature of the Sun's activity is the 11-or-so-year sunspot cycle. This is associated with rhythmic increases and decreases in solar radiation. The record of regular and frequent sunspot observations extends to about 1750; prior to those, less-frequent observations were made.[97]

These observations show that the Sun is variable. Between 1645 and 1715, the Maunder Minimum, there was greatly reduced sunspot activity, and climate was cooler than average. This period corresponded to the coldest portion of the Little Ice Age, during which Europe and North America experienced bitterly cold winters.[100] Less severe, the Dalton Minimum lasted from 1790 to 1830 and also corresponded to a period of lower-than-average temperatures.

Observations of the *Sunspot Number*, the longest scientific experiment still ongoing, is a crucial tool used to study the solar dynamo, space weather and climate change. Until recently, there was a general consensus that solar activity has been trending upward over the past 300 years (since the end of the Maunder Minimum), peaking in the late 20th century—called the *Modern Grand Maximum* by some. This trend had led some to conclude that the Sun has played a significant role in modern climate change.[101] However, a discrepancy between two parallel series of sunspot number counts has been found to be the apparent source of the Modern Grand Maximum.[102]

The Sunspot Number has now been recalibrated and shows a consistent history of solar activity over the past few centuries. The former two methods of counting the sunspot number—the Wolf Sunspot Number and the Group Sunspot Number—indicated significantly different levels of solar activity before about 1885 and also around 1945. The apparent upward trend of solar activity between the 18th century and the late 20th century has now been identified as a major calibration error in the Group Sunspot Number. Now that this error has been corrected, solar activity appears to have remained relatively stable since the 1700s.

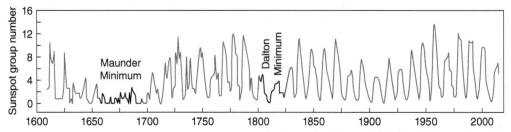

FIGURE 4.22 This graph shows the sunspot *group number* as measured over the past 400 years.[98] The group number is a standardized method for counting the presence of sunspots.[99] The Maunder Minimum is a time of low group number between 1645 and 1715, when sunspots were scarce and the winters harsh. Another period of low solar activity was in the early 1800s called the Dalton Minimum. Satellite measurements, begun in 1978, reveal that the Sun's output fluctuates by about 1 W/m² (or one-tenth of 1 percent) on approximately 11-year cycles.

Source: N. Hulbirt

[96] Spotty Sunspot Record Gets a Makeover, http://www.nature.com/news/spotty-sunspot-record-gets-a-makeover-1.18145 See also the papers Clette, F. and Lefèvre, L. (2016) The new sunspot number: Assembling all corrections, *Solar Physics* 291, 2629, doi:10.1007/s11207-016-1014-y. See also Svalgaard, L. and Schatten, K.H. (2016) Reconstruction of the sunspot group number: The backbone method, *Solar Physics* 291, 2653, doi:10.1007/s11207-015-0815-8

[97] Solar Variation, *Wikipedia,* http://en.wikipedia.org/wiki/Solar_variation

[98] Corrected Sunspot History Suggests Climate Change Not due to Natural Solar Trends, https://astronomynow.com/2015/08/08/corrected-sunspot-history-suggests-climate-change-not-due-to-natural-solar-trends

[99] Clette and Lefèvre (2016) Svalgaard and Schatten (2016)

[100] IPCC, Observed climate variability and change, In *Climate Change 2001: Working Group I: The Scientific Basis* (Cambridge, UK, Cambridge University Press, 2001). 2.3.3: "Was There a 'Little Ice Age' and a 'Medieval Warm Period'?" http://www.grida.no/publications/other/ipcc_tar/?src=/climate/ipcc_tar/wg1/070.htm

[101] Stott, P.A., et al. (2003) Do models underestimate the solar contribution to recent climate change? *Journal of Climate*, 16(24), 4079–4093.

[102] International Astronomical Union: https://www.iau.org/news/pressreleases/detail/iau1508/#1

With these discrepancies now eliminated, there is no longer any substantial difference between the two historical records. The new correction of the sunspot number, called the *Sunspot Number Version 2.0*,[103] nullifies the claim that there has been a Modern Grand Maximum. The new record has no significant long-term upward trend in solar activity since 1700, as was previously indicated. As a result, global warming since the industrial revolution cannot be attributed to increased solar activity—as there has not been any increased solar activity.

4.6 Did Global Warming End After 1998?

Between 1998 and 2012, a period in which the United Nations was engaged in political negotiations for preventing the most severe effects of climate change, Earth's surface appeared to scarcely warm. This period, termed the **global warming hiatus**, raised public doubt regarding the reliability of scientific projections of warming trends and impacts.[104] Did global warming end after 1998?[105]

Figure 4.23 shows a plot of the HadCRUT3 global surface temperature data set developed and maintained by the United Kingdom Met Office. The black dashed line shows the temperature trend for the period 1998 to 2008 in Figure 4.23(a). It is easy to see why some observers, especially those who do not accept the scientific consensus on climate change, would take the opportunity to declare that global warming had ended.[106]

Widely viewed by scientists as a typical example of climate variability,[107] the period in question begins with 1998, a very strong El Niño year that set a record at the time as the warmest year ever recorded. Subsequent years were cooler, a characteristic pattern following a strong El Niño year that had been anticipated by the research community. So it came as something of a surprise that so much fuss was being made by some media out of a decade trend that, to scientists, was consistent with normal climate behavior.[108] Researchers know from observations and models that global temperatures fluctuate on timescales of years to decades[109] producing longer periods of reduced warming rates.[110] A system with variability will always show a range

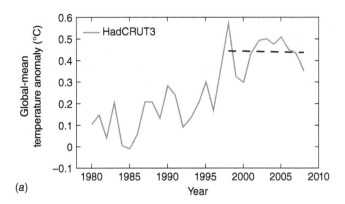

(a)

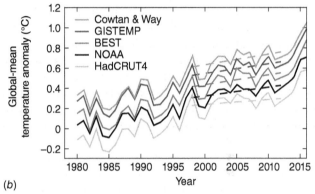

(b)

FIGURE 4.23 (a) Global surface temperature from the United Kingdom Met Office.[111] The Met Office dataset (HadCRUT3) shows global mean surface temperature from 1980 to 2008 (light line) together with the 1998 to 2008 linear trend (dashed line). (b) Global surface temperature (solid lines, top to bottom) from various datasets, 1980 to 2015: Cowtan and Way, NASA-GISTEMP, Berkeley-BEST, NOAA, and HadCRUT4 relative to 1961 to 1990. The data have been shifted by 0.1°C (from –0.2 to 0.2) relative to each other (no offset for BEST) for clarity. The dashed lines indicate the linear trend for 1998 to 2012 for the different datasets.

Source: https://www.nature.com/nature/journal/v545/n7652/full/nature22315.html[112]

of trends on various timescales.[113] So it wasn't until after a few years that the research community began publishing the results of investigations into the nature of the hiatus.[114]

Figure 4.23(b) plots datasets from several research groups: UK Met Office (HadCRUT4), NASA (GISTEMP),

[103] World Data Center for the production, preservation and dissemination of the international sunspot number: http://www.sidc.be/silso/newdataset

[104] Carter, B. (2006) There IS a problem with global warming...It stopped in 1998, *The Telegraph*, http://www.telegraph.co.uk/comment/personal-view/3624242/There-IS-a-problem-with-global-warming...-it-stopped-in-1998.html

[105] VIDEO: Reshuffling heat on a warming planet. https://science.nasa.gov/sciencecasts/reshuffling-heat-warming-planet

[106] Kerr, R. (2009) What happened to global warming? Scientists say just wait a bit, *Science*, 326, 28–29.

[107] Lean, J.L. and Rind, D.H. (2009) How will Earth's surface temperature change in future decades? *Geophysical Research Letters*, 36, L15708.

[108] Lewandowsky, S., et al. (2015) On the definition and identifiability of the alleged 'Hiatus' in global warming, *Scientific Reports*, 5, 16784.

[109] The UK Met Office, the National Meteorological Agency for the United Kingdom, has a climate research unit that collects climate observations from around the world.

[110] Medhaug, I., et al. (2017) Reconciling controversies about the "global warming hiatus", *Nature*, 545, 41–47, doi:10.1038/nature22315

[111] Mann, M.E. and Park, J. (1994) Global modes of surface temperature variability on interannual to century time scales, *Journal of Geophysical Research*, 99, 25819–25833.

[112] Watts, R.G. and Morantine, M.C. (1991) Is the greenhouse gas-climate signal hiding in the deep ocean? *Climatic Change*, 18, iii–vi.

[113] Medhaug et al. (2017)

[114] Fyfe, J.C., et al. (2013) Overestimated global warming over the past 20 years, *Nature Climate Change*, 3, 767–769.

NOAA, Berkeley Earth (BEST),[115] and Cowton and Way.[116] The graphed datasets are shifted vertically in order to clearly see their trends, and in every case the period 1998 to 2012 shows an upward trend, although at a somewhat slower rate than the longer time series; warming continued over the interval, but had slowed.

Note that Figure 4.23(a) plots the HadCRUT3 dataset, whereas Figure 4.23(b) plots an updated version, Had-CRUT4. Over the period 1998 to 2008, the HadCRUT3 data shows a mild negative trend (cooling). Since then, this and other observational datasets have undergone substantial, well-documented improvements.[117] The revised data show a positive temperature trend over the same period. As a result, it could be argued that the hiatus was not as substantial as first reported.

In 2009, the Associated Press conducted a blind test. They gave unidentified temperature data to four independent statisticians and asked them to look for trends. The experts found no true declines over time. One climate scientist said the following: "To talk about global cooling at the end of the hottest decade the planet has experienced in many thousands of years is ridiculous."[118]

In one study,[119] scientists showed that naturally occurring periods of no warming or even slight cooling could easily be a part of a longer-term pattern of global warming. These researchers conclude, "Claims that global warming is not occurring that are derived from a cooling observed over short time periods ignore natural variability and are misleading."[120] It is clear that global warming did not end in 1998 and that the climate in the following decade was simply experiencing an episode of natural variability.

Although global warming deniers argued that warming had been absent after 1998, a look at the temperature of the ocean reveals that warming had not relented, not even a little; the excess heat in Earth's climate system was being stored in the ocean. We learned earlier in **Chapter 3** that 93 percent of the heat trapped by increasing greenhouse gases goes into warming the ocean, not the atmosphere. So taking the ocean's temperature is the most comprehensive way to monitor global warming.

In 2012, a group of National Oceanic and Atmospheric Administration (NOAA) scientists published a revised and updated version of their decade-old compilation

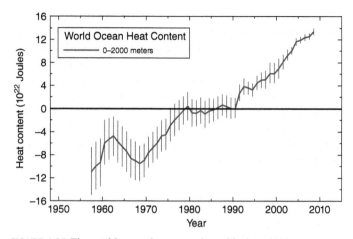

FIGURE 4.24 The world ocean has warmed steadily since 1990.

Source: Figure from "ScienceShot: No Letup in World's Warming," *Science*, http://www.sciencemag.org/news/2012/04/scienceshot-no-letup-worlds-warming

of temperature measurements from the upper 2,000 m (6,560 ft) of the world's oceans.[121] Ocean warmth (**Figure 4.24**) steadily increased in the decades since 1990, and the upper ocean warmed so much in the past 50 years that its additional heat, if released, would be enough to warm the lower atmosphere by about 36 degrees Celsius (65 degrees Fahrenheit).

Although it is no surprise that the rate of atmospheric warming declined somewhat after the super El Niño of 1998, scientists have nonetheless struggled to learn why. Several attempts to answer the question have been put forward: reflection of sunlight by aerosols in the stratosphere; processes related to the storage of heat by ocean circulation; natural variability; differences between observational datasets; and incomplete observations have all been forwarded.

In 2010, researchers[122] determined that the concentration of aerosols in the stratosphere over the previous decade was somewhat higher than assumed, and they calculated that the reflection of sunlight by particles might have had a cooling effect about 20 percent more than would be expected without them. *Aerosol particles* are produced by burning coal, wood, and animal dung. There is an especially high production of aerosols by power plants in Asia burning sulfur-rich coal whose influence had not previously been recognized. Aerosols are also produced by volcanic eruptions, and relatively small volcanic eruptions such as the 2006 eruptions of Soufriere Hills in Montserrat and Tavurvur in Papua New Guinea may have contributed more aerosols than previously realized.

[115] Berkeley Earth: http://berkeleyearth.org

[116] Cowtan, K. and Way, R.G. (2014) Coverage bias in the HadCRUT4 temperature series and its impact on recent temperature trends, *Quarterly Journal of the Royal Meteorological Society*, 140, 1935–1944.

[117] Cowtan and Way (2014).

[118] The Associated Press article describing the statistical testing contains this quote and can be found here: http://news.yahoo.com/s/ap/20091026/ap_on_bi_ge/us_sci_global_cooling

[119] Easterling, D. and Wehner, M. (2009) Is the climate warming or cooling? *Geophysical Research Letters*, 36, L08706

[120] NASA: http://climate.nasa.gov/news/index.cfm?FuseAction=ShowNews&NewsID=175

[121] Levitus, S., et al. (2012) World ocean heat content and thermosteric sea level change (0–2000), 1955–2010, *Geophysical Research Letters*, 39, L10603, doi: 10.1029/2012GL051106

[122] Solomon, S., et al. (2011) The persistently variable "background" stratospheric aerosol layer and global climate change, *Science*, 333(6044), 866–870, doi: 10.1126/science.1206027

Another team of researchers[123] looked at heat storage in the uppermost ocean (0 to 700 m, 0 to 2,300 ft) during the period 2003 to 2010 and found it had not gained any heat. By using an ensemble of computer climate models to trace heat budget variations, they learned that an 8-year period without upper ocean warming is not unusual. They probed the history of El Niño, which releases heat to the atmosphere, and the thermohaline circulation, which buries heat deep in the oceans. Models suggested that both processes combined starting in 2003, preventing excess energy from accumulating in the shallow ocean as usual. Approximately 45 percent of the missing heat was instead released to space, and 35 percent was stored below 700 m depth in the North Atlantic Ocean. How long would this pattern continue? The researchers point to recently observed changes in these two modes of climate variability and predicted an upward trend in upper ocean heat content.

Eventually they were proved correct as 2014 to 2016 set records as the warmest years in 136 years of record keeping. It was clear that whatever the hiatus had been, it had come to an end. A new round of analyses by scientists began that focused on observations of global mean surface temperature and how they were interpreted.

Scientists at NOAA published a study[124] that found the rate of global warming over the period 2000 to 2015 was as fast as or faster than that seen during the latter half of the 20th century. The study refuted the very existence of a hiatus and challenged the notion that there was a slowdown in the rate of global warming. In addition to being able to compute new trends in global mean surface temperature using additional years, the scientists applied corrections to both sea surface temperature and land surface air temperature datasets.

One of the most substantial improvements was a correction that accounts for the difference in data collected from buoys and ship-based data. Prior to the mid-1970s, ships were the predominant way to measure sea surface temperatures, and since then buoys have been used in increasing numbers. Compared to ships, buoys provide measurements of significantly greater accuracy and the researchers showed that data collected from buoys are biased cooler than ship-based data. The study authors developed a method to correct the difference between ship and buoy measurements and used this in their analysis.

Despite the improvements to the surface temperature datasets, global warming deniers attacked this work and a House of Representatives committee subpoenaed the scientists' emails.[125] NOAA agreed to provide data and respond to any scientific questions but refused to comply with the subpoena, a decision supported by scientists who feared the "chilling effect" of political inquisitions.

By January 2017, an independent analysis was published[126] that validated their work. The new study, which used independent data from satellites and robotic floats as well as buoys, concluded that the NOAA results were correct (**Figure 4.25**).

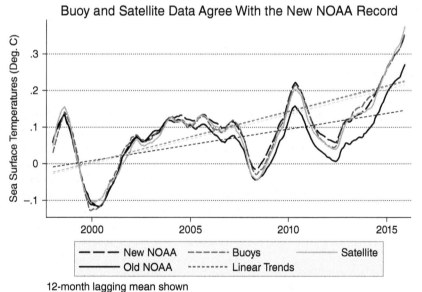

FIGURE 4.25 Analysis of ocean buoy and satellite data show that ocean temperatures increased steadily since 1999, as NOAA concluded in 2015 after adjusting for a cold bias in buoy temperature measurements. NOAA's earlier assessment had underestimated sea surface temperature changes, falsely suggesting a hiatus in global warming. Dashed lines show the general upward trend in ocean temperatures verifying that no hiatus in warming has been in existence over the period.[127]

Source: http://news.berkeley.edu/2017/01/04/global-warming-hiatus-disproved-again

Source: http://berkeleyearth.org

[123] Katsman, C.A. and van Oldenborgh, G.J. (2011) Tracing the upper ocean's "missing heat", *Geophysical Research Letters*, 38, L14610, doi: 10.1029/2011GL048417

[124] Karl, T.R., et al. (2015) Possible artifacts of data biases in the recent global surface warming hiatus, *Science*, doi: 10.1126/science.aaa5632

[125] New York Times: http://www.newyorker.com/news/news-desk/the-house-science-committees-anti-science-rampage

[126] Hausfather, Z., et al. (2017) Assessing recent warming using instrumentally homogeneous sea surface temperature records, *Science Advances*, 3(1), e1601207, doi: 10.1126/sciadv.1601207

[127] Figure 4.25 from Hausfather, Z., et al. (2017), http://advances.sciencemag.org/content/3/1/e1601207

4.7 Do Scientists Disagree on Global Warming?

According to the vast majority of climate scientists,[128] the planet is heating up. In a survey[129] of 3,146 Earth scientists, among the most highly qualified climatologists (those who wrote more than 50 percent of their peer-reviewed publications in the past 5 years on the subject of climate change) more than 95 percent agreed "human activity is a significant contributing factor in changing mean global temperatures." The survey found that as the level of active research and peer-reviewed publication in climate science increases, so does agreement that humans are significantly changing global temperatures (**Figure 4.26**). The divide between expert climate scientists (97.4 percent) and the general public (58 percent) is particularly striking.[130]

Studies into scientific agreement on human-caused global warming

(a)

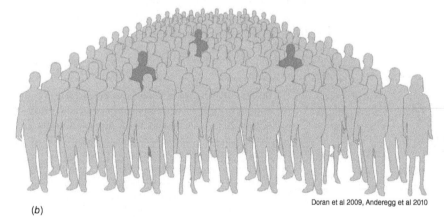

(b)

Doran et al 2009, Anderegg et al 2010

FIGURE 4.26 (a) This graphic summarizes studies of the scientific consensus on human-caused global warming. These studies examine either the expert opinion of climate scientists who have published peer-reviewed climate research, or the content of peer-reviewed climate papers.[131] (b) Response of scientists who publish in Earth science to the question "Do you think human activity is a significant contributing factor in changing mean global temperatures?"

Source: Figure from SkepticalScience.com

[128] Anderegg, W.R.L., et al. (2010) Expert credibility in climate change, *PNAS*, 107(27), 12107–12109, doi: 10.1073/pnas.1003187107

[129] Doran, P.T. and Zimmerman, M.K. (2009) Examining the scientific consensus on climate change, *Eos* 90(3), http://tigger.uic.edu/~pdoran/012009_Doran_final.pdf

[130] The Yale Program on Climate Change Communication conducts regular surveys on American attitudes toward climate change http://climate-communication.yale.edu

[131] Citations for these studies: Oreskes, N. (2004) Beyond the Ivory Tower: The scientific consensus on climate change, *Science*, 306(5702), 1686, http://www.sciencemag.org/cgi/content/full/306/5702/1686#. Doran and Zimmerman (2009). Anderegg, et al. (2010). Cook, J., et al. (2013) Quantifying the consensus on anthropogenic global warming in the scientific literature, *Environmental Research Letters*, 8(2), 031003. Verheggen, B., et al. (2014) Scientists views about attribution of global warming, *Environmental Science and Technology*, 48(16), 8963–8971. Stenhouse, N., et al. (2014) Meteorologists views about global warming: A survey of American Meteorological Society Professional Members, *Bulletin of the American Meteorological Society*: 1029–1040, http://journals.ametsoc.org/doi/abs/10.1175/BAMS-D-13-00091.1. Carlton, J.S., et al. (2015) The climate change consensus extends beyond climate scientists, *Environmental Research Letters*, 10.9, http://iopscience.iop.org/article/10.1088/1748-9326/10/9/094025/meta

An earlier study, published in 2004,[132] came to a similar conclusion: There is strong scientific consensus on global warming, and there is agreement that humans are the primary cause. The study analyzed all peer-reviewed scientific papers between 1993 and 2003 using the keyword phrase "climate change." Here is what its authors concluded:

> The 928 papers were divided into six categories: explicit endorsement of the consensus position, evaluation of impacts, mitigation proposals, methods, paleoclimate analysis, and rejection of the consensus position. Of all the papers, 75% fell into the first three categories, either explicitly or implicitly accepting the consensus view, and 25% dealt with methods or paleoclimate, taking no position on current anthropogenic climate change. Remarkably, none of the papers disagreed with the consensus position.

It is strikingly obvious, from any rational point of view, that there is very strong scientific consensus that global warming is real and that humans are the primary cause. What is concerning, especially to scientists and policy experts, is the lack of strong consensus among the average American. In May 2017, the Yale Center for Climate Change Communication published[133] the results of a survey of American attitudes toward climate change.[134] They found the following:

- 13% understand that nearly all climate scientists (more than 90%) are convinced that human-caused global warming is happening.
- 70% think global warming is happening.
- 13% think global warming is not happening.
- 58% of Americans believe climate change is mostly human caused.
- 30% say it is due mostly to natural changes in the environment.
- 39% think the odds that global warming will cause humans to become extinct are 50% or higher.
- 58% think the odds of human extinction from global warming are less than 50%.
- 24% say providing a better life for their children and grandchildren is the most important reason to reduce global warming.
- 16% said preventing the destruction of most life on the planet is the most important reason to reduce global warming.
- 13% said protecting God's creation is the most important reason to reduce global warming.

It is easy to conclude that either the scientific community or the media that report scientific results are not doing a sufficient job of educating the American public on the reality and consequences of climate change.

4.8 Are Climate Data Faulty?

Climate measurements come from many sources. Satellites measure Earth's temperature from a distance, weather balloons measure vertical profiles of the atmosphere as they ascend, a fleet of over 4,000 ocean robots collect temperature data in every corner of the seven seas, and ground-based weather stations located at thousands of sites across the globe measure the temperature of the lower troposphere and the ground. The agencies that collect temperature data take pains to remove factors that might skew the data artificially, relative to the true temperature. One such factor is known as the *urban heat island effect*; basically, urbanized regions tend to be hotter than the adjacent countryside.

In assembling climate data, NASA compares long-term urban temperature trends to nearby rural trends. They then adjust the urban trend so it matches the rural trend. The methodology that NASA Goddard Institute for Space Studies uses in assembling its dataset is explained in detail on their website.[135] Contrary to popular belief, because most urban climate stations are located in parks and other nonindustrial areas, NASA found that 42 percent of city trends are *cooler* relative to their country surroundings.

This is consistent with a study[136] finding that no statistically significant impact of urbanization could be identified in annual temperatures. Researchers found that industrial sections of towns may well be significantly warmer than rural sites, but urban meteorological observations are more likely to be made within parks that are cool islands compared to industrial regions.

Another study[137] analyzed 50-year records of temperatures on calm nights and on windy nights. It concluded "temperatures over land have risen as much on windy nights as on calm nights, indicating that the observed overall warming is not a consequence of urban development." The reasoning for this conclusion is that windy nights will circulate air from cool surroundings into the warm city, and thus warming should be suppressed on those nights if it is due to the urban effect.

Claims that global warming is based on unreliable data have been rigorously tested[138] and are simply not true. In any case, satellite data and ocean measurements confirm

---

[132] Oreskes (2004)

[133] Leiserowitz, A., et al. (2017) *Climate Change in the American Mind: May 2017.* Yale University and George Mason University. Yale Program on Climate Change Communication, New Haven, CT.

[134] Climate change attitudes: http://climatecommunication.yale.edu/publications/climate-change-american-mind-may-2017

[135] Temperature methodology: http://pubs.giss.nasa.gov/docs/2001/2001_Hansen_etal.pdf. This methodology is periodically updated and discussed at this site http://data.giss.nasa.gov/gistemp/updates.

[136] Peterson, T.C. (2003) Assessment of urban versus rural in situ surface temperatures in the contiguous United States: No difference found, *Journal of Climate*, 16(18), 2941–2959.

[137] Parker, D. (2006) A demonstration that large-scale warming is not urban, *Journal of Climate*, 19(12), 2882–2895.

[138] Menne, M., et al. (2010) On the reliability of the U.S. surface temperature record, *Journal of Geophysical Research*, 115, D11108, doi: 10.1029/2009JD013094

that global warming is not an artifact of biased ground-based measurements.

4.9 Is Today's Warming Simply a Repeat of the Recent Past?

It has been claimed that two periods in recent geologic history were warmer than today, and, therefore, today's warming is no big deal and probably a natural event. Until recently, one of these periods, known as the Eemian Interglacial, was thought to have been warmer than present as a result of the orbital parameters at the time. However, more recent research shows that today's surface temperature at 1 degree Celsius above background is equivalent to surface temperature during the Eemian. What is important to note is that global mean sea level in Eemian time was over 6.6 meters (22 feet) higher than present day.

The other warm period, called the **Medieval Climate Anomaly** (MCA), might have been warmer at various times in various places within the North Atlantic region, but this does not mean it was a global phenomenon. The MCA was followed by the *Little Ice Age* (LIA), suggestive to some of a temperature process that is completing one full cycle, MCA-LIA-Modern warming, and thus is natural in origin. Let's look into the MCA first.

4.9.1 Medieval Climate Anomaly

It has been claimed that the MCA was a time of warmer temperatures prior to industrial greenhouse gas production.[139] However, the IPCC AR5 concludes "The timing and spatial structure of the MCA and the LIA are complex, with different reconstructions exhibiting warm and cold conditions at different times for different regions and seasons."[140] (**Figure 4.27**) Southern Hemisphere reconstructions of the MCA are sparse and characterized by uncertainties that are not clearly separable from climate variability. In the Northern Hemisphere, the median of temperature reconstructions indicates mostly warm conditions from about 950 CE[141] to about 1250CE and colder conditions from about 1450CE to about 1850CE. These time intervals are taken by IPCC AR5 to represent the MCA and LIA, respectively.

The MCA has not been well documented outside of the North Atlantic region, hence one cannot assume it was a global phenomenon. Additionally, even in the North Atlantic, modeling and geologic proxy data reviewed by the

IPCC indicate that temperatures were very unlikely to have been higher than present temperatures and that they rose and fell at different times in different places. This scientific evidence makes the event fundamentally different from today's global warming pattern, which is nearly everywhere synchronous and statistically significant.

Reconstructions of global temperature (using climate proxies) during the past 1,300 years reveal that global temperatures today are the highest in the past 1,300 years. Scientists have shown that local temperatures from one place to another may have been as warm as or even warmer than today, but these do not represent a global trend, only a local pattern.

An important study[142] was published in 2006 on the MCA and it concludes as follows:

- Dramatic global warming has occurred since the 19th century.
- The record warm temperatures in the last 15 years are indeed the warmest temperatures in at least the last 1,000 years and possibly in the last 2,000 years.
- Comparisons between modern global warming and the MCA can only be made at the local and regional scale.

Researchers[143] examined tree ring data recording land temperatures and found clear MCA (warm), LIA (cool), and recent (warm) episodes preserved in North American and Eurasian tree ring records. They conclude that MCA temperatures were nearly 0.7 degrees Celsius (1.26 degrees Fahrenheit) cooler than in the late 20th century, with an amplitude difference of 1.14 degrees Celsius (2.05 degrees Fahrenheit) from the coldest (1600 to 1609) to warmest (1937 to 1946) decades. The study also stresses that "presently available paleoclimate reconstructions are inadequate for making specific inferences, at hemispheric scales, about MCA warmth relative to the present anthropogenic period and that such comparisons can only still be made at the local/regional scale." This means that it is wrong to use the MCA in a discussion of global warming, because the MCA was not likely a global event.

The Little Ice Age was a period of cooling that occurred after the MCA. The event has been depicted[144] as having three maxima beginning about 1650, 1770, and 1850, each separated by slight warming intervals. Scientific consensus[145] is that this cooling occurred (like the MCA) with

---

[139] VIDEO: The Medieval Warming Crock. http://www.youtube.com/watch?v=vrKfz8NjEzU

[140] Masson-Delmotte, V. et al. (2013)

[141] CE stands for "common era." It refers to years since the year 0. For instance, as I write this it is the year 2017CE. Years before the year 0 are refered to as BCE or "before the common era."

[142] D'Arrigo, R., et al. (2006) On the long-term context for late twentieth century warming, *Journal of Geophysical Research–Atmospheres*, 111(D3), D03103, doi: 10.1029/2005JD006352; NOAA: http://www.ncdc.noaa.gov/paleo/pubs/darrigo2006/darrigo2006.html

[143] D'Arrigo et al. (2006)

[144] NASA Glossary: http://earthobservatory.nasa.gov/Glossary/?mode=alpha&seg=l&segend=n

[145] IPCC (2007) *Climate Change 2007: The Physical Science Basis.* Contribution of Working Group I to the Fourth Assessment Report of the Intergovernmental Panel on Climate Change [Solomon, S., et al. (2007) (eds.)]. Cambridge University Press, Cambridge, United Kingdom and New York, NY, p. 996.

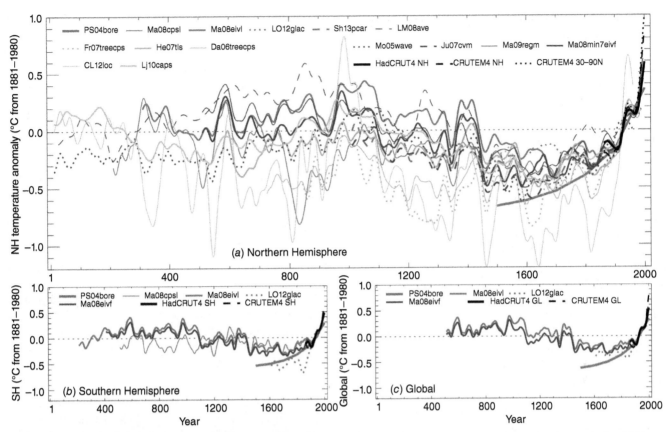

FIGURE 4.27 Reconstructed (a) Northern Hemisphere and (b) Southern Hemisphere, and (c) global annual temperatures during the last 2,000 years. Individual published reconstructions are shown in the legend grouped according to their spatial representation and instrumental temperatures shown in black (Hadley Center Climatic Research Unit (CRU)).[146] All series represent anomalies (degrees Celsius) from the 1881–1980 mean (horizontal dashed line) and have been smoothed with a filter that reduces variations on timescales less than about 50 years. The IPCC AR5 concludes that there is high confidence that the mean Northern Hemisphere temperature of the last 30 or 50 years very likely exceeded any previous 30- to 50-year mean during the past 800 years. Even accounting for uncertainties, almost all reconstructions agree that each 30-year (50-year) period from 1200 to 1899 was very likely colder in the Northern Hemisphere than the 1983–2012 (1962–2012) instrumental temperature.[147]

Source: https://www.ipcc.ch/pdf/assessment-report/ar5/wg1/WG1AR5_Chapter05_FINAL.pdf

varying degrees of intensity, at different times, in different places. Paleoclimatologists no longer expect to agree on either the start or end dates of this event, which varied according to local conditions.[148] Another researcher[149] studying the MCA established, "The Medieval period is found to display warmth that matches or exceeds that of the past decade in some regions, but which falls well below recent levels globally."

4.9.2 Eemian Interglacial

One hundred and twenty-five thousand years ago, the Eemian Interglacial was a natural episode of global warming that has been intensely studied[150] to improve understanding of what we might expect as the modern climate system comes into equilibrium with the amount of excess warming that anthropogenic greenhouse gases have produced. A number of studies[151] assert that during Eemian time, global sea level peaked 5.5 to 9 meters (18 to 30 feet) above present as a result of global temperatures that were

[146] IPCC AR5, Chapter 5, Appendix 5.A.1 for further information about each one (p. 456): https://www.ipcc.ch/pdf/assessment-report/ar5/wg1/WG1AR5_Chapter05_FINAL.pdf
[147] Source: Figure 5.7 from Masson-Delmotte, V. et al. (2013)
[148] Little Ice Age, *Wikipedia*, http://en.wikipedia.org/wiki/Little_Ice_Age
[149] Mann, M.E., et al. (2009) Global signatures and dynamical origins of the little ice age and medieval climate anomaly, *Science*, 326(5957), 1256–1260, doi: 10.1126/science.1177303

[150] Muhs, D.R., et al. (2002) Timing and warmth of the last interglacial period: New U-series evidence from Hawaii and Bermuda and a new fossil compilation for North America, *Quaternary Science Reviews*, 21, 1355–1383.
[151] Dutton, A. and Lambeck, K. (2012) Ice volume and sea level during the last interglacial, *Science*, 216–219. Hoffman et al. (2017) Regional and global sea surface temperatures during the last interglaciation, *Science*, 355, 276–279.

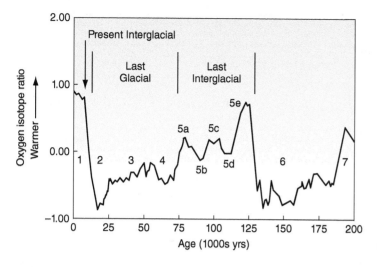

FIGURE 4.28 The last interglacial consisted of five stadials and interstadials, named MIS5a–e. The last glacial consisted of two stadials, MIS4 and MIS2, as well as one interstadial, MIS3. The present interglacial is MIS1, also known as the Holocene Epoch. The acronym MIS stands for Marine Isotopic Stage, because these detailed records were first identified using oxygen isotopes in seafloor sediments.

1 degree Celsius (1.8 degree Fahrenheit) above preindustrial levels. In other words, the same temperature produced by present-day global warming.

The last interglacial, broadly defined, occurred between approximately 130,000 and 75,000 years ago. Climate during this 55,000-year period was not continuously warm. Rather, researchers have identified five major phases consisting of three interstadials (warmings) and two stadials (coolings). These show up clearly in the ice-core records as well as the deep-sea record. **Figure 4.28** shows these phases, using the scientific naming system that has been adopted. The last interglacial is named after the oxygen isotope proxy that was used to first identify it in cores of marine sediments: Marine Isotopic Stage 5 (MIS5), and the stadials and interstadials are labeled MIS5a–e. Of these, MIS5e (the Eemian) was the warmest, and most like the modern Holocene Epoch.

MIS5e offers a geologically recent example of a warm period with characteristics similar to those of the Holocene; however, it differs from the Holocene in that the warmth at the time was driven by orbital parameters that were very different from the Holocene Epoch. Because it is also a relatively recent event, many geologic proxies that record climate conditions from that time have not been lost to erosion.

MIS5e lasted approximately 12,000 years, from 130,000 to 118,000 years ago, and the average age of fossil corals from around the world that grew at that time is 125,000 years. For example, **Figure 4.29** shows a fossil reef on the Hawaiian island of Oahu that illustrates another important feature of MIS5e: Sea level was higher than present, estimated by different researchers to have been from 6.6 to 8 meters[152] (22 to 26.5 feet) and 5.5 to 9 meters[153] (18 to 30 feet). Researchers have therefore concluded that because climate was warmer, melted ice contributed to the

FIGURE 4.29 This rocky shoreline in Hawaii is composed of limestone formed by a fossil reef that grew under higher-than-present sea levels during MIS5e.[154]

Courtesy of Chip Fletcher

higher sea level. Deep cores of ice in Greenland and Antarctica that preserve records from MIS5e show that it was a time of lower ice volume as a result of melting significant parts of both ice sheets.

The Eemian has been cited as a possible analogue for present climate[155] as a result of global warming. Studies have shown that CO_2 concentrations in the atmosphere were relatively high[156] (though not as high as they are today owing to the contribution of industrial greenhouse gases), temperatures were approximately the same as present, and sea level was significantly higher. Scientists study MIS5e for several reasons: 1) to improve understanding of the duration of the last interglacial period and global characteristics

[152] Kopp, R.E., et al. (2009) Probabilistic assessment of sea level during the last interglacial stage, *Nature*, 462, 863–868, doi: 10.1038/nature08686
[153] Dutton and Lambeck (2012).

[154] Fletcher, C.H., et al. (2008) Geology of Hawaii reefs. In: *Coral Reefs of the U.S.A.* (New York: Springer), pp. 435–488.
[155] Hoffman et al. (2017)
[156] Müller, U.C. (2005) Cyclic climate fluctuations during the last interglacial in Central Europe, *Geology*, 33(6), 449–452.

at the time, 2) to improve understanding of how sea level may rise as a result of modern global warming, and 3) as a basis for testing and advancing computer models that can be used to predict modern climate and how it may evolve over this century.

Computer models of climate change during MIS5e indicate that sea-level rise started with melting of the Greenland ice sheet and not the Antarctic ice sheet.[157] Research also suggests that ice sheets across both the Arctic and Antarctic could melt more quickly than expected this century because temperatures are likely to rise substantially higher than they did during MIS5e, especially in the polar regions.

If these predictions are correct, by 2100 the Arctic could warm by 3 to 5 degrees Celsius (5.4 to 9 degrees Fahrenheit) in summer. During MIS5e, researchers have concluded there is a 95 percent probability that meltwater from Greenland and other Arctic sources raised sea level by as much as 4 meters (13 feet). However, because global sea level actually rose significantly higher, researchers have concluded that Antarctic melting and thermal expansion of warm seawater must have produced the remainder of the rise in sea level.

The rise in sea levels produced by Arctic warming and melting could have floated, and thus destabilized, ice shelves at the edge of the Antarctic ice sheet. This would have led to their collapse, a positive feedback to sea-level rise. If such a process occurred today, global warming year-round would accelerate it. Over the past two decades, sea-level rise has accelerated,[158] and now it is rising at a rate of more than 3 centimeters per decade (12 inches per century), which is three times faster than it was rising over the 20th century. Studies have also found accelerated rates[159] of annual melt of both the Greenland and Antarctica ice sheets.[160]

During MIS5e, the amount of warming needed to initiate this melting was equivalent to modern day global warming, 1 degree Celsius (1.8 degrees Fahrenheit) above preindustrial levels.[161] The amount of Greenland ice sheet melting that produced higher sea levels is shown in **Figure 4.30**. This reconstruction predicts that sea level rose at a rate exceeding 1.6 meters (5.3 feet) per century, a rate that would be

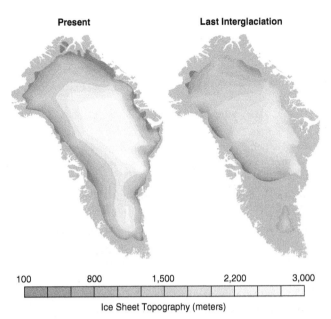

Present　　　　**Last Interglaciation**

100　　　800　　　1,500　　　2,200　　　3,000

Ice Sheet Topography (meters)

FIGURE 4.30 Computer models that simulate climate during MIS5e indicate that melting of the Greenland ice sheet was responsible for a global sea-level rise of approximately 4 m (13 ft). Additional melting came from Antarctica.

Source: Figure from UCAR, "Arctic, Antarctic Melting May Raise Sea Levels Faster than Expected," http://www.ucar.edu/news/releases/2006/melting.shtml; Figure 15.39 from C. Fletcher *Physical Geology, 3rd ed.*

potentially catastrophic for coastal communities worldwide if it were to happen today. Climate modeling[162] predicts several other features of the Eemian that are relevant to understanding our current situation: Global carbon dioxide rose by 1 percent per year (half the current rate of rise); 2100 will be significantly warmer than MIS5e, so Greenland is already headed toward a state similar to that depicted in Figure 4.30; and the West Antarctic ice sheet will also contribute significantly to global sea-level rise by 2100.

The end of the Eemian was characterized by precipitous changes in global climate. Although the period has been studied intensively, global climate during MIS5a–d is poorly understood. Researchers speculate that temperatures during MIS5d and 5b were significantly cooler than present temperatures, that global ice volume expanded, and that global sea level dropped perhaps by as much 25 meters (82 feet) below the present level. These were, in effect, mini ice ages that lasted a few thousand to 10,000 years each. The interstadials MIS5a and 5b were likely periods during which global temperature was cooler and ice volume greater than they are today.

4.10 In Conclusion

The origin of global warming has been vigorously tested for decades. There are still many details to work out, and there are occasional surprises in ongoing work, but the

[157] Overpeck, J.T., et al. (2006) Paleoclimatic evidence for future ice-sheet instability and rapid sea-level rise, *Science*, 311, 1747–1750.

[158] Dangendorf, S., et al. (2017) Reassessment of 20th century global mean sea level rise, *Proceedings of the National Academy of Sciences*, doi: 10.1073/pnas.1616007114

[159] Joughlin, I., et al., (2014) Marine ice sheet collapse potentially underway for the Thwaites Glacier Basin, West Antarctica, *Science*. See also Rignot, E., et al., (2014) Widespread, rapid grounding line retreat of Pine Island, Thwaites, Smith and Kohler Glaciers, West Antarctica from 1992 to 2011, *Geophysical Research Letters*.

[160] Velicogna, I. (2009) Increasing rates of ice mass loss from the Greenland and Antarctic ice sheets revealed by GRACE, *Geophysical Research Letters*, 36, L19503, doi: 10.1029/2009GL040222. See also Rignot, E., et al. (2011) Acceleration of the contribution of the Greenland and Antarctic ice sheets to sea level rise, *Geophysical Research Letters*, 38, LO5503, doi: 10.1029/2011GL046583

[161] Hoffman et al. (2017)

[162] Overpeck et al. (2006)

hypothesis that human production of heat-trapping gas has led to global warming is widely accepted, no natural process has been identified to account for it, and it represents a consensus opinion of the scientific community.

Animations and Videos

1. Carbon cycle. https://www.khanacademy.org/science/biology/ecology/biogeochemical-cycles/v/carbon-cycle

2. Biogeochemical cycles. https://www.khanacademy.org/science/biology/ecology/biogeochemical-cycles/v/biogeochemical-cycles

3. Tectonic plates. https://www.youtube.com/watch?v=7nxITuot-ko

4. Renewable energy. http://video.renewableenergyworld.com/#category/videos/solar/1

5. Climate change 101. https://www.youtube.com/watch?v=EtW2rrLHs08

6. Why reducing our carbon emissions matters. https://www.youtube.com/watch?v=rivf479bW8Q

7. Sir David Attenborough. http://www.youtube.com/watch?v=S9ob9WdbXx0&feature=related

8. ICE STORIES: Working to Reconstruct Past Climate. http://www.youtube.com/watch?v=r81MtDimgSg

9. Richard Alley's Global Warming. http://www.youtube.com/watch?v=T4GThA35s1s&feature=relmfu

10. The Thermohaline Circulation - The Great Ocean Conveyor Belt. https://pmm.nasa.gov/education/videos/thermohaline-circulation-great-ocean-conveyor-belt

11. AMOC Shutdown Potential and Implications. https://www.youtube.com/watch?v=IQ8j6jFKSLQ

12. Reshuffling Heat on a Warming Planet. https://science.nasa.gov/sciencecasts/reshuffling-heat-warming-planet

13. The Medieval Warming Crock. http://www.youtube.com/watch?v=vrKfz8NjEzU

Comprehension Questions

1. Why is it important to understand how carbon moves through the natural world?

2. Explain what a carbon "process" is.

3. What is a climate proxy? Identify two climate proxies and describe how they work.

4. Explain how the orbital parameters influence climate.

5. Why do ice cores and ocean cores tell the same story about paleoclimate?

6. What was the role of carbon dioxide in Earth's climate system at the end of the last ice age?

7. How do climate feedbacks work? Describe one positive and one negative feedback.

8. Is global warming caused by the Sun?

9. Describe the MCA.

10. Describe natural climate cycles. Are these responsible for modern global warming?

11. Did global warming end after 1998? Why or why not?

12. What is the Eemian? Why do researchers study the Eemian?

Thinking Critically

1. How are the geologic and biologic carbon cycles related and how are they different?

2. What climate processes are recorded in ice cores and deep-sea cores and why are they related?

3. You are asked to appear before a congressional hearing into climate change. Explain how paleoclimate improves our understanding of certain aspects of the global warming issue.

4. What is a Dansgaard–Oeschger event and what role does it play in the discussion of modern climate change?

5. The bipolar seesaw has been used to explain why Dansgaard–Oeschger events are not responsible for global warming. Elaborate.

6. How do oxygen isotopes reveal paleoclimate patterns?

7. How do we know that the orbital parameters are not responsible for modern global warming?

8. Explain how climate feedbacks play a critical role in understanding the origin of modern climate change.

9. The MCA has been used to explain global warming as a natural event. What is the logic behind this and why is it wrong?

10. Describe why attributing global warming to natural climate cycles is not supported by the evidence.

11. What aspects of the Eemian make it useful for understanding the impacts of global warming?

Activities

1. Visit the "Powers of 10" website http://www.ncdc.noaa.gov/paleo/ctl/index.html and answer the following questions.

   (a) How is climate related to the water cycle?
   (b) Describe the timescales of climate change.
   (c) What is a climate proxy and what do different proxies tell us about climate variability?
   (d) Compare and contrast climate variability on the timescale of 100 years versus 10,000 years.

2. Watch the video "Climate Denial Crock of the Week: That 1500 Year Thing" http://www.youtube.com/watch?v=G0HGFSUx2a8&feature=view_all&list=PL029130BFDC78FA33&index=55 and answer the following questions.

   (a) How are paleoclimate data used to address the question of natural cycles as a cause of global warming?
   (b) How does heat play a role in this issue?

   (c) Describe the methods used by climate deniers as outlined in this video and the potential impacts.
   (d) Describe the difference between 1,500-year climate cycles and modern global warming.

3. Watch the video "Climate Denial Crock of the Week: The Urban Heat Island" http://www.youtube.com/watch?v=B7OdCOsMgCw&feature=view_all&list=PL029130BFDC78FA33&index=54 and answer the following questions.

   (a) Describe the urban heat island effect.
   (b) Explain why the urban heat island effect is not a real source of error in global warming data. Use information from the video as well as from this chapter.
   (c) What are some of the impacts of global warming on the natural world?
   (d) Describe the methods used by "Climate Denial Crock of the Week" to improve understanding of climate change issues.

Key Terms

AMOC	Dansgaard–Oeschger	Mid-Pliocene Warm Period	Rock cycle
Anthropogenic hypothesis	Eccentricity	Milankovitch, Milutin	Rocks
Biogenic ooze	Eemian	(1879–1958)	Silica rocks
Bipolar seesaw	Glacials	Obliquity	Soil respiration
Calcification	Global biogeochemical cycling	Orbital parameters	Spreading center
Carbonate	Global warming hiatus	Paleoclimatology	Stadials
Carbonic acid	Hydrolysis	Photosynthesis	Subduction zone
Clay	Insolation	Pleistocene Epoch	Uplift Weathering Hypothesis
Climate feedbacks	Interstadials	Precession	Weathering
Climate proxies	Limestone	Processes	Younger Dryas
Conveyor belt hypothesis	Lithification	Reservoirs	
Crystallization	Medieval Climate Anomaly	Respiration	

CHAPTER 5

Sea-Level Rise

LEARNING OBJECTIVE

Today, global mean sea level is rising at an accelerating rate.[1] It is rising three times faster than in the 20th Century and, depending on the nature of future greenhouse gas emissions, will continue rising in the decades and centuries ahead. Global warming causes ice to melt and ocean water to warm and expand; these two processes are the main causes of global sea-level rise.[2] There is broad scientific consensus that accelerating sea-level rise will increase the damage caused by hurricanes, tsunami, high waves, and other coastal hazards.

National Oceanic and Atmospheric Administration (NOAA)

Devastation of Bolivar Peninsula, Texas, following Hurricane Ike. As sea level continues to rise because of global warming, the damage resulting from coastal hazards such as hurricanes, tsunamis, high waves, and extreme tides will increase.

Source: NOAA; http://www.noaa.gov/features/protecting_1208/coastalmanagement.html.

5.1 Chapter Summary

Today, rising seas threaten human communities and natural ecosystems in the coastal zone.[2] As a result, neighborhoods, commercial districts, and other types of development on the coast are subject to a broad range of flood types. These include flooding by rainstorms that are coincident with high tides, accelerated coastal erosion, saltwater intrusion into streams and aquifers, marine flooding by extreme tides, saltwater flooding of storm drains and other types of buried public infrastructure, and inundation due to rising groundwater that forms new wetlands. Sea-level rise

threatens cities, ports, and coastal environments with passive inundation due to rising waters and with damaging flooding that will increase in magnitude when hurricanes, tsunamis, and seasonal high waves strike. Because sea-level rise has enormous economic and environmental consequences, it is important to understand how global warming is creating this threat and how bad it may become later in the century and beyond.

In this chapter, you will learn that:

1. Global mean sea level has been rising for over a century, with thermal expansion of seawater and melting of glaciers principally contributing to the rate of rise.
2. The rate of global mean sea level rise is accelerating.
3. Tide gauges in the world's ports measure local sea-level change, and satellite missions that use radar altimetry are mapping the level of the world's oceans. Together these are used to understand changes in global and local sea level.

---

[1] Nerem, R.S., et al. (2018) Climate-change–driven accelerated sea-level rise detected in the altimeter era. *PNAS*, DOI: 10.1073/pnas.1717312115

[2] Causes of Sea Level Rise, Union of Concerned Scientists https://www.ucsusa.org/sites/default/files/legacy/assets/documents/global_warming/Causes-of-Sea-Level-Rise.pdf

4. According to satellite altimeter measurements, global mean sea level has risen over 8 cm (3 in) since 1993.[3]

5. The mean rate of global sea level rise over the period 1993–2017 is 3.2 mm/yr.[4] At the present rate of acceleration, the world's oceans could rise by about 75 cm (2.5 ft) over the next century.

6. It is widely believed by climate scientists that global sea level has the potential to increase its rate of acceleration and reach or exceed 3.3 ft (1 m) by the end of the century.

7. Global mean sea level will continue rising for many centuries even if greenhouse gas emissions cease rising and some excess CO_2 is removed from the atmosphere.

8. The last time it was this warm 1°C (1.8°F) was during the Eemian (ca. 125,0000 years ago) when global mean sea level was approximately 6.6 m (20 ft) above present.

9. Records of past sea levels indicate that the world is committed to a sea-level rise of approximately 2.3 m/°C (7.5 ft/°F) of warming over the next 2,000 years.

Sea level is rising today[5] and will continue to rise in the centuries ahead.[6] Greenhouse gas–induced global warming melts glaciers and warms the ocean that causes ocean water to expand. Melting ice and expanding ocean water are the main causes of global sea-level rise. Inertia in the climate and global carbon systems causes the global mean temperature to decline slowly even after greenhouse gas emissions have ceased. As a result, climate models[7] predict that, because of thermal expansion of deep ocean water, sea-level rise will persist for many centuries, even if greenhouse gas emissions cease rising and some excess CO_2 is removed from the atmosphere.[8]

IPCC-AR5 emission scenario modeling indicates that mean sea level may rise from 0.53 to 0.98 meters (1.74 to 3.21 feet) under RCP8.5 and from 0.28 to 0.61 meters (0.92 to 2.00 feet) under RCP2.6 by 2100.[9] However, a review of research since the publication of IPCC-AR5 reveals that the Greenland and Antarctic Ice sheets are decaying faster than expected. Under "high" and "extreme" scenarios proposed by NOAA,[10] it is physically plausible that global mean sea level could rise from 2.0 to 2.5 meters (6.56 to 8.20 feet) by the end of this century. Under business as usual greenhouse gas emission scenarios, there is a 17 percent probability of exceeding 0.34 meters (1.1 feet) by the year 2050.[11]

Global warming has already reached at least 1 degree Celsius (1.8 degrees Fahrenheit) above the preindustrial temperature of the nineteenth century. The last time it was this warm was during the Eemian interglacial (ca. 125,0000 years ago)[12] when global mean sea level was approximately 6.6 meters (20 feet) above present.[13] Records of past sea levels indicate that the world is committed to a sea-level rise of approximately 2.3 meters per degree Celsius (7.5 feet per 1.8 degree Fahrenheit) of warming over the next 2,000 years.[14]

Global warming is continuing and is unlikely to stop before reaching 3 to 4 degrees Celsius (5.4 to 7.2 degrees Fahrenheit).[15] We discuss this conclusion at length in **Chapter 8**. Simulations show that under these conditions, melting of the Antarctic ice sheet could raise global sea level 3 meters (10 feet) by the year 2300 and continue for thousands of years thereafter.[16] Research indicates that on multiple occasions over the past three million years, when global temperatures increased 1 to 2 degrees Celsius (1.8 to 3.6 degrees Fahrenheit), melting polar ice sheets caused global sea levels to rise at least 6 meters (20 feet) above present levels.[17]

Sea level rise is not uniform across the oceans. Vertical movement of the sea surface is highly variable, and although the global mean rate of change is positive, there are many locations where local sea-level change does not correlate to the global mean.[18]

[3] Chen, X., et al. (2017) The increasing rate of global mean sea level rise during 1993-2014. *Nature Climate Change*, v. 7, p. 492–495, doi:10.1038/nclimate3325

[4] Tollefson, J. (2017) Satellite Error Hid Rising Seas, 20 July, v. 547, *Nature*, pp. 265–266.

[5] Dangendorf, S., et al. (2017) Reassessment of 20th Century global mean sea level rise, *PNAS*, doi: 10.1073/pnas.1616007114

[6] Levermann, A., et al. (2013) The multimillenial sea-level commitment of global warming, *PNAS*, August 20, vol. 110, no. 34, p. 13745–13750.

[7] Meehl, G.A., et al. (2012) Relative outcomes of climate change mitigation related to global temperature versus sea-level rise, *Nature Climate Change*, doi:10.1038/nclimate1529.

[8] Zickfeld, K., et al. (2017) Centuries of thermal sea level rise due to anthropogenic emissions of short-lived greenhouse gases, *PNAS*, v. 114.4, p. 657–662, doi: 10.1073/pnas.1612066114

[9] Church J.A., et al. (2013) Sea Level Change. Climate Change 2013: The Physical Science Basis. Contribution of Working Group I to the Fifth Assessment Report of the Intergovernmental Panel on Climate Change, Chapter 13, eds Stocker TF, et al. (Cambridge Univ. Press).

[10] Sweet, W.V., et al. (2017a) Global and Regional Sea Level Rise Scenarios for the United States. NOAA Technical Report NOS CO-OPS 083, Silver Spring, 56p. plus Appendices

[11] Sweet et al. (2017a)

[12] Hoffman, J.S., et al. (2017) Regional and global sea surface temperatures during the last interglaciation, *Science*, 355(6322), 276–279, doi: 10.1126/science.aai8464

[13] Kopp, R.E., et al. (2009) Probabilistic assessment of sea level during the last interglacial stage, *Nature*, 462, 863–867, doi: 10.1038/nature08686.

[14] Levermann et al. (2013)

[15] Raftery, A.E., et al. (2017) Less than 2°C warming by 2100 unlikely, *Nature Climate Change*, 7, 637–641, DOI: 10.1038/nclimate3352.

[16] Golledge, N.R., et al. (2015) The multi-millennial Antarctic commitment to future sea-level rise, *Nature*, 2015; 526 (7573): 421 DOI: 10.1038/nature15706.

[17] Dutton, A., et al. (2015) Sea-level rise due to polar ice-sheet mass loss during past warm periods, *Science*, 10 Jul., v. 349, Is. 6244, DOI: 10.1126/science.aaa4019

[18] Hamlington, B. D., et al. (2016) An ongoing shift in Pacific Ocean sea level, *J. Geophys. Res. Oceans*, 121, 5084–5097, doi:10.1002/2016JC011815.

Globally, sea-level rise will increase coastal erosion, magnify inundation from storms and tsunamis, raise the groundwater table on low elevation lands and cause drainage problems, lead to saltwater intrusion among coastal ecosystems and aquifers, and threaten coastal economic activity among coastal communities.[19]

Today, rising seas threaten to forever change coastal wetlands, estuaries, reefs, islands, beaches, and all types of coastal environments. Coastal communities[20] are subject to flooding by rainstorms that are coincident with high tides, accelerated coastal erosion, and saltwater intrusion into streams and aquifers. Sea-level rise threatens cities, ports, coastal communities, and other areas with passive inundation due to rising waters, damaging storm surge associated with hurricanes, and destructive flooding by tsunamis. Because sea-level rise has enormous economic[21] and environmental consequences, it is important to understand how global warming is causing this threat.

5.2 Rate of Sea-Level Rise

By 2017, the rate of global mean sea-level rise was three times the rate of the 20th century,[22] and over the period 1993–2014, there was a 50 percent increase in the rate at which global mean sea level rose—from 2.2 millimeters per year to 3.3 millimeters per year.[23] The biggest increase in contributions to that sea-level rise came from the melting of the Greenland Ice Sheet, which rose from contributing 5 percent to global sea-level rise in 1993 to 25 percent in 2014.

Scientists use several types of instruments and geologic proxies to study the behavior of sea level. Geologic records of past sea-level positions include coastal sediments containing fossil organisms formerly inhabiting ecosystems that are tightly defined by a position between low and high tides. When these fossils are recovered in cores, say from the tidal wetlands lining an estuary, a geologist can interpret their age and depth as a past position of sea level.[24]

Physical oceanographers study sea-level change using tide gauges[25] in the calm waters of ports and harbors. Tide gauges measure water-level changes through time. Even

satellites are used to measure water-level position. **Satellite altimetry**[26] is a method of mapping the ocean surface with radar traveling at the speed of light. Information from tide gauges and that from satellite altimetry are used together to improve our understanding of sea-level change and to measure the long-term rate of sea-level rise that is occurring as a result of global warming.

5.2.1 How Can I Keep Track of the Changing Rate of Global Sea-Level Rise?

There are several scientific organizations that keep track of the global rate of sea-level rise. The University of Colorado Sea Level Research Group[27] provides analyses of satellite altimetry, an up-to-date list of the latest peer-reviewed research on sea-level rise, a sea-level blog, and a list of the rates of sea-level rise calculated by other institutions. In 2017, this group uncovered a mistake in the original calibration of a sensor on the first satellite altimetry mission, TOPEX-Poseidon.[28] Upon recalibrating the data, the estimate of the global rate of sea-level rise jumped from about 3.4 millimeters per year to 3.9 millimeters per year. If sea-level continues to accelerate at the newly calculated pace, global sea level will reach 75 centimeters (2.5 feet) above the present level by the end of the century.

The French Space Agency (CNES) maintains a site called AVISO (Archiving, Validation, and Interpretation of Satellite Oceanographic data),[29] which offers a number of satellite oceanography data products including updated calculations of global mean sea-level rise.

In Australia, the Commonwealth Scientific and Industrial Research Organization (CSIRO) maintains a "sea-level rise" research site[30] that offers discussion of historical and projected future sea-level changes, publications, and other relevant information. The NASA Jet Propulsion Laboratory (NASA-JPL) at the California Institute of Technology provides a "Physical Oceanography Distributed Active Archive Center" site that provides various products of satellite missions.[31]

The NOAA Center for Satellite Applications and Research provides plots of global mean sea-level rise at their Laboratory for Satellite Altimetry/Sea Level Rise.[32] All of these scientific organizations access and analyze the satellite altimetry data and apply slightly different methods of calculating the long-term rate of change (**Figure 5.1**).

[19] VIDEO: How much will sea level rise? http://vimeo.com/5188725

[20] Strauss, C., et al. (2012) Tidally Adjusted Estimates of Topographic Vulnerability to Sea-Level Rise and Flooding for the Contiguous United States, *Environmental Research Letters* 7, no. 1: 014033 doi: 10.1088/1748-9326/7/1/014033.

[21] National Oceanic and Atmospheric Administration (2010) *Adapting to Climate Change: A Planning Guide for State Coastal Managers* (Silver Spring, Md., NOAA Office of Ocean and Coastal Resource Management). http://coastalmanagement.noaa.gov/climate/adaptation.html

[22] Dangendorf et al. (2017)

[23] Chen, X., et al. (2017) The increasing rate of global mean sea-level rise during 1993–2014, *Nature Climate Change*. DOI: 10.1038/nclimate3325

[24] Kemp, A., et al. (2011) Climate Related Sea-Level Variations over the Past Two Millennia, *PNAS* 108, no. 27 (2011): 11017–11022, www.pnas.org/cgi/doi/10.1073/pnas.1015619108.

[25] National Ocean Service "What is a tide gauge" https://oceanservice.noaa.gov/facts/tide-gauge.html

[26] NASA Ocean Surface Topography https://podaac.jpl.nasa.gov/OceanSurfaceTopography

[27] Colorado Sea Level Group site: http://sealevel.colorado.edu

[28] Tollefson (2017)

[29] AVISO site: https://www.aviso.altimetry.fr/en/my-aviso.html

[30] CSIRO Sea Level Rise site: http://www.cmar.csiro.au/sealevel/sl_hist_last_decades.html

[31] Jet Propulsion Laboratory site: https://podaac-ftp.jpl.nasa.gov

[32] NOAA Laboratory for Satellite Altimetry/Sea Level Rise: https://www.star.nesdis.noaa.gov/sod/lsa/SeaLevelRise/LSA_SLR_timeseries_global.php

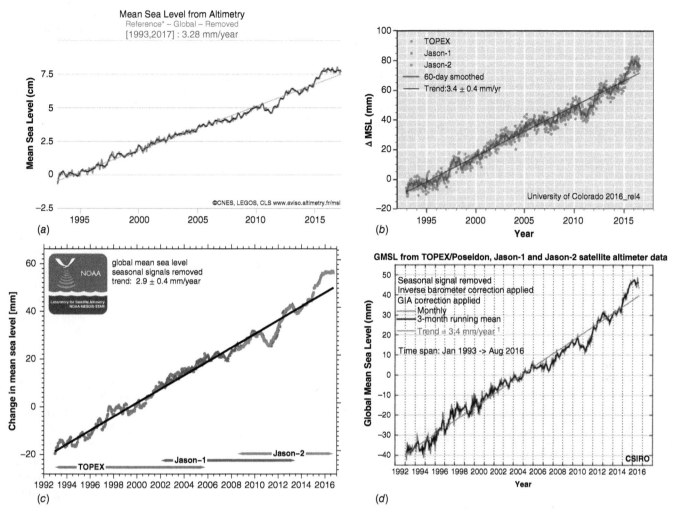

FIGURE 5.1 Global mean sea-level rise in the period 1992–2017, as measured by satellite detection (altimetry) of the ocean surface. Each graph is produced by a different scientific organization using the same dataset (however, these 4 datasets do use slightly different time periods). Using different time periods and different analytical techniques, the rates of change do not exactly match (although they do agree within statistical uncertainty). (a) French Space Agency–AVISO (3.28 ± 0.6 mm/yr); (b) Colorado Sea Level Research Group (3.4 ± 0.4 mm/yr); (c) NOAA Laboratory for Satellite Altimetry/Sea Level Rise (2.9 ± 0.4 mm/yr); (d) Commonwealth Scientific and Industrial Organization (3.4 ± 0.4 mm/yr).

Source: http://www.aviso.oceanobs.com/en/news/ocean-indicators/mean-sea-level/; http://sealevel.colorado.edu/content/2016rel4-global-mean-sea-level-time-series-seasonal-signals-removed; https://www.star.nesdis.noaa.gov/sod/lsa/SeaLevelRise/; http://www.cmar.csiro.au/sealevel/sl_hist_last_decades.html

5.2.2 Satellite Altimetry

Using the time it takes for radar to travel to Earth's surface and back, radar altimeters[33] on satellites can measure the sea surface from space to better than 5 cm (2 in).[34] The

TOPEX/Poseidon mission (1992–2006) and its successors Jason-1 (2001–2013), Jason-2 (2008), and Jason-3 (2016) have mapped the global extent of the sea surface approximately every 10 days for more than two decades. These missions have led to major advances in understanding the influence of winds, temperature, and atmospheric pressure on the ocean surface.[35]

Altimeter measurements indicate that global mean sea level has risen about 6.5 centimeters (2.5 inches) since 1993 at a mean rate of approximately 3.4 millimeters per year (0.13 inches per year).[36] However, this rise is not uniform across the oceans. In some locations, regional sea level has risen faster than the global average (e.g., prior to 2011, the

[33] Satellite altimetry measures the time taken by a radar pulse to travel from a satellite to Earth's surface and back to the satellite receiver. Combined with precise satellite location data, altimetry measurements yield sea-surface heights. See "Ocean Surface Topography Mission," *Wikipedia,* https://en.wikipedia.org/wiki/Ocean_Surface_Topography_Mission

[34] Leuliette, E.W., et al. (2004) Calibration of TOPEX/Poseidon and Jason altimeter data to construct a continuous record of mean sea level change. *Marine Geodesy* 27 (2004): 79–94. See also,Beckley, B.D., et al. (2007) A Reassessment of Global and Regional Mean Sea Level Trends from TOPEX and Jason-1 Altimetry Based on Revised Reference Frame and Orbits, *Geophysical Research Letters* 34, no. 14: L1-4608. See also the NASA entry on "Rising Water: New Map Pinpoints Areas of Sea Level Increase," http://climate.nasa.gov/news/index.cfm?FuseAction=ShowNews&NewsID=16

[35] Jet Propulsion Laboratory, Ocean Surface Topography from Space, http://sealevel.jpl.nasa.gov/.

[36] Sea Level http://sealevel.colorado.edu/

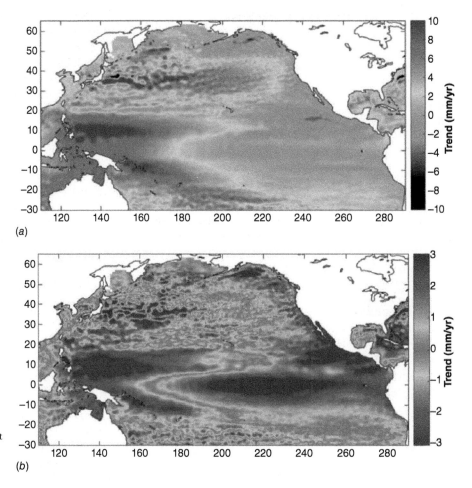

FIGURE 5.2 Two decades of high sea level that threatened island communities in the western tropical Pacific (A, 1993 through 2011), anomalously changed configuration (B, 2011 through 2015) and now threatens the west coast of North, Central, and South America.

Source: Hamlington et al. (2016)

western tropical Pacific[37]), and in other locations, regional sea level has risen slower than the global average (e.g., glacioisostatically uplifting[38] regions such as Alaska and Scandinavia[39]), and it might have even fallen over the period (e.g., the west coast of the United States[40]). However, two decades of high sea level that threatened low-lying island communities in the western tropical Pacific[41] anomalously changed configuration[42] and now threatens the west coast of North America.[43]

Altimeter measurements (**Figure 5.2**) reveal the shifting pattern of sea level since 1993 in the Pacific region.

In Figure 5.2,[44] rates are contoured by color: cool colors (blues and greens) indicate regions where sea level has been either falling or relatively stable; hot colors (yellow and red) show areas of sea-level rise.

This complex surface pattern largely reflects wind-driven changes in the thickness of the upper layer of the ocean and, to a lesser extent, changes in upper ocean heat content driven by surface circulation.[45] Most noticeable on the map is the dark red area (Figure 5.2A), which over the period 1993 through 2011 reached more than 10 millimeters per year (0.4 inches per year); the fastest rate of sea level rise yet observed. This pool of rising water has the signature shape of certain phases of quasi-periodic Pacific climate patterns; namely the **La Niña** phase of the El Niño Southern Oscillation (ENSO) and the negative phase of the **Pacific Decadal Oscillation** (PDO). La Niña conditions, and the negative phase of the PDO, are characterized by pronounced trade winds in the tropical western Pacific. The sea-level buildup in the western Pacific coincides with the absence of strong El Niño events over the period 1998–2011 and the development of persistent trade winds.[46]

However, in 2010, a moderate strength El Nino occurred and in 2014–2016 a major El Niño, with an unusual 1-year

[37] Merrifield, M.A. (2011) Shift in Western Tropical Pacific Sea Level Trends during the 1990s, *Journal of Climate* 24, 4126–4138, doi: 10.1175/2011JCLI3932.1

[38] Glacioisostatic uplift is a geologic phenomenon that occurs when Earth's crust relaxes upwards after a large mass of ice has melted. The ice, a glacier, would have been massive enough to depress the crust under its weight. The uplift can take place for thousands of years after the ice melts. Portions of Scandinavia and Canada are still uplifting after the removal of ice following the last ice age (which ended approximately 18,000 yrs ago).

[39] Sweet et al. (2017a)

[40] Bromirski, P., et al. (2018) Dynamical Suppression of Sea-Level Rise along the Pacific Coast of North America: Indications for Imminent Acceleration, *Journal of Geophysical Research* 116, no. C7: C07005, doi: 10.1029/2010JC006759

[41] Chowdhury, M. R., et al. (2010) Sea-level variability and change in the US-affiliated Pacific Islands: understanding the high sea levels during 2006–2008, *Weather, 65*(10), 263–268. https://doi.org/10.1002/wea.468

[42] Hamlington et al. (2016)

[43] Hamlington et al. (2016)

[44] Hamlington et al. (2016)

[45] Merrifield (2011)

[46] For a description of the El Niño Southern Oscillation (ENSO) phenomenon see "El Niño Southern Oscillation," *Wikipedia,* http://en.wikipedia.org/wiki/El_Ni%C3%B1o-Southern_Oscillation

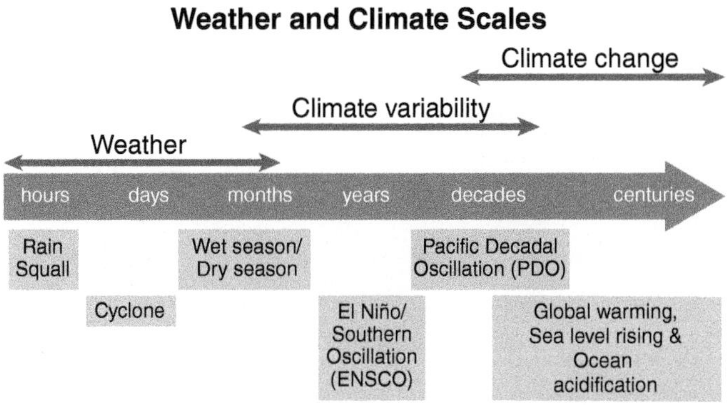

Weather and Climate Scales

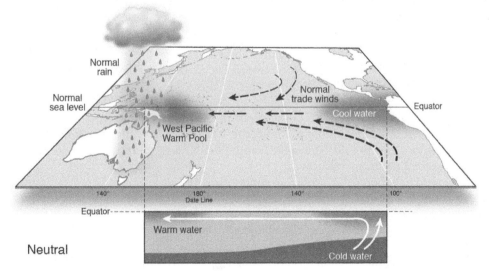

FIGURE 5.3 Weather, climate, and climate variability all operate over different lengths of time. Weather refers to hours, days, and months; climate variability refers to months, years, and decades; and climate change refers to decades and centuries.

Source: N. Hulbirt

ramp up period, developed in the Pacific ocean. Figure 5.2B shows a change in the pattern of sea level rise with a shift of the center of strongest rise moving to the eastern margin of the basin (note scale change between Figures 5.2A and 5.2B). This might be taken as a signal that the dominant characteristic of the PDO is shifting from a negative to a positive phase. Of interest, positive phases of the PDO are associated with periods of rapid atmospheric warming, and, accordingly, the years 2014–2016 each set record highs for global mean surface temperature.[47]

5.2.3 El Niño Southern Oscillation

As discussed in **Chapter 1**, weather, climate, and climate variability are all different aspects of the same question "How is global warming affecting Earth?" Weather is a common scientific term that is used by everyone. It describes the rain, wind, temperature, cloudiness, and other characteristics of the atmosphere at a particular place and time. *Climate* is the long-term average weather pattern in a specific place or region. When scientists describe the climate in a place, they use measurements and observations of the weather that have been made over periods of 30 years or longer. However, although the climate is the long-term average condition of the weather, it can change from one season to another, from one year to another, and over even longer time periods such as decades. This is referred to as *climate variability* (**Figure 5.3**). *Climate change* is a long-term shift in climate toward either a cooler or warmer state.

The east-to-west trade winds play a large role in the climate variability of the equatorial Pacific ocean (and this, in turn, can control the global mean surface temperature). These winds can change during a climate pattern that we briefly introduced in **Chapter 1** - the *El Niño Southern Oscillation* (ENSO). There are three types of ENSO patterns:

1. When the winds are normal, it is called a **neutral year**.
2. When the trade winds are stronger than normal, scientists call it a *La Niña* year.
3. When trade winds are weaker than normal, or absent, scientists say that it is an *El Niño* year.[48]

In a neutral year (normal winds), trade winds will blow seawater from the eastern part to the western part of the Pacific ocean. This water forms the **Western Pacific Warm Pool**, which straddles the equator.[49] Seawater in this region is much warmer than the seawater in the central or eastern portions of the Pacific ocean (**Figure 5.4**). The warm water leads to strong evaporation, and there is usually abundant rain. In the far eastern Pacific, off the coast of Central and South America, the surface water that has been blown to the west by the trade winds is replaced by cold water that rises from the deep sea floor. This water is rich in nutrients

FIGURE 5.4 In neutral ENSO years, the trade winds push seawater to the western tropical Pacific and create the Western Pacific Warm Pool, an area of warm water that readily evaporates and provides abundant rainfall in the region. The eastern Pacific region is characterized by cool, nutrient-rich water that rises from the deep sea and supports an important fishing industry.

Source: N. Hulbirt

[47] Trenberth, K. (2015) Has there been a hiatus? *Science*, 14 Aug, Vol. 349, Issue 6249, pp. 691–692, DOI: 10.1126/science.aac9225

[48] VIDEO: Ocean's role in El Nino. https://www.youtube.com/watch?v=gaFjlZxM7S4#action=share

[49] National Science Foundation-funded Pacific Islands Climate Education Partnership for resources that describe Pacific climate variability: http://pcep.wested.org/index.html

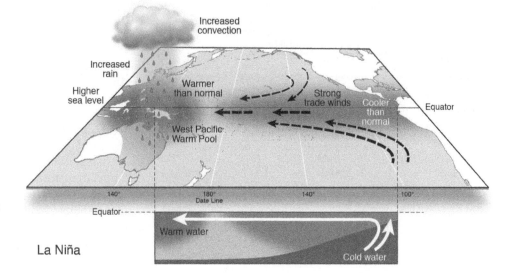

FIGURE 5.5 In a La Niña year, the trade winds are strong. They push water to the western tropical Pacific and increase the size of the Western Pacific Warm Pool. There is abundant rain throughout the region and higher than normal sea levels that can cause coastal flooding and erosion.

Source: N. Hulbirt

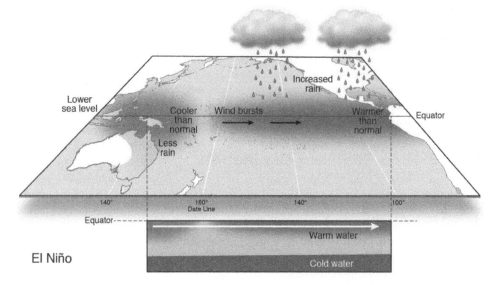

FIGURE 5.6 In an El Niño year, the trade winds are weak and may be replaced by winds that blow from west to east. This pushes warm seawater into the central and eastern tropical Pacific. There is abundant rain in the central and eastern tropical Pacific and higher than normal sea levels. The western Pacific, including the Philippines, Palau, and Micronesia, may experience drought and a fall in sea levels.

Source: N. Hulbirt

and feeds an economically important fishing industry in the eastern Pacific Ocean.

In a La Niña year (**Figure 5.5**), strong trade winds blow across the ocean surface into the western tropical Pacific. The Western Pacific Warm Pool grows larger and deeper, and the level of the ocean surface is higher. This raises the sea level and can cause coastal erosion (land loss due to wave action) and damaging king tides (nuisance flooding by unusually high tides).[50] La Niña years also tend to be very rainy in the western Pacific and support a strong monsoon (rainy) season in nations throughout Southeast Asia and around the Indian ocean. This region is globally important for food production.

In an El Niño year (**Figure 5.6**), trade winds are weaker than normal (or absent). There may be westerly wind bursts (gusts of wind in the tropics that blow from the west toward

the east). This causes warm seawater to migrate into the central and eastern Pacific ocean. The West Pacific Warm Pool is the source of this water, and it largely disappears and is replaced by deeper cool water. Abundant rainfall and flooding occur in the Americas and the fishing industry collapses. In the Western Pacific, the year following an El Niño year is usually drier and there is a greater chance of drought.

Table 5.1 summarizes the differences between neutral years, El Niño years, and La Niña years.

ENSO climate variability plays an important role in year-to-year global mean surface temperature. Typically, warmer than average years correlate with El Niño episodes and cooler than average years correlate to La Niña episodes. This is because the additional heat that is stored in the Western Pacific Warm Pool during a La Niña comes out of the atmosphere, thus causing a year with a lower than average global mean surface temperature. During an El Niño

[50] Chowdhury (2010)

TABLE **5.1**

ENSO Conditions and the Effects of ENSO Changes

Feature	Neutral ENSO Year	El Niño ENSO Year	La Niña ENSO Year
Wind	Normal east-to-west trade winds	Weak-to-absent trade winds; winds can blow from west to east	Stronger east-to-west trade winds
Rainfall	Usual amounts of rainfall with normal variability	Western Pacific drier than usual and can have long droughts. Eastern Pacific wetter than usual, can experience flooding and landslides.	Western Pacific wetter than usual and can experience flooding and landslides. Eastern Pacific drier than usual and can experience drought.
Sea level	Usual sea level with normal tide variability	Western Pacific experiences lower sea level so high tides tend to cause less flooding. Eastern Pacific experiences higher than normal sea level and can experience coastal erosion and flooding.	Western Pacific experiences high sea level so high tides tend to cause more coastal erosion and flooding. Eastern Pacific experiences lower than normal sea level so high tides cause less flooding and erosion.

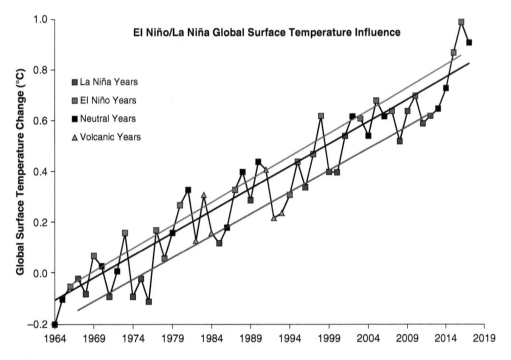

FIGURE 5.7 1964–2017 global surface temperature data from NASA, divided into El Niño, La Niña, and neutral years, with linear trends added. 2017 was the hottest year on record without an El Niño, thanks to global warming.

Source: Dana Nuccitelli, https://www.theguardian.com/environment/climate-consensus-97-per-cent/2018/jan/02/2017-was-the-hottest-year-on-record-without-an-el-nino-thanks-to-global-warming

event, waters from the Western Pacific Warm Pool spread across the Central and Eastern Pacific, heat is released, warming the air and causing a year with a warmer than average global mean surface temperature.

In a strong El Niño year, air temperature may set a record, such as in 1997/1998, 2009/2010, and 2015/2016. In fact, 2017 surface temperature set a record in that it was the hottest year to that point that was not an El Nino year (**Figure 5.7**). It is appropriate to note here that ENSO years are like academic years; they start in the summer and end in the following spring (thus, El Niño events are usually identified using 2 years, such as 2015/2016). ENSO is one type of interannual climate process that plays an important role in the year-to-year temperature of the atmosphere.

Another climate process that exerts strong variability on global climate is the Pacific Decadal Oscillation or PDO.

5.2.4 Pacific Decadal Oscillation (PDO)

The PDO consists of two phases, each historically lasting an irregular number of years.[51] In a positive (or warm) phase of the PDO, surface waters in the western Pacific above 20°N latitude tend to be cool, and equatorial waters in the

---

[51] Pacific Decadal Oscillation, *Wikipedia,* http://en.wikipedia.org/wiki/Pacific_decadal_oscillation

Pacific Decadal Oscillation

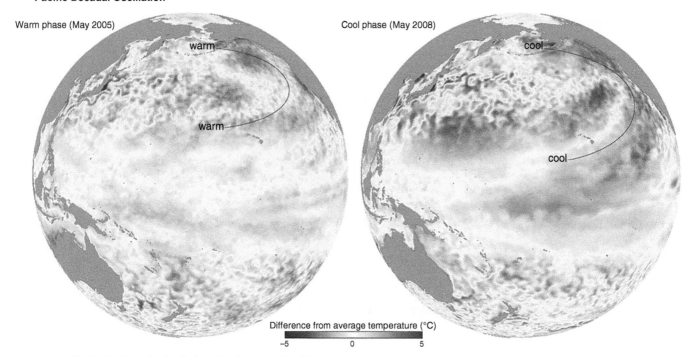

FIGURE 5.8 The Pacific Decadal Oscillation describes periodic shifts in ocean temperatures that occur over the span of years to decades. The classic signal of the cool phase is a crescent of cool water in the eastern North Pacific that surrounds a pocket of warm water in the central North Pacific.

Source: https://www.climate.gov/sites/default/files/HR_PDO2005-2008.jpg

central and eastern Pacific tend to be warm. In a negative (or cool) phase, the opposite pattern develops (**Figure 5.8**).

Natural climate variability such as the Pacific Decadal Oscillation influences the climate on an uneven interannual basis, occurring against a backdrop of global warming that has been developing over a period of decades to centuries. As such, these patterns may be changing over time because the background state of the climate is getting warmer. However, the period of satellite observations of the ocean surface is so short (since 1970s) that it is hard to separate natural patterns from those that are influenced by anthropogenic greenhouse gas emissions. Thus, the PDO is a poorly understood process.

Rapid sea-level rise in the western Pacific following the 1997/1998 El Niño matched the negative phase of the PDO.[52] Studies[53] of the history of sea level change in the western tropical Pacific reveal a strong historical correlation between phases of regionally rapid sea-level rise to negative phases of the PDO. The degree to which this pattern contributes to the global mean rate of sea-level rise observed in satellite altimetry is not known, but the PDO has been recognized as a major factor controlling sea-level change in

the Pacific, the world's largest ocean and a key player in the year-to-year fluctuation of global mean surface temperature.

With the rapid, record-setting warming of 2014–2016, there is speculation that the PDO has once again changed phase.[54] If the warming of 2014–2016 signals a positive phase of the PDO, it is not known how long it will prevail nor what its effects may be. Some researchers even view the PDO pattern of decadal timing as having broken down in favor of shorter-duration events.[55] If indeed the Pacific has moved into a positive phase of the PDO (**Figure 5.9**), it does not bode well for low-lying coastal communities in the eastern Pacific.[56] A positive phase is likely to be accompanied by increased sea levels and storminess that promote flooding and erosion in that region.

Although our understanding of the PDO is incomplete, at least one researcher may have successfully predicted the latest phase change.[57] Peter Bromirski forecast that persistent trade winds would decrease, and the recent historical stabilization of sea level along the U.S. west coast would come to an end. This would usher in a period of accelerated sea-level rise in the eastern Pacific, and indeed if the

[52] In April 2008, scientists at NASA's Jet Propulsion Laboratory announced that the Pacific Decadal Oscillation had shifted to its cool (or negative) phase: http://earthobservatory.nasa.gov/IOTD/view.php?id=8703
[53] Merrifield, M.A., et al. (2013) Multidecadal sea level anomalies and trends in the western tropical Pacific, *Geophysical Research Letters*, 39, 13, doi:10.1029/2012GL052032

[54] Trenberth (2015)
[55] NOAA, http://www.nwfsc.noaa.gov/research/divisions/fed/oeip/ca-pdo.cfm
[56] Barnard, P.L., et al. (2017) Extreme oceanographic forcing and coastal response due to the 2015–2016 El Niño, *Nature Communications*, v. 8, pp. 1–8, doi: 10.1038/ncomms14365
[57] Bromirski (2011)

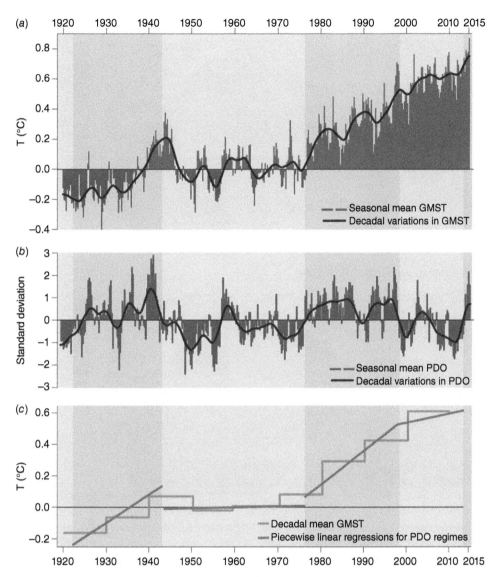

FIGURE 5.9 The Pacific Decadal Oscillation (PDO) is thought to have switched to a negative (or cool) phase following the 1997/1998 El Niño. Cool phases (light shading) tend to enhance La Niña conditions and suppress global mean surface warming. Significant El Nino events in 2010, and again in 2014–2016, may signal another switch, this time to a positive (or warm) phase (dark shading) of the PDO.

Source: http://science.sciencemag.org/content/349/6249/691.full?sid=583a9ed8-7641-491a-8715-463fd9b6d516

PDO is now in a positive phase, he would be correct. The sea surface topography illustrated in Figure 5.2B appears to bear this out.

Because winds play an important role in regional sea-level change, it is worthwhile asking, "Are the winds changing as a result of global warming?" Young et al.[58] address this question. They used a 23-year database of satellite altimeter measurements to investigate global changes in oceanic wind speed and wave height (**Figure 5.10**).

They discovered a general global trend of increasing wind speed and, to a lesser degree, wave height. The rate of wind speed increase is greater for extreme events compared to the mean condition and indicates the intensity of extreme events is increasing at a faster rate than that of the mean conditions. At the mean and 90th percentile, wind speeds over the majority of the world's oceans have increased by at least 0.25%–0.5% per year (a 5%–10% net increase over the past 20 years). The trend is stronger in the Southern Hemisphere than in the Northern Hemisphere. The only significant exception to this positive trend is the central north Pacific, where there are smaller localized increases in wind speed of approximately 0.25% per year and some areas where there is a weak negative trend.

Climate is changing throughout the Pacific, and studies[59] indicate that winds exert an important control on sea-level behavior in the Pacific basin. But how the ENSO, the PDO, and global warming are linked and how they will continue to interact in the future is somewhat speculative. For instance, models show that the tropics have expanded, and this has been verified by observations.[60] Presumably related

[58] Young, I., et al. (2012) Global Trends in Wind Speed and Wave Height, *Science* 332, 451–455

[59] Timmermann, A., et al. (2010) Wind Effects on Past and Future Regional Sea Level Trends in the Southern Indo-Pacific, *Journal of Climate* 23, 4429–4437, doi: 10.1175/2010JCLI3519.1

[60] Lu, J., et al. (2009) Cause of the Widening of the Tropical Belt Since 1958, *Geophysical Research Letters* 36, L03803, doi: 10.1029/2008GL036076

mean wind speed (1991–2008)

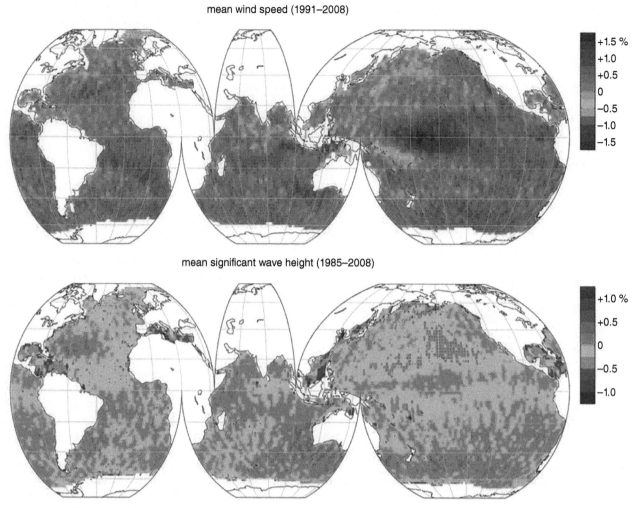

mean significant wave height (1985–2008)

FIGURE 5.10 Global contour plots of mean trend (percent per year); wind speed (top) and wave height (bottom). Points that are statistically significant are shown with dots. Researchers have found that since 1990, there is a general global trend of increasing wind speed and, to a lesser degree, wave height. The rate of increase is greater for extreme events as compared to the mean condition.

Source: Young et al. (2012)

to this is widening of the Hadley cell,[61] the convective system that governs tropical winds, and as seen, it is winds that are currently influencing the rates of sea-level change in the low-latitude Pacific.

While it is unknown how any of these processes are likely to change in a warmer future, global climate models do project that as we move into a warmer world, the frequency and intensity of both La Niña[62] and El Niño[63] events will increase, that extreme El Niño events will increase in frequency long after temperatures stabilize

at 1.5 degrees Celsius (2.7 degrees Fahrenheit),[64] and that other types of climate variability will become more prominent.[65]

5.2.5 Tide Gauges

In addition to satellite altimetry, sea level is measured around the world using tide gauges.[66] A tide gauge is a water-surface measurement device (**Figure 5.11**) mounted on a

[61] Johanson, C.M. and Fu, Q. (2009) Hadley Cell Widening: Model Simulations versus Observations, *Journal of Climate* 22, 2713–2725.

[62] Cai, W., et al. (2015) Increased frequency of extreme La Niña events under greenhouse warming, *Nature Climate Change*, v. 5, p. 132–137, doi:10.1038/nclimate2492

[63] Cai, W., et al. (2014) Increasing frequency of extreme El Niño events due to greenhouse warming. *Nature Climate Change*, 4, 111–116. http://dx.doi.org/10.1038/nclimate2100

[64] Wang, G., et al. (2017) Continued increase of extreme El Niño frequency long after 1.5°C warming stabilization, *Nature Climate Change*, doi:10.1038/nclimate3351

[65] Cai et al. (2015). See also, Widlansky, M.J., et al. (2015) Future extreme sea level seesaws in the tropical Pacific, *Science Advances*, 25 Sept., V. 1, No. 8, DOI: 10.1126/sciadv.1500560

[66] NOAA website on tide gauges: http://oceanservice.noaa.gov/education/kits/tides/tides10_oldmeasure.html

USGS

FIGURE 5.11 A tide gauge and weather station in New Jersey operated by the U.S. Geological Survey. The instruments measure, store, and transmit water levels and meteorological data—rainfall, wind speed and direction, air and water temperature, relative humidity, and barometric pressure. The data are transmitted by telephone and satellite to specific critical decision-making centers to track the potential for flooding and high winds. Precise land-surface elevations are established for tide gauge locations. The elevation is part of a network of benchmarks, which are permanent markers on which a known elevation is established. The benchmarks are referenced to the North American Vertical Datum of 1988.

Source: https://pubs.usgs.gov/fs/2007/3064/

pier, seawall, or other coastal foundation fixed to the land in order to monitor vertical motion of the ocean surface.

The global sea-level record from tide gauges is an important indicator of the evolution and impact of global climate change. Tide gauge data also capture a variety of local and regional phenomena related to decadal climate variability, tides, storm surges, tsunamis, swells, and other oceanographic processes.

Tide gauge data are used to validate models of ocean circulation and to detect errors and drifts in satellite altimetry. Compared to satellite altimetry data, tide gauge data offer a longer record and finer temporal resolution but coarser spatial resolution. The calculation of global mean sea level from tide gauges is not straightforward due to a number of considerations, including local and regional changes in winds and ocean circulation that impact sea level, the impact of atmospheric pressure changes on sea level, the relative lack of long, continuous records, and the lack of a common datum across tide gauge sites.

Tides Tides are very long-period ocean waves that move in response to the gravitational force exerted by the Moon and Sun (and other planets). Tides originate in the oceans and progress toward the coastlines where they appear as the regular rise and fall of the sea surface. When the highest part, or crest, of the tide wave reaches a location, high tide occurs; low tide corresponds to the lowest part of the wave, or its trough. The difference in height between the high tide and the low tide is called the **tidal range**.[67]

Tides are caused mainly by the pull of gravity from the Moon and the Sun acting on Earth's surface. The Moon pulls more on Earth's water than the Sun because the Moon is much closer to our planet (**Figure 5.12**a). When the Moon is in one of its half-moon phases, the gravitational attraction of the Moon and that of the Sun are pulling in different directions. During those times of the month, the high tides are not very high and the low tides are not very low—these are called **neap tides**. When the Moon is bright and full or is nearly completely dark, the gravitational attraction of the Moon and that of the Sun are both pulling on the ocean in the same direction. During those times of the month, the high tides are higher than average and the low tides are lower than average—these are called **spring tides** (**Figure 5.12**b).

Tide Gauge Studies In addition to measuring the tide, over time a tide gauge measures long-term water level trends or **local relative sea-level change**. Local relative sea-level change can reflect a number of factors: rising land will produce a record of falling local relative sea level in the tide gauge data; subsiding land will produce a record of rising local relative sea level. Global warming will cause local relative sea level rise. But it is the relative difference between the rate of **vertical land motion** and rate of sea-level rise that will determine whether the local relative sea level is rising or falling and by how much each year. Tide gauges are distributed across the globe and are maintained, and their data archived, by a number of research and monitoring organizations (**Figure 5.13**).[68]

Many processes affect the water level history at a tide gauge. Changes in atmospheric pressure, storms and storm surge, episodes of high waves, oceanographic circulation phenomena known as **mesoscale eddies**, multimonthly bulges of water that move through the ocean (e.g., Rossby Waves), shifts in the location and velocity of currents, and changes in sea surface temperature and winds associated

---

[67] NOAA: http://oceanservice.noaa.gov/facts/tides.html
[68] For instance the University of Hawaii Sea Level Center is one such institution with responsibility for maintaining tide gauges across the Pacific, Southern, and Indian Oceans: http://uhslc.soest.hawaii.edu

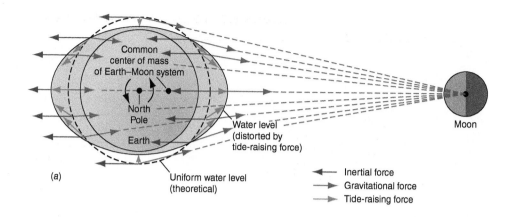

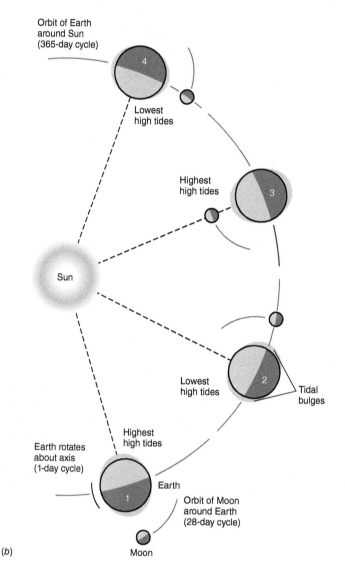

FIGURE 5.12 (a) Two tide-raising forces are responsible for the lunar ocean tide: the gravitational attractive force (red arrows) and the inertial force (blue arrows). The balance between these two opposing forces (green arrows) defines the tide-raising force. On the side of Earth facing the Moon, the gravitational attraction force is greater than the inertial force, and a tidal bulge is created on the ocean surface. On the side facing away from the moon, the inertial force is greater and similarly creates a tidal bulge. Earth rotates beneath this watery envelope and encounters these two bulges (high tides) twice every day, as land areas rotate into the tidal bulges. This water flows within, throughout, and around coastlines in the form of a series of tidal currents. (b) Tidal bulges on the ocean surface related to the gravitational and inertial forces of the Sun–Earth system modulate the lunar tide. When the Sun, Earth, and Moon are aligned, the spring tide is established. When the Moon is at 90° to the Sun–Earth alignment, neap tide is established.

with climate variability processes (e.g., PDO, ENSO, etc.) all influence the vertical position of the sea surface recorded at a tide gauge (and, for that matter, in altimetry data as well).

If some or all of these processes converge at the same time, extreme high water events can be produced known as king tides. King tides are most likely to occur during times of the year when the Moon and/or Sun are closest to the Earth, such as at solstices, and during spring tides. As sea level rises with continued global warming, king tides will lead to flooding in low-lying communities that will become more damaging with time.

All in all, tide gauge data provide information on the relative rate of change between the land that the gauge is attached to and the ocean surface it measures. To isolate the tide gauge so that the influence of rising or sinking land does not control the long-term history of water level,

GLOSS Real Time Sea Level Monitoring Network at VLIZ

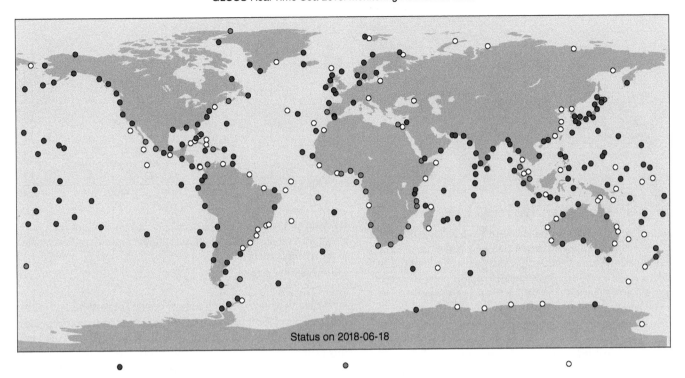

Status on 2018-06-18

●	●	○
Updated in past 31 days (181)	Has some data (28)	No data (81)

FIGURE 5.13 The Global Sea Level Observing System (GLOSS) is an international program established by the Intergovernmental Oceanographic Commission of UNESCO. GLOSS provides oversight and coordination of global and regional sea-level networks in support of scientific research. The main component of GLOSS is the 'Global Core Network' (shown here) of 290 sea level stations. Maintaining a large array of data collecting sites requires the time and effort of a large network of partners.[69]

Source: http://www.gloss-sealevel.org

modern Global Positioning Systems[70] (GPS) are used to monitor the movement of the gauge. This information is used to resolve a true water level history separately from vertical land motion. Networks of tide gauges provide information on sea-level rise and fall at coastal sites around the world.[71]

It is common for researchers to use the global network of tide gauges to analyze patterns of recent sea-level change resulting from global warming,[72] to project future patterns of sea-level change,[73] to identify important patterns of climate variability,[74] and to analyze other oceanographic, atmospheric, and geologic influences on the sea surface.

One study[75] identified acceleration in the global rate of sea-level rise that occurred in approximately 1990 (**Figure 5.14**). The study recognized an average global sea-level trend over the period 1962–1990 of 1.56 millimeters per year (0.06 inches per year); however, after 1990, the global trend increased to a rate of 3.2 millimeters per year (0.13 inches per year), matching estimates obtained from satellite altimetry. Increased rates in the tropical and southern oceans primarily account for the acceleration. The timing of the global acceleration corresponds to similar trend changes in upper ocean heat content and ice melt.

Another study[76] used the global network of tide gauges in combination with satellite altimetry data to establish that

---

[69] The University of Hawaii Sea Level Center is one such partner: http://uhslc.soest.hawaii.edu

[70] The Global Positioning System (GPS) is a constellation of up to 32 satellites that orbit at a height of 26,600 km above Earth. The satellites are owned by the U.S. Department of Defense, but anyone can use the signals from those satellites, provided they have a receiver. For the receiver to work, it needs to be able to "see" four of the satellites. When you turn on your receiver, it may take a minute or so to locate these satellite signals, then to download data from the satellite before software in the receiver can identify its exact position. Fundamentally, two things need to happen for this to work effectively: 1) The GPS receiver measures the distance from itself to a satellite by measuring the time a signal takes to travel that distance at the speed of light; 2) When the satellite's position is known, the GPS receiver knows it must lie on a sphere that has the radius of this measured distance with the satellite at its center. The receiver need only intersect three such spheres. This process, known as trilateration, is an effective means of determining absolute or relative locations.

[71] NOAA page for sea-level trends: http://tidesandcurrents.noaa.gov/sltrends/sltrends.shtml

---

[72] Dieng, H. B., et al. (2017), New estimate of the current rate of sea level rise from a sea level budget approach, *Geophys. Res. Lett.*, 44

[73] Kopp, R.E., et al. (2014) Probabilistic 21st and 22nd century sea-level projections at a global network of tide-gauge sites. *Earths Future* 2, 383–406.

[74] Widlansky (2015)

[75] Merrifield, M.A., et al. (2009) An Anomalous Recent Acceleration of Global Sea-Level Rise. *Journal of Climate* 22, 5772–5781

[76] Church, J.A. and White, N.J. (2006) 20th Century Acceleration in Global Sea-Level Rise, *Geophysical Research Letters* 33, no. 1: L01602.

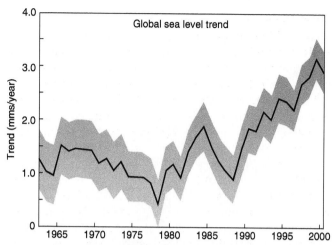

FIGURE 5.14 A study of tide gauges identified acceleration in the rate of global sea-level rise from approximately 1.56 mm/yr (0.06 in/yr) over the period 1962–1990 to 3.2 mm/yr (0.13 in/yr) between 1990 and 2000. The timing of the acceleration corresponds to similar trend changes in upper ocean heat content and ice melt; it also matches measurements made with satellites over the time period.

Source: Figure from Merrifield et al. (2009)

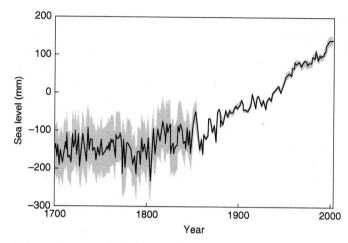

FIGURE 5.15 Using long tide-gauge records, researchers reconstructed global sea level since 1700. The shaded portion represents uncertainties in the reconstruction.

Source: Jevrejeva et al. (2008)

global mean sea level rose 19.5 centimeters (7.7 inches) between 1870 and 2004 at an average rate of about 1.44 millimeters per year (0.05 inches per year). Over the 20th-century portion of the record, sea level averaged 1.7 millimeters per year (0.07 inches per year). This acceleration provided important confirmation of climate models predicting that the rate of sea-level rise will accelerate in response to global warming. If the same acceleration continues, then the amount of rise from 1990 to 2100 will range from 28 to 34 centimeters (11–13 inches), which is consistent with IPCC AR4 projections of 18–59 centimeters (7–23 inches) of sea-level rise by 2100.

More recent analysis of tide gauge observations has refined the average global sea-level trend before 1990 to 1.1 ± 0.3 millimeters per year (0.05 inches per year), lower than the previous estimates. From 1993 to 2012, the trend has increased by three times to 3.1 ± 1.4 millimeters per year (0.12 inches per year).[77]

Using a combination of tide gauge and altimeter data, Hamlington[78] applied statistical techniques to define the primary components of sea-level change in the satellite data set and used these to unravel the characteristics of sea-level change in the tide gauge era. The combined data capture sea-level change over the period 1950–2009. The computed rate of global mean sea-level rise from the reconstructed time series is 1.97 millimeters per year (0.077 inches per year) from 1950 to 2009 and 3.22 millimeters per year (0.126 inches per year) from 1993 to 2009.

Work[79] with long tide-gauge records (**Figure 5.15**) reveals that sea-level acceleration might have started earlier than widely thought, more than 200 years ago. By reconstructing global mean sea level since 1700 from long tide-gauge records, researchers concluded that sea-level acceleration began at the end of the 18th century. Sea level rose by 6 centimeters (2.4 inches) during the 19th century and 19 centimeters (7.5 inches) in the 20th century. On the basis of this analysis, they conclude that if the conditions that established the acceleration continue, then sea level will rise 34 centimeters (13.4 inches) over the 21st century.

Tide gauges also provide information on the regional behavior of sea level. The National Research Council[80] studied sea level rise for the coasts of California, Oregon, and Washington. They found that because of vertical land motion resulting from plate tectonics and the ongoing response of Earth's surface to disappearance of North American ice sheets, future sea-level rise is likely to vary along the U.S. west coast. The study projected that, relative to 2000 levels, global sea level will reach 8–23 centimeters (3–9 inches) by 2030, 18–48 centimeters (7–19 inches) by 2050, and 50–140 centimeters (20–55 inches) by 2100. South of Cape Mendocino on the California coast, sea levels are projected to rise by an amount similar to that estimated by global calculations. But north of the Cape, along the coasts of northern California, Oregon, and Washington, future sea-level projections are lower than those to the south. This difference in sea level trends along the California coastline is related to the fact that the coast is the approximate location of the boundary between the North American and Pacific lithospheric plates. Stresses along the plate boundary cause

[77] Dangendorf et al. (2017)

[78] Hamlington, B. et al. (2011) Reconstructing Sea Level using Cyclostationary Empirical Orthogonal Functions, *Journal of Geophysical Research* 116, C12015, doi: 10.1029/2011JC007529.

[79] Jevrejeva, S. et al. (2008) Recent Global Sea Level Acceleration Started over 200 Years Ago? *Geophysical Research Letters* 35, LO8715, doi: 10.1029/2008GL033611.

[80] Committee on Sea Level Rise in California, Oregon, and Washington; Board on Earth Sciences and Resources; Ocean Studies Board; Division on Earth and Life Studies; National Research Council *Sea level rise for the coasts of California, Oregon, and Washington: Past, Present, and Future*, (2012), National Academies Press, Washington, D.C.

vertical land motion that influence the relative rate of sea-level change from north to south along the U.S. west coast.

On the U.S. east coast, a study[81] of tide-gauge records revealed a "hotspot" of rapid sea-level rise along the highly populated coast between Cape Hatteras (North Carolina) and New England. Between 1950–1979 and 1980–2009, sea-level rise rate increases were approximately three to four times higher than the global average and were consistent with a slowdown of North Atlantic (thermohaline) circulation. This conjecture has since been confirmed by at least two studies that found decreases in North Atlantic circulation.[82]

5.2.6 Coastal Sediment Studies

To extend the record of sea-level changes beyond the era of tide gauges and satellite altimetry, coastal geologists use natural archives of shoreline sediment to reconstruct the past history of sea level. Certain types of sediment can serve as sea-level proxies, including environmental features that grow and collect at the edge of the sea, such as beach sands, shallow-water corals, mud deposited on tidal flats, mangrove roots, and others.

One of the most precise sea-level proxies is a type of plankton (**foraminifera**,[83] a tiny protist) that collects on salt marshes that are only flooded by the highest tides. Because different species of foraminifera live at different levels of the tide, a survey of the types of remains buried in marsh mud tells researchers where the level of the tide was in that particular spot at the time the sediment layer was laid down.[84] As sea level rises (or the land subsides) these microscopic animals are buried by mud that collects in the salt marsh as it maintains its position between rising high and low tides.

Geologists take cores of salt marshes, analyze the entombed forams (and plant fragments and other types of remains) that indicate the position of past sea level through time and use radiocarbon (or other methods) to date the sediments and assess the age of the samples. Radiocarbon dating is a method that permits age dating of organic samples up to an age of about 50,000 years old.

For instance, a cored sample from 1.5 meters (5 feet) below the marsh surface might contain plankton and plant fossils known to grow only in mud inundated by the full-moon high tide; these are very good indicators of sea level. The same sample might provide a radiocarbon date of, say, 2100 years (or so) before present. Hence, by building a record of changing tide level out of these geologic materials, a sea-level history can be assembled that predates the instrumental record.

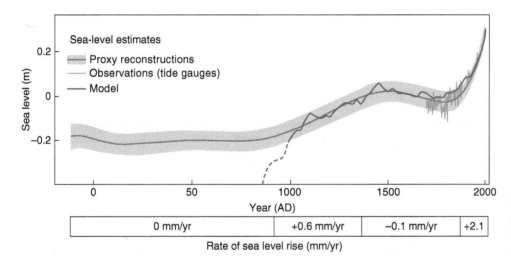

FIGURE 5.16 Sea-level history along the North Carolina coast. The rate of sea-level rise along the U.S. Atlantic coast is greater now than at any time in the past 2,000 years and has shown a consistent link between changes in global mean surface temperature and sea level for the past 1,000 years.

Source: Figure after Kemp et al. (2011)

[81] Sallenger, A.H. et al. (2012) Hotspot of accelerated sea-level rise on the Atlantic coast of North America, *Nature Climate Change*, 24 June, DOI: 10.1038/NCLIMATE1597

[82] Thornalley, D.J.R., et al. (2018) Anomalously weak Labrador Sea convection and Atlantic overturning during the past 150 years, *Nature*, 556 (7700): 227 DOI: 10.1038/s41586-018-0007-4. See also Caesar, L., et al. (2018) Observed fingerprint of a weakening Atlantic Ocean overturning circulation, *Nature*, 556 (7700): 191 DOI: 10.1038/s41586-018-0006-5

[83] Foraminifera ("forams" for short) are single-celled protists with shells. Their shells are also referred to as "tests". In some forams, the protoplasm covers the exterior of the shell. The shells are commonly divided into chambers which are added during growth, though the simplest forms are open tubes or hollow spheres. Depending on the species, the shell may be made of organic compounds, sand grains and other particles cemented together, or crystalline calcite. There are an estimated 4,000 species alive in the world's oceans. Of these, 40 species are planktonic, that is they float in the water. The remainder live on or in the sand, mud, rocks and plants at the bottom of the ocean and in coastal environments. Foraminifera are found in all marine environments, from the intertidal to the deepest ocean trenches, and from the tropics to the poles, but species of foraminifera can be very particular about the environment in which they live. Some are abundant only in the deep ocean, others are found only on coral reefs, and still other species live only in brackish estuaries or intertidal salt marshes.

[84] Kemp, A. et al. (2009) Timing and Magnitude of Recent Accelerated Sea-Level Rise (North Carolina, United States), *Geology* 37, 1035–1038, doi: 10.1130/G30352A.1

It was just such a research effort that produced a reconstructed sea level history of the North Carolina coast extending 2,100 years into the past[85] (**Figure 5.16**). Analyzing cores of tide marsh mud, researchers found four phases of persistent sea-level change, a history that applies only to the North Carolina coastal plain because of the unique behavior of Earth's crust (localized vertical land movement) from one region to another. (The crust rises in some places and subsides in others, making detailed sea-level records only representative of their home region.)

The four phases of sea-level behavior consisted of: 1) stable sea level from at least 100 BCE until 950 CE,[86] 2) rising sea level from 950 to 550 CE at a rate of 0.6 millimeters per year (0.023 inches per year), 3) a period of stable or slightly falling sea level from 550 CE until the late 19th century, and 4) most recently rising sea level to the present at an average rate of 2.1 millimeters per year (0.082 inches per year), representing the steepest century-scale increase of the past two millennia. This rate was initiated between 1865 and 1892 CE, toward the end of the Little Ice Age in the North Atlantic region and the beginning of a clear signal of human-induced global warming.

In summary, research shows that today's rate of sea-level rise is the most rapid of the past 2,000 years and that the rate of global mean sea-level rise has accelerated (approximately doubling) over the 20th and 21st centuries[87] and reached more than 3 millimeters per year (0.13 inches per year).

5.3 Sea-Level Components

Global sea-level rise results from a combination of factors. As the oceans absorb heat, the water molecules tend to separate and produce **thermal expansion**; having no other direction to go but upward, they contribute to a rise in sea level. Thermohaline circulation in the North Atlantic, and in the Southern Ocean near Antarctica, circulates warm water downward, leading to thermal expansion in the deep ocean as well (a process that will play out over several centuries). The melting of three forms of ice also contributes to sea-level rise: alpine glaciers in the valleys of mountain systems, ice caps that cover larger surface area than alpine glaciers, and continental ice sheets, of which there are only two, Greenland and Antarctica, left over from the last ice age (which culminated about 21,000 years ago).[88]

One study[89] considered the various components that go into determining the rate of global mean sea-level change. Using tide-gauge data only, researchers calculated that global mean sea level rose between 1972 and 2008 at an average rate of 1.8 ±0.2 millimeters per year (0.071 ±0.008 inches per year). Using a combination of tide gauges and altimeter observations, they calculated a rate of 2.1 ± 0.2 millimeters per year (0.083 ± 0.008 inches per year). The largest contributors to sea-level rise over the period include ocean thermal expansion (0.8 millimeters per year; 0.031 inches per year) and melting of various types of ice (0.7 millimeters per year; 0.027 inches per year), with Greenland and Antarctica contributing about 0.4 millimeters per year (0.016 inches per year). Contributions from melting ice increased throughout the study period, as did contributions from thermal expansion, although less rapidly.

The Intergovernmental Panel on Climate Change Assessment Report 5[90] calculated the components of sea-level rise in their 2013 analysis. They consider contributions to sea-level rise over the periods 1971–2010 and 1993–2010 (the era of satellite altimetry observation). Between 1971 and 2010, thermal expansion contributed a mean of 0.8 millimeters per year (0.031 inches per year). Between 1993 and 2010, thermal expansion had increased to 1.1 millimeters per year (0.043 inches per year).

The contributions of glaciers not located in Greenland or Antarctica totaled a mean 0.62 millimeters per year (0.024 inches per year) between 1971 and 2010; between 1993 and 2010, the contribution increased to 0.76 millimeters per year (0.029 inches per year). Glaciers in Greenland other than the ice sheet contributed 0.06 millimeters per year (0.002 inches per year) between 1971 and 2010 and 0.10 millimeters per year (0.003 inches per year) between 1993 and 2010. Changes in the Greenland ice sheet between 1971 and 2010 were not sufficiently well observed to quantify. However, between 1993 and 2010, the Greenland ice sheet contributed a mean 0.33 millimeters per year (0.012 inches per year) to global sea level.

Contributions of the Antarctic ice sheet to global sea level were also only known for the period 1993–2010 and equaled 0.27 millimeters per year (0.010 inches per year). Another contribution to global sea-level change comes from the storage (and release) of water on land in the form of reservoirs, groundwater pumping, lakes, wetlands, rivers, the vadose zone,[91] aquifers, and snow pack. Between 1971 and 2010, release of stored water on land (largely by human pumping) contributed 0.12 millimeters per year (0.004 inches per year), and between 1993 and 2010, had more than tripled to 0.38 millimeters per year (0.014 inches per year).

[85] Kemp, A., et al. (2011) Climate Related Sea-Level Variations over the Past Two Millennia, *PNAS* 108, no. 27, 11017–11022, www.pnas.org/cgi/doi/10.1073/pnas.1015619108

[86] BCE – before Common Era, and CE-Common Era, are standard terms for referring to time before the year "0" (BCE) and after the year "0" (CE). These terms replaced the previously widely used terms "BC" and "AD".

[87] Church, J.A. and White, N.J. (2011) Sea-Level Rise from the Late 19th to the Early 21st Century, *Surveys in Geophysics* 32, 585–602, doi: 10.1007/s10712-011-9119-1.

[88] VIDEO: Melting ice, rising seas. http://www.youtube.com/watch?v=gbnW3MK8wgY

[89] Church, J., et al. (2011) Revisiting the Earth's Sea-Level and Energy Budgets from 1961 to 2008, *Geophysical Research Letters*, 38, L18601, doi: 10.1029/2011GL048794

[90] Church et al. (2013)

[91] The vadose zone, also termed the unsaturated zone, is the part of the crust between the land surface and the water table, the position at which the groundwater (the water in the soil's pores) is at atmospheric pressure ("vadose" is from the Latin for "shallow").

TABLE 5.2

IPCC AR5[92] Sea-Level Budget (mm/yr)

Source	1971–2010	1993–2010
Thermal expansion	0.8 (0.5 to 1.1)	1.1 (0.8 to 1.4)
Glaciers except in Greenland and Antarctica	0.62 (0.25 to 0.99)	0.76 (0.39 to 1.13
Glaciers in Greenland	0.06 (0.03 to 0.09)	0.10 (0.07 to 0.13)
Greenland ice sheet	No data	0.33 (0.25 to 0.41)
Antarctica ice sheet	No data	0.27 (0.16 to 0.38)
Land water storage	0.12 (0.03 to 0.22)	0.38 (0.26 to 0.49)
TOTAL of contributions	Incomplete data	2.8 (2.3 to 3.4)
Observed global mean sea-level rise	2.0 (1.7 to 2.3)	3.2 (2.8 to 3.6)

Observed global mean sea-level rise over the period 1971–2010 equaled 2.0 millimeters per year (0.078 inches per year). Contributions from thermal expansion, glacier loss, and water stored on land equal 1.6 millimeters per year (0.062 inches per year). Contributions from the Greenland and Antarctic ice sheets were not known for this period. Over the period 1993–2010, estimated global mean sea-level rise ranged from 2.8 to 3.6 millimeters per year (0.110 to 0.141 inches per year) and contributions ranged from 2.3 to 3.4 millimeters per year (0.090 to 0.133 inches per year). Thus, within uncertainties, observations of global mean sea-level rise match the observed contributions (Table 5.2).

The effort to delineate contributions to global mean sea-level change is based on the need to identify accelerations or decelerations in the various components. For instance, a paper in 2017[93] revealed that global mean sea level rose 50 percent faster in 2014 than in 1993, with meltwater from the Greenland ice sheet supplying 25 percent of total sea-level increase compared with just 5 percent 20 years earlier.

These findings add to growing concern among scientists that global sea level is climbing more rapidly than forecast only a few years ago, with potentially devastating consequences. Overall, the pace of sea-level rise went up from about 2.2 millimeters per year (0.086 inches per year) in 1993, to 3.3 millimeters per year (0.129 inches per year) two decades later. In the early 1990s, thermal expansion accounted for fully half of the added rise. Two decades later, that figure was only 30 percent. Accelerations in the contribution of meltwater from Greenland can be useful in focusing research programs to spend more effort in understanding the behavior of the Greenland ice sheet as a research priority.

As mentioned at the start of this chapter, global mean sea level is rising now at 3.9 millimeters per year (0.153 inches per year). If it maintains the current pace of acceleration, this will result in about 75 centimeters (2.5 feet) of sea-level rise by the end of the century.[94] This is an increase over the roughly 3 millimeters per year (0.118 inches per year) rate that was previously calculated from the 20-year history of satellite altimetry. The increase results from the discovery of a calibration error in the first satellite, TOPEX/Poseidon, which had led to overestimates of the global rate. Researchers are reacting that the correction clears up a source of confusion over the fact that during the past 20 years, global warming has increased, the rate of ocean thermal expansion has increased, and the melting of the world's ice has increased, yet the rate of global mean sea-level rise had not reflected these increases. In the words of one researcher, bringing the observations of the causes of sea-level rise in line with the measured rate ". . .gives us much more confidence that we understand what is happening."[95]

5.3.1 Ocean Warming

The world ocean is so immense that it dominates Earth's heat budget, storing more than 90 percent of the excess heat in Earth's climate system that has been trapped by anthropogenic greenhouse gases. The capacity of the oceans to store heat is enormous. For instance: the upper 2.5 meters (8.2 feet) of ocean water stores as much heat as the entire atmosphere![96] Although an increase in the average temperature of the ocean of only 0.01 degrees Celsius (0.018 degrees Fahrenheit) seems small, it is a very large amount of heat. In fact, if this energy were released all at once, the average temperature of the atmosphere would increase by about 10 degrees Celsius (18 degrees Fahrenheit).[97] Thus, a small change in the mean temperature of the ocean represents a very large change in the total heat content of the climate system. It also contributes to sea-level rise, because warming water expands.

The trend of ocean heating has been quite strong over the longer term,[98] but research indicated that the period 2003–2010 showed no net warming of the ocean, and investigators wanted to know why. Using an ensemble of global climate models, one study[99] concluded that an 8-year period without upper ocean warming was not unusual and occurs as a normal event in model scenarios of the climate system. Another study,[100] looking into whether

[92] Church et al. (2013)
[93] Chen et al. (2017)
[94] Tollefson (2017)

[95] Tollefson (2017)
[96] Bindoff, N.L., et al. (2007) Observations: Oceanic Climate Change and Sea Level. In Solomon, S., et al., eds, *Contribution of Working Group I to the Fourth Assessment Report (AR4) of the Intergovernmental Panel on Climate Change* (Cambridge, U.K., Cambridge University Press)
[97] Levitus, S., et al. (2005) Ocean Warming 1955–2003. Poster presented at the U.S. Climate Change Science Program Workshop, November 14–16, Arlington Va., Climate Science in Support of Decision-Making.
[98] Lyman, J.M., et al. (2010) Robust Warming of the Global Upper Ocean, *Nature*, 465 no. 7296, 334–337, doi: 10.1038/nature09043
[99] Katsman, C.A. and van Oldenborgh, C.J. (2011) Tracing the Upper Ocean's 'missing heat,' *Geophysical Research Letters* 38, L14610, doi: 10.1029/2011GL048417
[100] Loeb, N.G., et al. (2012) Observed Changes in Top-of-the-Atmosphere Radiation and Upper-Ocean Heating Consistent within Uncertainty, *Nature Geoscience*, doi: 10.1038/ngeo1375

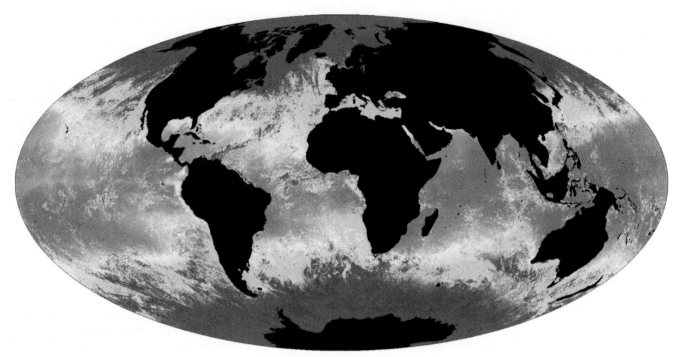

FIGURE 5.17 This sea-surface temperature map was produced using data from MODIS (Moderate Resolution Imaging Spectroradiometer, a satellite operated by NASA). The data were acquired daily over the whole globe.

Source: http://visibleearth.nasa.gov/view.php?id=54229

the rate of surface warming had slowed in the first decade of the 21st century (the so-called global warming "hiatus" discussed in **Chapter 4**), concluded that when uncertainties in measurement systems were considered, there was, in fact, no missing heat and that between January 2001 and December 2010, Earth accumulated heat in the ocean at a rate of 0.5 Watts per square meter (10.8 square foot), with no sign of a decline. This extra energy, they inferred, would eventually find its way back into the atmosphere and lead to rising surface temperatures on Earth.

In 2012, scientists provided estimates[101] of global ocean warming and its influence on sea level. Over the period 1955–2010, the heat content of the world ocean from 0 to 2000 meters (0–6560 feet) depth increased by 0.09 degrees Celsius (0.16 degrees Fahrenheit) and from 0 to 700 meters (0–2300 feet) depth, it increased by 0.18 degrees Celsius (0.32 degrees Fahrenheit). On this basis, the global ocean accounts for approximately 90 percent of the warming of the entire Earth climate system that has occurred since 1955. The ocean warming component of the sea-level trend is 0.54 millimeters per year (0.019 inches per year) for depths 0–2000 meters (0–6500 feet) and 0.41 millimeters per year (0.016 inches per year) for depths 0–700 meters (0– 2300 feet).

Understanding how ocean warming (**Figure 5.17**), and the resulting thermal expansion, contributes to sea-level rise

is important to forecast future sea-level impacts. Researchers[102] found that from 1961 to 2003, ocean temperatures to a depth of about 700 meters (2300 feet) contributed to an average rise in sea level of about 0.5 millimeters per year (0.02 inches per year). Although recent warming is greatest in the upper ocean, observations[103] also indicate that the deep ocean below 700 meters is warming.

One study[104] combined observations and modeling to conclude that deep ocean warming might have contributed 1.1 millimeters per year to the global mean sea-level rise, or one third of the altimeter-observed rate of 3.11 ± 0.6 millimeters per year (0.122± 0.02 inches per year) over the period 1993–2008. In the IPCC AR4,[105] researchers calculated that thermal expansion of ocean water is responsible for an average 5 millimeters per decade (0.2 inches per decade) of sea-level rise over the 20th century, compared to 18 millimeters per decade (0.7 inches per decade) in the first decade of the 21st century.

[102] Domingues, C.M., et al. (2008) Improved Estimates of Upper-Ocean Warming and Multi-decadal Sea-Level Rise, *Nature* 453, 1090–1093, doi: 10.1038/nature07080

[103] Johnson, G.C. and Doney, S.C. (2006) Recent Western South Atlantic Bottom Water Warming., *Geophysical Research Letters* 33, L14614, doi: 10.1029/2006GL026769. See also Johnson, G.C., et al. (2007) Recent Bottom Water Warming in the Pacific Ocean., *Journal of Climate*, 13, 2987–3002; Song, Y.T. and Colberg, F. (2011) Deep Ocean Warming Assessed from Altimeters, Gravity Recovery and Climate Experiment, in situ Measurements, and a non-Boussinesq Ocean General Circulation Model, *Journal of Geophysical Research* 116, C02020, doi:10.1029/2010JC006601

[104] Song and Colberg (2011)

[105] AR4 (2007), http://www.wmo.int/pages/partners/ipcc/index_en.html

[101] Levitus, S., et al. (2012) World Ocean Heat Content and Thermosteric Sea Level Change (0–2000), 1955–2010, *Geophysical Research Letters* 39, L10603 doi: 10.1029/2012GL051106

According to AR4, global ocean temperature has increased by 0.1 degrees Celsius (0.18 degrees Fahrenheit) from 1961 to 2003 from the surface to a depth of 700 meters (2300 feet). Ocean heat content has increased over the upper 3000 meters (9842 feet) over the same period, equivalent to absorbing a heating of 0.21 ±0.04 Watts per square meter.

Studies indicate that the temperature of the world's oceans have been trending upward for more than 100 years. In an innovative analysis of temperature records, researchers[106] compared the modern temperature of the ocean as measured by a global deployment of 3000 free-floating probes (the Argo Array[107] now enlarged to over 4000 probes) with 300 measurements taken during the historic global voyage of the *HMS Challenger* (1872–1876), the first systematic exploration of the seas.

The study shows a mean warming of the ocean surface of 0.59±0.12 degrees Celsius (1.06±0.22 degrees Fahrenheit) over the past century. Below the surface, the mean warming decreases to 0.39±0.18 degrees Celsius (0.7±0.32 degrees Fahrenheit) at 366 meters (1200 feet) and 0.12±0.07 degrees Celsius (0.22±0.13 degrees Fahrenheit) at 914 meters (3000 feet). The 0.33±0.14 degrees Celsius (0.59±0.25 degrees Fahrenheit) average temperature difference from 0 to 700 meters (2300 feet) is twice the value that has been observed globally in that depth range over the past 50 years by previous studies,[108] implying a centennial timescale for the present rate of global warming. Warming in the Atlantic Ocean is stronger than in the Pacific.

Techniques to extend ocean temperature measurements from Argo buoys back to the decades when other methods were used are important because they provide information on changes in ocean heat absorption. One study[109] shows that the rate of global warming has changed significantly in the past 60 years and that oceans are now warming about 13 percent faster than previously thought. Not only that but the warming has accelerated.

The warming rate from 1992 is almost twice as great as the warming rate from 1960 (**Figure 5.18**). Moreover, it is only since about 1990 that the warming has penetrated to depths below about 700 meters (2300 feet). The study also revealed that Earth's five oceans are sequestering heat at different rates. The Atlantic Ocean had the largest heat increase from 1960 to 2015; it was 3.5 times as high as that

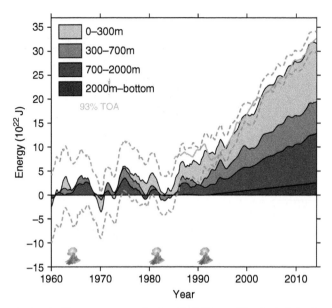

FIGURE 5.18 The ocean energy budget (10 billion trillion joules) for four depth layers: 0–300 m, 300–700 m, 70–2000 m, and 2000 m to the bottom. The warming rate from 1992 is almost twice as great as the warming rate from 1960.

Source: http://advances.sciencemag.org/content/3/3/e1601545.full.

of the Pacific Ocean, which covers twice the area. Circulation within and between oceans is the likely cause of the differences, which may themselves change if global warming alters circulation patterns.

It is interesting to compare sea-level rise today to conditions during the Eemian interglacial 120,000–130,000 years ago. At that time, average sea-surface temperatures were equivalent to the present,[110] and, remarkably, there is abundant evidence that global average sea level at that time was several meters higher than today. By analyzing[111] geologic proxies of global sea-surface temperature during the last interglacial, and comparing the data to results of global climate models simulating ocean temperatures over a 200-year period, investigators were able to calculate the contributions to sea-level rise from thermal expansion of seawater and from the melting of Greenland and Antarctica. The study revealed that the thermal expansion component of last interglacial sea-level rise was small, contributing no more than 40 centimeters (15.7 inches) to global sea level during the two-century period; Antarctic ice sheets must have contributed 2.8 to 4.5 meters of sea-level rise. Hence, researchers conclude that polar ice sheets may be sensitive to small changes in global temperature.

These results have implications for what we can expect in the warmer world we have created. The study suggests that even small amounts of warming today might have committed us to more ice sheet melting than we previously thought. The ocean temperature during the last interglacial was equal to the ocean temperatures of today, yet sea level at the time

[106] Roemmich, D., et al. (2012) 135 Years of Global Ocean Warming between the Challenger Expedition and the Argo Programme, *Nature Climate Change* 2, no. 6, 425–428, doi: 10.1038/nclimate1461

[107] Argo is a global array of over 4,000 free-drifting profiling floats that measures the temperature and salinity of the upper 2000 m (6561.6 ft) of the ocean. This allows, for the first time, continuous monitoring of the temperature, salinity, and velocity of the upper ocean, with all data being relayed and made publicly available within hours after collection. See the ARGO home page http://www.argo.ucsd.edu/index.html

[108] Levitus, S., et al. (2009) Global ocean heat content 1955–2008 in light of recently revealed instrumentation problems, *Geophysical Research Letters* 36, L07608.

[109] Cheng, L., et al. (2017) Improved estimates of ocean heat content from 1960 to 2015, *Science Advances*, DOI: 10.1126/sciadv.1601545

[110] Hoffman et al. (2017)

[111] McKay, N.P., et al. (2011) The Role of Ocean Thermal Expansion in Last Interglacial Sea-Level Rise, *Geophysical Research Letters* 38, L14605, doi: 10.1029/2011GL048280.

Greenland

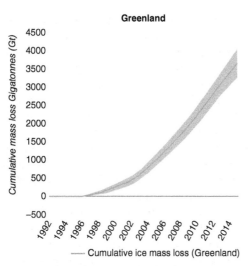

— Cumulative ice mass loss (Greenland)

Antarctica

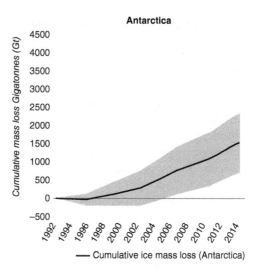

— Cumulative ice mass loss (Antarctica)

FIGURE 5.19 Both the Greenland and Antarctic ice sheets are melting. Together, they have contributed approximately 15 mm (0.6 in) to global sea-level rise over the period 1992–2015.[116]

Source: https://www.eea.europa.eu/data-and-maps/indicators/greenland-ice-sheet-3/assessment.

peaked several meters higher than present. One study[112] found a 95 percent probability that global sea level peaked at least 6.6 meters (21.6 feet) higher than today during the last interglacial; it is likely (67 percent probability) to have exceeded 8.0 meters (26.2 feet) but is unlikely (33 percent probability) to have exceeded 9.4 meters (30.8 feet).

If the research is correct, it indicates that even if we stopped greenhouse gas emissions right now, the troposphere would keep warming, the oceans would keep warming, the ice sheets would keep shrinking, and global sea level would keep rising for a long time.

Researchers have concluded that the climate system must experience a series of time lags; greenhouse gas buildup leads to atmospheric warming, ocean warming lags behind the atmosphere, ice melting lags further still, and last is thermal expansion. Researchers argue[113] that humans, by warming the atmosphere and oceans, are pushing Earth's climate toward a threshold where we will be committed to at least 4–8 meters (13–26 feet) of sea-level rise in coming centuries, with the bulk of the water coming from the melting of the great polar ice sheets.

5.3.2 Melting Ice

Excess heat in Earth's climate system produces thermal expansion of seawater, and it leads to melting of glaciers and sea ice, a decrease in the extent of snow cover, changes in the extent and timing of snowfall and rainfall, and shifts in seasonal patterns. Glacier melt and snowmelt contribute to sea-level rise, especially from Greenland and Antarctica,[114]

the two largest ice-covered regions on the planet. Both of these locations are experiencing accelerating melting (**Figure 5.19**).[115]

The cumulative loss of ice on Greenland from 1992 to 2015 was 3,600 billion tons and contributed approximately 10 millimeters (0.4 inches) to global sea-level rise over the period. The cumulative loss of ice on Antarctica over the same period was 1,500 billion tons, which corresponds to approximately 5 millimeters (0.2 inches) of global sea-level rise since 1992.[117]

Global climate models project that Greenland and Antarctic a will experience further declines in the future, but the uncertainties are large. It is estimated that over the course of the 21st century, melting of these ice sheets will contribute up to 50 centimeters (1.6 feet) of global sea-level rise.[118] Very long-term projections (until the year 3000) suggest potential sea-level rise of several meters with continued melting of the ice sheets. As we discussed in **Chapter 4**, there are also model projections suggesting that continued disruption of the North Atlantic conveyor belt will lead to rapid polar ice sheet melting if greenhouse gas emissions are not promptly curtailed.[119]

---

[112] Kopp et al. (2009)

[113] Rising Oceans – Too late to turn the tide? https://uanews.arizona.edu/story/rising-oceans—too-late-to-turn-the-tide

[114] Velicogna, I. (2009) Increasing rates of ice mass loss from the Greenland and Antarctic ice sheets revealed by GRACE, *Geophysical Research Letters* 36, L19503, doi: 10.1029/2009GL040222.

[115] NASA tracking of ice mass on Greenland and Antarctica at: https://climate.nasa.gov/vital-signs/land-ice/

[116] European Environment Agency, Greenland and Antarctic Ice Sheets https://www.eea.europa.eu/data-and-maps/indicators/greenland-ice-sheet-3/assessment

[117] Clark, P.U., et al. (2015) Recent Progress in Understanding and Projecting Regional and Global Mean Sea Level Change, *Current Climate Change Reports 1*, no. 4 (10 October): 224–46, doi:10.1007/s40641-015-0024-4.

[118] European Environment Agency, Greenland and Antarctic Ice Sheets https://www.eea.europa.eu/data-and-maps/indicators/greenland-ice-sheet-3/assessment

[119] Hansen, J.M., et al. (2016) Ice melt, sea level rise and superstorms: Evidence from paleoclimate data, climate modeling, and modern observations that 2°C global warming could be dangerous. *Atmos. Chem. Phys.*, 16, 3761-3812, doi:10.5194/acp-16-3761-2016.

Antarctica Antarctica consists of three main geographic regions: the Antarctic Peninsula, West Antarctica, and East Antarctica. In West Antarctica, which warmed 0.17 degrees Celsius per decade (0.3 degrees Fahrenheit per decade) at the same time that mean global temperature was increasing by about 0.11 degrees Celsius per decade (0.2 degrees Fahrenheit per decade), ice loss increased by 59 percent in the early 21st century.[120]

The East Antarctic ice sheet, by far the largest region of the continent, is also melting. It is losing mass at the rate of approximately 57 billion tons per year, apparently caused by increased ice loss since 2006.[121] The East Antarctic ice sheet is experiencing melting along the coastal margin in warming seas and snow accumulation in the hinterlands.

Dramatic increases in the yearly loss from ice streams along the Antarctic Peninsula have also occurred. The ice streams were broadly stable up until 2009, since then they have been losing on the order of 56 billion tons of ice each year to the ocean.[122] Researchers conclude that warm waters from the deep sea may be driving the changes by melting floating ice shelves from beneath. These ice shelves serve an important role in that they slow down, or limit, the rate of ice stream discharge into the ocean.[123]

The contribution of ice melt solely from West Antarctica has been reported as having the potential to contribute more than 1 meters (3.3 feet) of sea-level rise by 2100.[124] New observations of melting in the Amundsen Sea sector are six times higher than reported in previous research.[125] Probabilistic projections of high-end SLR that include rapid Antarctic ice sheet loss indicate that end of century ice mass-loss strongly depends on emission scenario and thereby global temperature change. Under a business as usual emission scenario, extreme SLR by end of the century could reach more than 3 meters (10 feet).[126]

Overall, the entire continent of Antarctica is experiencing increasing air temperatures.[127] All three regions are warming, and the overall rate of ice loss increased by 75 percent in the late 20th and early 21st centuries.[128] Current ice loss across the entire continent is 125+/−39 billion tons of ice per year (**Figure 5.20**).[129, 130]

Glaciers that flow into the sea around Antarctica, Greenland, and Canada can form thick (100–1000 meter; 330–3300 feet) floating platforms of ice called ice shelves. Ice shelves constitute the seaward extension of grounded glacier ice. Ice shelves slow the advance of glaciers into the ocean and when ice shelves melt or fracture, there is a possibility that the adjoining glacier can accelerate.

With direct measurements using laser altimetry on NASA's ICESat satellite, researchers[131] have learned that 20 of the 54 ice shelves studied around Antarctica are experiencing melting by warm ocean currents on their undersides, and as a result, adjoining grounded glaciers have accelerated their rate of flow.[132] This form of melting is the dominant cause of recent ice loss from the continent (**Figure 5.21**). Melting is dramatic in some cases, with some shelves thinning by a few meters per year leading to billions of tons of ice draining into the sea.

The Antarctic Peninsula gave one of the earliest warnings on the impact of a changing climate when warming air and ocean temperatures led to the dramatically fast breakup of the Larsen B ice shelf in 2002. Within approximately 1 month, 3237 square kilometers (1,250 square miles) of floating ice that had been stable for over 10,000 years were gone. In the following years, other ice shelves in the Peninsula, including the last remainder of Larsen B, collapsed, speeding up the flow of the glaciers that they were buttressing.

In 2014, West Antarctica grabbed the spotlight when two studies[133] focusing on the acceleration of glaciers in the Amundsen Sea sector showed that its collapse is underway and probably unstoppable. While one of the studies said that the demise could take 200–1,000 years, depending on how rapidly the ocean heats up, both studies concurred that the collapse is unstoppable and will add up to 3.6 meters (12 feet) of sea-level rise.

Pine Island Glacier is one of the main avenues for ice to flow from West Antarctica into the ocean. As the tip

[120] Rignot, J., et al. (2008) Recent Mass Loss of the Antarctic Ice Sheet from Dynamic Thinning," *Nature Geoscience* 1, 106–110, doi: 10.1038/ngeo102

[121] Chen, J.L., et al. (2009) Accelerated Antarctic ice loss from Satellite Gravity Measurements, *Nature Geoscience* 2, 859–862, doi: 10.1038/NGEO694

[122] Wouters, B., et al. (2015) Dynamic thinning of glaciers on the Southern Antarctic Peninsula, Science, *Science*, 22 May, v. 348, iss. 6237, p. 899-903, DOI: 10.1126/science.aaa5727

[123] Scambos, T.A. (2004). Glacier acceleration and thinning after ice shelf collapse in the Larsen B embayment, Antarctica, *Geophysical Research Letters*, 31(18).

[124] DeConto, R.M. and Pollard, D. (2016) Contribution of Antarctica to past and future sea-level rise, *Nature* 531(7596): 591–597.

[125] Khazendar, A., et al. (2016) Rapid submarine ice melting in the grounding zones of ice shelves in West Antarctica, *Nature Communications* 7, 13243. See also Scheuchl, B., et al. (2016) Grounding line retreat of Pope, Smith, and Kohler Glaciers, West Antarctica, measured with Sentinel-1a radar interferometry data. *Geophysical Research Letters*, 43 (16): 8572 DOI: 10.1002/2016GL069287

[126] Le Bars, D., Drijfhout, S., de Vries, H., 2017, A high-end sea level rise probabilistic projection including rapid Antarctic ice sheet mass loss, *Environmental Research Letters*, v. 12, no. 4, April 3.

[127] Steig, E.J., et al. (2009) Warming of the Antarctic Ice-Sheet Surface Since the 1957 International Geophysical Year, *Nature* 457, 459–462

[128] Rignot, E., et al. (2008) Recent Antarctic Ice Mass Loss from Radar Interferometry and Regional Climate Modeling, *Nature Geoscience* 1, no. 2, 106–110. doi: 10.1038/ngeo102

[129] NASA Vital Signs of the Planet page at https://climate.nasa.gov

[130] VIDEO: Runaway glaciers in West Antarctica. https://www.youtube.com/watch?v=UXBKZPWuUdE

[131] Pritchard, H.D., et al. (2012) Antarctic Ice-Sheet Loss Driven by Basal Melting of Ice Shelves. *Nature* 484, no. 7395, 502, doi: 10.1038/nature10968

[132] ThinkProgress https://thinkprogress.org/nature-antarctica-is-melting-from-below-which-may-already-have-triggered-a-period-of-unstable-7aa42de18fa3

[133] Joughlin, I., et al. (2014) Marine ice sheet collapse potentially underway for the Thwaites Glacier Basin, West Antarctica, *Science*, May 12. See also Rignot, E., et al. (2014) Widespread, rapid grounding line retreat of Pine Island, Thwaites, Smith and Kohler glaciers, West Antarctica from 1992 to 2011, *Geophysical Research Letters*.

GRACE Observations of Antarctic Ice Mass Changes

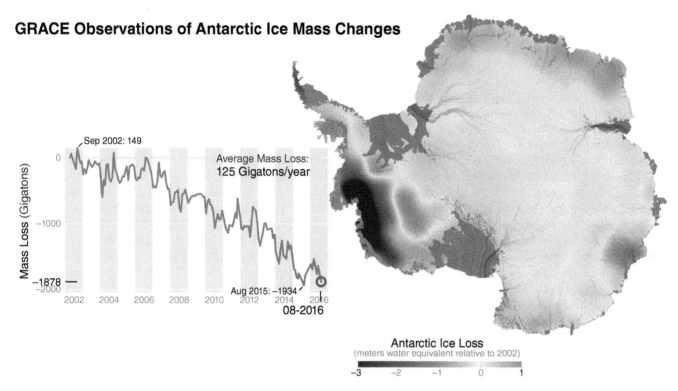

FIGURE 5.20 The NASA/German Aerospace Center Gravity Recovery and Climate Experiment (GRACE) satellite indicates that between 2002 and 2016, Antarctica shed approximately 125 gigatons of ice per year, causing global sea level to rise by 0.35 mm (0.014 in) per year.

Source: NASA/Goddard CGI Lab.

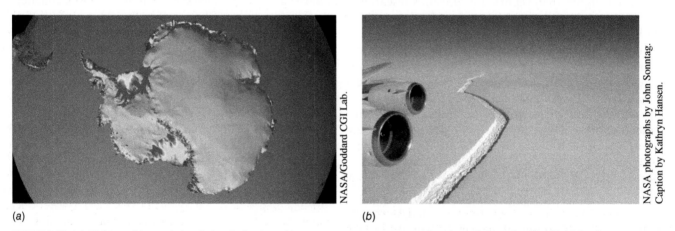

(a)   (b)

FIGURE 5.21 (a) Melting of Antarctic ice shelves is dominated by warm ocean currents sweeping along their undersides:[134] thicker ice (dark; greater than 550 m [1,800 ft]), and thinner ice (light; less than 200 m [656 ft]). (b) Ice shelves not only melt on their undersides, they can fracture and release large icebergs into the sea. Larsen C, a floating platform of glacial ice on the east side of the Antarctic Peninsula, is the fourth largest ice shelf ringing Antarctica. In 2014, a crack that had been slowly growing into the ice shelf for decades started to spread northward. In July of 2017, a massive iceberg broke off the ice shelf, causing the Larsen C shelf area to shrink by approximately 10%.[135]

Source: NASA/Goddard CGI Lab.

of the glacier melts and thins, the glacier is discharging more ice into the sea. The glacier has been losing about 20 billion tons of ice each year for the last two decades, but scientists see this rising to 100 billion tons per year in coming decades. Pine Island Glacier accounts for about one-fourth of the total ice melt from the West Antarctic Ice Sheet.

If the entire West Antarctic Ice Sheet retreated, it could cause sea level to rise by more than 3 meters (10 feet). The entire Antarctic Ice Sheet, covering an area bigger

---

[134] NASA, http://www.nasa.gov/topics/earth/features/currents-ice-loss.html

[135] NASA https://www.nasa.gov/feature/goddard/2017/massive-iceberg-breaks-off-from-antarctica

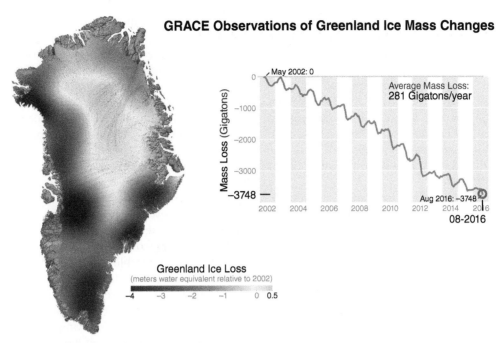

GRACE Observations of Greenland Ice Mass Changes

FIGURE 5.22 The mass of the Greenland ice sheet has rapidly declined in the 21st century due to surface melting and iceberg calving. The NASA/German Aerospace Center Gravity Recovery and Climate Experiment (GRACE) satellite indicates that between 2002 and 2016, Greenland shed approximately 280 gigatons of ice per year, causing global sea level to rise by 0.8 m (0.03 in) per year.

Source: https://svs.gsfc.nasa.gov/30879.

than the continental United States, contains nine-tenths of the ice on Earth and could raise sea level by over 200 meters (656 feet) if it melted.

Greenland The contribution of the Greenland continental glacier to sea-level rise has also been measured. Increased melting of the Greenland ice sheet has been observed, and it is known that the glacier is becoming smaller.[136] The balance between annual ice gained and lost is in deficit, and the deficiency tripled between 1996 and 2007 (**Figure 5.22**).[137, 138]

In Greenland, 2007 marked a rise to record levels of the summertime melting trend over the highest altitudes of the ice sheet. Melting in areas above 2000 meters (6,560 feet) rose 150 percent above the long-term average, with melting occurring on 25–30 more days in 2007 than the average in the past 19 years.[139] Crevasses, fractures in the ice that promote sliding, melting, and faster movement, have

been seen to grow in some areas over the past two decades, suggesting that mechanical processes are speeding up ice movement into the ocean.[140] Scientists have found that glaciers in southern Greenland are melting faster than they were 10 years ago, and the overall amount of ice discharged into the sea has increased from 20 cubic kilometers (5 cubic miles) in 1996 to 54 cubic kilometers (13 cubic miles) in 2005, an increase of 25 percent.

These observations agree with studies of global mean sea-level rise that found that the contribution of the Greenland ice sheet to global sea level increased from 5 percent in 1992 to 25 percent in 2014.[141] The study is one of only a few works to confirm acceleration in sea-level rise during recent decades (the era of satellite altimetry). There had been greater uncertainty about this before, with climate deniers latching onto that and arguing that such acceleration has not, in fact, been occurring. However, by using calculations of the various contributing factors to sea-level rise, such as melting ice sheets, thermal expansion, and other factors, researchers found that global mean sea level increased from about 2.2 millimeters (0.09 inches) per year in 1993 to 3.3 millimeters (0.13 inches) per year in 2014. This is consistent with other studies that found recent sea-level rise as measured by satellite altimetry is occurring three times faster than in the 20th century.[142]

[136] Schrama, E.J.O. and Wouters, B. (2011) Revisiting Greenland Ice Sheet Mass Loss Observed by GRACE, *Journal of Geophysical Research* 116, B02407, doi: 10.1029/2009JB006847

[137] Rignot, E., et al. (2008) Mass Balance of the Greenland Ice Sheet from 1958 to 2007, *Geophysical Research Letters* 35, L20502, doi: 10.1029/2008GL035417

[138] VIDEO: Greenland's thinning ice. https://www.youtube.com/watch?v=Rl7mPdZCRKg

[139] NASA, Earth Observatory, Melting Anomalies in Greenland in 2007, http://earthobservatory.nasa.gov/Newsroom/NewImages/images.php3?img_id=17846

[140] Colgan, W., et al. (2011) An Increase in Crevasse Extent, *West Greenland: Hydrologic Implications, Geophysical Research Letters* 38, L18502, doi: 10.1029/2011GL048491

[141] Chen et al. (2017)

[142] Dangendorf et al. (2017)

In 2010, the melting of Greenland ice broke previous records,[143] with melting in some regions extending up to 50 days longer than average. Ice melting on Greenland also spread to previously stable portions of the northwest coast.[144] Researchers reported[145] that melting in 2010 started exceptionally early at the end of April and ended quite late in mid-September; summer temperatures up to 3 degrees Celsius (5.4 degrees Fahrenheit) above average were combined with reduced snowfall; and Nuuk, the capital of Greenland, had the warmest spring and summer since records began in 1873.

How vulnerable is the Greenland ice sheet to irreversible melting? Research suggests that it may be more susceptible than previously assumed. A model simulation of Greenland ice in a warmer world[146] reveals that the temperature threshold for melting the ice sheet completely is in the range of 0.8–3.2 degrees Celsius (1.4–5.7 degrees Fahrenheit) of global warming, with a best estimate of 1.6 degrees Celsius (2.8 degrees Fahrenheit) above preindustrial levels. Global warming has already exceeded the minimum of this range (1.1 degrees Celsius [1.98 degrees Fahrenheit]), and the Arctic is on track to double this amount before mid-century.[147]

The time it takes before most of the ice in Greenland is lost strongly depends on the level of warming. The more we exceed the threshold, the faster it melts. According to the simulation,[148] in a business-as-usual scenario of greenhouse gas emissions, in the long run, humanity might be aiming at 8 degrees Celsius (14.4 degrees Fahrenheit) of global warming. This would result in one-fifth of the ice sheet melting within 500 years and a complete loss in 2000 years.

Under certain conditions, the melting of the ice sheet becomes irreversible because of climate feedbacks. For instance, Greenland's ice is more than 3,000 meters (9,842 feet) thick, and much of its surface is located at cooler high elevations. Prolonged melting will lower the surface of the ice to warmer elevations, preventing it from rebuilding again. Also, the loss of sunlight-reflecting ice cover and its replacement with heat-absorbing seawater and dark rock will prevent the ice sheet from future growth, even if climate returned to its preindustrial state. Melting the total ice on Greenland would result in a sea-level rise of about 6.5 meters[149] (21 feet) and affect many of the world's major cities, which are located on coastlines because of historical ties to shipping.

Greenland's contribution to average sea-level rise is accelerating: Ice losses quickened in 2006–2008 to the equivalent of 0.75 millimeters per year (0.03 inches per year) of sea-level rise, from an average 0.46 millimeters per year (0.018 inches per year) for 2000–2008.[150] Icebergs breaking away and meltwater runoff are equally to blame for the shrinking ice sheet. Greenland contributions account for between 20 and 38 percent of the observed yearly global sea-level rise.[151] As glacier acceleration continues to spread northward from its original focus in southern Greenland, the global sea-level rise contribution from the world's largest island continues to increase.

The northeast corner of Greenland was thought to be the last remaining stable portion of the ice sheet. However, a group of researchers has identified a long "river" of ice—known as the Zachariae ice stream—that has retreated about 20 kilometers (12.4 miles) over the last decade.[152] For comparison, one of the fastest moving glaciers, the Jakobshavn ice stream in southwest Greenland, has retreated 35 kilometers (21.7 miles) over the last 150 years. Ice streams drain large regions; the same way the Mississippi River drains central North America. The Zachariae ice stream drains over 16 percent of the Greenland ice sheet—an area twice as large as the one drained by Jakobshavn. Researchers found that the northeast Greenland ice sheet lost about 10 billion tons of ice per year from April 2003 to April 2012. Increased ice flow in this region is particularly troubling because the northeast ice stream stretches more than 600 kilometers (about 373 miles) into the center of the ice sheet, where it connects with the heart of Greenland's ice reservoir.

Many glaciers draining the Greenland ice sheet end in the ocean. These are especially vulnerable to melting as ocean water warms and causes them to retreat. Although several marine-terminating glaciers are known to have retreated over the past decade, the extent and magnitude of retreat relative to past history were unknown until a study[153] in mid-2011. Scientists used changes in the front positions of 210 marine-terminating glaciers using Landsat imagery spanning nearly four decades, and they compared rates of change with earlier observations to the early 20th century. They found that 90 percent of the observed glaciers retreated between 2000 and 2010, approaching 100 percent in the northwest, with rapid retreat observed

[143] Tedesco, M., et al. (2011) The Role of Albedo and Accumulation in the 2010 Melting Record in Greenland, *Environmental Research Letters* 6, 014005, doi: 10.1088/1748-9326/6/1/014005

[144] Khan, S.A., et al. (2010) Spread of Ice Mass Loss into Northwest Greenland Observed by GRACE and GPS, *Geophysical Research Letters* 37, L06501, doi: 10.1029/2010GL042460

[145] Early melting http://www.sciencedaily.com/releases/2011/01/110121144011.htm

[146] Robinson, A., et al. (2012) Multistability and Critical Thresholds of the Greenland Ice Sheet, *Nature Climate Change* 2, 429–432, doi: 10.1038/NCLIMATE1449

[147] Smith, S.J., et al. (2015) Near-term acceleration in the rate of temperature change, *Nature Climate Change*, March 9, DOI: 10.1038/nclimate2552

[148] Robinson et al. (2012)

[149] Poore, R.Z., et al. (2000) Sea Level and Climate, U.S. Geological Survey Fact Sheet 002–00, http://pubs.usgs.gov/fs/fs2-00/

[150] van den Broeke, M., et al. (2009) Partitioning Recent Greenland Mass Loss, *Science* 326, no. 5955, 984, doi: 10.1126/science.1178176

[151] Rignot, E. and Kanagaratnam, P (2006) Changes in the Velocity Structure of the Greenland Ice Sheet, *Science* 311 986–989

[152] Khan, S.A., et al. (2016) Sustained mass loss of the northeast Greenland ice sheet triggered by regional warming. *Nature Climate Change*, DOI: 10.1038/NCLIMATE2161

[153] Howat, I.M. and Eddy, A. (2011) Multi-decadal Retreat of Greenland's Marine-Terminating Glaciers, *Journal of Glaciology* 57, no 203, 389–396

throughout the entire ice sheet. The retreat today is accelerating and likely began between 1992 and 2000, which was the onset of warming in the region. Previously, during the middle of the 20th century, glaciers were largely stable, and they even advanced, coincident with a cooling period. The early 20th century was warm, and although there was extensive glacier retreat at that time, the current retreat is more widespread.

Research now indicates that both Greenland and Antarctica are experiencing melting at faster rates than originally expected by scientists. Greenland's summer melt season now lasts 70 days longer than it did in the early 1970s. Every summer, warmer air temperatures cause melt over about half of the surface of the ice sheet—although recently, 2012 saw an extreme event where 97 percent of the ice sheet experienced melt at its top layer.

Alpine Glaciers and Ice Caps

Alpine (or mountain) glaciers worldwide are melting.[154] These glaciers contribute to global sea-level rise.[155] Additionally, for the 200 million people in communities that depend on seasonal snow and ice melting as a source of freshwater (Himalayan, western North America, Andes mountains regions, and others), the retreat and eventual loss of these ice centers deliver a fundamental blow to sustainability.[156]

Thermal expansion was the main driver of global sea level rise for 75–100 years after the start of the Industrial Revolution, though its relative contribution has declined as the shrinking of land ice has accelerated. Land ice—glaciers, ice caps, and ice sheets—is shrinking at a faster rate in response to rising temperatures, adding water to the world's oceans. As the rate of ice loss has accelerated, its contribution to global sea-level rise has increased from a little more than half of the total increase from 1993–2008 to 75–80 percent of the total increase between 2003 and 2007.[157]

The GRACE satellite mission detects changes in the gravitational field on Earth's surface. Researchers completed a comprehensive study using data from GRACE[158] documenting the contribution of the world's melting glaciers and ice caps to global sea-level rise. The study concluded that Earth's glaciers and ice caps outside of the regions of Greenland and Antarctica are shedding roughly 150 billion tons (or 163 cubic kilometers; 39 cubic miles) of ice annually, and this is sufficient to raise global mean sea level to approximately 0.4 millimeters per year (0.016 inches per year).

The data on melting glaciers, ice caps, and ice sheets were collected by the GRACE satellite (Gravity Recovery and Climate Experiment), a joint effort of space programs in NASA and Germany. The GRACE mission consists of two satellites (launched in 2002) that orbit Earth together 16 times per day at an altitude of about 482 kilometers (300 miles). Traveling together, the two satellites sense subtle variations in Earth's mass and gravitational pull caused by regional changes in the planet's mass, including changes in ice sheets, oceans, water stored in the soil and underground aquifers, and in Earth's mantle.

Data indicate that Earth has lost a total of 4.3 trillion tons of ice between 2003 and 2010.[159] Greenland and Antarctica lost the bulk of the ice, but nearly a quarter of the losses came from glaciers in Alaska, Canada, and Patagonia (**Figure 5.23**). Southern Alaska lost 46 billion tons of ice per year, or 8 percent of the total. Islands in northern Canada lost 67 billion tons of ice, or 12 percent of the total. Patagonia lost 23 billion tons of ice per year, or 4 percent of the total.

Warming temperatures lead to the melting of alpine glaciers, and the total volume of glaciers on Earth is declining sharply.[160] In fact, worldwide, current glaciers are out of balance with current climatic conditions, indicating that glaciers will continue to shrink in the future even without further temperature increase.[161] Glaciers have been retreating worldwide for at least the last century, and the rate of retreat has increased in the past decade.[162] Only a few glaciers are actually advancing (in locations that are well below freezing and where increased precipitation has outpaced melting). The progressive disappearance of glaciers has implications not only for a rising global sea level but also for water supplies in regions of Asia, South America, and western North America.

Mountain glaciers are retreating and thinning in nearly every mountainous region of the planet. For instance, over the period 2007–2009, a sharp increase in the rate of ice mass loss due to melting made the Canadian Arctic Archipelago, the single largest contributor to global sea-level rise outside of Greenland and Antarctica.[163] Researchers have documented that melting there was due largely to warmer summertime temperatures, to which rates of ice loss are highly sensitive.

The cumulative mean thinning of the world's mountain glaciers has accelerated from about −1.8 to −4 meet (−6 to −13 feet) between 1965 and 1970 to about −12 to −14 meet (−40 to −46 feet) of change in the first decade of

[154] Jacob, T., et al. (2012) Recent contributions of glaciers and ice caps to sea level rise, *Nature*, 482, p. 514-518, Feb. 23, doi:10.1038/nature10847

[155] Radić, V. and Hock, R. (2012) Regionally differentiated contribution of mountain glaciers and ice caps to future sea-level rise, *Nature Geoscience*, 4, 91-94, doi: 10.1038/ngeo1052

[156] Melting glaciers http://www.ipsnews.net/2011/09/200-million-depend-on-melting-glaciers-for-water/

[157] Fact Sheet, Union of Concerned Scientists http://www.ucsusa.org/global_warming/science_and_impacts/impacts/causes-of-sea-level-rise.html#.WX1gvcaZOV4

[158] Jacob et al. (2012)

[159] Jacob et al. (2012)

[160] NOAA website at http://www.ncdc.noaa.gov/indicators/

[161] IPCC-AR5, Vaughan, D.G., et al. (2013) Observations: Cryosphere. In: *Climate Change 2013: The Physical Science Basis. Contribution of Working Group I to the 5th Assessment Report of the Intergovernmental Panel on Climate Change* [Stocker, T.F., et al. (eds.)]. Cambridge University Press, Cambridge, UK and New York, NY.

[162] Chen et al. (2013)

[163] Gardner, A.S., et al. (2011) Sharply increased mass loss from glaciers and ice caps in the Canadian Arctic Archipelago, *Nature* 473, 357-360

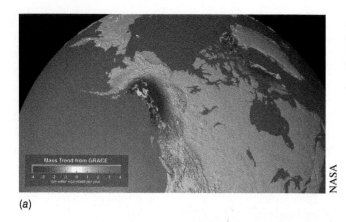
(a)

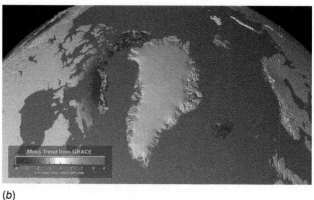
(b)

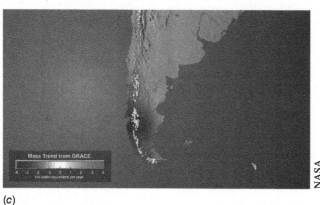
(c)

FIGURE 5.23 Between 2003 and 2010, Earth has lost a total of 4.3 trillion tons of ice. This image shows the average yearly change in mass, in centimeters of water, from glaciers and ice caps in three regions: (a) Alaska, (b) Canada, and (c) Patagonia. These three regions lost a combined 136 billion tons of ice, or 24% of the total. The remainder was lost by Greenland and Antarctica. Locations with large ice loss stand out clearly with dark shading. The bright circles mark the locations of glaciers.

Source: https://svs.gsfc.nasa.gov/10892.

the 21st century.[164] Over the period 1961–2003, mountain glaciers contributed an estimated 0.5 millimeters per year (0.02 inches per year) to global sea-level rise, increasing to 0.8 millimeters per year (0.03 inches per year) for the period 1993–2003.[165]

Researchers show in a 2018 study[166] that further melting of glaciers cannot be prevented in the current century - even if all emissions were stopped now. However, due to the slow reaction of glaciers to climate change, human behavior has a massive impact beyond the 21st century. Whether the average temperature rises by 2 degrees Celsius or only 1.5 degrees Celsius makes no significant difference for the development of glacier mass loss over the next 100 years. Around 36 percent of the ice still stored in glaciers today

would melt even without further emissions of greenhouse gases. That means: more than one-third of the glacier ice that still exists today in mountain glaciers can no longer be saved even with the most ambitious measures.

5.3.3 Groundwater

Another important source of sea-level rise is groundwater extraction. As human population increases and the use of water for manufacturing, agriculture, and all types of industrial and domestic purposes grow; the natural renewal rate of groundwater stores has not been able to keep pace with the rate of human use. The vast majority of water drawn from the ground is ultimately released to become runoff to the sea. Hence, human withdrawal of groundwater has become a small but measurable component of global sea-level rise.

Using calibrated hydraulic models, analysis of observational data, and inferences about human use, groundwater depletion during the period 1900–2008 is estimated[167] to equal approximately 4,500 cubic kilometer (1,080 cubic miles).

[164] Meier et al. (2007). See also WWF Nepal Program, *An Overview of Glaciers, Glacier Retreat, and Subsequent Impacts in Nepal, India and China* (Katmandu, World Wildlife Fund, 2005), assets.panda.org/downloads/himalayaglaciersreport2005.pdf

[165] Dyurgerov, M.B. and M. F. Meier (2005) Glaciers and the Changing Earth System: A 2004 Snapshot, *Occasional Paper 58*) (Boulder, Colo., Institute of Arctic and Alpine Research, University of Colorado), http://instaar.colorado.edu/other/occ_papers.html

[166] Marzeion, B., et al. (2018) Limited influence of climate change mitigation on short-term glacier mass loss. *Nature Climate Change*, DOI: 10.1038/s41558-018-0093-1

[167] Konikow, L.F. (2011) Contribution of Global Groundwater Depletion since 1900 to Sea-Level Rise, *Geophysical Research Letters* 38, L17401, doi: 10.1029/2011GL048604.

Researchers calculate that this is equivalent to a sea-level rise of 12.6 millimeters (0.5 inches), or more than 6 percent of the total sea-level rise of the entire period; however, groundwater withdrawal has increased substantially since 1950, and more recently (2000–2008), the rate of withdrawal averaged approximately 145 cubic kilometers per year (35 cubic miles per year) or about 0.40 millimeters per year (0.016 inches per year) of sea-level rise. This number is 13 percent of the reported rate of sea-level rise during the period, 3.1 millimeters per year (0.12 inches per year).

More recent studies[168] have found that groundwater extraction and other land water uses contribute about three times less to sea level rise than previously estimated. This information does not change the overall picture of future sea level rise, but it does provide a much more accurate understanding of the interactions between water on land, in the atmosphere, and the oceans, which could help to improve future models of sea level rise.

5.3.4 Future Commitment to Sea-Level Rise

It is important to realize that past greenhouse gas emissions have committed the world to continued sea level rise for several centuries, and our present and future emissions choices affect the rise in seas and the pace at which it unfolds beyond 2050.[169]

Even if global warming emissions were to drop to zero before 2020 (an impossibility), scientists project another 0.36 to 0.80 meters (1.2 to 2.6 feet) of global sea-level rise by 2100 as oceans and land ice adjust to the changes we have already made to the atmosphere.[170] Projections for global sea-level rise by 2100 range from 0.2 to 2.0 meters (0.66 to 6.6 feet) above 1992 levels, though the lowest end of this range is a simple extension of historic sea level rise—and recent data indicate this rate has doubled in recent years.

It may take until the 2060s to know how much the sea level will rise by the end of this century. Research[171] indicates that Earth faces a broad range of possible outcomes with climate change. At the less severe end, 0.6 meters (2 feet) of global-average sea-level rise by 2100 would submerge land that's currently home to about 100 million people. Toward the high end, 1.8 meters (6 feet) of rise would swamp the current homes of more than 150 million. Either scenario has drastic implications for the world's major cities, most of which are located at ports and harbors established centuries ago.

The world can make lower sea-level rise outcomes much more likely by meeting the 2015 Paris Agreement goal of bringing net greenhouse gas emissions to zero in the second half of this century.[172] The IPCC-AR5 results have been confirmed by other studies. They found that global sea-level rise in a high-emissions future would likely be between 0.6 and 1.1 meters (2 and 3.5 feet) by 2100. However, these studies fail to consider an important physical process, marine ice-cliff collapse.

Marine ice-cliff collapse becomes increasingly likely as ice shelves retreat. Ice shelves are the seaward ends of huge ice streams that flow toward the ocean. In West Antarctica, currently dominated by rapidly melting ice shelves, the ocean floor gets deeper in the landward direction, so each new iceberg that breaks away exposes taller and taller cliffs. Ice gets so heavy that these taller cliffs can't support their own weight. Once they start to crumble, the destruction would be unstoppable. When this process is included in climate models, the resulting sea-level rise ranges up to 1.2–2.1 meters (4 to 7 feet). By contrast, marine ice-cliff instability doesn't have much effect if we meet the Paris Agreement emissions goal. That keeps the likely global rise in the range of about 0.6 to 1.1 meters (2 and 3.5 feet) by 2100.

Even with declining emissions, sea level is predicted to rise by at least 80 centimeters (2.6 feet) at the end of this century and is expected to continue rising for at least the next two hundred years. The rate and magnitude of the loss of ice sheets, primarily in Greenland and West Antarctica, will have the greatest effect on long-term sea-level rise. Thus, the choices we make today will determine how high sea level rises this century, how fast it occurs, and how much time we have to protect our communities.[173]

Researchers have assessed the role that anthropogenic carbon emissions lock in long-term sea-level rise. Long-term sea level rise over centuries greatly exceeds projections for this century.[174] This poses a profound challenge for human communities located in coastal areas. Unabated carbon emissions up to the year 2100 would commit the world to an eventual global sea-level rise of 4.3–9.9 meters (14.1 to 32.5 feet).[175]

Within the United States, land that is home to more than 20 million people is threatened by this rise and is widely distributed among different states and coasts. The total area exposed to sea-level impacts includes 1,185–1,825 municipalities that are home to more than half of the current population, among them at least 21 cities exceeding 100,000 residents. Under aggressive carbon cuts, more than half of these municipalities would avoid this commitment if the West Antarctic Ice Sheet remains stable.

[168] Wada, Y., et al. (2016) Fate of water pumped from underground and contributions to sea-level rise. *Nature Climate Change*, DOI: 10.1038/NCLIMATE3001

[169] Schaeffer, M.W., et al. (2012) Long-term sea-level rise implied by 1.5° C and 2° C warming levels, *Nature Climate Change* doi:10.1038/nclimate1584.

[170] Zecca, A., and Chiari, L. (2012) Lower bounds to future sea-level rise. *Global and Planetary Change* 98–99:1–5. Abstract online at *http://www.sciencedirect.com/science/article/pii/S0921818112001579*

[171] Kopp, R.E., et al. (2017) Evolving understanding of Antarctic ice-sheet physics and ambiguity in probabilistic sea-level projections. *Earth's Future*, DOI: 10.1002/2017EF000663

[172] Sea-level rise projections made hazy by Antarctic instability https://www.sciencedaily.com/releases/2017/12/171213095545.htm

[173] Sea level rise and global warming: An infographic from the Union of Concerned Scientists: http://www.ucsusa.org/sites/default/files/legacy/assets/documents/global_warming/Methodology-and-Assumptions-UCS-Sea-Level-Rise-and-Global-Warming-Infographic.pdf

[174] Levermann et al. (2013)

[175] Strauss et al. (2015)

BOX 5.1 | Spotlight on Climate Change

Changes in the Cryosphere[176]

Careful scientific observations show that Earth's ice system (known as the **cryosphere** among scientists), has experienced strong and significant changes in recent decades. Melting is widespread and, in many cases accelerating. The IPCC-AR5 published a summary of the observed variations in the cryosphere that provides a basis for understanding this new planetary phenomenon.

Frozen Ground: Permafrost temperatures have risen by as much as 2°C (3.6°F) and the thickness of the "active" layer (zone of permafrost melt) has increased by up to 90 cm (3 ft) since the early 1980s. In the Northern Hemisphere, the southern limit of permafrost has been moving north since the mid-1970s, and there has been decreasing thickness of seasonal frozen ground by 32 cm (1 ft) since the 1930s.

Snow Cover: Between 1967 and 2012, satellite data show decreases in snow cover throughout the year, with the largest decreases (53%) in June. Most locations report strongest decreases in snow in spring.

Lake and River Ice: The period of winter ice is contracting with delays in autumn freeze-up proceeding more slowly than advances in spring breakup. There is evidence of recent acceleration in both across the Northern Hemisphere.

Glaciers: Alpine glaciers are a major contributor to global sea-level rise. Ice mass loss from glaciers has increased since the 1960s. Rates of ice loss from glaciers outside Greenland and Antarctica were equivalent to raising sea level 0.76 mm/yr (0.029 in/yr) from 1993 to 2009 and 0.83 mm/yr (0.032 in/yr) from 2005 to 2009.

Sea Ice: Between 1979 and 2012, Arctic sea ice extent declined at a rate of 3.8% per decade with larger losses in summer and autumn. Over the same period, the extent of thick multiyear ice in the Arctic declined at a higher rate of 13.5% per decade. Mean sea ice thickness decreased by 1.3 to 2.3 m (4.3 to 7.5 ft) between 1980 and 2008.

Ice Shelves and ice tongues: The continuing retreat and collapse of regions of the Larsen ice shelf along the Antarctic Peninsula threaten the stability of their adjoining ice streams. Progressive thinning has been observed among other ice shelves/ice tongues in Antarctica and Greenland.

Ice Sheets: Both Greenland and Antarctic ice sheets lost mass and contributed to sea-level change over the last 20 years. The rate of total loss and discharge from a number of major outlet glaciers in Antarctica and Greenland has increased over this period (**Figure 5.24**).

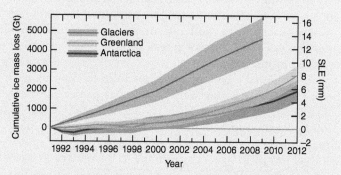

FIGURE 5.24 The cumulative ice mass loss from glaciers and ice sheets (in sea level equivalent) is 1.0 to 1.4 mm/yr (0.039 to 0.055 in/yr) over the period 1993-2009. Over the period 2005-2009, it was higher at 1.2 to 2.2 mm/yr (0.047 to 09.086 in/yr). Thus, the rate of global sea-level rise is accelerating.

Source: p. 367, https://www.ipcc.ch/pdf/assessment-report/ar5/wg1/WG1AR5_Chapter04_FINAL.pdf.

5.4 Sea-Level Impacts

5.4.1 End of the Century

Sea-level rise is one of the main consequences of global warming.[177] Knowing how fast sea-level rise can develop given a scenario of future greenhouse gas emissions is crucial to inform both mitigation and adaptation choices. According to the IPCC AR5,[178] global mean sea-level rise is projected to reach approximately 98 centimeters (3.2 feet) by the end of the century as the outer bound of the modeled worst-case emissions scenario (RCP 8.5). However, subsequent studies of melting in Antarctica,[179] Greenland,[180] and alpine ice

[176] IPCC-AR5, Vaughan et al. (2013)

[177] Arnell, N.W., et al. (2016) Global-scale climate impact functions: the relationship between climate forcing and impact. *Climatic Change*, 134(3), 475–487.

[178] Church et al. (2013)

[179] DeConto and Pollard (2016). Khazendar et al. (2016). Scheuchl et al. (2016).

[180] Tedesco, M., et al. (2016) The darkening of the Greenland ice sheet: trends, drivers, and projections (1981-2100) *The Cryosphere*, 10, 477–496.

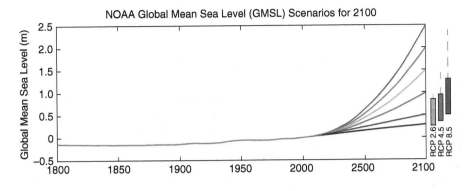

(a)

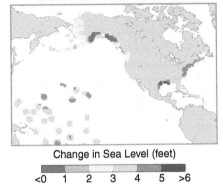

Projected Relative Sea Level Change for 2100
under the Intermediate Scenario

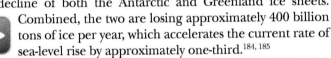

Change in Sea Level (feet)

(b)   <0 1   2   3   4   5 >6

FIGURE 5.25 (a) Global mean sea level (GMSL) rise from 1800 to 2100, the six Interagency Global Mean Sea Level (GMSL) scenarios the very likely ranges in 2100 for different RCPs (shaded boxes), and lines augmenting the very likely ranges by the difference between the median Antarctic contribution of Kopp et al. (2017) and the various median Antarctic projections of DeConto and Pollard (2016). (b) Relative sea level (RSL) rise (ft) in 2100 projected for the Interagency Intermediate Scenario (1 m [3.3 ft] GMSL rise by 2100).[187]

Source: https://science2017.globalchange.gov/chapter/12/

systems,[181] and the rate of ocean warming,[182] suggest that it is now physically plausible to see more than double this amount.[183] A large contributor to this is the accelerating decline of both the Antarctic and Greenland ice sheets.

> Combined, the two are losing approximately 400 billion tons of ice per year, which accelerates the current rate of sea-level rise by approximately one-third.[184, 185]

A comprehensive 2017 report[186] by the National Oceanic and Atmospheric Administration (NOAA) presents six scenarios of future sea-level rise (**Figure 5.25**) on the basis of peer-reviewed information regarding the Greenland and Antarctic ice sheets. While the NOAA scenarios are not model projections, they are physical assessments based on peer-reviewed research observations, and they do take into account new understanding of the behavior of the ice sheets that have not been considered in previous studies.

Notably, the IPCC AR5 *worse-case* scenario is depicted in the NOAA report as an *intermediate* scenario of future sea level rise (aqua line in Figure 5.24). The NOAA report breaks the six scenarios into levels of sea-level rise associated with particular time periods this century (Table 5.3).

These values offer planners a time frame for assessing the duration of major infrastructure projects in the context of accelerations of global sea-level rise.

Sea-level rise presents challenges to coastal communities and ecosystems. Modeled estimates of sea-level rise by 2100, for a temperature rise of 4 degrees Celsius (7.2 degrees Fahrenheit) or more over the same time frame, is between 0.5 and 2 meters (1.64 to 6.6 feet); potentially a devastating result globally placing up to 187 million people at risk of forced displacement.[188] In the United States, approximately 32,000 square kilometers (12,355 square miles) of land lies within one vertical meter of the high tide line, encompassing 2.1 million housing units where 3.7 million people live.[189]

5.4.2 Impact Analysis

It is important that community managers, resource officials and other decision-makers, and community groups concerned with natural hazards and environmental conservation begin the process of planning for the impacts

[181] Ciraci, E., et al. (2015) Mass Loss of Glaciers and Ice Caps from GRACE during 2002-2015. Presented at the American Geophysical Union - Fall 2015 Meeting.

[182] Glecker, P.J., et al. (2016) Industrial era global ocean heat uptake doubles in recent decades. *Nature Climate Change.* doi:10.1038/nclimate2915

[183] Sweet et al. (2017a). See also Le Bars (2017).

[184] Ciraci et al. (2015)

[185] VIDEO: Major sea level rise in the near future. https://www.youtube.com/watch?v=f7sEhuSbQo8

[186] Sweet et al. (2017a)

[187] Sweet, W.V., et al. (2017b) Sea level rise. In: Climate Science Special Report: Fourth National Climate Assessment, Volume I[Wuebbles, D.J., et al., eds.]. U.S. Global Change Research Program, Washington, DC, USA, pp. 333-363, doi: 10.7930/J0VM49F2.

[188] Nicholls, R., et al. (2011) Sea-Level Rise and its Possible Impacts Given a Beyond 4C World in the Twenty-First Century, *Philosophical Transactions of the Royal Society A* 369: 161–181 doi: 10.1098/rsta.2010.0291

[189] Tebaldi, C., et al. (2012) Modeling sea-level rise impacts on storm surges along US coasts, *Environmental Research Letters* 7, 014032, doi: 10.1088/1748-9326/7/1/014032.See also Strauss et al. (2012)

TABLE 5.3												

Global Mean Sea-Level Rise Scenario Heights in Meters for 19-Year Averages Centered on Decade Through 2200 (Showing Only a Subset after 2100) Initiating in Year 2000. Only Median Values Are Shown.[190]

Global Mean Sea Level (meters)	2010	2020	2030	2040	2050	2060	2070	2080	2090	2100	2120	2150	2200
Low	0.03	0.06	0.09	0.13	0.16	0.19	0.22	0.25	0.28	0.30	0.34	0.37	0.39
Intermediate low	0.04	0.08	0.13	0.18	0.24	0.29	0.35	0.4	0.45	0.50	0.60	0.73	0.95
Intermediate	0.04	0.10	0.16	0.25	0.34	0.45	0.57	0.71	0.85	1.0	1.3	1.8	2.8
Intermediate high	0.05	0.10	0.19	0.30	0.44	0.60	0.79	1.0	1.2	1.5	2.0	3.1	5.1
High	0.05	0.11	0.21	0.36	0.54	0.77	1.0	1.3	1.7	2.0	2.8	4.3	7.5
Extreme	0.04	0.11	0.24	0.41	0.63	0.90	1.2	1.6	2.0	2.5	3.6	5.5	9.7

of sea-level rise.[191] Because sea level is rising now, and is very likely to rise at increasingly accelerated rates in the future,[192] coastal communities cannot avoid the immediate impacts by mitigating global warming. Some amount of sea-level rise has already been set into unstoppable motion, and it is thus incumbent upon communities to begin the process of adapting to the inevitable impacts.

Coastal communities around the world are beginning to realize that sea-level rise is not a hypothetical hazard. Year after year, valid, peer-reviewed scientific studies confirm that sea-level rise is a serious threat that is growing more ominous as studies reveal that the ice sheets are melting faster than anticipated and the ocean is absorbing heat faster than originally thought probable. As an example of a coastal community that is preparing for a future of higher ocean levels, the island state of Hawaii has engaged in sea-level rise planning by modeling the impacts of higher oceans and assessing the community assets that will be affected.

Hawaiian Islands Of the threats posed by sea-level rise, there are two fundamental categories of negative impacts: 1.) high-frequency events that typically are not life-threatening and do not constitute disasters (i.e., coastal erosion, groundwater inundation including drainage problems, and seasonal high wave flooding) and 2.) low-frequency events that will become more damaging because of sea-level rise and that may be life-threatening and may constitute disasters (i.e., hurricane storm surge and tsunamis). It is important to plan for both categories.

Low-frequency events that will become more damaging because of sea-level rise include hurricane storm surge and tsunami. Both of these processes occur on the Hawaiian coast. When a hurricane strikes, storm surge poses an enormous threat to life and property. For instance, when New Orleans was hit by Hurricane Katrina in 2005, at least 1500 persons lost their lives and many of those deaths occurred directly, or indirectly, as a result of storm surge.

Storm Surge Storm surge is a rapid rise of water that is pushed onto the coast by a hurricane. Three mechanisms contribute to storm surge: 1) wind blowing onto the coast piles up water (potentially more than 85 percent of the surge); 2) waves push water inland faster than it can drain off (called wave setup, potentially 5–10 percent of the surge); low pressure of a hurricane sucking water higher into the air near the eye (potentially 5–10 percent of the surge). The stage of tide at the time the surge hits plays an important role because at high tide the water depth can increase an additional 0.3 to 1.0 meters (1 to 3 feet), this can spell the difference between moderate damage and catastrophic damage.

The level of flooding associated with storm surge depends on several factors: the size and intensity of a hurricane; the angle that it approaches the coast; the slope of the seabed at the coastline; the slope of the land at the coast; and how fast the hurricane is moving. Winds in a Northern Hemisphere hurricane rotate counter clockwise. Hence, the forward right quarter of a hurricane approaching a coastline is the most dangerous area. The wind is blowing onshore, the forward speed of the storm adds to the wind speed, and the storm surge is greatest.

Storm surge moves with the forward speed of the hurricane—typically 16 to 24 kilometers per hour (10 to 15 miles per hour). This wind-driven water has tremendous power because 0.76 cubic meters (1 cubic yard) of sea water weighs 784 kilograms (1,728 pounds)—almost a ton. Hence, a 30-centimeters (1 feet) deep storm surge can sweep a car off the road. Compounding the destructive power of the rushing water is the large amount of floating debris that typically accompanies the surge. Trees, pieces of buildings, and other debris float on top of the

---

[190] Sweet et al. (2017a)

[191] Rahmstorf, S. (2012) Sea-Level Rise: Towards Understanding Local Vulnerability, *Environmental Research Letters* 7, 021001, doi: 10.1088/1748-9326/7/021001

[192] Hansen, J.E. (2007) Scientific Reticence and Sea-Level Rise, *Environmental Research Letters* 2, 024002, doi: 10.1088/1748-9326/2/2/024002

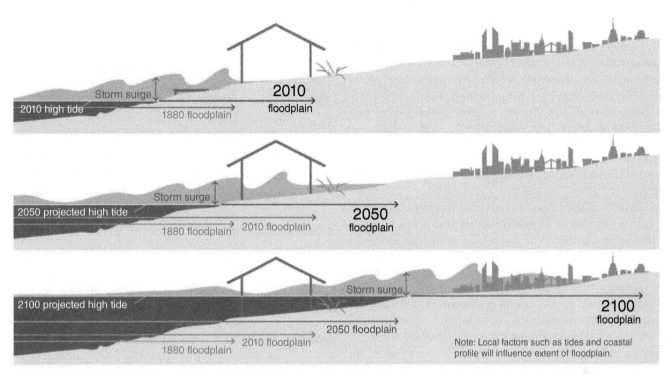

FIGURE 5.26 As sea level rises, storm surge causes more damage by penetrating farther inland. With higher global sea levels in 2050 and 2100, areas farther inland will be at risk of being flooded.[194]

Source: https://www.ucsusa.org/sites/default/files/legacy/assets/documents/global_warming/Causes-of-Sea-Level-Rise.pdf.

storm surge and act as battering rams that can cause severe damage to any buildings unfortunate enough to stand in the way.[193]

Since the 19th century, global sea level has risen an average of 0.20 meters (8 inches). Sea-level rise is expected to accelerate through this century as the climate continues to warm. With rising seas, storm surge is amplified. That means a storm today could create more extensive flooding than an identical storm a century ago, with sometimes catastrophic damage to our homes and critical infrastructure (**Figure 5.26**). Even if there were no immediate increase in the strength of hurricanes from climate change, continued sea-level rise means storm surge will be more destructive.

These impacts will affect human coastal communities as well as coastal ecosystems. Careful planning based on the best available science that identifies the location, magnitude, and timing of these impacts, is needed for communities to adapt effectively, and avoid hasty and environmentally damaging emergency interventions. Modeling that provides spatially accurate visualizations of impacts is especially important for developing implementable policy approaches and for initiating long-term planning.

Especially, given the uncertainty that still exists around the timing and severity of impacts, scenario-based modeling can be a useful tool to allow planners and policy makers to evaluate a range of risk scenarios and select from them as appropriate to the nature and scale of the project at hand. For instance, it will be important for major infrastructure projects with long life spans to plan for both low- and high-frequency impacts of sea-level rise. While, depending on the specifics of the project, it might be acceptable for short-term, minor development projects to only plan for high-frequency impacts.

Groundwater Inundation One of the impacts of sea-level rise that is not widely appreciated is that flooding will not only cross the shoreline from the sea, it will come from beneath the ground.[195] This phenomenon is known as **groundwater inundation**.[196] In most coastal settings, the water table lies at an elevation of approximately mean sea level or somewhat above. It rises and falls with the tides and even with weather patterns; thus, it is intimately and immediately connected to the surface of the sea.

---

[193] Description from Weather Underground https://www.wunderground.com/prepare/storm-surge

[194] From Union of Concerned Scientists https://www.ucsusa.org/sites/default/files/legacy/assets/documents/global_warming/Causes-of-Sea-Level-Rise.pdf

[195] Bjerklie, D.M., et al. (2012) *Preliminary Investigation of the Effects of Sea-Level Rise on Groundwater Levels in New Haven, Connecticut,* U.S. Geological Survey Open-File Report 2012–1025, http://pubs.usgs.gov/of/2012/1025/

[196] Rotzoll, K. and Fletcher, C.H. (2013) Assessment of groundwater inundation as a consequence of sea-level rise. *Nature Climate Change* 3, p. 477–481

This connectivity decays with distance from the shoreline, but within several blocks of the ocean, the position of the water table can be strongly correlated to marine processes (i.e., tides, waves, etc.). Thus, a rise in sea level related to global warming means that the coastal water table will rise (**Figure 5.27**), possibly to the point of breaking through the ground surface and creating new wetlands; not a desirable feature in an urban setting at the foundation of a building or road, or in an ecosystem that serves as a refuge for endangered species that are unaccustomed to saturated soil and free-standing water bodies.[197]

Analysis[198] of the impacts of groundwater inundation in the urban corridor of Honolulu found that it will threaten $5 billion of taxable real estate, flood nearly 48.2 kilometers (30 miles) of roadway, and impact pedestrians, commercial and recreation activities, tourism, transportation, and infrastructure. Researchers found that the flooding will occur regardless of seawall construction and thus will require innovative planning and intensive engineering efforts to accommodate standing water in the streets.

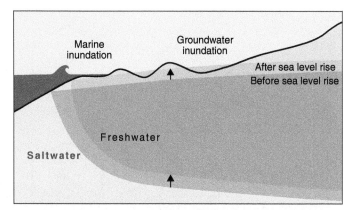

FIGURE 5.27 Sea-level rise lifts the water table, causing groundwater inundation in low-lying areas.

Source: N. Hulbirt

Figure 5.28 is a model[199] of sea-level vulnerability of the central urban core of Honolulu, Hawaii—a major coastal city of over 400,000 people. Mapped in blue is the extent of standing surface water at high tide. Mapped in yellow are

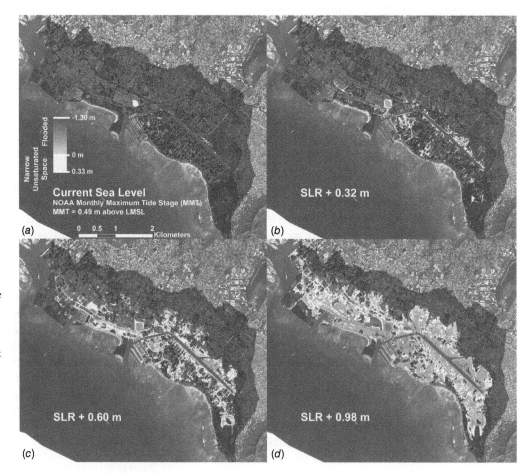

FIGURE 5.28 Groundwater model of Honolulu, Hawaii. Dark shading indicates standing surface water, light shading indicates vadose zone 30 cm (1 ft) or less in thickness. Black defines the study area. Flooding and changes in the vadose zone due to sea-level rise are indicated at (a) current levels, (b) 32 cm (1 ft), (c) 60 cm (2 ft), and (d) 98 cm (3 ft). Surface flooding originates from groundwater inundation, storm drain flooding, and flow across the shoreline.

Source: Habel et al. (2017).

---

[197] Habel, S., et al. (2017) Development of a model to simulate groundwater inundation induced by SLR and high tides in Honolulu, Hawaii. *Water Research*, 114, 122–134.

[198] Habel et al. (2017)

[199] Habel et al. (2017)

FIGURE 5.29 As sea level rises, coastal erosion threatens homes, roads, and other beachfront development. Protecting these assets with rock walls is called armoring. On a retreating shoreline, armoring leads to beach narrowing and beach loss. Beach loss, as shown here, represents the destruction of a critical natural ecosystem, the loss of a tourism asset, and a blow against the lifestyle that coastal residents seek in their lives.

Source: Brad Romine.

locations where dry soil is only 30 centimeters (1 foot) thick or less. The figure shows model results for present-day high tide and results for 30 centimeters (1 foot) increments of sea-level rise to 98 centimeters (3.2 feet) above present level.

Most of the flooded area is not connected to the ocean, and inundation is a result of seawater flowing out of storm drainage pipes along road sides, the water table breaking through the land surface to create wetlands, and the collection of runoff into pools because it cannot drain into the ground or the ocean. In nonstorm conditions, these lands are more vulnerable to the rise of the water table as a main cause of flooding than direct marine inundation. An aspect of a high water table is that rainfall and runoff of all types will experience restricted drainage due to the saturation of the ground surface by groundwater and the inundation of the storm drainage system by seawater.

Coastal Erosion Communities located in most coastal settings engage in some form of tourism economy, and beaches are the most desirable natural asset these communities have. However, as sea level rises, coastal erosion increasingly threatens beachfront land.[200] During sea-level rise, beaches, which are by definition environments located at the edge of the ocean, are forced to retreat landward. In so doing, beachfront land is eroded—this is called coastal erosion.

Seawalls are often used to protect homes, roads, and other types of developed infrastructure on beachfront lands from the effects of erosion. This practice is known as **armoring,** and while it provides a temporary reprieve from

coastal erosion, armoring on retreating beaches eventually leads to beach narrowing and **beach loss** (**Figure 5.29**).[201]

If left alone to migrate landward with the rising ocean levels, most beaches will survive sea-level rise just fine, as they have for 20,000 yrs since global sea level was 130 meters (400 feet) lower at the culmination of the last ice age. Beach loss represents the destruction of a critical natural ecosystem, the loss of a tourism asset, and a blow against the lifestyle that coastal residents seek in their lives. Armoring the shoreline temporarily solves one problem, but it creates another that may be equally or more serious.

Because coastal erosion can lead to negative outcomes, it is useful to model future erosion to identify where it is likely to strike, which beaches are most threatened, and to provide data for strategic planning. Simply responding to coastal erosion by building seawalls is reactive and does not constitute good planning. For example, on the Hawaiian island of Oahu, sections of world-famous Sunset Beach are characterized by chronic erosion.[202] In 2015, there were 18 beachfront homes (10 percent of the beachfront homes) directly threatened by erosion. Using a model of erosion associated with sea-level rise, it is possible to estimate the rising threat to beachfront homes at Sunset.[203]

With only 30 centimeters (1 feet) of sea-level rise, the number of buildings threatened by erosion at Sunset Beach swells to 109, comprising 60 percent of the beachfront homes.

[200] Romine, B.M., et al. (2016) Beach erosion under rising sea-level modulated by coastal geomorphology and sediment availability on carbonate reef-fringed island coasts. *Sedimentology*, 63(5), 1321–1332.

[201] Fletcher, C.H., et al. (1997) Beach loss along armored shorelines on Oahu, Hawaiian islands. *Journal of Coastal Research*, v. 13, p. 209–215.
[202] Fletcher, C.H., et al. (2012) National assessment of shoreline change: Historical shoreline change in the Hawaiian Islands: U.S. Geological Survey Open-File Report 1051.
[203] Anderson, T.R., et al. (2015) Doubling of coastal erosion under rising sea level by mid-century in Hawaii. *Natural Hazards*, 78(1), 75–103. https://doi.org/10.1007/s11069-015-1698-6.

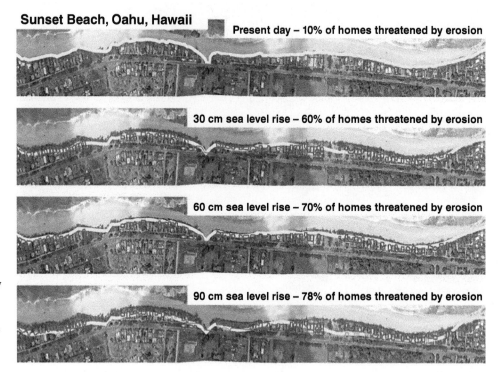

Sunset Beach, Oahu, Hawaii

Present day – 10% of homes threatened by erosion

30 cm sea level rise – 60% of homes threatened by erosion

60 cm sea level rise – 70% of homes threatened by erosion

90 cm sea level rise – 78% of homes threatened by erosion

FIGURE 5.30 Model depiction of erosion threat with sea-level rise on Sunset Beach Oahu, Hawaii. The light band is the 50th percentile location of the shoreline (the landward edge of the beach). Today (top panel), the beach has a relatively small number of homes (10% of the beachfront homes) threatened by erosion (squares). With 30 cm (1 ft) of sea-level rise (second panel), the number of homes threatened by erosion swells to 109, constituting 60% of the buildings. At 60 cm (2 ft) of sea-level rise (third panel), erosion threatens 138 buildings (70% of the homes), and with 98 cm (3.2 ft) of sea-level rise (fourth panel), 175 buildings and 1 km of roadway are threatened. Notable in this trend is the rapid increase in threatened homes associated with only 30 cm of sea-level rise.

Source: C. Fletcher.

If a significant number of these buildings were to build sea-walls, it would spell the swift end of Sunset Beach, an iconic stretch of shore that is world famous and a critical part of the tourism allure of the Hawaiian Islands. Sunset Beach is also home to endangered monk seals that haul onto the sand to rest and to give birth to their pups. Green sea turtles, another endangered species, also use the beach for egg laying and resting.

The modeling sends a clear message that unless authorities prepare soon with comprehensive policies designed to protect the beach, near-term sea-level rise is going to trigger a cascade of seawall construction, which will cause serious environmental and economic problems. It is incumbent upon authorities that are responsible for protecting natural coastal environments to consider this future and develop planning strategies to avoid the worst effects (**Figure 5.30**).

Seasonal High Waves Impacts related to 98 centimeters (3.2 feet) of sea-level rise include flooding by predictable high waves that approach Hawaiian waters every year. These seasonal high waves are responsible for the renowned surfing culture originated by ancient Hawaiians and for which Hawaii is world famous today. On shores facing to the south, high waves produced by storminess in the southern hemisphere arrive every summer, and on shores facing to the north, high waves produced by storminess in the North Pacific arrive every winter.

Presently, these seasonal high waves may be responsible for occasional flooding of the first row of beachfront homes. But modeling the characteristics of these annual waves under higher sea levels reveals that between 60 and 98 centimeters (2 to 3.2 feet) of sea level rise, wave flooding moves from being damaging, to catastrophic. This threshold has

been called a **critical point**.[204] It is the sea level height beyond which flooding rapidly accelerates and threatens an entire beachfront community. A critical point can be thought of as a "tipping point," a limit at which some aspect of the climate system irretrievably shifts to a new state (**Figure 5.31**).

Exposed Assets Hawaiian authorities have assessed the combined assets that are exposed to high-frequency hazards as a result of sea-level rise.[205] With only 30 centimeters (1 foot) of sea-level rise, the developed assets and land value that are exposed to erosion, groundwater inundation, and/or high seasonal waves sums to over $4 billion, including 22 kilometers (13.7 miles) of roadway. Approximately 2000 residents will be displaced. At 98 centimeters (3.2 feet) of sea-level rise, a total $12.9 billion of developed assets and land value are exposed with over 13,000 residents displaced, and 111 kilometers (69.3 miles) of roadway at risk (Table 5.4).

Hawaiian authorities have also assessed the potential impact of low-frequency hazards by modeling the 100-year flood, a flood event that has a 1 percent chance of happening in any year (this includes storm surge, tsunami, and runoff types of floods). With 30 centimeters (1 foot) of sea level rise, $70 billion of structure and land value are exposed, 330 kilometers (205 miles) of roads are at risk, and over

[204] Kane, H.H., et al. (2015) Critical elevation levels for flooding due to SLR in Hawaii. *Regional Environmental Change*, 15(8), 1679-1687.

[205] Hawai'i Climate Change Mitigation and Adaptation Commission (2017) Hawai'i Sea Level Rise Vulnerability and Adaptation Report. Prepared by Tetra Tech, Inc. and the State of Hawai'i Department of Land and Natural Resources, Office of Conservation and Coastal Lands, under the State of Hawai'i Department of Land and Natural Resources Contract No: 64064. https://climateadaptation.hawaii.gov/wp-content/uploads/2017/12/SLR-Report_Dec2017.pdf

(a)

Chip Fletcher

(b)

Chip Fletcher

FIGURE 5.31 Models of annual wave run-up on the south shore of Oahu indicate that a critical point of flooding is passed at 60 cm (2 ft) of sea-level rise. A visual estimation of the difference of wave flooding clearly identifies the transition from (a) 60 cm of sea-level rise to (b) 98 cm (3.2 ft) of sea-level rise as a significant jump in flooded area. This critical point reflects a major hazard threshold that should be considered in all forms of community planning and sea-level rise adaptation.

55,000 residents are displaced. A total 13,500 structures are flooded. With 98 centimeters (3.2 feet) of sea-level rise, the potential impacts rise to $120 billion in structure and land value, 500 kilometers (310 miles) of roads, 73,700 residents displaced, and 17,700 structures flooded.

5.4.3 Adapting to Sea-Level Rise

Following an analysis[206] and calculation of global sea-level projections to the end of the century, study leader Dr. Anthony

[206] Committee on Sea Level Rise in California, Oregon, and Washington; Board on Earth Sciences and Resources; Ocean Studies Board; Division on Earth and Life Studies; National Research Council *Sea level rise for the coasts of California, Oregon, and Washington: Past, Present, and Future*, (2012), National Academies Press, Washington, D.C.

TABLE 5.4

Oahu Island Hawaii—Sea-Level Rise Impacts

Assets at risk from coastal erosion, seasonal wave flooding, and groundwater inundation

	30-cm sea-level rise	98-cm sea-level rise
Structures flooded	650	3800
Residents displaced	2000	13,300
Km of road flooded	22.1	111.5
Structure and land loss	$4.1 billion	$12.9 billion

Assets at risk from 1% annual chance coastal flood (storm surge, tsunami, or runoff)

	30-cm sea-level rise	98-cm sea-level rise
Structures flooded	13,500	17,700
Residents displaced	55,100	73,700
Km of road flooded	330	500
Structure and land loss	$70 billon	$120 billion

Dalrymple stated "There will be about 1 meter of sea level rise by 2100." As you might imagine, for a human community located on the edge of the sea, the prospect of 1 meter (3.3 feet) or more of sea-level rise represents a significant challenge, possibly affecting as many as 3.7 million people.[207] Since the 2012 date of Dalrymple's report, as we have seen, there are reasons to believe that sea level by the end of the century may exceed 1 meter (3.3 feet) above present.[208]

As a result of sea-level rise, marine inundation from storm surge and tsunamis will increase in severity and frequency, coastal erosion and flooding will increase, and ecosystems near intertidal elevations will be affected. Studies[209] estimate that by 2050, one-third of coastal communities in the United States will see an increase in the frequency of extreme high water levels that currently only happen once per century.

Small storms that previously had little impact on the coast will begin, over time, to cause greater damage. Buildings and roads located at the water's edge will experience wave-related flooding and structural damage, placing demands on emergency workers, public works crews, and utility companies. Tourism and private land ownership will be threatened as beach erosion spreads and increases in severity. Ecosystems, human communities, infrastructure, ports and harbors, and other coastal assets will all come under increasing attack during the course of the 21st century. Lands previously dry throughout a tidal cycle will experience flooding at high tide, even in the absence of storms.[210]

[207] Tebaldi et al. (2012); See also Strauss et al. (2012)
[208] Rahmstorf, S. (2012) Sea-level rise: towards understanding local vulnerability, *Environmental Research Letters*, v. 7, no. 2, http://iopscience.iop.org/article/10.1088/1748-9326/7/2/021001
[209] Tebaldi et al. (2012)
[210] VIDEO: Sea level rise accelerating. http://climatecrocks.com/2011/05/06/sea-level-rise-accelerating/

Because it is too late to stop the global warming happening today, communities are faced with a dual challenge: 1.) Adapting to the unavoidable consequences of present warming and 2.) Mitigating the worst effects of future warming. Engaging in the process of adaptation is no reason to cease efforts to limit the production of greenhouse gases; working to prevent the worst impacts of future warming is more important than ever. If we love the generation of our grandchildren as much as we love ourselves, we must limit the worst effects of future warming. For the present generation, however, adaptation has become a reality that must be embraced even though it is expensive and time-consuming and will slowly begin to take over as a core mission of government agencies in the next few decades.

Adapting to sea-level rise is not a one-time event for a coastal community; it will become an unending series of expenses, decisions, and construction projects lasting through this century and into the next. Adaptation is, in other words, a process[211] requiring community participation; technical skill in geographic information systems (digital cartography); data on the spatial distribution of developed infrastructure, ecosystems, and the built environment; scientific knowledge to build future scenarios of climate change; leadership; and decision-making. Fundamentally, a community must engage in a planning process leading to judgments about what assets are at risk and about what assets are to be protected, moved, or sacrificed.

The National Oceanographic and Atmospheric Administration (NOAA) has developed guidance[212] on this issue on behalf of coastal communities around the United States and elsewhere. They articulate a planning process that begins with identifying a planning team, scoping the level of effort, and educating and involving stakeholders. This early phase in sea-level adaptation requires leadership from a small number of knowledgeable individuals, an agency office, or an elected official. NOAA recommends that a new state law or executive order authorizing the climate change adaptation planning process would help ensure it has adequate resources, support, and legitimacy. This could require educating elected officials, which should be done early in the planning process.

Once a planning team is assembled, stakeholders are involved, and responsibilities are established, it is important to assess community vulnerability to sea-level rise. This begins with a technical effort to determine the impacts of sea-level rise, such as accelerated coastal erosion, increased inundation due to tsunami and hurricane storm surge, and drainage problems related to tidal flooding of runoff infrastructure such as suburban storm sewer systems.

A focused analysis involves modeling inundation, mapping high-resolution topography in the coastal zone, determining erosion patterns under high rates of sea-level rise, and other technical steps. Planners typically use a geographic information system of map layers depicting community assets, including vulnerable ecosystems, social phenomena, transportation assets, and other key features of the coastal zone. By this stage, the planning team will have identified the sea level–related physical processes likely to affect the coast, examined the associated impacts, and assessed what it is about the coast that could affect its vulnerability to climate change. Armed with this knowledge, the planning team can develop scenarios that illustrate (**Figure 5.32**) potential impacts and consequences of sea-level rise.

After assessing the vulnerability of community and ecosystem assets to sea-level rise, planners should define the steps needed to adapt to and mitigate negative impacts. This phase requires developing an adaptation strategy characterized by features such as:

1. Reducing the vulnerability of the built environment to sea-level rise,
2. Monitoring and maintaining healthy coastal ecosystems,
3. Reducing the expense and building the capacity of disaster response and recovery,
4. Protecting critical infrastructure,
5. Minimizing economic losses attributable to the impacts of climate change,
6. Adapting to sea-level rise in a manner that minimizes harm to the natural environment and loss of public access, and others.

FIGURE 5.32 Under 1 m (3.3 ft) of sea-level rise, the tourist mecca of Waikiki, Hawaii, will be severely affected by seawater inundation at high tide, and the negative impacts of this will ripple throughout the tourism-based economy of the entire state of Hawaii. This image shows an analysis of sea-level rise impact on the built environment. *Blue* indicates flooding by 90 cm (3 ft) of sea-level rise. Buildings are color-coded by vulnerability to higher sea level based on their elevation. *Red* indicates buildings located at modern high tide. *Orange* indicates buildings vulnerable to 30 cm (1 ft) of sea-level rise. *Yellow* indicates buildings vulnerable to 60 cm (2 ft) sea-level rise. *Green* indicates buildings vulnerable to 90 cm (3 ft) sea-level rise.

[211] Marcy, D., et al. (2012) Incorporating Sea Level Change Scenarios at the Local Level, NOAA Coastal Services Center, http://www.csc.noaa.gov/digitalcoast/_/pdf/slcscenarios.pdf

[212] National Oceanic and Atmospheric Administration, "Adapting to Climate Change: A Planning Guide for State Coastal Managers," http://coastalmanagement.noaa.gov/climate/adaptation.html

Prioritizing these many steps, funding them, and executing the many projects that embody these goals require the development of an action plan that needs implementation and ongoing management and revision throughout the 21st century.

Some planners have begun the arduous process of assessing management options and developing new policies. For instance, state agencies in California have mapped the impact zone of a 1.4 meters (4.6 feet) rise in sea level after having calculated that by the end of the century lands at this elevation and lower are vulnerable to negative impacts. They have identified the land and development that is vulnerable to inundation,[213] including 480,000 people, $100 billion in property, 140 schools, 34 police and fire stations, 55 health-care facilities, 330 EPA hazardous waste sites, 3,500 miles of roads and highways, 280 miles of railroads, 30 power plants, 28 wastewater treatment plants, and more.[214] For each one of these assets, a decision-making process must be engaged to decide on the specific adaptation strategy to be employed (even if it is to abandon the asset), a budget scoped and funded, and a construction project detailed and completed.

Adaptation is, in short, a massive operation.

5.4.4 SLR Refugees

Unmitigated SLR is expected to reshape the distribution of human communities, potentially stressing landlocked areas unprepared to accommodate a wave of coastal migrants.[215] By the year 2100, rising sea levels could force up to 2 billion people inland, creating a refugee crisis among one-fifth of the world's population.[216] Worse yet, there won't be many places for those migrants to go.

Researchers have identified three obstacles, or "barriers to entry," that stand in the way of people driven inland from their homes by rising seas.[217] The first is that drought and desertification make some areas uninhabitable at worst, and incapable of sustaining a large influx of migrants at best. Second: if climate refugees flock to cities, the increasing urban sprawl might require land formerly used to grow food. Those communities could lose the ability to feed their inflated populations. Third: regions and municipalities might erect walls and post guards to prevent climate migrants from entering and settling down. This phenomenon is dubbed the "no-trespass zone."

This discussion highlights a problem; too much of the conversation around SLR adaptation is focused on building sea walls, learning to live with regular flooding, and relocating communities inland. SLR greatly exceeding 1 m (3.3 feet), barriers to entry, and the occurrence of rapid onset events such as hurricanes and extreme flooding that force residents to relocate ahead of planning, mean that these overly simplistic ideas of adaptation could leave human communities unprepared for a mass migration that could dwarf the current refugee crisis in Europe.[218]

5.4.5 How High?

As we have discussed already, to properly design community adaptation strategies, it is desirable to have an estimate of sea-level rise this century. An approximation of the amount of expected sea-level rise, by certain years that serve as benchmarks during the course of this century, can allow planners to develop a number of key assessments: vulnerability to coastal hazards (such as increased risk from tsunamis, storm surge, and coastal erosion); flooding and drainage threats to coastal assets such as the built environment, coastal ecosystems, and others; development of climate risk-management policies; and development of urban planning and ecosystem conservation modeling scenarios.[219]

Geologic observations shed light on the natural rate and magnitude of sea-level change. Researchers[220] have reconstructed sea-level fluctuations over the past 22,000 years, spanning the period from the last glacial maximum to the present interglacial warm phase (beginning about 11,500 years ago when the Younger Dryas[221] came to an end). From this work, it is apparent that changing climate, in the form of shifts in ice volume and global temperature (the kind of processes operating today), is responsible for driving sea-level changes over this period.

Researchers who reconstructed the relationship between climate and sea level predict 4 to 24 centimeters (1.6 to 9.4 inches) of sea-level rise over the 20th century, in agreement with other peer-reviewed findings. When used to forecast sea-level heights over the 21st century on the basis of

---

[213] California Executive Order S-13-08 2008 at http://gov.ca.gov/executive-order/11036/

[214] Pacific Institute at: http://www.pacinst.org/reports/sea_level_rise/index.htm

[215] Hauer, M.E. (2017) Migration induced by sea-level rise could reshape the US population landscape, *Nature Climate Change*, 7, 321-325, doi:10.1038/nclimate3271

[216] Geisler, C. and Currens, B. (2017) Impediments to inland resettlement under conditions of accelerated sea level rise, *Land Use Policy, v.* 66 (July), p. 322-330, https://doi.org/10.1016/j.landusepol.2017.03.029

[217] Geisler and Currens (2017)

[218] Geisler and Currens (2017)

[219] VIDEO: The Sea Also Rises: Prepare for a Changing Shoreline. https://www.youtube.com/watch?v=pKyK_YpGdn8

[220] Siddall, M., et al. (2009) Constraints on Future Sea-Level Rise from Past Sea-Level Change, *Nature Geoscience*, 2, 571–575, doi: 10.1038/ngeo587.

[221] We discussed the Younger Dryas (YD) in Chapter 4. The YD was an abrupt climate event consisting of pronounced cooling during the transition from the last ice age to the current interglacial. Paleoclimate proxies indicate the YD occurred ca. 12,800 to 11,500 years ago. It was most strongly expressed in the North Atlantic region leading to speculation that it was associated with a slow-down, perhaps a full cessation, of the Atlantic Meridional Overturning Circulation (AMOC). Generally the YD is not considered an example of global climate change, although slowing of the AMOC can have global consequences.

modeled temperature projections of 1.1 to 6.4 degrees Celsius (2 to 11.5 degrees Fahrenheit), the reconstruction predicts 7 to 82 centimeters (2.8 to 32.3 inches) of sea-level rise by the end of this century. This range overlaps with IPCC-AR5 estimates, increasing confidence in the projections.[222]

Another study[223] used a similar approach, developing an equation that estimates the relationship between climate and sea-level change over the past 2,000 years. The researchers found that sea-level rise by the end of the century will be roughly three times higher than predictions by the IPCC-AR5. They also conclude that even if temperature rise were stopped today, sea level will still rise another 20 to 40 centimeters (7.9–15.7 inches) and that actual cooling would be needed to stop the ongoing rise. According to this model, the most optimistic emissions scenario, one that produces a temperature rise of less than 2 degrees Celsius (3.6 degrees Fahrenheit) by 2100, will nonetheless result in a sea-level rise of 80 centimeters (31.5 inches). The most pessimistic scenario, one that produces a temperature rise of 4 degrees Celsius (7.2 degrees Fahrenheit) by 2100, results in a sea-level rise of 1.35 meters (4.4 feet). The study estimates that sea level will rise 0.9 to 1.3 meters (3 to 4.3 feet) by 2100.

These estimates are consistent with a study published in the Proceedings of the *National Academy of Sciences* in late 2009[224] and subsequently tested and verified using multiple modeling and statistical tests.[225] This work concludes that a simple relationship links global sea-level variation on timescales of decades to centuries to global mean temperature, namely about 1 meter (3.3 feet) of global sea-level rise will result from warming of 1.8 degree Celsius (3.24 degrees Fahrenheit).

This provides a basis for predicting sea level by the end of the century using IPCC emission scenarios. The model projects a sea-level rise of 0.75 to 1.90 meter (2.5 to 6.2 feet) for the period 1990–2100 (**Figure 5.33**). The study authors note that to limit the amount of global sea-level rise to a maximum of 1 meter (3.3 feet) in the long run, reductions in greenhouse gas emissions would likely have to be deeper than those needed to limit global warming to 2 degrees Celsius (3.6 degrees Fahrenheit), which is the policy goal now supported by many countries and signed into international law by the Paris Agreement.[226]

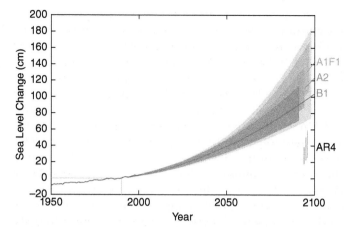

FIGURE 5.33 Sea-level rise from 1990 to 2100 based on IPCC AR4 temperature projections for three different emission scenarios (labeled on *right*).[227] The sea-level range projected in the IPCC AR4[228] for these scenarios is shown for comparison in the bars on the right. Also shown is the observations-based annual global sea-level data[229] (*red*) including artificial water reservoir correction.[230]

Source: Vermeer and Rahmstorf (2009) http://www.pnas.org/content/106/51/21527.full.

Several other researchers have published estimates of sea-level rise by 2100: 0.5 to 1.4 meters[231] (1.6–4.6 feet), 0.8 to 2.0 meters[232] (2.6–6.6 feet), 1.6 meters[233] (5.2 feet), the NOAA report[234] that we discussed earlier, and there are others.

5.4.6 Modeling Issues

There are two problems with the so-called semi-empirical modeling used in Figure 5.33.[235] 1) This approach relies on relatively simple relationships that are statistically established between global temperature change and global sea-level change that might not hold true in the warming future, rather than attempting to reproduce the actual physical relationships and feedbacks that result in sea-level changes. 2) Semi-empirical modeling does not account for the high regional variability that characterizes real sea-level change.

[222] Church et al. (2013)

[223] Grinsted, A., et al. (2010) Reconstructing Sea Level from Paleo and Projected Temperatures 200 to 2100 AD, *Climate Dynamics* 34, no. 4, 461–472, doi: 10.1007/s00382-008-0507-2

[224] Vermeer, M, and Rahmstorf, S. (2009) Global Sea Level Linked to Global Temperature, *PNAS* 106, no. 51, 21527–21532, http://www.pnas.org/content/106/51/21527.full

[225] Rahmstorf, S., et al. (2011) Testing the Robustness of Semi-Empirical Sea Level Projections, *Climate Dynamics*, doi: 10.1007/s00382-011-1226-7

[226] United Nations Framework Convention on Climate Change homepage http://unfccc.int/2860.php

[227] Vermeer and Rahmstorf (2009)

[228] Solomon et al. (2014)

[229] Church and White (2006)

[230] Chao, B.F., et al. (2008) Impact of Artificial Reservoir Water Impoundment on Global Sea Level, *Science* 320, 212–214.

[231] Rahmstorf (2007)

[232] Pfeffer, W.T., et al. (2008) Kinematic Constraints on Glacier Contributions to 21st Century Sea-Level Rise, *Science* 321, no. 5894, 1340–1343.

[233] Rohling, E.J., et al. (2008) High Rates of Sea-Level Rise during the Last Interglacial Period, *Nature Geoscience* 1, 38–42.

[234] Sweet et al. (2017a)

[235] Willis, J. and Church, J.A. (2012) Regional Sea-Level Projections, *Science* 336, 550–551.

Regional variability[236] is a term describing the fact that the ocean surface is not flat. As was shown in Figure 5.2, sea level can rise and fall simultaneously in different parts of the world owing to forcing by winds, currents, differences in heating, vertical changes in the seafloor and land, and shifts in Earth's gravity field due to melting ice and the weight of water on the sea floor.

Global climate models (GCMs), such as used in IPCC assessment reports, produce estimates of future global mean sea-level rise that are typically significantly less than the results of semi-empirical modeling. This is because GCMs, to date, have not been able to account for the dynamics of surging continental-scale ice sheets on Greenland and Antarctica. Especially difficult is modeling the behavior of marine-based segments of ice sheets such as found in the Amundsen Sea sector of Antarctica. For instance, IPCC-AR4 projected a global mean sea-level rise of 18 to 59 centimeters (7.1–23.2 inches) by 2100, but lacked an assessment of ice dynamics (such as ice acceleration, fracturing, and collapse) in Greenland and Antarctica because these processes were too poorly understood to be adequately modeled.

More-recent global climate modeling using IPCC-AR5 climate forcing scenarios has been geared toward improving understanding of regional variability in sea-level rise, but has not advanced understanding of ice sheet dynamics. To account for this gap, researchers may choose to insert a fixed value for ice sheet decay based on satellite observations and expert judgment.[237] This is especially important given that melting ice sheets are now the largest contributor[238] to sea-level rise and the rate at which they are melting is accelerating.[239] An 18-year satellite study found that from 1 year to the next, the Greenland ice sheet lost mass faster than it did the year before, by an average of 21.9 billion tons per year. In Antarctica, the year-over-year speedup in ice mass lost averaged 14.5 billion tons.

In one study of sea level, researchers[240] used a coupled model[241] that included regional estimates of thermal expansion, glacioisostatic land-level changes,[242] and local effects of small glaciers and ice caps to project regional variations in sea level by the end of the century. For the AR4 A1B scenario (moderate economic growth), regional variability in sea level was found to range from –3.91 meters (–12.8 feet, sea level fall) to +0.79 meters (2.6 feet, sea level rise), with a global mean of +0.47 meters (1.5 feet), which is significantly less than the projections of the semi-empirical models. However, as is commonly the case, this projection lacked an assessment of ice dynamics in Greenland and Antarctica. When a fixed value for ice dynamics[243] is included in the projection, 0.41 meters (1.3 feet), sea level equivalent for the Antarctic Ice Sheet and 0.22 meters (0.7 feet) sea level equivalent for the Greenland Ice Sheet, the global mean sea-level rise by 2100 increased to 1.02 meters (3.3 feet).

Because GCMs do not (yet) account for the physics of sliding ice sheets, some researchers are relying on expert judgment. One researcher[244] asked experts for their estimate of the ice contribution to sea-level rise by 2100. A survey of 28 colleagues (half of whom responded) produced a best estimate of 32 centimeters (1 feet) of sea-level rise resulting from ice sheet losses. That results in a total rise of 61 to 73 centimeters (2 to 2.4 feet) from all sources by the end of the current century. Any calculation of a global mean such as this must, however, take into account local and regional variability of sea level when planning for the impacts (an area of research where the GCMs excel).

An important outcome of GCM sea-level projections is a picture of the regional variability of future sea level. As Greenland continues to lose ice during the 21st century, the land rebounds upward because there is not as much weight pushing down on it. As Greenland rebounds upward, the North Atlantic seafloor around the Greenland coast flexes downward in a seesaw-like action called lithospheric flexure, a form of glacioisostatic land level change. This (ironically) results in falling sea level near Greenland (and other coastal areas experiencing melting and upward flexing).

Another important phenomenon is regional differences in thermal expansion. For instance, as Arctic sea ice continues to retreat during the 21st century, the Arctic Ocean—more so than other parts of the ocean—will absorb heat and experience thermal expansion. There are also shifts in atmospheric pressure associated with continued global warming—these will cause regional sea level differences. Air pressure is projected to increase over the subtropics and mid-latitudes

[236] For a thorough discussion of regional variability in the context of media reports of sea-level vulnerability by oceanic islands, see Donner, S. (2012) Sea-Level Rise and the Ongoing Battle of Tarawa, *Eos*, 93, no. 17, 169.

[237] Katsman, C.A., et al. (2010) Exploring High-End Scenarios for Local Sea-Level Rise to Develop Flood Protection Strategies for a Low-Lying Delta–the Netherlands as an Example, *Climatic Change* 109, nos. 3–4, 617–645, doi: 10.1007/s10584-011-0037.

[238] Chen et al. (2017)

[239] Khazendar et al. (2016)

[240] Slangen, A., et al. (2011) Towards Regional Projections of Twenty-First Century Sea-Level Change Based on IPCC SRES Scenarios, *Climate Dynamics* 38, nos. 5–6, 1191–1209, doi: 10.1007/s00382-011-1057-6.

[241] A coupled model uses two or more separate models of the Earth climate system, say a model of ice behavior and a model of atmospheric circulation, to determine an outcome. A coupled model uses the two more focused models to project behavior under some condition (e.g., higher CO_2 concentrations). Coupled models are considered more powerful and more likely to simulate natural behavior.

[242] Glacio-isostatic land level changes occur when the land rebounds upward after a glacier melts and its weight is removed from the crust. Much of the land throughout northern Europe and Canada is uplifting in this way because the weight of the continental glaciers during the last ice age has been removed.

[243] Katsman et al. (2010)

[244] Meeting Briefs, Climate outlook looking much the same or even worse, *Science* 23 Dec 2011: Vol. 334, Issue 6063, pp. 1616, DOI: 10.1126/science.334.6063.1616-a

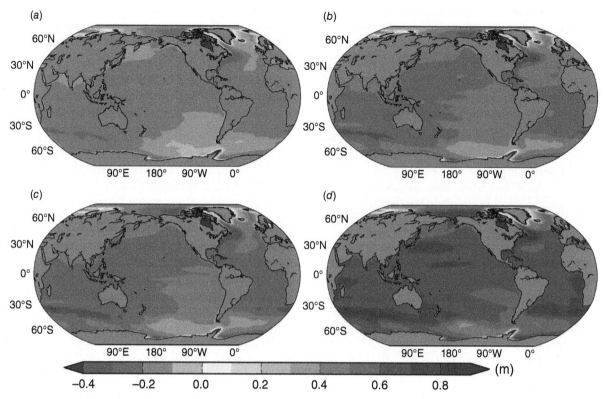

FIGURE 5.34 Regional variability (meters) by 2081 to 2100 as modeled for four scenarios of sea level change. Each scenario, tied to an RCP (representative concentration pathway), depicts sea level response to greenhouse gas concentrations necessary to achieve a certain radiative forcing.[248] RCP2.6 models an increase of energy in the climate system due to anthropogenic greenhouse gas emissions, equivalent to 2.6 W/m² by the period 2081–2100. (a) RCP2.6, (b) RCP4.5, (c) RCP6.0, and (d) RCP8.5. Across all RCPs, global mean sea level is projected to rise 0.26 to 0.82 m (0.85 to 2.69 ft) by the late-21st century. These sea level scenarios, however, reflect low confidence in projecting large-scale Antarctic ice sheet instability.

Source: https://www.ipcc.ch/report/graphics/index.php?t=Assessment%20Reports&r=AR5%20-%20WG1

(depressing sea level) and decrease over high latitudes (raising sea level), especially over the Arctic, by the end of the 21st century. These are associated with expansion of the Hadley Circulation toward the poles, and a shift of storm tracks away from the equator of several degrees latitude.

There are more causes of regional sea level variations. As the ocean accumulates heat, and ice around the world melts, Earth's crust under the oceans and at ice covered regions vertically deforms. In some places it rises and in other places it subsides in response to the removed or added weight. These changes alter the rotational characteristics of planet Earth and they cause regional shifts in Earth's gravity field. Altogether, these are referred to as *deformational, rotational, and gravitational responses to mass redistribution.* Unique combinations of these three, along with the localized effects of changing atmospheric pressure and differences in thermal expansion mentioned above, produce distinctive regional departures from a single global mean sea level value.[245] These are called **sea level fingerprints**.[246]

A characteristic of sea level fingerprints[247] is that regions where ice is melting experience relative sea level fall of about an order of magnitude greater than the equivalent global mean sea-level rise from these mass contributions. Away from the region influenced by ice sheet melting, known as the **far field**, the resulting sea level rise is larger (up to about 30 percent) than the global average rise. IPCC-AR5 concluded that it is very likely that over 95 percent of the world ocean, regional relative sea level rise will be positive. While most regions that will experience a sea level fall are located near current and former glaciers and ice sheets.

How does regional variability affect sea level by the end of the century? **Figure 5.34** shows the projected sea level by the end of the century under four scenarios by IPCC-AR5.

Note in Figure 5.34, the IPCC has modeled global mean sea level change and the regional variability of sea level for the end of the century (2081 to 2100) under four different global warming scenarios. In each scenario, the global mean sea level reflects the effects of anthropogenic warming

---

[245] IPCC-AR5, Working Group I, Climate Change 2013-The Physical Science Basis, Chapter 13 "Sea-Level Change", has a discussion of regional sea level changes on pages 1191-1199. https://www.ipcc.ch/report/ar5/wg1/
[246] Spada G., et al. (2012), The gravitationally consistent sea-level fingerprint of future terrestrial ice loss, *Geophys. Res. Lett.*, 40, 482–486, doi:10.1029/2012GL053000.

[247] Slangen et al. (2012)
[248] Recall from Chapter 2 that the term "radiative forcing" means the difference between sunlight absorbed by Earth and energy radiated back out to space. An increase, a positive radiative forcing, causes global warming. When it is caused by human greenhouse gas emissions, it is called the anthropogenic greenhouse effect.

TABLE 5.5				
Median Values and Likely (66–100% Probability) Ranges for Projections of Global Mean Sea-Level Rise in 2081–2100 Relative to 1986–2005.[249]				
	RCP2.6	**RCP4.5**	**RCP6.0**	**RCP8.5**
Global mean sea level rise in 2100 (m)	0.44 (0.28 to 0.61)	0.53 (0.36 to 0.71)	0.55 (0.38 to 0.73)	0.74 (0.52 to 0.98)

including thermal expansion, changes in land storage of water, and melting of the cryosphere. But regional variability determines whether a location will experience greater than or less than the global mean. Vertical land movement, changes in ocean circulation, geographic differences in thermal expansion, differences in atmospheric pressure, and other complex processes control regional variability.

The global mean sea level rise for the end of the century is shown in Table 5.5 for each greenhouse gas scenario modeled by the IPCC-AR5.

The authors of IPCC-AR5, **Chapter 13**, Sea-Level Change[250] express high confidence that ocean thermal expansion and alpine glacier melting have been the dominant contributors of 20th Century global mean sea level rise. The contribution of the Greenland and Antarctic ice sheets has increased since the early 1990s, partly from increased outflow induced by warming of the adjacent ocean. In modeling future sea level, they have high confidence in projections of thermal expansion and surface contributions from Greenland. But they have only medium confidence in projections of alpine glacier contributions and Antarctic surface contributions. Notably, there is low confidence in projections of contributions from marine-terminating glaciers and marine-based sectors of the Antarctic ice sheet. It is the future behavior of these floating ice shelves that has been of greatest concern to those who study the Antarctic contribution to future sea level.[251]

5.4.6 Consensus

If Greenland and Antarctica continue to undergo the observed high rate of decay observed to date,[252] a global sea-level rise of approximately 1 meter (3.3 feet) over the 21st century is emerging as a consensus[253] of the scientific community.[254] This elevation, as a planning target for the

sea-level rise adaptation groups around the world, makes a robust guideline for identifying risk and vulnerability of coastal assets.

If greenhouse-gas concentrations were stabilized today, sea level would nonetheless continue to rise for hundreds of years.[255] After 500 years, sea-level rise from thermal expansion alone would have reached only half of its eventual level, which models suggest could lie within ranges of 0.5 to 2 meters (1.6 to 6.6 feet). Glacier retreat will continue and the loss of a substantial fraction of Earth's total glacier mass is likely. Areas that are currently marginally glaciated are likely to become ice-free. But, as discussed in **Chapter 8**, it is unlikely that greenhouse gases will be stabilized soon, so we can very likely count on additional atmospheric heating and sea-level rise.

In one study,[256] researchers found that if current warming trends continue, by 2100 Earth will likely be at least 4 degrees Celsius (7.2 degrees Fahrenheit) warmer than present, with the Arctic at least as warm as it was nearly 130,000 years ago, when the Greenland ice sheet was a mere fragment of its present size. The study leader said, "The last time the Arctic was significantly warmer than present day, the Greenland Ice Sheet melted back the equivalent of about 2 to 3 meters (6.6 to 9.8 feet) of sea level." The research also suggests the Antarctic ice sheet melted substantially, contributing another 2 to 3 meters (6.6–9.8 feet) of sea-level rise. The ice sheets are melting already. The new research suggests melting could accelerate, thereby raising sea level as fast as, or faster, than 1 meter (3.3 foot) per century.[257]

Global sea-level rise has accelerated in response to warming of the atmosphere and the ocean and melting of the world's ice environment. Projections indicate that a 1 meter (3.3 foot) rise by the end of this century is likely. It has been pointed out[258] that observed sea-level rise has exceeded the best-case projections thus far (**Figure 5.35**). This has been independently confirmed by other studies as well; observed global average sea level rose at a rate near the upper end of projections of the IPCC Third and Fourth assessment reports.[259]

[249] Church et al. (2013)

[250] Church et al. (2013)

[251] Interview http://www.huffingtonpost.com/todd-r-miller/the-sunset-of-antarcticas_b_7921112.html

[252] Khazendar et al. (2016)

[253] Committee on Sea Level Rise in California, Oregon, and Washington; Board on Earth Sciences and Resources; Ocean Studies Board; Division on Earth and Life Studies; National Research Council *Sea level rise for the coasts of California, Oregon, and Washington: Past, Present, and Future,* (2012), National Academies Press, Washington, D.C.

[254] Linwood, P., et al. (2011) Estimating the Potential Economic Impacts of Climate Change on Southern California Beaches, *Climatic Change* 109, no S1, 277, doi: 10.1007/s10584-011-0309-0.

[255] Solomon, S., et al. (2009) Irreversible Climate Change due to Carbon Dioxide Emissions, *PNAS* 106, no., 1704–1709.

[256] Overpeck et al. (2006)

[257] VIDEO: Sea level rise accelerating. http://climatecrocks.com/2011/05/06/sea-level-rise-accelerating/

[258] Pielke, R.A. (2008) Climate Predictions and Observations. *Nature Geoscience* 1, 206

[259] Church and White (2011)

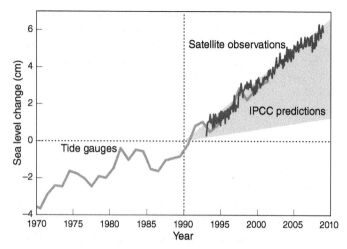

FIGURE 5.35 Sea-level change during 1970–2010 showing tide-gauge data[260] (thick light line) and satellite data[261] (dark line). Shading shows the high and low projections of the IPCC Third Assessment report for comparison.[262]

Source: The Copenhagen Diagnosis, 2009: Updating the World on the Latest Climate Science. Sydney, University of New South Wales Climate Change Research Centre (2009)

The map of sea-level change produced by satellite altimetry (Figure 5.2) suggests that sea-level rise will have significant local variability (rising in some areas, falling in others, but as a global mean, rising overall) that is worthy of continued research to improve our understanding of localized impacts and adaptation needs. Planners should consider this variability, as impacts to coastal assets will be scaled to localized sea-level change. The scientific community is converging on a consensus that it is appropriate to plan for at least 1 meter (3.3 feet) rise in mean sea level over the 21st century,[263] with some research indicating it may be reached well before the end of the century.[264] As stated in an update to the IPCC in late 2009,[265]

> By 2100, global sea-level is likely to rise at least twice as much as projected by AR4, for unmitigated emissions it may well exceed 1 m (3.3 ft). The upper limit has been estimated as ~2 m (6.6 ft) sea-level rise by 2100. Sea-level will continue to rise for centuries after global temperatures have been stabilized and several meters of sea-level rise must be expected over the next few centuries.

5.5 The Eemian Analogue and a Warning of Our Future

As the science of sea-level analysis develops, projections of how high global mean sea level will rise by 2100 continue to change. Thus far, there is general agreement about sea level rising 1 meter (3.3 feet) over this century, possibly more.[266] One source of information for improving our understanding of how high and how fast sea-level could rise is to look to the natural analogue of the last interglacial climate period, known by its European name, the Eemian (see discussion in **Chapter 4**).

The Eemian interglacial, which lasted from about 129,000 to 116,000 years ago, was apparently warmer than the preindustrial climate of 150 years ago by about 0.5 degrees Celsius (0.9 degrees Fahrenheit).[267] This is indistinguishable from the 1995–2014 global mean surface temperature. Hence, the Eemian is useful as an analogue for the ice volume and sea level position of a warmer world because it was the most recent time in Earth history when global sea level was substantially higher than it is at present.[268]

Researchers[269] found a 95 percent probability that global sea level during the Eemian peaked at least 6.6 meters (21.6 feet) higher than today. They found that it is likely (67 percent probability) to have exceeded 8.0 meters (26 feet), but it is unlikely (33 percent probability) to have exceeded 9.4 meters (30.9 feet). They also found that the rate of sea-level rise to peak heights ranged between 5.6 and 9.3 millimeters per year (22 and 36.6 inches per century).

It is well known that ice sheets are very sensitive to climate, and because global ice volume takes some time to equilibrate with climate, ice sheets have a long-term vulnerability to even relatively low levels of sustained global warming. Hence, excess heat in the climate system today has the potential to drive ice sheet melting, and thus sea level rise,[270] far into the future. A rise of over 6 meters (20 feet) would, of course, be devastating to coastal communities, and the hundreds of millions of individuals who live in them.

5.6 Conclusion

Sea-level rise could affect coastal communities in a number of ways. Coastal erosion will increase, marine inundation will worsen, coastal ecosystems will evolve or go extinct, and the coastal water table will rise causing severe

[260] Church and White (2006)

[261] Cazenave, A., et al. (2009) Sea Level Budget over 2003–2008: A Reevaluation from GRACE Space Gravimetry, *Satellite Altimetry and ARGO, Global and Planetary Change* 65, 83–88.

[262] Figure Allison, I., et al. (2009) *The Copenhagen Diagnosis, 2009: Updating the World on the Latest Climate Science.* Sydney, University of New South Wales Climate Change Research Centre

[263] Rahmstorf, S. (2012) Sea-Level Rise: Towards Understanding Local Vulnerability, *Environmental Research Letters* 7, 021001, doi: 10.1088/1748-9326/7/021001.

[264] Sweet et al. (2017a)

[265] Allison et al. (2009)

[266] Rahmstorf, S. (2010) A New View of Sea-Level Rise, *Nature Reports Climate Change* 4, no. 1004, 44–45, doi: 10.1038/climate.2010.29.

[267] Hoffman et al. (2017)

[268] Kopp et al. (2009)

[269] Kopp et al. (2009)

[270] Levermann et al. (2014)

drainage problems. Although these problems have not yet become endemic in much of the developed world, southeast Florida, the tropical western Pacific, and the islands of Micronesia are flooding at today's high tides, and age-old communities are suffering impacts to their food and drinking water. Many are considering migrating to new lands.

Although coastal communities in developed nations are still awakening to the inevitabilities of sea-level rise and planning for adaptation is not widespread, as impacts grow (and expenses with them), sea-level rise is likely to become the leading issue in coastal community planning within just a few years.[271]

Animations and Videos

1. How Much Will Sea-level rise? http://vimeo.com/5188725
2. Oceans Role in El Nino's https://www.youtube.com/watch?v=gaFjlZxM7S4#action=share
3. Melting Ice, Rising Seas, http://www.youtube.com/watch?v=gbnW3MK8wgY
4. Greenland's Thinning Ice, https://www.youtube.com/watch?v=Rl7mPdZCRKg
5. Runaway Glaciers in West Antarctica, https://www.youtube.com/watch?v=UXBKZPWuUdE
6. Major Sea Level Rise in Near Future, https://www.youtube.com/watch?v=f7sEhuSbQo8
7. Sea-level rise Accelerating, http://climatecrocks.com/2011/05/06/sea-level-rise-accelerating/
8. The Sea Also Rises: Prepare for a Changing Shoreline, https://www.youtube.com/watch?v=pKyK_YpGdn8

Comprehension Questions

1. What are the sources of sea-level rise?
2. Melting ice is a source of sea-level rise. Where is ice melting occurring?
3. How high is sea level projected to rise by mid-century?
4. How high is sea level projected to rise by the end of the century?
5. What factors should you take into consideration in assessing how fast sea level will rise? That is, when will sea level rise reach a certain level – such as 1 meter for instance?
6. Describe the process of planning for community adaptation to sea level rise.
7. Why do scientists study sea-level history during the Eemian?
8. What does the Eemian teach us about future sea level rise?
9. Describe how model projections of sea level rise compare to observations of sea-level rise.
10. Is today's sea level-rise unusual in recent geologic history? Explain your answer.
11. Describe the patterns of sea-level rise across the globe as revealed by satellite altimetry.
12. How is sea-level rise influenced by winds, air temperature, and ocean temperature?

Thinking Critically

1. Describe how global sea level has changed over the past 2000 years.
2. Compare the maps in Figures 5.2a and 5.3b.
   (a) How are they the same?
   (b) How are they different?
   (c) The maps show the same portion of the Pacific ocean, what caused the changes shown by the two maps?
   (d) Why is it important to understand the changes in these two maps?
3. Global mean sea level is rising over 3 mm/yr.
   (a) What causes global sea level rise?
   (b) Describe the data you would need to create a sea level "budget"; that is, to assign portions of this rate to various processes that contribute to global sea-level rise.
   (c) How would you get these data?
4. You are asked to appear before a congressional hearing about sea-level rise.
   (a) Explain how global warming causes sea-level rise.
5. Satellite altimetry measurements began approximately 20 years ago.
   (a) Has the rate of sea level rise accelerated over the past 20 years?
   (b) Since earlier in the twentieth century, has sea level rise accelerated?
6. Compare the history of global surface temperature and the history of global sea level over the past 20 years.
   (a) How are they different?
   (b) How are they the same?
   (c) Would you expect them to change at the same rate? Why or why not? That is, why might there be a difference in the rate at which global temperature changes and global sea level changes?
7. Imagine you are in charge of a city-planning team asked to rank the vulnerability of transportation assets (roads, harbors,

[271] Fletcher, C.H., and Richmond, B.R. (2010) Climate Change in the Federated States of Micronesia: Food and Water Security, Climate Risk Management, and Adaptive Strategies, http://seagrant.soest.hawaii.edu/climate-change-federated-states-of_micronesia

rail, airports, bus lines and terminals, maintenance yards, etc.) in the face of sea-level rise. In fact this is a real-life issue and such planning exactly like this is taking place today in dozens of cities across the United States.

(a) Who should be on your team?

(b) How will you organize the team?

(c) What sort of resources will you need to do a thorough job in this study?

(d) What data do you need and how will you get the data?

(e) What will you do with this data? What is the final product of your work?

(f) Describe the ultimate purpose and outcome of this study.

8. The Eemian offers an analogue to a future with higher seas. You live in a coastal region where there is evidence of an old Eemian shoreline.

(a) How will you use this evidence to improve your understanding of the pattern and impacts of sea level rise locally?

9. You are building a home on a beach.

(a) What design features will you use to mitigate the negative impacts of sea-level rise?

10. You live one block from the ocean.

(a) Describe how you are vulnerable to sea-level rise.

11. Describe the negative impacts of sea-level rise on a coastline.

(a) How would you design a road to mitigate each impact?

(b) A parking lot?

(c) A school building?

(d) A hospital?

Activities

1. Visit the NASA website "Global Climate Change," http://climate.nasa.gov/evidence/, and answer the following questions.

(a) What is the evidence for rapid climate change?

(b) Describe the rate of global sea-level rise?

(c) Describe some facts about Earth's climate that are not in dispute.

(d) Choose two of the key lines of evidence that Earth's climate is changing and describe their impact where you live.

2. Watch the video "Climate Denial Crock of the Week: Sea-Level Rise Accelerating," http://climatecrocks.com/2011/05/06/sea-level-rise-accelerating/, and answer the following questions.

(a) Do climate deniers on TV accurately portray the sea-level rise problem? What tactics do they use when communicating about sea level rise (or climate change in general)?

(b) What are the primary causes of global sea-level rise?

(c) What important cause of global sea-level rise is not included in the IPCC AR4? Why is this important?

(d) Study IPCC AR5. How has the modeling of ice sheet contributions to sea level rise improved since this publication? How is it ice sheet modeling still deficient?

(e) When "rapid dynamical changes in ice flow" are included in future sea-level rise estimates, how high could the sea level rise by the end of the century?

(f) How would a sea-level rise of 0.8 to 2 m by the end of the century impact coastlines?

3. Read the article "Rising Sea Levels Set to Have Major Impacts around the World," http://www.sciencedaily.com/releases/2009/03/090310104742.htm.

(a) Describe the major points of the article.

(b) Carefully summarize the wording used by the experts in describing sea level at the end of the century. What do they state in a definitive manner and what do they state in a tentative manner? Why do they use these styles of communication?

(c) What are some of the impacts of sea-level rise as described in the article?

(d) This website provides "Related Stories" on sea-level rise. Explore these and provide a summary of impacts, projections, research methods, and understanding of the causes of sea-level rise.

Key Terms

Armoring	Groundwater inundation	Mesoscale eddies	Thermal expansion
Beach loss	Ice shelves	Neap tides	Tidal range
Coastal erosion	Ice stream	Neutral year	Tide gauge
Critical point	King tides	Pacific Decadal Oscillation	Vertical land motion
Cryosphere	La Niña	Regional variability	Western Pacific Warm Pool
Far field	Lithospheric flexure	Satellite altimetry	
Foraminifera	Local relative sea level change	Sea level fingerprints	
Groundwater extraction	Marine ice-cliff collapse	Spring tides	

CHAPTER 6

Modeling Climate

This global climate model simulation of the circulation, surface elevation, and water temperature of the North Atlantic is produced by the Los Alamos National Laboratory[1] Model for Prediction Across Scales Ocean (MPAS-O) component of the Accelerated Climate Model for Energy (ACME). Complex regions of warmer water adjacent to the Gulf Stream off the eastern coast of the United States indicate the model's capability to simulate the transport of heat by small-scale circulation cells called "eddys" within the ocean, a key component necessary to accurately simulate global climate change.

Source: http://www.lanl.gov/newsroom/picture-of-the-week/pic-week-2.php

LEARNING OBJECTIVE

Researchers represent the complex behavior and interaction of the oceans, land, biosphere, and atmosphere using mathematical relationships to simulate Earth's climate system. These models must account for detailed characteristics of Earth's heat engine by including total solar irradiance, certain human activities, the carbon cycle, ocean and atmospheric circulation, and other processes that affect climate. By projecting climate simulations into the future, these models provide useful large-scale (and increasingly regional-scale) estimates of future conditions resulting from the anthropogenic greenhouse effect. Human communities use these estimates of future climate conditions to develop adaptation and mitigation strategies.

6.1 Chapter Summary

Climate models successfully reproduce the past 100 years of global warming, but only when the rise of anthropogenic greenhouse gases is included. The International Panel on Climate Change Assessment Report 5 (IPCC-AR5)[2] concludes that continued emission of greenhouse gases will cause further warming and long-lasting changes in all components of the climate system, increasing the likelihood of severe, pervasive, and irreversible negative impacts for people and ecosystems. Limiting climate change will require substantial and sustained reductions in greenhouse gas emissions, which, together with adaptation, can limit climate change risks.

To provide guidance for developing mitigation[3] and adaptation[4] strategies, IPCC-AR5 identified four different

[1] The Los Alamos National Laboratory is a research component of the U.S. Department of Energy. Their mission is to deliver science and technology to protect the United States and promote World Stability. http://www.lanl.gov/index.php.
[2] IPCC (2014) Climate Change 2014: Synthesis Report. Contribution of Working Groups I, II and III to the Fifth Assessment Report of the Intergovernmental Panel on Climate Change [Core Writing Team, Pachauri, R.K. and Meyer, L.A. (eds.)]. IPCC, Geneva, Switzerland, 151 pp.

[3] Mitigation is the act of decreasing greenhouse gas emissions to avoid the worst future effects of climate change.
[4] Adaptation is the act of changing human behavior in order to reduce the negative impacts of climate change occurring today and in the future that cannot be avoided.

21st century future pathways of greenhouse gas emissions, air pollutants and land use. These pathways are called "RCPs" for **Representative Concentration Pathways**. The RCPs include a stringent mitigation scenario (RCP2.6), two intermediate scenarios (RCP4.5 and RCP6.0), and one scenario with very high greenhouse gas emissions (RCP8.5). Scenarios in which humans fail to engage in efforts to limit greenhouse gas emissions lead to pathways ranging between RCP6.0 and RCP8.5. RCP2.6 represents a scenario that attempts to keep global warming below 2 degrees Celsius (3.6 degrees Fahrenheit) above preindustrial temperatures.

Modeling of future climate by IPCC-AR5 indicates that the global mean surface temperature change for the period 2016–2035 relative to 1986–2005 is similar for the four RCPs and will likely be in the range 0.3 to 0.7 degrees Celsius (0.5 to 1.3 degrees Fahrenheit). By mid-century, the four RCPs begin to diverge and the likely range of warming under RCP2.6 is 0.4 to 1.6 degrees Celsius (0.7 to 2.9 degrees Fahrenheit) and under RCP8.5 is 1.4 to 2.6 degrees Celsius (2.5 to 4.7 degrees Fahrenheit).

At the end of the 21st Century, likely warming under RCP2.6 is 0.3 to 1.7 degrees Celsius (0.5 to 3.1 degrees Fahrenheit) and under RCP8.5 is 2.6 to 4.8 (4.7 to 8.7 degrees Fahrenheit). The other RCPs fall between these end members. Climate models provide important results for understanding future global climate. Their ability to project regional and localized climate is more limited but improving as computer processing capacity grows.

In this chapter, you will learn that:

1. Climate models rely on fundamental equations of physics combined with **parameterizations** of climate processes to simulate Earth's climate system.
2. Future climate will depend on committed warming caused by past anthropogenic emissions, as well as future anthropogenic emissions and natural climate variability.
3. By the mid-21st century, the magnitude of projected climate change is substantially affected by the choice of emissions scenarios.
4. ENSO (El Niño–Southern Oscillation) is a large-scale meteorological pattern that governs temperature, wind, and rainfall trends in the Pacific Ocean. ENSO exerts a global influence on weather patterns and annual mean temperature.
5. Global climate is influenced by explosive volcanic eruptions, the ice–albedo effect, clouds, anthropogenic sulfur emissions, and variations in solar radiation; models must take these into account. These, and other influences, cause climate variability.
6. Confidence in climate model projections is strengthened because of the agreement between model simulations of the past, and actual observed temperature increases.
7. If greenhouse gas and aerosol concentrations were kept at year 2000 levels, climate models project that a temperature rise of about 0.1°C (0.18°F) per decade would be expected for the next two decades.
8. Climate models project a temperature rise of about 0.3 to 0.7°C (0.5 to 1.3°F) for the next two decades for all potential future scenarios of greenhouse gas emissions.
9. The best model estimate for a low scenario of surface air warming in the 21st century is 1.0°C (1.8°F), with a likely range of 0.3 to 1.7°C (0.5 to 3.1°F). However, air temperature already reached this level of warming in 2016 and already emitted greenhouse gases commit the surface to significant additional warming.
10. The best estimate for a high scenario of surface air warming in the 21st century is 3.4°C (7.2°F), with a likely range of 2.6 to 4.8°C (4.7 to 8.6°F).
11. The difference in potential impacts between low and high scenarios is significant.
12. Surface temperature is projected to rise over the 21st century under all modeled emission scenarios.
13. It is very likely that heat waves will occur more often and last longer and that extreme precipitation events will become more intense and frequent in many regions.
14. The ocean will continue to warm and acidify, global ice to melt, and global mean sea level to rise.
15. Anthropogenic sulfur aerosols from Asian industrialization may have controlled global temperature variations over the past two decades.

6.2 Climate Models

Earth's climate system is very complex. Some climate processes operate on cycles (such as seasons and glacial–interglacial cycles), some occur with irregular timing (such as ENSO and Dansgaard–Oeschger events), and some are essentially unpredictable far in advance, but it is possible to say that they are "likely" or "unlikely" (such as various types of weather including hurricanes, snowfall, rain, and extremely hot days) depending on certain conditions.

In reality, climate processes all interact with one another over different lengths of time, sometimes enhancing and sometimes suppressing each other's effects. Climate complexity is enormous, and several fundamental processes continue to be the focus of intensive research (e.g., the role of cloud formation in future warming). Thus, important changes in the climate system can occur that catch researchers off guard, and climate is characterized by a great deal of **uncertainty** (random fluctuations in observations) and **variability** (the range of climate compared to its average state). Despite this complexity, climate does follow certain core principles of physics related to Earth's radiation budget, and these were expounded upon by scientists over a century ago.[5]

As we learned in **Chapter 1**, the chemist and physicist Svante Arrhenius[6] (1859–1927) and the engineer and inventor Guy Stewart Callendar[7] (1898–1964) calculated and documented (respectively) the process of heat trapping by CO_2. The planetary energy budget quantified in the work of these two scientists allows for retrospectively predicting global warming over the course of the 20th century. A model based on the core principles described by Arrhenius and Callendar was constructed by modern climate modelers[8] and used to calculate warming in response to increasing atmospheric greenhouse gases during the 21st century. The resulting model projects a temperature increase at the lower bound of results generated by the Intergovernmental Panel on Climate Change. Thus, although Earth's climate system is conceptually complex, it has at its core, physical laws of radiative transfer as originally conceived by Arrhenius and documented by Callendar. These core principles are not up for debate as they describe the fundamental thermodynamics of natural systems, thus neither should the obvious future they imply be a subject of debate.

The core physics alone leads to quantitatively robust projections of baseline warming. Modern climate models include not only these fundamental principles but also climate feedbacks that introduce uncertainty. Thus, the projections of end-of-century global warming by modern models are fundamentally trustworthy: robust baseline warming based on the well-understood physics of radiative transfer, with extra warming due to climate feedbacks. These projections provide a compelling case that global climate will continue to undergo significant warming in response to ongoing anthropogenic emissions of CO_2 and other greenhouse gases to the atmosphere.

Beyond the basic laws governing radiative equilibrium, understanding climate change can become exceedingly complicated. For example, in **Chapter 4**, we discussed the relative slowdown in warming that occurred during the first decade of this century, the so-called global warming hiatus. Subsequent research on this problem has resulted in several hypotheses to explain the pattern. One group of researchers[9] noted that the upper ocean stopped accumulating heat following 2003. They concluded that some of the missing energy had been lost to space, and the remainder had been delivered to the deep ocean by the thermohaline circulation system of the North Atlantic.

Another group[10] concluded that the lack of strong warming resulted from stratospheric aerosol particles that "persistently varied" rather than staying stable as researchers had assumed. They identified Asian power plants that burn sulfur-rich coal and moderately small volcanic eruptions in the Caribbean and Indonesian regions as two sources of these particles (later in this chapter, this idea remerges in a 2016 paper that provides a powerful narrative of anthropogenic manipulation of climate).

A third research team[11] concluded that the warming slowdown didn't occur at all. They applied corrections to both sea surface and land surface air temperature datasets and challenged the notion that there was a decrease in the rate of global warming. The apparent slowdown was simply a fixable bias in the dataset.

Regardless of these different points of view, one overarching message to be taken from this discussion is that modeling climate complexity[12] continues to be a difficult and challenging aspect of predicting the impacts of global warming.[13] But this does not mean that climate models serve no purpose. On the contrary, early climate modeling from the 20th century provided an amazingly accurate depiction of the world we live in today[14] (**Figure 6.1**).

[5] Anderson, T.R., et al. (2016) CO_2, the greenhouse effect and global warming: from the pioneering work of Arrhenius and Callendar to today's Earth System Models, In *Endeavour*, v. 40, Is. 3, p. 178-187, ISSN 0160-9327, https://doi.org/10.1016/j.endeavour.2016.07.002 http://www.sciencedirect.com/science/article/pii/S0160932716300308

[6] In 1895, Svante Arrhenius described an energy budget model of the radiative effects of carbon dioxide and water vapor on Earth's surface temperature. He argued that variations in CO_2 could influence the heat budget of Earth's surface and made calculations describing the relationship of carbon dioxide to temperature. He calculated that the "temperature of the Arctic regions would rise about 8° or 9°C, if the carbonic acid (carbon dioxide) increased 2.5 to 3 times its present value. In order to get the temperature of the ice age between the 40th and 50th parallels, the carbonic acid in the air should sink to 0.62 to 0.55 of present value (lowering the temperature 4° to 5°C)." Thus, he postulated that carbon dioxide acted as a thermostat responsible for temperature changes associated with ice ages and future global warming. Today this is accepted as a basic physical principal of Earth's climate system.

[7] In 1938, Guy Stewart Callendar was the first person to compile temperature measurements and propose that Earth's surface was warming. He correlated his findings to measurements of atmospheric CO_2 and concluded that over the previous 50 years the global land temperatures had increased, and this was an effect of the increase in carbon dioxide. His estimates have now been shown to be remarkably accurate. However, his findings were met with skepticism at the time as leading scientists depicted his results as coincidence. However, his papers throughout the 1940s and 1950s slowly convinced some scientists of the need to conduct an organized research program to collect CO_2 concentrations, leading eventually to Charles Keeling's iconic Mauna Loa measurements, which proved pivotal to advancing the theory of anthropogenic global warming. Callendar remained convinced of the accuracy of his theory until his death in 1964 despite continued mainstream skepticism. His postulation that warming was related to carbon dioxide emissions became known for a time as the "Callendar Effect." Callendar is still relatively unknown as a scientist, but his contribution was fundamental to climate science.

[8] Anderson et al. (2016)

[9] Katsman, C.A. and van Oldenborgh, G. (2011) Tracing the Upper Ocean's 'Missing Heat', *Geophysical Research Letters* 38: L14610, doi:10.1029/2011GL048417.

[10] Solomon, S., et al. (2011) The persistently variable 'background' stratospheric aerosol layer and global climate change, *Science* 333, no. 6044: 866–870, doi: 10.1126/science.1206027.

[11] Karl, T.R., et al. (2015) Possible artifacts of data biases in the recent global surface warming hiatus, *Science*, DOI: 10.1126/science.aaa5632

[12] Schultz, C. (2011) Interview with De-Zheng Sun, co-editor of "Climate Dynamics: Why does climate vary?" *Eos* 92, no. 34, 285–286.

[13] How reliable are climate models? https://www.skepticalscience.com/climate-models.htm

[14] See CarbonBrief Analysis: How well have climate models projected global warming? https://www.carbonbrief.org/analysis-how-well-have-climate-models-projected-global-warming

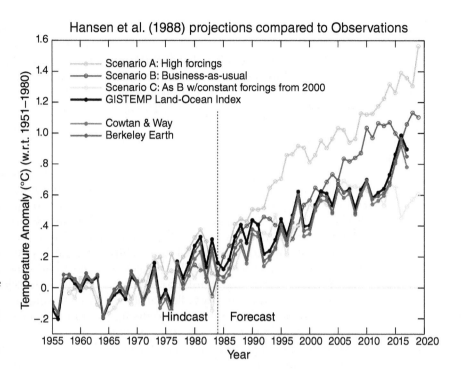

FIGURE 6.1 Hansen et al. (1988) published[15] projections of continued global warming assuming increases in human produced greenhouse gases. Hypothesizing future concentrations of the main greenhouse gases, Scenario A assumed exponential growth in emissions, Scenario B assumed a roughly linear increase, and Scenario C was similar to Scenario B, but assumed emissions remained constant after the year 2000. Hansen specifically stated that he thought the middle scenario (B) the "most plausible."[16] Observations of land–sea surface temperature through 2016 shown for comparison.

Source: http://www.realclimate.org/index.php/climate-model-projections-compared-to-observations/

6.2.1 General Circulation Models

Attempting to make order out of this complexity is the role of mathematical relationships constructed by researchers in the form of **general circulation models** (GCMs). GCMs reproduce the fundamental, physically predictable processes that operate in Earth's climate system: solar radiation and differences in heating Earth's surface, atmosphere and ocean circulation, and other well-understood natural phenomena linked to the climate.

GCMs are computer-based sets of mathematical calculations that simulate the behavior and interaction of Earth's oceans, land, and atmosphere under the influence of the Sun. They produce countless calculations on supercomputers that solve the equations of fluid dynamics. The equations are used to calculate the properties of fluids (made of gas, Earth's atmosphere, is considered a fluid) such as velocity, pressure, density, temperature, and how they change over space and time. These equations are used as laws to predict the behavior of the atmosphere and the ocean in various useful and applied settings.

There are various types of GCMs; separate atmospheric (AGCMs) and oceanic (OGCMs) models treat different parts of the climate system. In earlier generation GCMs, the ocean was treated in a very basic way, as a simple slab of water where the only thing accounted for was the transfer

of heat across the surface between the air and the water. However, climate models today include an ocean that plays a far more active role in the climate system. The major horizontal and vertical current systems are represented, as well as air–sea mixing, and their role in transporting heat toward the poles and influencing the atmosphere at all latitudes and longitudes.

When atmospheric and ocean GCMs are united, they form an AOGCM, also known as a **coupled general circulation model**. If other components are added, such as a model of sea ice or a model of evapotranspiration over land (for instance), the coupled GCM can become a full **climate model**.[17] The most actively researched use of climate models in recent years has been to project future temperature changes resulting from increases in atmospheric concentrations of anthropogenic greenhouse gases.

These models receive several types of input in the form of data on ocean currents and seawater temperature, the concentration of CO_2 and other greenhouse gases in the atmosphere at specific geographic locations; the amount of sunlight as distributed across the planet; the cover of vegetation, ice, and snow; the development of clouds; the amount of urbanized land cover; pollutants related to air travel, shipping, and biomass burning; a detailed carbon cycle; and many other factors that affect Earth's energy budget.[18] These inputs are used to guide how the equations treat the various factors that influence climate.

[15] Hansen, J., et al. (1988) Global climate changes as forecast by Goddard Institute for Space Studies three-dimensional model. *J. Geophys. Res.*, 93, 9341-9364, doi:10.1029/JD093iD08p09341. https://pubs.giss.nasa.gov/abs/ha02700w.html

[16] Hansen's 1988 Projections, see discussion http://www.realclimate.org/index.php/climate-model-projections-compared-to-observations/

[17] Description of a global climate model: http://www.sciencedaily.com/releases/2004/06/040623082622.htm

[18] Recall from our discussion in Chapter 2 that Earth's energy budget is determined by the net flow of energy into and out of the Earth system.

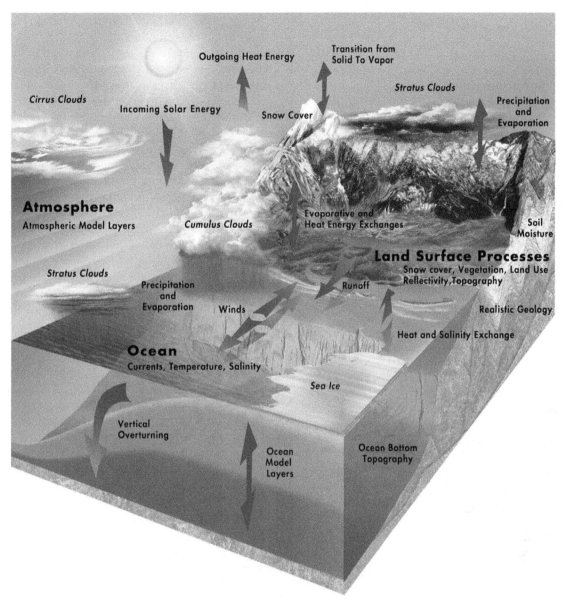

FIGURE 6.2 A climate model must take many factors into account such as how the atmosphere, the oceans, the land, ecosystems, ice, topography, and energy from the Sun all affect one another and Earth's climate.

It is also possible to model the impacts of, but not the exact timing, of essentially random events such as volcanic eruptions, ENSO, irregular variations in solar activity, and others. These processes exert important influences on the climate system, but modern science does not yet understand them enough to predict when they will occur and when they won't. It is this climate variability[19] that makes it essentially impossible for GCMs to project year-to-year climate. However, accurate prediction of general background trends and patterns is an attainable goal, and various research teams around the world continue to make important progress.

Building and running a climate model is a complex process of identifying and quantifying Earth system processes, representing them with mathematical equations, setting variables to initial conditions, and repeatedly solving the equations using powerful computers.[20] GCMs include equations that predict how greenhouse gases influence the climate (**Figure 6.2**).

[19] Recall from Chapter 5 that in our discussion of the El Niño Southern Oscillation, we said that although the climate is the long-term average condition of the weather, it can change from one season to another, from one year to another, and over longer time periods such as decades. This is referred to as climate variability.

[20] VIDEO: Recipe for a better climate model. http://www.youtube.com/watch?v=TLvCCHNCEdc&NR=1

6.2.2 Modeling Concept

Climate models are designed to simulate Earth's climate on a range of scales, from global to regional (thousands to hundreds of kilometers). But few models regularly tackle climate changes at the local level (tens of kilometers). Most break up the atmosphere into a stack of 10 to 30 layers between Earth's surface and outer space. The ocean is also characterized by a set of layers, breaking up the water column from the deepest parts of the seafloor to the surface into discrete levels. Each of these layers is further subdivided into individual cells that overall make a three-dimensional grid.

The basic building blocks of climate models are 3-D "grid cells" that contain climate-related physical information about a particular location. Each grid cell has information about whether it is covered by land or ocean and the proportion of each (**Figure 6.3**). Information is also included about what is on the land: mountains, forests, farmland, cities, and other features. The characteristics of the ocean are also described: temperature, currents, depth, and more. Within each grid cell, climate models simulate the movement of air, water, and heat. There are equations that calculate the pressure in the atmosphere, changes in the Sun's energy, the speed of ocean currents, and others.

Parameterization One of the problems with dividing the atmosphere up into lots of little grid cells is that there are many climate processes that are smaller than a cell: cloud formation, rainfall, the effect of topography on winds, storms, and others. This problem could potentially mean that individual clouds, which play an important role in the climate system, might not be represented in modeling, or they might be mischaracterized. Somehow the processes that form clouds and other small-scale climate factors (and their consequences) must be represented.

Researchers address this problem with expert estimations based on the fundamentals of climate science and observations of climate processes. For example, cloudiness and rainfall can be estimated based on knowledge of the temperature and humidity in a cell. Raindrops require a very small solid particle in the air to precipitate, and thus, modelers must also estimate how much dust (aerosol) is in the cell. This process of estimation is called parameterization, and most models run many parameterization schemes to approximate many climate processes. Some of these schemes are well defined and thought to be quite reliable, but others are far less well understood, and confidence in them is not high.

Examples of important processes that must be parameterized in climate models include turbulent mixing, heating/cooling through radiation, and small-scale physical processes such as cloud formation and precipitation, chemical reactions, and exchanges between the biosphere and atmosphere. For example, these models cannot represent every raindrop. However, they can simulate the total amount of rain that would fall over a large area, the size of a grid cell in the model. Parameterizations are usually derived from a set of measurements taken in a natural setting or higher resolution modeling in a restricted area with smaller grid cells. The problem with this approach is that the relationships simulated by a parameterization may not hold true for every location or under all possible conditions.

Climate in the real-world changes over time, and climate processes in one part of the world affects processes in other parts. These space and time realities need to be

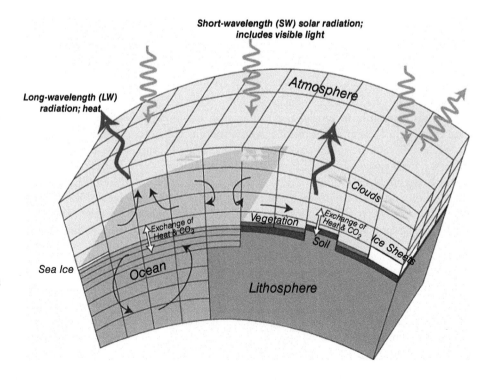

FIGURE 6.3 Climate models include land, air, and ocean components, and each of these is treated somewhat separately since different processes act within them. The more cells in a model, the closer it can approximate the real Earth, but too many cells would require more computing power than is available.

Source: https://www.e-education.psu.edu/
earth103/node/524

tracked by a climate model. For instance, imagine three model cells next to each other (**Figure 6.4**). In one cell, air currents are moving over flat farmland. In the neighboring cell, these currents encounter a mountain range and must climb to higher/cooler elevations causing water vapor in the air to condense and precipitate. In the third cell, it may rain over a city or a rain shadow may develop. This relatively simple sequence requires that the model continuously recalculate all equations in discrete steps of time and across discrete units of physical space. Simultaneously, the model must replicate processes elsewhere in the same grid cell related to Earth's rotation, ocean currents and tides, evaporation, cloud formation, heating and cooling of the land and water, industrial pollution, albedo, small-scale atmospheric mixing, runoff, and changes in all of these details over time. When we scale this back up to the size of the Earth where many thousands of grid cells are each engaged in similar activities, it becomes clear why supercomputers are needed to run these models (and why a major limitation to climate models is available computer power).

Models must include a calculation of time in their operations. For instance, it is critically important to allow sufficient time for realistic heat exchange between the ocean surface and the atmosphere or from one part of the globe to another. When a model starts, it begins with a set of initial conditions for the atmosphere and ocean and then calculates how they will have changed after one time step, say, half an hour. The time step must be chosen with care. For example, if you want to run a model through 50 years as quickly as possible, you want to use as large a time step as possible. However, past some critical threshold, the time step is so large that air (or, more accurately, heat) in the real world could have traveled farther than one grid box in one time step, and it becomes impossible to accurately determine how various elements of the climate develop.

Some aspects of the climate can change more rapidly than others, and so they need to be calculated more frequently. For example, the movement of the air needs to be calculated every half hour, but the balance of incoming and outgoing radiation can be calculated less frequently. In the ocean, the ratio of the horizontal grid size to the length of a time step must not exceed the largest flow speed of water. There are many similar detailed relationships in the climate system requiring careful attention when building and running a reliable climate model.

6.2.3 Model Resolution

The **spatial resolution** of a model determines whether the climate simulations it makes are specific to an area the size of the continental United States, the size of New England, or the size of Manhattan (**Figure 6.5**). Spatial resolution governs the size of grid cells in a model (in degrees of latitude and longitude or in kilometers), and **temporal resolution** refers to how often (in "model time") the model recalculates climate factors (say, every half hour, 6 hours, every week, etc.). Climate models are typically run with time steps of about 30 minutes. A climate

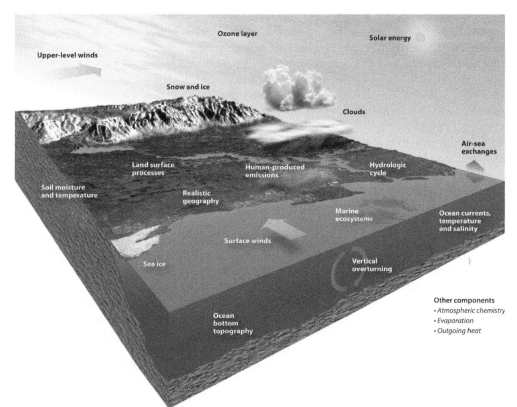

FIGURE 6.4 Climate models must track changing conditions in all aspects of a grid cell and neighboring cells.

Source: http://nas-sites.org/climate-change/climatemodeling/page_3_1.php

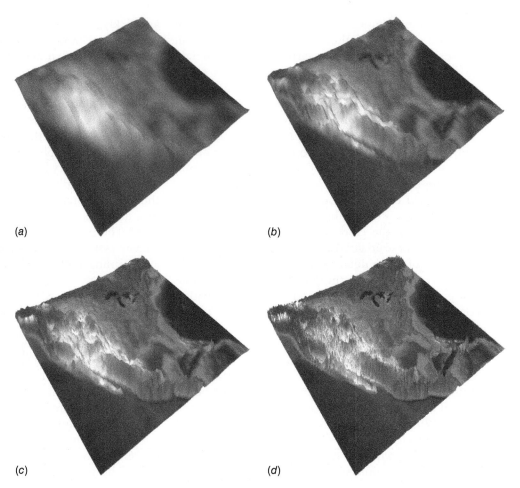

(a)

(b)

(c)

(d)

FIGURE 6.5 Illustration of climate model resolution from the National Center for Atmospheric Research. (a) 300 km or 2.8° resolution; (b) 111 km or 1° resolution; (c) 55 km or 0.5° resolution; (d) 11.13 km or 0.1° resolution. The decimal degree measurement is an expression of latitude and longitude, Earth's coordinate system.

Source: https://www.vets.ucar.edu/vg/ Climate_Model_Resolution%20%20/ index.html

model run for a century might, therefore, involve 1,753,152 (the number of half hours in a century) time steps. All model parameters (temperature, wind speed, humidity, etc.) need to be calculated at each of the thousands to millions of grid points in the model at each of those time steps.[21]

Climate models are achieving improved resolution all the time. A typical climate model might have horizontally spaced grid cells of 100 square kilometers (62 square miles). This is equivalent to saying "Calculate the temperature at a point, then move 100 km west and calculate temperature again, then move another 100 km west and repeat. Once you've gone all the way around the globe, move 100 km north and repeat the process; and so on."[22] In effect, the model is creating virtual weather stations at 100 km intervals around the planet surface and reporting climate characteristics that it calculated at each of them.

Supercomputer capabilities are the primary limitation on grid cell size. Grid cells can be made smaller for higher resolution (**Figure 6.6**), but this requires more

computing time, which, on supercomputers, can be very expensive. To pay for this, researchers typically seek special government grants. As a general rule, increasing the resolution of a model by a factor of 2 (say, going from a cell size of 100 square kilometer to one of 50 square kilometers) means that about 10 times more computing power will be needed (the model will take 10 times as long to run on the same computer).

6.2.4 Representing the Climate System

Climate models are constantly being enhanced as scientific understanding of the climate system improves and as computational power increases. For example, in 1990, the average model divided up the world into grid cells measuring approximately 500 kilometers (310 miles) per side. Today, leading models divide the world up into grid cells of about 30 kilometers (18.6 miles) per side, and some models analyzing the climate in localized areas are able to run short simulations with grid cells of only a few kilometers per side. With a grid resolution of 30 kilometers, the modeling in IPCC AR5 includes representations of Earth system processes that are much more extensive and improved compared to previous efforts. Simulations of

---

[21] VIDEO: Ultra high resolution climate model. https://www.youtube.com/watch?v=4794mgJLTbU

[22] After Russell, R.M. (2011) *Resolution of Climate Models,* http://eo.ucar.edu/staff/rrussell/climate/modeling/climate_model_resolution.html

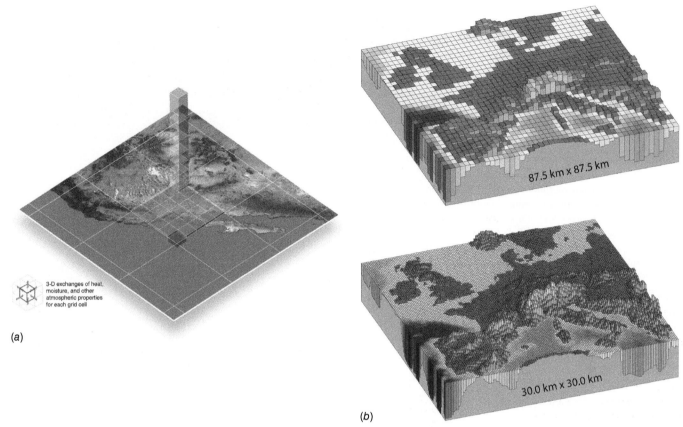

(a)

(b)

FIGURE 6.6 The resolution of climate models has increased over time. (a) Models reach high into a virtual atmosphere and deep into the ocean. They simulate climate by dividing the world into three-dimensional grid boxes, measuring physical processes such as temperature at each grid point. Such models can be used to simulate changes in climate over years, decades, or even centuries. (b) Example from IPCC AR5—Horizontal resolution considered in today's climate models and in the very high-resolution models now being tested: (top) Illustration of European topography at a resolution of 87.5 km²; and (bottom) at a resolution of 30.0 km².[23]

Source 6.6a: Figure from University Corporation for Atmospheric Research (UCAR) http://www2.ucar.edu/climate/faq/aren-t-computer-models-used-predict-climate-really-simplistic#mediaterms

Source 6.6b: https://www.ipcc.ch/report/graphics/index.php?t=Assessment%20Reports&r=AR5%20-%20WG1&f=Chapter%2001

radiation and aerosol–cloud interactions and of the cryosphere are particularly enhanced.[24]

A representation of the carbon cycle was added to a larger number of models and has been improved since AR4. A high-resolution stratosphere is now included in many models. Climate models now include enhanced representation of nitrogen effects on the carbon cycle. As new processes or treatments are added to the models, they are also evaluated and tested relative to available observations.[25]

Newer models incorporate more of the physical processes and components that make up Earth's climate system. The very first global climate models were designed to simulate only the circulation of the atmosphere (**Figure 6.7**). Over time, the ocean, clouds, land surface, ice, snow, and other features were added one by one. Most of these features were new modules that were developed by experts in those fields and then added into an existing climate model framework. Today, there are more than 35 climate models created and maintained by more than 20 modeling groups around the world. Some of the newest models are known as Earth System Models, or ESMs, which include all the previous components of a typical GCM but also incorporate modules that represent additional aspects of the climate system, including agriculture, vegetation, and the carbon cycle.[26]

6.2.5 Model Ensembles

Weather is chaotic, meaning that, while it does obey the laws of physics (every effect has a cause), there are so many possible causes affecting weather that it is impossible to

[23] Cubasch, U., et al. (2013) Introduction. In: *Climate Change 2013: The Physical Science Basis. Contribution of Working Group I to the Fifth Assessment Report of the Intergovernmental Panel on Climate Change* [Stocker, T.F., et al. (eds.)]. Cambridge University Press, Cambridge, United Kingdom and New York, NY, USA.

[24] Cubasch et al. (2013)

[25] Cubasch et al. (2013)

[26] From Supplemental Message 6, National Climate Assessment 3, 2014 http://nca2014.globalchange.gov/report/appendices/climate-science-supplement

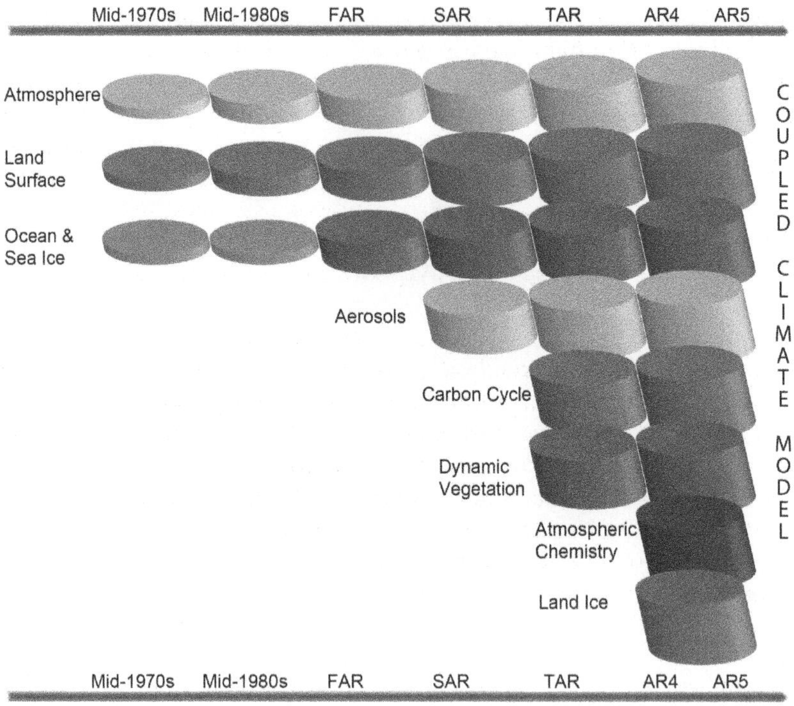

FIGURE 6.7 The development of climate models over the last 35 years showing how the different components were coupled into comprehensive climate models over time. In each aspect (e.g., the atmosphere, which comprises a wide range of atmospheric processes), the complexity and range of processes have increased (illustrated by growing cylinders). Note that during the same time, the horizontal and vertical resolution of models has increased considerably. FAR—IPCC First Assessment Report; SAR—IPCC Second Assessment Report; TAR—IPCC Third Assessment Report; AR4—IPCC Fourth Assessment Report; AR5—IPCC Fifth Assessment Report.

Source: https://www.ipcc.ch/report/graphics/index.php?t=Assessment%20Reports&r=AR5%20-%20WG1&f=Chapter%2001

Box 6.1 | Spotlight on Climate Change

Research Groups that Model Climate[27]

There are many groups in the United States and around the world at universities, government laboratories, and other organizations that work on climate modeling. A few of the larger U.S. efforts are listed here:

National Center for Atmospheric Research (NCAR) - NCAR created the Community Earth System Model (CESM), a fully coupled global climate model that provides state-of-the-art computer simulations of the Earth's past, present, and future climate states. http://www.cesm.ucar.edu

Geophysical Fluid Dynamics Laboratory (GFDL) - The mission of the GFDL is to be a world leader in the production of timely and reliable knowledge and assessments on natural climate variability and anthropogenic changes and in the development of the required Earth system models. https://www.gfdl.noaa.gov

Goddard Institute for Space Studies (GISS) - The climate modeling program at NASA's GISS focuses on decadal to centennial climate change with an emphasis on investigation of climate sensitivity, including the climate system's response to such forcings as solar variability and anthropogenic and natural emissions of greenhouse gases and aerosols. https://www.giss.nasa.gov/projects/gcm/

National Center for Environmental Modeling (NCEP) - The NCEP focuses on seasonal prediction; NCEP's Environmental Modeling Center is responsible for the enhancements, transitions-to-operations, and maintenance of more than 20 numerical prediction systems comprising NCEP's operational production suite. http://www.emc.ncep.noaa.gov

Goddard Global Modeling and Assimilation Office (GMAO) - The GMAO, a component of NASA's Goddard Space Flight Center, aims to maximize the impact of satellite observations in climate, weather, and atmospheric composition prediction using comprehensive global models and data assimilation. https://gmao.gsfc.nasa.gov

International Pacific Research Center (IPRC) - The IPRC is an international climate research center located at the University of Hawaii at Manoa with a focus on the Asia-Pacific region. Research at IPRC seeks to understand how the climate system may respond to human activity by conducting experiments with computer simulation models and by analyzing the many direct and remote observations related to climate. http://iprc.soest.hawaii.edu

Note that this is not a complete list of all U.S. climate modeling activities.

---

[27] After Climate Modeling 101, the National Academy of Sciences, Board on Atmospheric Sciences and Climate: http://nas-sites.org/climate-change/climatemodeling/page_1_1.php

know about all of them. To address this problem in climate modeling, researchers need to get an idea of all the possible ways that climate could change, and the likelihood, or probability, of each possible way. The probability of a certain climate outcome is developed by running groups of general circulation models (known as *ensembles*),[28] each of which uses different parameterizations and makes different types of assumptions. A **model ensemble** is a collection of model runs designed to identify the most probable future climate.

There are several ways to conduct ensemble experiments. One common method is to run several different models to discover all of their answers to the same question. For instance, let's say we wanted to know "How will central Pacific tropical sea-surface temperature (SST) change if global mean air temperature increases by 2 degrees Centigrade (3.6 degrees Fahrenheit)?" To get at the answer, we might choose 20 different climate models, each with slightly different parameterizations of various climate processes (e.g., ENSO, cloudiness, aerosols). If 15 out of the 20 (75 percent) agree that central Pacific tropical SST will rise between 1.7 and 2.3 degrees Centigrade (3.1 and 4.1 degrees Fahrenheit), and the rest offer answers that fall outside this range, we have confidence that the answer to our question is "SST will rise between 1.7 and 2.3 degrees Centigrade."

It is still important to assess the 25 percent of answers that fall outside this range and improve our understanding of why they do not agree, but we can conclude that the most agreed upon range of temperature in this case has the highest statistical probability of being correct. Often, in cases such as this, a **multimodel mean** is reported, which is the mean prediction of all 20 model runs. The range of model outcomes around the mean allows researchers to calculate the probability of an answer's falling within some range; scientists typically use the 95th percentile or "There is a 95 percent chance that the answer falls within a certain range of values."

6.2.6 Communicating Uncertainty[29]

Climate change exposes human communities and natural ecosystems to risk. **Risk** is the potential for consequences when something of value is at stake and the outcome is uncertain. Risks from climate change impacts arise from the interaction between three qualities: (1) **hazards** triggered by an event or a trend related to climate change, (2) **vulnerability**, which is the susceptibility to harm, and (3) **exposure** of people, assets, or ecosystems to impacts. Hazards include a range of events and trends; from brief but severe storms, to slow, multidecade droughts or multicentury sea-level rise.

Risk can be represented as the probability of occurrence of hazardous events or trends multiplied by the magnitude

of the consequences if these events occur. Therefore, high risk can result not only from high-probability outcomes but also from low-probability outcomes that carry very severe consequences. This makes it important to assess the full range of possible outcomes, from low-probability outcomes to very likely outcomes.

Because the **findings** published in Fifth Assessment Report of the IPCC[30] are intended to guide policy and management decisions (and therefore need to use consistent and specific language), the authors agreed to use a certain set of terms throughout the report to communicate the uncertainty or likelihood of the findings (Table 6.1). The terms in Table 6.1 were used in an exact fashion by all IPCC working groups (see section on IPCC Working Groups) and are referred to as **calibrated uncertainty language**.[31]

6.2.7 Regional Projections

For all their success at simulating climate at the global scale, GCMs are limited in their ability to replicate climate processes at more highly resolved scales (regionally or locally).[33] In regions where the land surface is flat for thousands of kilometers (hundreds of miles), there is no ocean or coastline nearby, land cover is simple, and land use by humans is either absent or uncomplicated, the coarse resolution of a GCM may be enough to acceptably simulate changes under future climate conditions. However, most land areas are affected by human development and have mountains, coastlines, and changing vegetation characteristics on much smaller scales. In these areas, GCM

TABLE 6.1	
Likelihood Scale, Calibrated Uncertainty Language	
Term[32]	Likelihood of the Outcome
Virtually certain	99% to 100% probability
Very likely	90% to 100% probability
Likely	66% to 100% probability
About as likely as not	33% to 66% probability
Unlikely	0% to 33% probability
Very unlikely	0% to 10% probability
Exceptionally unlikely	0% to 1% probability

[28] Kim, H.-M., et al. (2012) Evaluation of Short-Term Climate Change Prediction in Multi-Model CMIP5 Decadal Hindcasts, *Geophysical Research Letters* 39: L10701, doi:10.1029/2012GL051644.

[29] IPCC (2014) Climate Change 2014: Synthesis Report.

[30] IPCC (2014) Climate Change 2014: Synthesis Report.

[31] Mastrandrea, M.D., et al. (2010) Guidance Note for Lead Authors of the IPCC Fifth Assessment Report on Consistent Treatment of Uncertainties. Intergovernmental Panel on Climate Change, Geneva, Switzerland, 4 pp. http://www.ipcc.ch/pdf/supporting-material/uncertainty-guidance-note.pdf

[32] Additional terms that were used in limited circumstances in the AR4 (extremely likely—95% to 100% probability, more likely than not—>50% to 100% probability, and extremely unlikely—0% to 5% probability) may also be used in AR5 when appropriate.

[33] World Climate Research Program http://www.wcrp-climate.org/ and at Climate Prediction.net http://climateprediction.net/

simulations are not adequate for the practical purposes of planning water resources, changes in storminess, impacts on ecosystems, and individual community planning. In these cases, and others, information is required on a much more detailed scale than GCMs are typically able to provide.

Even with improved resolution, the ability of models to depict climate in the place where you live is limited. Global models are called that for a reason: Their output is averaged over time and over space because achieving even a global projection is a major chore for the fastest supercomputer on the planet. A commonly used model, the Community Climate System Model[34] (at the National Center for Atmospheric Research), is so complex it requires about 3 trillion computer calculations to simulate a single day of global climate (**Figure 6.8**).

Because the resolution of global models does not allow for fine-scale understanding of climate in a particular location, researchers engage in "downscaling." Two types of downscaling are typically used: dynamical and statistical. **Dynamical downscaling** applies the fundamental governing laws of physics, coupled with parameterizations, to run a regional climate model (RCM) on a high-resolution grid.[36] RCMs need a partner global model to specify

boundary conditions; boundary conditions are an existing set of climate parameters such as water vapor, winds, temperature, greenhouse gas content, and others that set the stage for the RCM calculations. The global model establishes the overarching climate conditions, the RCM applies them to a finer-scale grid network.

For instance, a researcher can use a GCM to project various climate parameters under a scenario of doubled CO_2 content. Then, using an RCM with a finer grid size over, say, the New England region, the global climate parameters are used as boundary conditions to calculate climate in the finer cells of the RCM. This is called **nested modeling**. The benefit is in the finer resolution of the RCM. For instance, in the global model, perhaps only a dozen data points (12 cells) represent the topography of the White Mountains, the Berkshires, the Green Mountains, the Catskills, and other significant topographic features that influence circulation, clouds, precipitation, and air temperature. In the RCM, perhaps this complex topography is represented with 60 to 120 cells, thus, improving the simulation of winds, snow cover, rainfall, biological systems, the coastline, and others.[37]

Another way to project future climate conditions at higher resolution is called **statistical downscaling**. Statistical

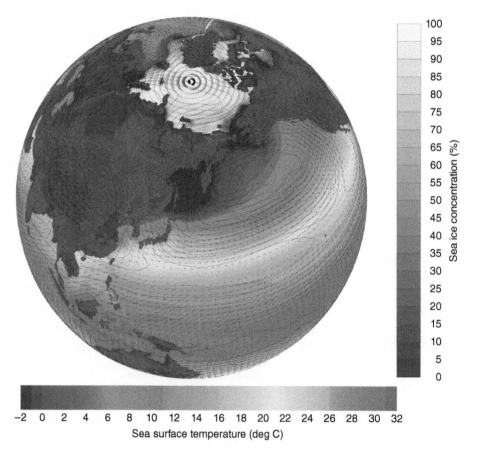

FIGURE 6.8 This image, taken from a simulation of 20th century climate, depicts several aspects of Earth's climate system. Sea surface temperatures and sea–ice concentrations are shown by the two scales. The figure also captures sea-level pressure and low-level winds, including warmer air moving north on the eastern side of low-pressure regions and colder air moving south on the western side of the lows. Such simulations, produced by the NCAR-based Community Climate System Model, can also depict additional features of the climate system, such as precipitation. Companion software, released as the Community Earth System Model, enables scientists to study the climate system in even greater complexity.[35]

Source: https://www2.ucar.edu/atmosnews/news/2366/ new-computer-model-advances-climate-change-research

Sea ice concentration (%)

Sea surface temperature (deg C)

---

[34] Community Climate System Model home page, http://www.cesm .ucar.edu/models/ccsm4.0/

[35] National Science Foundation Special Report Clouds: http://www.nsf .gov/news/special_reports/clouds/downloads.jsp

[36] Description of RCM at http://www.climateprediction.net/climate-science/climate-modelling/regional-models/

[37] VIDEO: NASA Supercomputing the climate. http://www.nasa.gov/ topics/earth/features/climate-sim-center.html

downscaling is a two-step process consisting of 1) the development of statistical relationships between local climate variables of interest (e.g., surface air temperature and precipitation) and large-scale predictors (e.g., pressure fields), and 2) the application of these relationships to the output of GCM experiments that simulate future climate characteristics in the area of interest. The accuracy of statistical downscaling relies on important assumptions. Chiefly, have the statistical relationships between variables and predictors been appropriately defined, and do they remain appropriate under the changing climate conditions of the future? As both dynamical and statistical downscaling experiments continue to be developed, improvements in both approaches are likely to provide increasingly reliable results.

Increasingly, as decision-makers responsible for community sustainability and environmental conservation grow more worried about the long-term impacts of global warming, RCMs are being used to clarify plans for the future. For instance, GCMs show[38] that in the future, water availability in the western United States will be increasingly tied to extreme events[39]; however, the degree to which this change will affect local communities is poorly defined by the scale of global models.

This problem has been improved with the use of regional climate modeling. One study[40] used an ensemble of eight RCMs embedded within the projections of GCMs to estimate future winter average and extreme precipitation in the western United States. Researchers found a consistent and statistically significant increase in the intensity of future extreme winter precipitation events over the western United States. For the years 2038 to 2070, 20-year return-period and 50-year return-period winter storms are modeled to increase across the entire west by 12.6 percent and 14.4 percent, respectively. Model results show that this increase in storminess will be accompanied by a 7.5 percent decrease in winter average precipitation in the southwestern United States. For water managers in communities faced with population growth (the southwestern United States is the fastest-growing region in the nation), this type of information provided by a regional climate model is a key element in effective planning for the future.

6.2.8 Historical Accuracy; No Guarantee of Future Success

Despite success in reproducing historical climate, it is nonetheless possible that models are achieving the right

results for the wrong reasons. That is, assumptions about climate processes represented in a model may be wrong, yet the combined effects of various processes could lead to a model successfully matching historical observations. Scientists tested[41] this possibility by running 11 atmosphere–ocean coupled GCMs. Instead of looking at the models' ability to reproduce 20th century temperatures, the study focused on model skill in recreating global average, Arctic, and tropical climates. Additionally, climate forcings (such as solar activity), feedback systems (such as Arctic ice melt or the effects of clouds), and representations of heat storage and transport mechanisms were analyzed.

Of the 11 models tested, eight successfully reproduced global average, Arctic, and tropical temperatures for the past century; most failed to capture warming that occurred in the 1920s and 1930s; three failed to achieve historical accuracy; and two unrealistically depicted either Arctic or tropical temperature change. All the models emphasize climate feedbacks and forcings to different degrees and therefore, point out the study authors, model skill in reproducing 20th century climate is not an indication that they will accurately predict future climate. The results suggest that researchers should focus on improving parameterization of the complex relationships among climate forcing and feedback processes that are poorly represented in GCMs.

6.3 Intergovernmental Panel on Climate Change— Assessment Report 5

In 2013, the IPCC[42] released the first of three "Working Group" reports constituting the Fifth Assessment Report (AR5) on Climate Change. AR5 was the product of thousands of scientists, criticizing each other's work, struggling to find common wording to describe results, and ultimately building a description of the impacts of global warming on all sectors of human interest. AR5 was an enormous effort summarizing the current understanding of climate change.

The final two Working Group reports were released in 2014, along with a "Synthesis Report". Hence, the Fifth Assessment is made up of four reports: three IPCC Working Groups' contributions dealing respectively with 1) *The Physical Science Basis* (2013), 2) *Impacts, Adaptation, and Vulnerability* (2014), and 3) *Mitigation of Climate Change* (2014), and a *Synthesis Report* (2014). Each report contains its own *Summary for Policymakers*, which is approved line by line by all member countries of the IPCC and constitutes a formally agreed statement on key findings and uncertainties.

[38] Meehl, G.A., et al. (2007) Current and Future US Weather Extremes and El Niño, *Geophysical Research Letters* 34: L20704, doi:10.1029/2007GL031027.
[39] Emori, S. and Brown, S. (2005) Dynamic and Thermodynamic Changes in Mean and Extreme Precipitation under Changed Climate, *Geophysical Research Letters* 32: L17706, doi:10.1029/2005GL023272.
[40] Dominguez, F., et al. (2012) Changes in Winter Precipitation Extremes for the Western United States under a Warmer Climate as Simulated by Regional Climate Models, *Geophysical Research Letters* 39: L05803, doi:10.1029/2011GL050762

[41] Crook, J.A. and Forster, P.M. (2011) A Balance between Radiative Forcing and Climate Feedback in the Modeled 20th Century Temperature Response, *Journal of Geophysical Research* 116: D17108, doi:10.1029/2011JD015924.
[42] IPCC description, http://www.ipcc.ch/pdf/press/ipcc_leaflets_2010/ipcc_ar5_leaflet.pdf

There is also a *Technical Summary* that accompanies each of the working group reports and the Synthesis Report.[43]

6.3.1 IPCC Working Groups

The IPCC Working Group I (WG I) assesses the physical scientific aspects of the climate system and climate change. The main topics assessed by WG I include: changes in greenhouse gases and aerosols in the atmosphere; observed changes in air, land, and ocean temperatures, rainfall, glaciers and ice sheets, oceans and sea level; historical and paleoclimate perspective on climate change; biogeochemistry, carbon cycle, gases and aerosols; satellite data and other data; climate models; and climate projections, causes and attribution of climate change.

Working Group II (WG II) assesses the vulnerability of socioeconomic and natural systems to climate change, negative and positive consequences of climate change, and options for adapting to it. It also takes into consideration the interrelationship between vulnerability, adaptation, and sustainable development. The assessed information is considered by sectors (water resources; ecosystems; food and forests; coastal systems; industry; and human health) and regions (Africa; Asia; Australia, and New Zealand; Europe; Latin America; North America; Polar Regions; and Small Islands).

Working Group III (WG III) assesses options for mitigating climate change through limiting or preventing greenhouse gas emissions and enhancing activities that remove them from the atmosphere. The main economic sectors are taken into account, both in a near-term and in a long-term perspective. The sectors include energy, transport, buildings, industry, agriculture, forestry, waste management. WG III analyzes the costs and benefits of the different approaches to mitigation, considering also the available instruments and policy measures. The approach is solution-oriented.

6.3.2 IPCC Findings

AR5 was the result of several years of work and involved more than 2,500 scientific expert reviewers and more than 800 authors from more than 130 countries. Key findings in AR5 provide a comprehensive summary of the state of change in Earth's climate system.[44]

Existing Impacts With regard to existing impacts of global warming, the report finds the following:

- Human influence on the climate system is clear, and recent anthropogenic emissions of greenhouse gases are the highest in history. Recent climate changes have had widespread impacts on human and natural systems.
- Warming of the climate system is unequivocal, and since the 1950s, many of the observed changes are unprecedented over decades to millennia. The atmosphere and ocean have warmed, the amounts of snow and ice have diminished, and sea level has risen.
- Anthropogenic greenhouse gas emissions have increased since the preindustrial era, driven largely by economic and population growth, and are now higher than ever. This has led to atmospheric concentrations of carbon dioxide, methane, and nitrous oxide that are unprecedented in at least the last 800,000 years. Their effects, together with those of other anthropogenic drivers, have been detected throughout the climate system and are extremely likely to have been the dominant cause of the observed warming since the mid-20th century.
- In recent decades, changes in climate have caused impacts on natural and human systems on all continents and across the oceans. Impacts are due to observed climate change, irrespective of its cause, indicating the sensitivity of natural and human systems to changing climate.
- Changes in many extreme weather and climate events have been observed since about 1950. Some of these changes have been linked to human influences, including a decrease in cold temperature extremes, an increase in warm temperature extremes, an increase in extreme high sea levels, and an increase in the number of heavy precipitation events in a number of regions.

Future Impacts The model projections for future climate change in AR5 are similar to projections made in AR4, published in 2007. The ranges (max to min) of projected temperature, precipitation, sea level, and other types of changes are less in AR5, but this is due to the nature of the scenarios used in the modeling. In contrast to the emission scenarios based on economic activity modeled in AR4, the scenarios modeled in AR5 are based on emissions resulting in certain levels of radiative forcing. Carbon cycle uncertainties affecting atmospheric CO_2 concentrations that were inherent in the AR4 modeling are not intrinsic to the AR5 simulations.

Projections of sea-level rise in AR5 are larger than in AR4, primarily because of improved modeling of land–ice. Temperature and precipitation projections for the next few decades show geographic patterns of change that are similar to the projections for later this century, but with smaller magnitude. According to the models in AR5, natural variability will continue to be a major influence on climate, particularly in the next few decades and at the regional scale. By the middle of the century, the magnitudes of various future changes are tied largely to the choice of emissions scenario being modeled.

---

[43] IPCC Working Groups: http://www.ipcc.ch/working_groups/working_groups.shtml

[44] IPCC (2014) Climate Change 2014: Synthesis Report.

With regard to future impacts of global warming, AR5 finds the following:[45]

- Continued emission of greenhouse gases will cause further warming and long-lasting changes in all components of the climate system, increasing the likelihood of severe, pervasive, and irreversible impacts for people and ecosystems. Limiting climate change would require substantial and sustained reductions in greenhouse gas emissions, which, together with adaptation, can limit climate change risks.

- Cumulative emissions of CO_2 largely determine global mean surface warming by the late 21st century and beyond. Projections of greenhouse gas emissions vary over a wide range, depending on both socioeconomic development and climate policy.

- Surface temperature is projected to rise over the 21st century under all assessed emission scenarios. It is very likely that heat waves will occur more often and last longer and that extreme precipitation events will become more intense and frequent in many regions. The ocean will continue to warm and acidify and global mean sea level to rise.

- Climate change will amplify existing risks and create new risks for natural and human systems. Risks are unevenly distributed and are generally greater for disadvantaged people and communities in countries at all levels of development.

- Many aspects of climate change and associated impacts will continue for centuries, even if anthropogenic emissions of greenhouse gases are stopped. The risks of abrupt or irreversible changes increase as the magnitude of the warming increases.

Reacting to Climate Change   Assessment Report 5 concludes that adapting to changes in the climate system and reducing emissions of greenhouse gases are the two most appropriate reactions to climate change. Identifying the most suitable adaptation and mitigation pathways should be assessed using criteria of sustainability and equity. That is, policies and projects designed to improve mitigation or adaptation should be sustainable in the long run and fair to all human communities.

It is important to recognize that different countries have very different histories of contributing to the accumulation of greenhouse gases in the atmosphere. It is also important to recognize that different countries face varying challenges and circumstances and thus have different capacities to address mitigation and adaptation. This raises issues of equity, justice, and fairness. Consideration should be given to the following: 1) Many of the most vulnerable populations have contributed very little to greenhouse gas emissions. 2) Delaying mitigation shifts the burden of dealing with the worst aspects of climate change from the present to the future. 3) The lack of sufficient adaptation steps to date erodes the basis for sustainable development. 4) Comprehensive strategies in response to climate change that are consistent with sustainable development will have to take into account the benefits, adverse side effects, and risks that may arise from both adaptation and mitigation options.

With regard to adaptation and mitigation activities, AR5 finds the following:[46]

- Adaptation and mitigation are complementary strategies for reducing and managing the risks of climate change. Substantial emissions reductions over the next few decades can reduce climate risks in the 21st century and beyond, increase prospects for effective adaptation, reduce the costs and challenges of mitigation in the longer term, and contribute to climate-resilient pathways for sustainable development.

- Effective decision-making to limit climate change and its effects can be informed by a wide range of analytical approaches for evaluating expected risks and benefits, recognizing the importance of governance, ethical dimensions, equity, value judgments, economic assessments and diverse perceptions and responses to risk and uncertainty.

- Without additional mitigation efforts beyond those in place today, and even with adaptation, warming by the end of the 21st century will lead to high to very high risk of severe, wide-spread and irreversible impacts globally. Mitigation involves some level of benefits and of risks due to adverse side effects, but these risks do not involve the same possibility of severe, wide-spread, and irreversible impacts as risks from climate change, increasing the benefits from near-term mitigation efforts.

- Adaptation can reduce the risks of climate change impacts, but there are limits to its effectiveness, especially with greater magnitudes and rates of climate change. Taking a longer-term perspective, in the context of sustainable development, increases the likelihood that more immediate adaptation actions will also enhance future options and preparedness.

- There are multiple mitigation pathways that are likely to limit warming to below 2.0°C (3.6°F) relative to preindustrial levels. These pathways would require substantial emissions reductions over the next few decades. Implementing such reductions poses substantial technological, economic, social and institutional challenges, which increase with delays in additional mitigation and if key technologies are not available. Limiting warming to lower or higher levels involves similar challenges but on different timescales.

[45] IPCC (2014) Climate Change 2014: Synthesis Report.

[46] IPCC (2014) Climate Change 2014: Synthesis Report.

- Many adaptation and mitigation options can help address climate change, but no single option is sufficient by itself. Effective implementation depends on policies and cooperation at all scales and can be enhanced through integrated responses that link adaptation and mitigation with other societal objectives.
- Adaptation and mitigation responses are underpinned by common enabling factors. These include effective institutions and governance, innovation and investments in environmentally sound technologies and infrastructure, sustainable livelihoods and behavioral and lifestyle choices.
- Adaptation options exist in all sectors, but their context for implementation and potential to reduce climate-related risks differs across sectors and regions. Some adaptation responses involve significant benefits, synergies, and trade-offs. Increasing climate change will increase challenges for many adaptation options.
- Mitigation options are available in every major sector. Mitigation can be more cost-effective if using an integrated approach that combines measures to reduce energy use and the greenhouse gas intensity of end-use sectors, decarbonize energy supply, reduce net emissions, and enhance carbon sinks in land-based sectors.
- Effective adaptation and mitigation responses will depend on policies and measures across multiple scales: international, regional, national, and subnational. Policies across all scales supporting technology development, diffusion, and transfer, as well as finance for responses to climate change, can complement and enhance the effectiveness of policies that directly promote adaptation and mitigation.
- Climate change is a threat to sustainable development. Nonetheless, there are many opportunities to link mitigation, adaptation, and the pursuit of other societal objectives through integrated responses. Successful implementation relies on relevant tools, suitable governance structures, and enhanced capacity to respond.

6.3.3 CMIP—The Coupled Model Intercomparison Project

The United Nations World Climate Research Program[47] sponsors the Coupled Model Intercomparison Project or CMIP, a global endeavor to compare and share data among the research groups who build and run climate models. Virtually the entire international climate modeling community has participated in CMIP since its inception in 1995. Originally, CMIP began by collecting and comparing output from participating research groups using control runs, in which climate was held constant under a set of unchanging conditions (constant carbon dioxide, solar brightness, and other external forcing). The main goal of the early project was to compare and analyze differences in the simulation of Earth's climate system from various models that use the same climate values.

Phase two of CMIP (CMIP2) collected and compared output from an idealized scenario of global warming, in which atmospheric CO_2 increased at the rate of 1 percent per year until it doubled at about Year 70 of a model run. No other anthropogenic climate forcing, such as anthropogenic aerosols (which have a net cooling effect), was included. The modeling experiment in CMIP2 did not include natural variations in climate forcing, such as from volcanic eruptions or changing solar brightness. The idealized scenarios under the CMIP2 design resulted in easy-to-identify differences among the various model results. This facilitated comparing, and thereby possibly explaining, different responses arising from different model designs. Understanding model performance, or skill, was the primary purpose of the CMIP2 effort.

Phase three (CMIP3) included scenarios for both past and present climate forcing to improve understanding of conditions that led to past climate changes and compared results to geologic proxies of paleoclimate. The research based on CMIP3 provided much of the material underlying projections of future climate in the Fourth Assessment Report (AR4) of the Intergovernmental Panel on Climate Change.[48]

CMIP5 provided modeling support for the most recent IPCC assessment, AR5. At a conference in 2008 attended by 20 modeling groups from around the world, the CMIP5 experiment was established with the objectives of 1) evaluating how realistic climate models are at simulating the recent past, 2) providing projections of future climate change on two timescales, near term (out to about 2035) and long term (to about 2100 and beyond), and 3) improving understanding of some of the factors responsible for differences in model projections, including quantifying some key feedbacks such as those involving clouds and the carbon cycle.

CMIP5 modeling has continued since the publication of AR5 in 2013. However, the various more recent results are not collated in a single report and the AR5 Working Group I Summary for Policymakers[49] offers a good reference point for the state of modeling at this date.

---

[47] The World Climate Research Program (WCRP) facilitates analysis and prediction of Earth system change for use in a range of practical applications of direct relevance, benefit, and value to society. WCRP aims to determine the predictability of climate and the effect of human activities on climate. https://www.wcrp-climate.org

[48] There was no CMIP4 model experiment. The project skipped from CMIP3 to CMIP5.

[49] IPCC (2013) Summary for Policymakers. In: *Climate Change 2013: The Physical Science Basis. Contribution of Working Group I to the Fifth Assessment Report of the Intergovernmental Panel on Climate Change* [Stocker, T.F., et al. (eds.)]. Cambridge University Press, Cambridge, United Kingdom and New York, NY, USA.

Coherence with Observed Changes General circulation model projections of the future are based on an ability to accurately simulate the past and current conditions of the climate system. An important step in validating climate models is testing their ability to replicate historical climate observations. CMIP5 models produced remarkable coherence among observed changes in climate over the past century and simulations of both anthropogenic and natural forcing. This was particularly strong for temperature-related variables. CMIP5 simulations of both surface temperature and ocean heat content show emerging anthropogenic and natural signals.

In **Figure 6.9**, there is a clear separation between natural climate variations, and anthropogenic influences on

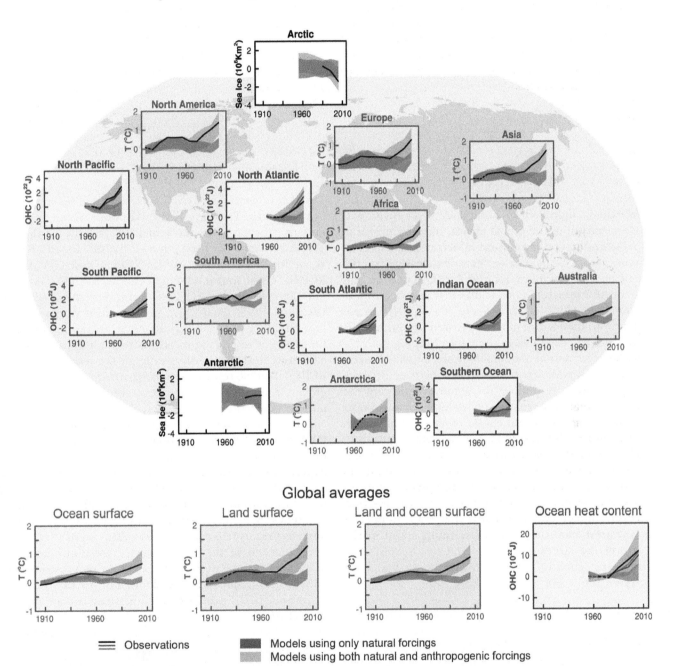

FIGURE 6.9 Comparison of observed and simulated change in the climate system, at regional scales (top panels) and global scales (bottom four panels). Panels depict land surface temperature time series, ocean heat content time series, and sea ice time series (decadal averages). Each panel shows observations (black or black and shades of gray), and the 5% to 95% range of the simulated response to natural forcings and natural and anthropogenic forcings, together with the corresponding ensemble means. The observation lines are either solid or dashed and indicate the quality of the observations and estimates. For land and ocean surface temperature panels and precipitation panels, solid observation lines indicate where spatial coverage of areas being examined is above 50% coverage and dashed observation lines where coverage is below 50%. For example, data coverage of Antarctica never goes above 50% of the land area of the continent. For ocean heat content and sea ice panels, the solid observations line is where the coverage of data is good and higher in quality, and the dashed line is where the data coverage is only adequate.

Source: http://www.ipcc.ch/report/graphics/index.php?t=Assessment%20Reports&r=AR5%20-%20WG1&f=Technical%20Summary

TABLE 6.2

Projected Change in Global Mean Surface Air Temperature and Global Mean Sea Level for the Mid- and Late 21st Century Relative to the Reference Period of 1985-2005.[52]

		2046-2065		2081-2100	
	Scenario	**Mean**	**Likely range**	**Mean**	**Likely range**
Global Mean Surface Temperature Change (°C)	RCP2.6	1.0	0.4 to 1.6	1.0	0.3 to 1.7
	RCP4.5	1.4	0.9 to 2.0	1.8	1.1 to 2.6
	RCP6.0	1.3	0.8 to 1.8	2.2	1.4 to 3.1
	RCP8.5	2.0	1.4 to 2.6	3.7	2.6 to 4.8
	Scenario	Mean	Likely range	Mean	Likely range
Global Mean Sea-Level Rise (m)	RCP2.6	0.24	0.17 to 0.32	0.40	0.26 to 0.55
	RCP4.5	0.26	0.19 to 0.33	0.47	0.32 to 0.63
	RCP6.0	0.25	0.18 to 0.32	0.48	0.33 to 0.63
	RCP8.5	0.30	0.22 to 0.38	0.63	0.45 to 0.82

climate, when compared to observations. These signals do not appear just in the global means, but also appear at regional scales on continents and in ocean basins in each of these variables. Sea ice extent emerges clearly from the range of natural variability for the Arctic. Based on this modeling, AR5 researchers conclude that at subcontinental scales, human influence is likely (66 to 100 percent probability) to have substantially increased the probability of occurrence of heat waves in some locations.[50]

Because of strong agreement between modeled simulations and observations, there is very high confidence that CMIP5 models reproduce the general features of the global and annual mean surface temperature changes over the historical period. This includes certain features such as the warming in the second half of the 20th century and the cooling immediately following large volcanic eruptions. However, most model simulations of the historical period do not reproduce the observed reduction in global mean surface warming that occurred over the period 1998–2012, the so-called global warming hiatus. Most (though not all) models overestimate the observed warming trend in the tropics over the last 30 years and tend to underestimate the long-term cooling trend taking place in the lower stratosphere.

Representative Concentration Pathways—RCPs To model[51] future global warming and its impacts, researchers participating in CMIP5 based their efforts on greenhouse gas concentrations (called Representative

Concentration Pathways, RCPs) that would force a range of radiative responses by the year 2100 (relative to the year 1795): 2.6 Watts per square meter (W/m²) for RCP2.6; 4.5 W/m² for RCP4.5; 6.0 W/m² for RCP6.0; and 8.5 W/m² for RCP8.5. These four modeling scenarios include a stringent mitigation scenario (RCP2.6), two intermediate scenarios (RCP4.5 and RCP6.0), and one scenario with very high greenhouse gas emissions (RCP8.5). Scenarios in which humans fail to engage in efforts to limit greenhouse gas emissions lead to pathways ranging between RCP6.0 and RCP8.5. RCP2.6 represents a scenario that attempts to keep global warming below 2 degrees Celsius (3.6 degrees Fahrenheit) above preindustrial temperatures.

Model projections of global mean surface temperature for the period 2016–2035 relative to 1986–2005 is similar for all four RCPs and will likely be in the range 0.3 to 0.7 degrees Celsius (0.5 to 1.3 degrees Fahrenheit). By midcentury, the four RCPs begin to diverge and the likely range of warming under RCP2.6 is 0.4 to 1.6 degrees Celsius (0.7 to 2.9 degrees Fahrenheit) and under RCP8.5 is 1.4 to 2.6 degrees Celsius (2.5 to 4.7 degrees Fahrenheit).

At the end of the 21st century, likely warming under RCP2.6 is 0.3 to 1.7 degrees Celsius (0.5 to 3.1 degrees Fahrenheit) and under RCP8.5 is 2.6 to 4.8 (4.7 to 8.7 degrees Fahrenheit). The other RCPs fall between these end members (Table 6.2)

6.3.4 Projected Climate in AR5

The future climate projected in AR4 (2007, CMIP3) would not have been substantially different if the CMIP5 models had been used instead. The global mean temperature response simulated by CMIP3 and CMIP5 models is very

---

[50] Stocker, T.F., et al. (2013) Technical Summary. In: *Climate Change 2013: The Physical Science Basis. Contribution of Working Group I to the Fifth Assessment Report of the Intergovernmental Panel on Climate Change* [Stocker, T.F., et al. (eds.)]. Cambridge University Press, Cambridge, United Kingdom and New York, NY, USA.

[51] VIDEO: CMIP5, RCP8.5. https://vimeo.com/70674503

[52] Table TS.1, Stocker et al. (2013)

similar, both in the mean and in the model range. In both cases, future climate will depend on warming caused by past greenhouse gas emissions, as well as future emissions and natural climate variability. The future temperatures projected by CMIP5 models assume that there will be no major volcanic eruptions or changes in natural emissions.

By the end of the 21st century, the Arctic region will continue to warm more rapidly than the global mean. As global mean surface temperature increases, it is virtually certain (99 to 100 percent probability) that there will be more frequent hot and fewer cold temperature extremes over most land areas on daily and seasonal timescales. It is very likely (90 to 100 percent probability) that heat waves will occur with a higher frequency and longer duration. Occasional cold winter extremes will continue to occur.

Changes in precipitation will not be uniform. The high latitudes and the equatorial Pacific are likely (66 to 100 percent probability) to experience an increase in annual mean precipitation under the RCP8.5 scenario. In many mid-latitude and subtropical dry regions, mean precipitation will likely decrease, while in many midlatitude wet regions, mean precipitation will likely increase under the RCP8.5 scenario. Extreme precipitation events over most of the midlatitude land masses and over wet tropical regions will very likely become more intense and more frequent.

The global ocean will continue to warm during the 21st century, with the strongest warming projected for the surface in tropical and Northern Hemisphere subtropical regions. Models project a global increase in ocean acidification for all RCP scenarios by the end of the 21st century, with a slow recovery after midcentury under RCP2.6. The decrease in surface ocean pH is in the range of 0.06 to 0.07 (15 to 17 percent increase in acidity) for RCP2.6, 0.14 to 0.15 (38 to 41 percent) for RCP4.5, 0.20 to 0.21 (58 to 62 percent) for RCP6.0, and 0.30 to 0.32 (100 to 109 percent) for RCP8.5.

It is virtually certain that near-surface permafrost extent at high northern latitudes will be reduced as global mean surface temperature increases, with the area of permafrost near the surface (upper 3.5 m [11.5 ft]) projected to decrease by 37 percent (RCP2.6) to 81 percent (RCP8.5) for the multimodel average.

The global glacier volume, excluding glaciers on the periphery of Antarctica (and excluding the Greenland and Antarctic ice sheets), is projected to decrease by 15 to 55 percent for RCP2.6 and by 35 to 85 percent for RCP8.5.

There has been significant improvement in understanding and projection of sea level change since the AR4. Global mean sea-level rise will continue during the 21st century, very likely at a faster rate than observed from 1971 to 2010. For the period 2081–2100 relative to 1986–2005, across regions. By the end of the 21st century, it is very likely that sea level will rise in more than about 95 percent of the ocean area. About 70 percent of the coastlines worldwide are projected to experience a sea-level change within ±20 percent of the global mean (**Figure 6.10**).

6.3.5 Future Global and Regional Climate Change

CMIP5 model results indicate the following future global and regional-scale climate patterns:

- Climate change over the next few decades will be similar to what is expected by the end of the century, but with smaller magnitude.
- Natural climate variability will continue to be a major influence, particularly in the near-term and at the regional scale.
- By the middle of the century, the amount of climate change will depend on the amount of carbon that is emitted by human activities.

Temperature The global mean surface temperature change over the next two decades, relative to 1986–2005, will likely be in the range of 0.3 to 0.7 degrees Celsius (0.5 to 1.3 degrees Fahrenheit). This assessment is based on multiple lines of evidence and assumes there will be no major volcanic eruptions or changes in total solar irradiance.

- Near-term increases in seasonal and annual temperatures are expected to be larger in the tropics and subtropics than in midlatitudes.
- The increase in global mean surface temperatures for the end of the century, relative to 1986–2005, is projected to be in the following range:
  a. RCP2.6—0.3 to 1.7°C (0.5 to 3.0°F);
  b. RCP4.5—1.1 to 2.6°C (2.0 to 4.7°F);
  c. RCP6.0—1.4 to 3.1°C (2.5 to 5.6°F);
  d. RCP8.5—2.6 to 4.8°C (4.7 to 8.6°F).
- The Arctic region will warm more rapidly than the global mean, and warming over land will be larger than over the ocean.
- It is virtually certain that there will be more frequent hot and fewer cold temperature extremes over most land areas on daily and seasonal timescales as global temperatures increase.
- It is very likely that heat waves will occur with a higher frequency and duration. Occasional cold winter extremes will continue to occur.

Water Cycle Projected changes in the water cycle over the next few decades show similar large-scale patterns to those toward the end of the century, but with smaller magnitude.

- Changes in the near-term water cycle, and at the regional scale, will be strongly influenced by natural variability and may be affected by anthropogenic aerosol emissions.
- The high latitudes and the equatorial Pacific Ocean are likely to experience an increase in annual mean precipitation by the end of this century under the RCP8.5 scenario.

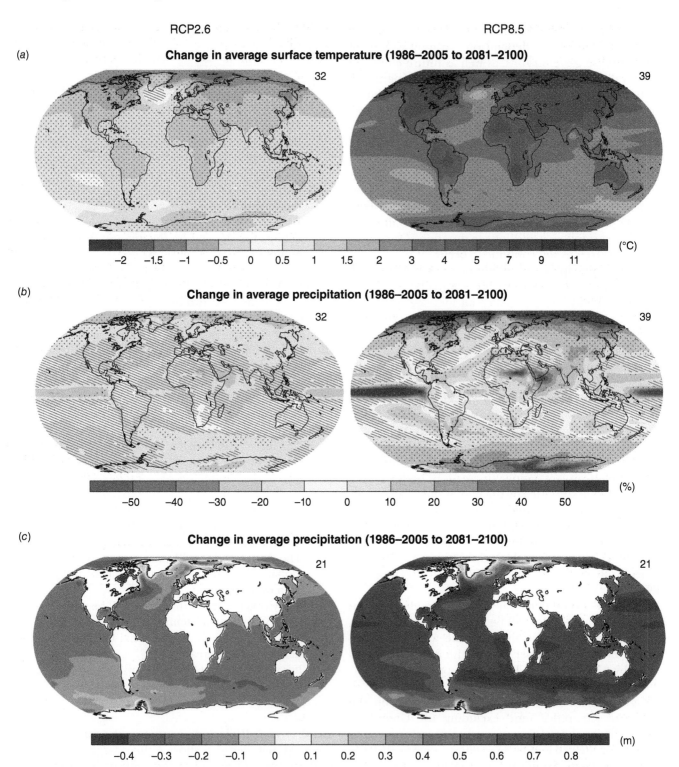

FIGURE 6.10 Change in (a) average surface temperature, (b) average precipitation, and (c) average sea level. These CMIP5 model results are based on ensemble mean projections for 2081–2100 relative to 1986–2005 under the RCP2.6 (left) and RCP8.5 (right) scenarios. The number of models used to calculate the ensemble mean is indicated in the upper right corner of each panel. Stippling (i.e., dots) shows regions where the projected change is large compared to natural internal variability and where at least 90% of models agree on the sign of change. Hatching (i.e., diagonal lines) shows regions where the projected change is less than one standard deviation of the natural internal variability (indicating results are not statistically different than natural variability).[53]

Source: http://www.ipcc.ch/report/graphics/index.php?t=Assessment%20Reports&r=AR5%20-%20Synthesis%20Report

[53] Figure SPM.7 from IPCC (2014) Climate Change 2014: Synthesis Report.

- In many midlatitude and subtropical dry regions, mean precipitation will likely decrease, while in many midlatitude wet regions, mean precipitation will likely increase by the end of this century under the RCP8.5 scenario.
- Extreme precipitation over most of the midlatitude land masses and over wet tropical regions will very likely become more intense and more frequent by the end of this century.
- Globally, it is likely that the area encompassed by **monsoon** systems will increase over the 21st century.
- While monsoon winds are likely to weaken, monsoon precipitation is likely to intensify due to the increase in atmospheric moisture.
- Monsoon onset dates are likely to become earlier or not to change much. Monsoon retreat dates will likely be delayed, resulting in a longer monsoon season in many regions.
- There is high confidence that the El Niño–Southern Oscillation (ENSO) will remain the dominant mode of interannual variability in the tropical Pacific, with global effects in the 21st century.
- Due to the increase in moisture availability, ENSO-related precipitation variability on regional scales will likely intensify.

Air Quality Evidence indicates that locally higher temperatures in polluted regions will trigger regional feedbacks in chemistry and local emissions that will increase peak levels of ozone and particulate matter.

Ocean The strongest ocean warming is projected for the surface in tropical and Northern Hemisphere subtropical regions. At greater depth, the warming will be most pronounced in the Southern Ocean.

- Best estimates of ocean warming by the end of the 21st century in the top 100 m (328 ft) are about 0.6°C (1.0°F) for RCP2.6, to 2.0°C (3.6°F) for RCP8.5.
- At a depth of about 1000 m (3281 ft), best estimates of ocean warming are about 0.3°C (0.5°F) for RCP2.6, to 0.6°C (1.1°F) for RCP8.5.
- It is very likely that the Atlantic Meridional Overturning Circulation (AMOC) will weaken over the 21st century. The best estimate is 11% in RCP2.6 and 34% in RCP8.5.
- It is likely that there will be some decline in the AMOC by about 2050, but there may be some decades when the AMOC increases due to large natural variability.
- It is very unlikely that the AMOC will undergo an abrupt transition or collapse in the 21st century for the scenarios considered. A collapse beyond the 21st century for large sustained warming cannot be excluded.

Cryosphere Year-round reductions in Arctic sea ice extent are projected by the end of the 21st century from

multimodel averages. These reductions range from 43% for RCP2.6 to 94 percent for RCP8.5 in September and from 8 percent for RCP2.6 to 34 percent for RCP8.5 in February.

- By the end of the 21st century, the global glacier volume, excluding glaciers on the periphery of Antarctica, is projected to decrease by 15% to 55% for RCP2.6 and by 35% to 85% for RCP8.5.
- The area of Northern Hemisphere spring snow cover is projected to decrease by 7% for RCP2.6 and by 25% in RCP8.5 by the end of the 21st century for the model average.
- It is virtually certain that near-surface permafrost extent at high northern latitudes will be reduced as global mean surface temperature increases. By the end of the 21st century, the area of permafrost near the surface (upper 3.5 m [11.5 ft]) is projected to decrease by between 37% (RCP2.6) and 81% (RCP8.5) for the model average.

Sea Level Global mean sea-level rise for 2081–2100 relative to 1986–2005 will likely be in the ranges of 0.26 to 0.55 meters (0.85 to 1.8 feet) for RCP2.6, 0.32 to 0.63 meters (1.0 to 2.1 feet) for RCP4.5, 0.33 to 0.63 meters (1.1 to 2.1 feet) for RCP6.0, and 0.45 to 0.82 meters (1.5 to 2.7 feet) for RCP8.5.

- For RCP8.5, the rise by the year 2100 is 0.52 to 0.98 m (1.7 to 3.2 ft), with a rate during 2081 to 2100 of 8 to 16 mm/yr (0.3 to 0.63 in/yr) (**Figure 6.11**).

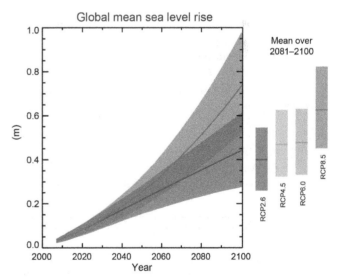

FIGURE 6.11 Projections of global mean sea-level rise over the 21st century relative to 1986–2005 from the combination of the CMIP5 ensemble with process-based models, for RCP2.6 and RCP8.5. The assessed likely range is shown as a shaded band. The assessed likely ranges for the mean over the period 2081–2100 for all RCP scenarios are given as vertical bars, with the corresponding median value given as a horizontal line.

Source: https://www.ipcc.ch/report/graphics/index.php?t=Assessment%20 Reports&r=AR5%20-%20WG1&f=SPM

- In the RCP projections, thermal expansion accounts for 30% to 55% of 21st century global mean sea level rise and glaciers for 15% to 35%.
- The increase in surface melting of the Greenland ice sheet will exceed the increase in snowfall, leading to a positive contribution from changes in surface mass balance to future sea level.
- While surface melting will remain small, an increase in snowfall on the Antarctic ice sheet is expected, resulting in a negative contribution to future sea level from changes in surface mass balance.
- Changes in outflow from both ice sheets combined will likely make a contribution in the range of 0.03 to 0.20 m (0.01 to 0.65 ft) by 2081–2100.
- Based on current understanding, only the collapse of marine-based sectors of the Antarctic ice sheet, if initiated, could cause global mean sea level to rise substantially above the likely range during the 21st century. However, there is medium confidence that this additional contribution would not exceed several tenths of a meter of sea-level rise during the 21st century.
- The basis for higher projections of global mean sea-level rise in the 21st century has been considered, and it has been concluded that there is currently insufficient evidence to evaluate the probability of specific levels above the assessed likely range. Many semi-empirical model projections of global mean sea-level rise are higher than process-based model projections (up to about twice as large), but there is no consensus in the scientific community about their reliability and there is thus low confidence in their projections.
- Sea-level rise will not be uniform. By the end of the 21st century, it is very likely that sea level will rise in more than about 95% of the ocean area. About 70% of the coastlines worldwide are projected to experience sea-level change within 20% of the global mean sea-level change.

Carbon and Other Biogeochemical Cycles Ocean uptake of carbon dioxide will continue under all RCPs through 2100, with higher uptake for higher concentration pathways.

- The future evolution of land carbon uptake is less certain. A majority of models project a continued land carbon uptake under all RCPs, but some models simulate a land carbon loss due to the combined effect of climate change and land use change.
- Based on Earth System Models, there is high confidence that the feedback between climate and the carbon cycle is positive in the 21st century; that is, climate change will partially offset changes in land and ocean carbon sinks caused by rising atmospheric CO_2. As a result, more of the emitted CO_2 will remain in the atmosphere. A positive feedback between climate and the carbon cycle on century to millennial timescales is supported by paleoclimate observations and modeling.
- Earth system models project a global increase in ocean acidification for all RCP scenarios. The corresponding decrease in surface ocean pH by the end of the 21st century is in the range of 0.06 to 0.07 for RCP2.6, 0.14 to 0.15 for RCP4.5, 0.20 to 0.21 for RCP6.0, and 0.30 to 0.32 for RCP8.5.
- Cumulative CO_2 emissions for the 2012 to 2100 period compatible with the RCP atmospheric CO_2 concentrations, as derived from 15 Earth System Models, range from 140 to 410 GtC for RCP2.6, 595 to 1005 GtC for RCP4.5, 840 to 1250 GtC for RCP6.0, and 1415 to 1910 GtC for RCP8.5.
- By 2050, annual CO_2 emissions derived from Earth System Models following RCP2.6 are smaller than the 1990 emissions (by 14% to 96%). By the end of the 21st century, about half of the models infer emissions slightly above zero, while the other half infer a net removal of CO_2 from the atmosphere.
- The release of carbon dioxide or methane to the atmosphere from thawing permafrost carbon stocks over the 21st century is assessed to be in the range of 50 to 250 GtC for RCP8.5.

Climate Stabilization, Climate Change Commitment and Irreversibility Cumulative total emissions of carbon dioxide and global mean surface temperature response are approximately linearly related.

- Limiting the warming caused by anthropogenic CO_2 emissions alone to less than 2°C (3.6°F) since the period 1861–1880, with a probability of
  - >33%, will require cumulative CO_2 emissions to stay between 0 and about 1570 GtC (5760 $GtCO_2$);
  - >50%, will require cumulative CO_2 emissions to stay between 0 and about 1210 GtC (4440 $GtCO_2$); and
  - >66%, will require cumulative CO_2 emissions to stay between 0 and about 1000 GtC (3670 $GtCO_2$) since that period.
- An amount of 445 to 585 GtC (1630 to 2150 $GtCO_2$) was already emitted by 2011.
- A lower warming target, or a higher likelihood of remaining below a specific warming target, will require lower cumulative CO_2 emissions. Accounting for warming effects of increases in non-CO_2 greenhouse gases, reductions in aerosols, or the release of greenhouse gases from permafrost will also lower the cumulative CO_2 emissions for a specific warming target.
- A large fraction of anthropogenic climate change resulting from CO_2 emissions is irreversible on a multicentury to millennial timescale, except in the case of

a large net removal of CO_2 from the atmosphere over a sustained period.

- Surface temperatures will remain approximately constant at elevated levels for many centuries after a complete cessation of net anthropogenic CO_2 emissions.
- Due to the long timescales of heat transfer from the ocean surface to the deep ocean, ocean warming will continue for centuries.
- Depending on the scenario, about 15% to 40% of emitted CO_2 will remain in the atmosphere longer than 1,000 years.
- It is virtually certain that global mean sea-level rise will continue beyond 2100, with sea-level rise due to thermal expansion to continue for many centuries.
- The few available model results that go beyond 2100 indicate global mean sea-level rise above the preindustrial level by 2300 to be less than 1 m (3.2 ft) for a radiative forcing that corresponds to CO_2 concentrations that peak and decline and remain below 500 ppm, as in the scenario RCP2.6.
- For a radiative forcing that corresponds to a CO_2 concentration that is above 700 ppm but below 1500 ppm, as in the scenario RCP8.5, the projected rise is 1 m (3.2 ft) to more than 3 m (9.8 ft).
- Sustained mass loss by ice sheets would cause larger sea level rise, and some part of the mass loss might be irreversible.
- There is high confidence that sustained warming greater than some threshold would lead to the near-complete loss of the Greenland ice sheet over a millennium or more, causing a global mean sea-level rise of up to 7 m (23 ft). Current estimates indicate that this threshold is greater than about 1.0°C (1.8°F) but less than about 4.0°C (7.2°F) global mean warming with respect to preindustrial temperature.
- Methods that aim to deliberately alter the climate system to counter climate change, termed geoengineering, have been proposed. Limited evidence precludes a comprehensive quantitative assessment of both Solar Radiation Management (SRM) and Carbon Dioxide Removal (CDR) and their impact on the climate system.
- CDR methods have biogeochemical and technological limitations to their potential on a global scale. There is insufficient knowledge to quantify how much CO_2 emissions could be partially offset by CDR on a century timescale.
- Modeling indicates that SRM methods, if realizable, have the potential to substantially offset a global temperature rise, but they would also modify the global water cycle and would not reduce ocean acidification.
- If SRM were terminated for any reason, there is high confidence that global surface temperatures would rise very rapidly to values consistent with the greenhouse gas forcing.
- CDR and SRM methods carry side effects and long-term consequences on a global scale.

6.4 Conclusion

This rather exhaustive summary of AR5 has been provided as an update on the status of climate knowledge when the IPCC Working Group I cut off the consideration of new research on March 15 of 2013. The WGI report on the physical science basis for climate change was finalized in 2014. However, since these dates, new understanding of the climate system has been produced by the research community in several areas: melting in the cryosphere, the extent that anthropogenic climate change is influencing extreme weather, climate variability, the rate of sea-level rise, impacts related to heat waves, and many others. Thus, it is appropriate to recognize that newer information may conflict with some of the findings in AR5.

Box 6.2 | Spotlight on Climate Change

Reasons for Concern Regarding Climate Change[54]

In AR5, the IPCC authors identified five overarching categories where there is reason for us to be concerned about the potential impacts of anthropogenic climate change. These provide a framework for summarizing key risks, they illustrate the implications of warming, and they point out potential limits of adaptation for people, economies, and ecosystems across sectors and regions.

All warming levels in the text of this box are relative to the 1986–2005 period. Adding ~0.6°C to these warming levels roughly gives warming relative to the 1850–1900 period, used here as a proxy for preindustrial times.

1. **Unique and threatened systems:** Some ecosystems and cultures are already at risk from climate change. With additional warming of around 1.0°C (1.8°F), the number of unique and threatened

---

[54] Box 2.4, IPCC (2014) Climate Change 2014: Synthesis Report.

systems at risk of severe consequences increases. Many systems with limited adaptive capacity, particularly those associated with Arctic sea ice and coral reefs, are subject to very high risks with additional warming of 2.0°C (3.6°F). In addition to risks resulting from the *magnitude* of warming, terrestrial species are also sensitive to the *rate* of warming, and marine species are sensitive to the rate and degree of ocean acidification and coastal systems to sea level rise (**Figure 6.12**).

2. **Extreme weather events:** Climate-change-related risks from extreme events, such as heat waves, heavy precipitation, and coastal flooding, are already moderate. With 1.0°C (1.8°F) additional warming, risks are high. Risks associated with some types of extreme events (e.g., extreme heat) increase progressively with further warming.

3. **Distribution of impacts:** Risks are unevenly distributed between groups of people and between regions; risks are generally greater for disadvantaged people and communities everywhere. Risks are already moderate because of regional differences in observed climate change impacts, particularly for crop production. Based on projected decreases in regional crop yields and water availability, risks of unevenly distributed impacts are high under additional warming of above 2.0°C (3.6°F).

4. **Global aggregate impacts:** Risks of global aggregate impacts are moderate under additional warming of between 1 and 2°C (1.8 and 3.6°F), reflecting impacts on both Earth's biodiversity and overall global economy. Extensive biodiversity loss, with associated loss of ecosystem goods and services, leads to high risks at around 3.0°C (5.4°F) additional warming. Aggregate economic damages accelerate with increasing temperature, but few quantitative estimates are available for additional warming of above 3.0°C (5.4°F).

5. **Large-scale singular events:** With increasing warming, some physical and ecological systems are at risk of abrupt and/or irreversible changes. Risks associated with such tipping points are moderate between 0 and 1°C (1.8°F) additional warming, since there are signs that both warm-water coral reefs and Arctic ecosystems are already experiencing irreversible regime shifts. Risks increase at a steepening rate under an additional warming of 1 and 2°C (1.8 to 3.6°F) and become high above 3°C (5.4°F), due to the potential for large and irreversible sea-level rise from ice sheet loss. For sustained warming above some threshold greater than about 0.5°C (0.9°F), additional warming but less than about 3.5°C (6.3°F), near-complete loss of the Greenland ice sheet would occur over a millennium or more, eventually contributing up to 7 m (23 ft) to global mean sea-level rise.

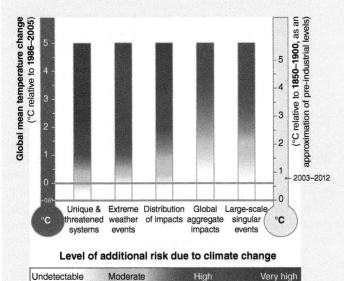

FIGURE 6.12 Reasons for concern regarding climate change. Shading indicates the additional risk due to climate change when a temperature level is reached and then sustained or exceeded. From lightest to darkest - no associated impacts are detectable and attributable to climate change; associated impacts are both detectable and attributable to climate change with at least medium confidence; severe and widespread impacts; very high risk is indicated by all key criteria.

Source: IPCC AR5, Synthesis Report, Topic 2, Box 2.4 Fig 1_Rev1-01. https://www.ipcc.ch/report/graphics/

6.5 Additional Considerations

Models do a good job of simulating air and ocean circulation, solar heating, and the role of greenhouse gases. But there are also factors whose influence on global climate can be modeled, but when they will occur is essentially unpredictable: major volcanic eruptions that throw sulfur compounds into the high atmosphere that absorb and scatter sunlight; ENSO events with magnitudes that are unknowable beforehand; and of course, the political decisions that will determine future fossil fuel consumption and land surface changes. There is also an ongoing effort to understand the role of clouds: Do they block sunlight? Or do they trap heat?

Let's examine some of these and other factors and how scientists are working to understand their effect on climate.[55]

6.5.1 If We Can't Predict Weather, How Can We Predict Climate?

In fact, GCMs do predict climate with accuracy, but at a large scale. Your TV weather forecaster has the harder task of predicting detailed weather, at specific localities, in very short time periods. Predicting climate and predicting weather are very different from each other.

Weather is the short-term (up to a week) state of the atmosphere at a given location. It affects the well-being of humans, plants, and animals and the quality of our food and water supply. Weather is somewhat predictable because of our understanding of Earth's global climate patterns. Climate is the long-term (about 30 years) average weather pattern and is the result of interactions among land, ocean, atmosphere, ice, and the biosphere. Climate is described by many weather elements, such as temperature, precipitation, humidity, sunshine, and wind. Both climate and weather result from processes that accumulate and move heat within and between the atmosphere and the ocean.

As a general rule, global warming will produce more hot days and fewer cool days in most places.[56] Warming will be greatest over land, and longer, more-intense heat waves will become more common (**Figure 6.13**). We will see an increase in the severity of storms, floods, and droughts as rain and snowfall patterns change. It has been difficult for the meteorology community to reach agreement on how hurricanes will change with global warming. But there is general agreement that because of warmer ocean surface temperatures, hurricanes could decrease in frequency yet increase in intensity[57] as a global average.

It is impossible to pin any single unusual weather event on global warming, but emerging evidence suggests that global warming is already influencing the weather.[58] [59] Heat waves, droughts, and intense rain events have

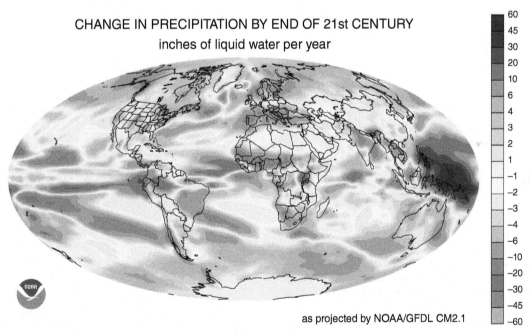

FIGURE 6.13 The change in annual average precipitation projected by the Geophysical Fluid Dynamics Laboratory (NOAA) CM2.1 model for the 21st century. These results are from a model simulation in which atmospheric carbon dioxide levels increase from 370 to 717 ppm (a pathway we are currently traveling). The plotted precipitation differences were computed as the difference between the 2081 to 2100 20-year average and the 1951 to 2000 50-year average.

Source: NOAA GFDL Climate Research Highlights Image Gallery: Will the Wet Get Wetter and the Dry Drier? http://www.gfdl.noaa.gov/will-the-wet-get-wetter-and-the-dry-drier

[55] VIDEO: Yale Climate Connections. https://www.yaleclimateconnections.org/2017/10/humans-experimenting-with-climates-playing-nice/
[56] Meehl, G.A., et al. (2009) The Relative Increase of Record High Maximum Temperatures Compared to Record Low Minimum Temperatures in the U.S., *Geophysical Research Letters* 36: L23701, doi:10.1029/2009GL040736

[57] Knutson, T.R., et al. (2010) Tropical cyclones and climate change, *Nature Geoscience* 3: 157–163
[58] Pall, P., et al. (2011) Anthropogenic Greenhouse Gas Contribution to Flood Risk in England and Wales in Autumn 2000, *Nature* 470: 382–385, doi:10.1038/nature09762
[59] VIDEO: How climate change makes hurricanes worse. https://www.youtube.com/watch?v=_0TCrGtTEQM

increased in frequency during the last 50 years, and human-induced global warming more likely than not contributed to the trend.[60] Climate change is neither proved nor disproved by individual warming or cooling spells. It's the longer-term trends, of a decade or more, that place less emphasis on single-year variability, that count. Nonetheless, unusual bouts of weather, and other weather changes that are a result of naturally occurring patterns, are still consistent with a globally warming world.

6.5.2 Rapid Warming by Midcentury

How high will global temperature rise by the middle of the century? One study[61] ran almost 10,000 climate simulations on volunteers' home computers to increase the horsepower needed to calculate global climate change using a GCM. The study was the first to run so many simulations using a coupled ocean–atmosphere climate model. Using so many simulations improves definition of some of the uncertainties of previous forecasts that used simpler models or only a few dozen simulations. The modeling experiment found that a global warming of 3 degrees Celsius (5.4 degrees Fahrenheit) by 2050 is as equally plausible as a rise of 1.4 degrees Celsius (2.5 degrees Fahrenheit). The results suggest that the world is very likely to cross the "2°C barrier" at some point in this century if emissions continue unabated. Thus, those planning for the impacts of climate change need to consider the possibility of warming of up to 3 degrees Celsius (above the 1961–1990 average) by 2050 even on a mid-range emissions scenario. This is a faster rate of warming than most other models predict.

6.5.3 Variability in the Pacific

The El Niño event of 2015–2016[62] broke many records for flooding, excessive heat, and other examples of punishing weather. Much of South America, southern United States, and East Africa experienced intense rainfall and flooding. Australia and southeast Asia endured severe drought. Studies indicate that similar "extreme" El Niño events could become more frequent as global temperatures rise.[63] If global warming reaches 1.5 degrees Celsius (2.7 degrees Fahrenheit), the limit of warming aspired to by signatories of the 2015 Paris Accord[64]—extreme El Niño events could

happen twice as often. That means seeing an extreme El Niño on average every 10 years, rather every 20 years,[65] an increase of 130 percent over natural frequency.

ENSO is the El Niño–Southern Oscillation, which we studied in **Chapter 5**. It is a large-scale, quasiperiodic meteorological pattern historically characterized by two conditions: La Niña and El Niño. Recently, however, a new third pattern has emerged known as a central-Pacific El Niño. These conditions govern sea-surface and air temperature trends as well as rainfall patterns throughout the tropical Pacific Ocean. ENSO also exerts a global influence on weather and temperature patterns. In fact, the year-to-year variation in global average temperature shown in **Figure 6.14** is largely a reflection of whether the year was dominated by La Niña (years tend to be cool) or El Niño (years tend to be warm).

ENSO[66] is related to the atmospheric pressure difference between a body of dry air (a high-pressure system, the descending limb of the Southern Hemisphere Hadley Cell) located in the southeast Pacific over Easter Island and a body of wet air (a low-pressure system) located over Indonesia in the southwest Pacific. Under normal conditions in the southern hemisphere, air flows from the high pressure to the low pressure and creates the trade winds. These winds blow east to west across the surface of the Pacific and drive a warm surface current of water into the western Pacific. The resulting accumulation of warm tropical water in the western Pacific is known as the Western Pacific warm pool; it extends well below the surface, and it has the highest sea-surface temperatures on the planet.

Seawater in the warm pool evaporates readily and produces lush rainfall throughout Southeast Asia, India, Africa, and other areas associated with the monsoon, the rainy season storms that nourish food production and ecosystems from Indonesia to Africa.[67] In the eastern Pacific, this displaced seawater is replaced by nutrient-rich cold ocean water that rises from the deep sea, a process called upwelling. The upwelling current is loaded with nutrients fueling an important fishing industry off the coast of South America.

On occasion, the pressure difference between the two centers decreases and the trade winds respond by weakening (**Figure 6.15**). This condition is known as El Niño. As a result, the warm pool in the west Pacific surges to the east; it shallows and spreads across the surface, releasing its heat to the atmosphere and causing a broad area of the Pacific to experience warmer, wetter conditions than normal. The ocean surface in the central and eastern Pacific warms substantially and heats the lower troposphere. This temporarily

[60] Min, S.K., et al. (2011) Human Contribution to More-Intense Precipitation Extremes, *Nature* 470: 378–381, doi:10.1038/nature09763

[61] Rowlands, D., et al. (2012) Broad Range of 2050 Warming from an Observationally Constrained Large Climate Model Ensemble, *Nature Geoscience* 5, no 4: 256 doi: 10.1038/ngeo1430

[62] Australian Bureau of Meteorology http://www.bom.gov.au/climate/updates/articles/a018.shtml

[63] Wang, G. (2017) Continued increase of extreme El Niño frequency long after 1.5C warming stabilization, *Nature Climate Change*, doi:10.1038/nclimate3351

[64] United Nation Framework Convention on Climate Change http://unfccc.int/paris_agreement/items/9485.php

[65] CarbonBrief https://www.carbonbrief.org/extreme-el-ninos-double-frequency-under-one-point-five-celsius-warming-study

[66] VIDEO: El Nino, What is it? https://www.youtube.com/watch?v=WPA-KpldDVc

[67] VIDEO: NASA, Monsoons - Wet, Dry, Repeat. https://svs.gsfc.nasa.gov/12255

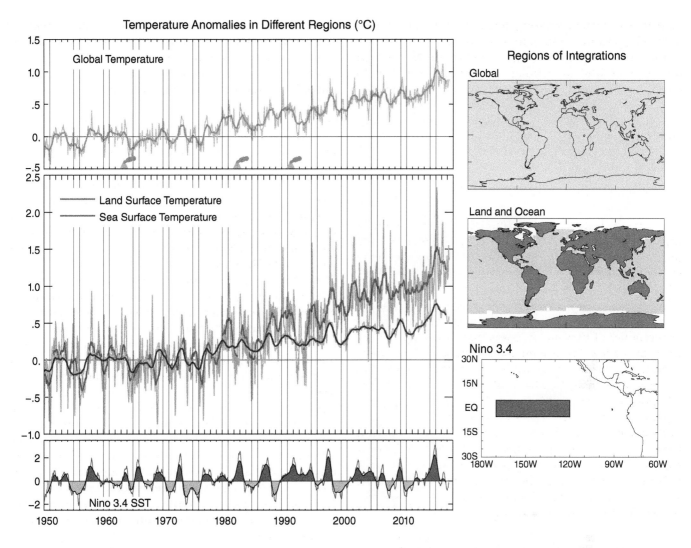

FIGURE 6.14 Monthly (thin lines) and 12-month running mean (thick lines in the case of Niño 3.4[69] index) global land–ocean temperature anomaly, global land and sea surface temperature, and El Niño index. All have a base period 1951–1980. Data are through April 2018. Year-to-year global mean temperature variability largely correlates to the prevalence of El Niño (warm) and La Niña (cool). Some temperature variability is dominated by large-scale volcanic emissions that tend to produce cooling of a year or two. Shown are three volcanic eruptions with global impact; from left to right, Mount Agung (Indonesia, 1963), El Chichon (Mexico, 1982), and Mount Pinatubo (Philippines, 1991).

Source: http://www.columbia.edu/~mhs119/Temperature/T_moreFigs/

raises global mean temperature[68] for any year characterized by El Niño. When a very strong El Niño develops, so much heat is released to the atmosphere that it usually sets a record for the warmest year in the more than century-long history of air temperature observations. This occurred in the El Niños of 1982/83, 1997/98, and 2015/16.[70]

El Niño can have devastating social and economic consequences. The eastern movement of the tropical warm pool takes with it a critical source of rainfall; as a result, seasonal rains in Indonesia collapse, leading to drought, famine, and forest fires in Southeast Asia. The monsoon, the life-sustaining, crop-nourishing rains that overcome the summer drought in India, is known to fail in the onset

[68] Thompson, D., et al. (2009) Identifying Signatures of Natural Climate Variability in Time Series of Global-Mean Surface Temperature: Methodology and Insights, *Journal of Climate* 22: 6120–6141, doi:10.1175/2009JCLI3089.1.

[69] How do researchers follow the ENSO system? Climatologists have learned that the pattern of sea surface temperature (SST) in the tropical Pacific, where the ENSO system operates, provides a tell-tale to the development of El Niño and La Niña conditions. One pattern that is closely followed is the Niño 3.4 index. The Niño 3.4 index tracks changes in sea surface temperature in a box centered on the equator between 5°N and 5°S, and reaching from 170°W to 120°W. This region has large variability on El Niño time scales. An El Niño or La Niña event is identified

if the 5-month running-average of the NINO3.4 index exceeds +0.4°C for El Niño or -0.4°C for La Niña for at least 6 consecutive months. For more information, see the discussion (and the links) at the National Center for Atmospheric Research https://climatedataguide.ucar.edu/climate-data/nino-sst-indices-nino-12-3-34-4-oni-and-tni

[70] El Niño information at http://ggweather.com/enso/oni.htm

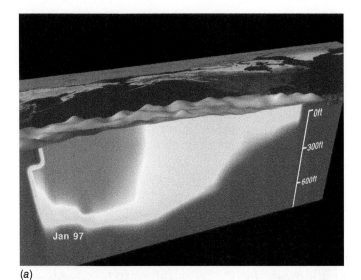

(a)

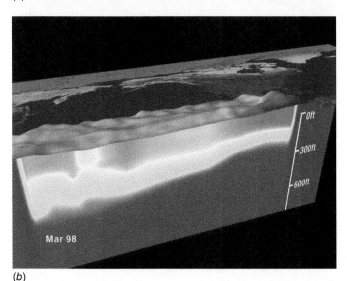

(b)

FIGURE 6.15 El Niño is a large-scale phenomenon in the Pacific Ocean that influences global temperature and precipitation patterns. In normal conditions (a), trade winds blow from east to west across the tropical Pacific and create the Western Pacific Warm Pool. The warm pool has deep roots and stores heat that would otherwise be in the atmosphere. On occasion, the winds break down and are replaced by westerly wind gusts, heralding the development of an El Niño. The warm pool surges to the east (b), shallows, releases its heat to the atmosphere, and increases the sea surface temperature in the central and eastern Pacific Ocean. Heat released to the air by a very strong El Niño may cause a record-setting warm year. Shown here are sea surface temperatures and sea level heights observed during the record-setting El Niño of 1997–1998.

Source: https://svs.gsfc.nasa.gov/cgi-bin/details.cgi?aid=280

year of an El Niño event, thus leading to famine and water shortages.

Precipitation in the east Pacific increases with the arrival of the warm seawater, causing higher (often catastrophic) rainfall on the Pacific coasts of both North and South America. In the northern United States, winters are warmer and drier than average and summers are wetter than average. Torrential rains and damaging floods can occur across

the southern United States. There is also a clear statistical relationship between El Niño and hurricanes. Whereas the number of hurricanes in the Atlantic basin tends to drop by about 50 percent during El Niño years, the number of hurricanes tends to increase in the Pacific.

6.5.4 How Will ENSO Change as the World Warms?

A new type of El Niño has emerged in the past two decades, one that has its warmest waters in the tropical central Pacific Ocean rather than in the eastern Pacific. This new type of El Niño is known by several names, including *central-Pacific El Niño, warm-pool El Niño,*[71] *dateline El Niño,* and *El Niño Modoki* (Japanese for "similar but different"). The intensity of these central-Pacific El Niño events has nearly doubled in 20 years.[72] They have been observed in 1991–1992, 1994–1995, 2002–2003, and 2004–2005, and the most intense occurred in 2009–2010. How the central Pacific El Niño will evolve as warming continues is not well understood. However, modeling by CMIP3 indicates that central-Pacific El Niños will increase as much as five times under global warming.[73]

CMIP5 models predict that very large El Niño events will become twice as frequent as today as global warming continues.[74] On average, very large El Niños occur today every 20 years, this is expected to increase in frequency, occurring every 10 years in the future. A number of researchers have concluded that ENSO in general will become stronger as a result of global warming,[75] that strong El Niño events will become more frequent,[76] and that strong La Niña[77] events will as well.

6.5.5 How Will Hadley Circulation Change as the World Becomes Warmer?

Trade winds breakdown in the tropical Pacific when an El Niño is underway, and they accelerate during La Niña. The atmospheric Hadley circulation we discussed in **Chapter 3** is

[71] NOAA discussion of "warm-pool" and "cold tongue" El Niño types https://www.climate.gov/news-features/blogs/enso/enso-flavor-month
[72] Lee, T. and McPhaden, M. (2010) Increasing Intensity of El Niño in the Central-Equatorial Pacific, *Geophysical Research Letter* 37, no. 14: L14603, doi: 10.1029/2010GL044007
[73] Yeh, S.-W., et al. (2010) El Niño in a changing climate. *Nature* 461, 511–514, DOI: 10.1038/nature08316
[74] Cai, W., et al. (2014) Increasing frequency of extreme El Niño events due to greenhouse warming. *Nature Climate Change*, DOI: 10.1038/nclimate2100
[75] Cai, W., et al. (2015) ENSO and Greenhouse Warming, *Nature Climate Change*, 5, p. 849–859, doi:10.1038/nclimate2743
[76] Wang, G. (2017) Continued increase of extreme El Niño frequency long after 1.5C warming stabilization, *Nature Climate Change*, doi:10.1038/nclimate3351
[77] Cai, W., et al. (2015) Increased frequency of extreme La Niña events under greenhouse warming. *Nature Climate Change*, DOI: 10.1038/nclimate2492

the source of these winds. How this circulation will change in a warmer world is tied closely to the ENSO process. Although several reports indicate that both El Niño and La Niña conditions will strengthen and become more frequent in the future, the situation with Hadley circulation is less clear.

Hadley circulation influences the north–south distribution of rainfall, clouds, and relative humidity over half of Earth's surface. Consequently, it controls the geographic distribution of the world's dry and wet regions and the prosperity of associated economies. Hadley circulation can expand or contract in a warmer or colder global climate, leading to major floods and droughts that might have triggered the collapse of ancient civilizations in the past.[78] Hadley circulation correlates strongly with global mean surface temperature and at least one modeling experiment[79] indicates that Hadley circulation widens and weakens as the climate warms, but the results are only statistically valid in the Northern Hemisphere.

Over the past decade or two, the Northern Hemisphere Hadley circulation has indeed expanded, but at a rate faster than what is predicted by global climate models.[80] Also unexpected is that the circulation has intensified and contributed to increased droughts over many subtropical regions.[81] Intensification of the Hadley circulation is expressed as stronger trade winds, which have led to accelerating sea-level rise in the western tropical Pacific as well as increases in surface currents.[82] Understanding the mechanisms that control the variability of the Hadley circulation is central to determining changes in rainfall, tropical cyclones, and other key climate elements where much of the world lives.[83] However, a number of studies, including results from RCP8.5 model experiments, suggest that Hadley circulation should weaken as atmospheric CO_2 content increases throughout this century.[84] The Hadley cell appears to be evolving in an opposite direction to what the models predict (with the exception of at least one result[85]). Accurately projecting how Hadley circulation will change as the world gets warmer is fundamental to guiding human communities striving to adapt to climate change.

Box 6.3 | Spotlight on Climate Change

Record carbon dioxide levels in 2015

According to measurements taken at the Mauna Loa Observatory in Hawaii,[86] global carbon dioxide in 2015 grew by 3.03 ppm (or about 6.3 GtC), the highest single year accumulation in over 50 years of record-keeping. In recent years, the annual increase has been about 2 ppm or 4 GtC. The spike in 2015 was unexpected, given the strong growth in renewable energy use in most of the world's major economies.

To sleuth out the source of this unforeseen CO_2, NASA's Orbiting Carbon Observatory-2,[87] or OCO-2, mapped various locations where carbon dioxide was released that year and compared them to earlier years. They found that the total amount of carbon released to the atmosphere from all land areas increased by 3 GtC. About 80% of that, or 2.5 GtC, came from three tropical

forests locations—South America, Africa, and Southeast Asia[88] (**Figure 6.16**). For comparison, the United States released 6.59 GtC into the atmosphere in 2015.

Researchers have long known these forests act like sponges, soaking up carbon dioxide through photosynthesis and thus providing a buffer against its buildup in the atmosphere. But 2015 revealed that in some years, we cannot count on this process as fully as we had assumed. Why was 2015 so different? You will recall from our earlier discussions that 2015 was a very strong El Niño year and in all three of these forested regions, the weather changes dramatically, and in each case differently, during an El Niño.

In South America, the Amazon rainforest and other tropical regions experienced severe drought, the worst in 30 years. Combined with higher than normal

[78] Hodell D.A., et al. (1995) Possible role of climate in the collapse of classic Maya Civilization, *Nature*, 375 (6530). 391–394

[79] D'Agostino, R., et al. (2017) Factors controlling Hadley circulation changes from the Last Glacial Maximum to the end of the 21st century. *Geophysical Research Letters*, 44, 8585–8591. doi: 10.1002/2017GL074533.

[80] Nguyen H., et al. (2013) The Hadley circulation in reanalysis: Climatology, variability, and change. *Journal of Climate*, 26(10):3357–3376

[81] Fu, R. (2015) Global warming-accelerated drying in the tropics, *PNAS USA*.

[82] Merrifield, M. A., and Maltrud, M.E. (2011), Regional sea level trends due to a Pacific trade wind intensification, *Geophys. Res. Lett.*, 38, L21605, doi:10.1029/2011GL049576

[83] Fu (2015)

[84] Seo, K.H., et al. (2014) A mechanism for future changes in Hadley circulation strength in CMIP5 climate change simulations. *Geophysical Res Lett*, 41(14): 5251–5258

[85] Lau, W.K.M., and Kim, K.-M. (2015) Robust Hadley Circulation changes and increasing global dryness due to CO_2 warming from CMIP5 model projections. *PNAS USA* 112:3630-3635

[86] Mauna Loa Observatory homepage, part of the NOAA global monitoring network https://www.esrl.noaa.gov/gmd/obop/mlo/

[87] Orbiting Carbon Observatory-2 home page https://oco.jpl.nasa.gov

[88] NASA Pinpoints Cause of Earth's Recent Record Carbon Dioxide Spike https://www.jpl.nasa.gov/news/news.php?feature=6973

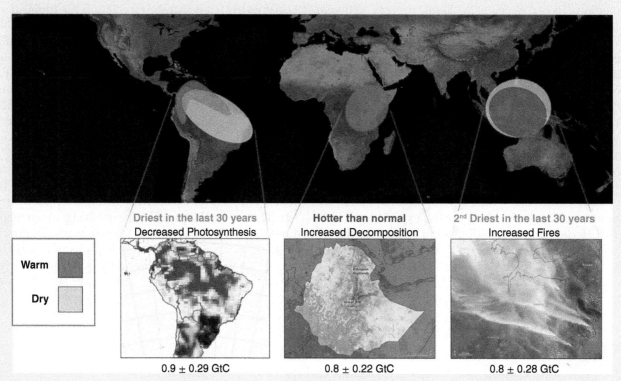

Warm

Dry

Driest in the last 30 years
Decreased Photosynthesis

Hotter than normal
Increased Decomposition

2ⁿᵈ Driest in the last 30 years
Increased Fires

0.9 ± 0.29 GtC

0.8 ± 0.22 GtC

0.8 ± 0.28 GtC

FIGURE 6.16 During the El Niño of 2015/16, tropical forests released about 2.5 Gt more carbon to the atmosphere than they did in 2011, a cooler and wetter La Niña year. This contributed to a record spike of CO_2 in 2015.

Source: https://www.jpl.nasa.gov/news/news.php?feature=6973

temperatures, this stressed vegetation and reduced photosynthesis, leaving more carbon in the air than a normal year.

In Africa, the rainfall remained at typical levels, but the temperature grew hotter. This meant that dead trees and plants decomposed rapidly, releasing carbon to the atmosphere in unusually large quantities.

Southeast Asia, meanwhile, had the second driest year of the past 3 decades. This fueled widespread fires, largely set by farmers who wanted to clear land for agriculture, and releasing vast quantities of carbon.[89]

Prior to the detailed picture of carbon exchange with the atmosphere afforded by the OCO-2, scientists had not understood what the most important processes were in governing the carbon cycle during

El Niño events. Researchers have known for decades that El Niño influences the productivity of tropical forests' and their contributions to atmospheric carbon dioxide, but they had very few direct observations of the process in action. Now, OSO-2 data can be used to test understanding of whether the response of tropical forests is likely to make climate change worse or not.

These findings imply that if future climate brings more frequent or more intense El Niño events, more carbon dioxide may remain in the atmosphere leading to additional warming. This is an unexpected positive feedback to global warming that is likely to grow worse because climate models indicate that strong El Nino events are going to grow more common as the world warms.

6.5.6 Aerosols

Aerosols[90] are tiny particles suspended in the atmosphere (**Figure 6.17**). If they are large enough, they will scatter and absorb sunlight and in the lower atmosphere can modify the size of cloud particles, changing how the clouds reflect

and absorb sunlight. Their impact on sunlight and cloud formation affects Earth's energy budget.

Types of Aerosols Three types of aerosols significantly affect Earth's climate. The first is aerosols that collect in the stratosphere after major volcanic eruptions such as

---

[89] ScienceNews, During El Niño, the tropics emit more carbon dioxide https://www.sciencenews.org/article/during-el-nino-tropics-emit-more-carbon-dioxide

[90] Aerosols at Union of Concerned Scientists http://www.ucsusa.org/global_warming/science_and_impacts/science/aerosols-and-global-warming-faq.html#.WfZd7EyZOV5

William Putman, NASA/Goddard

FIGURE 6.17 Global aerosols simulated by NASA. Dust is lifted from the surface, sea salt swirls inside cyclones, smoke rises from fires, and sulfate particles stream from volcanoes and fossil fuel emissions.

Source: https://www.nasa.gov/multimedia/imagegallery/image_feature_2393.html

Mt. Pinatubo in 1992. **Volcanic aerosols** are formed by sulfur dioxide gas from the volcano, which is converted to droplets of sulfuric acid (H_2SO_4) and solid particles of sulfate (SO_4). These sulfur aerosols are blown by winds in the stratosphere until they mostly cover the globe and may stay in the stratosphere as long as two years. They reflect sunlight, reducing the amount of energy reaching the lower atmosphere and Earth's surface, cooling them.

Volcanic eruptions come in all sizes and shapes. Eruptions that are relatively passive, in which the lava quietly effuses from a volcanic vent and flows across the ground, are not known to influence the climate. But some eruptions are explosive. These can send thick columns of sulfuric aerosols and ash high into the atmosphere. If the aerosols reach the stratosphere, they can scatter and absorb incoming sunlight and change Earth's radiation balance, temporarily offsetting warming with global cooling.

It is the major sulfur-rich eruptions that cause the greatest global effects. For example, Mount Pinatubo (Philippines) erupted in 1991[91] and ejected almost 15 million tons of sulfur dioxide into the stratosphere. For many months, a satellite tracked the sulfur cloud produced by the eruption as it lowered average global temperature by about 0.6 degrees Celsius (1 degree Fahrenheit).

Research[92] has revealed that the Little Ice Age, a period of regional cooling in the North Atlantic, North America, and Europe (and perhaps beyond) that lasted for hundreds of years until the late 19th century may have been triggered by an unusual 50-year-long episode of four massive tropical volcanic eruptions between 1275 and 1300 CE. The stratospheric aerosols associated with this sequence of eruptions apparently produced persistent cold summers that generated a positive feedback in the form of expanding sea ice and weakened Atlantic Ocean currents that transport heat.

Computer simulations paint a picture of Earth's climate system being hit time and time again by cold conditions over a short period, all leading to a cumulative cooling effect that culminated in the start of the Little Ice Age. The study relied on a convergence of data from ice and sediment cores, patterns of dead vegetation (from the start of the Little Ice Age) that were recently revealed by receding glacier ice in the Arctic, and computer simulations of climate feedbacks. The research indicates that the start of the cold period was prevalent throughout the North Atlantic region and involved major components of the climate system through a series of feedbacks that amplified the original impacts of explosive volcanic aerosols.

---

[91] NASA page on Mt. Pinatubo https://earthobservatory.nasa.gov/IOTD/view.php?id=1510

[92] Miller, G., et al. (2012) Abrupt Onset of the Little Ice Age Triggered by volcanism and Sustained by Sea-Ice-Ocean feedbacks. *Geophysical Research Letters* 39: L02708, doi: 10.1029/2011GL050168

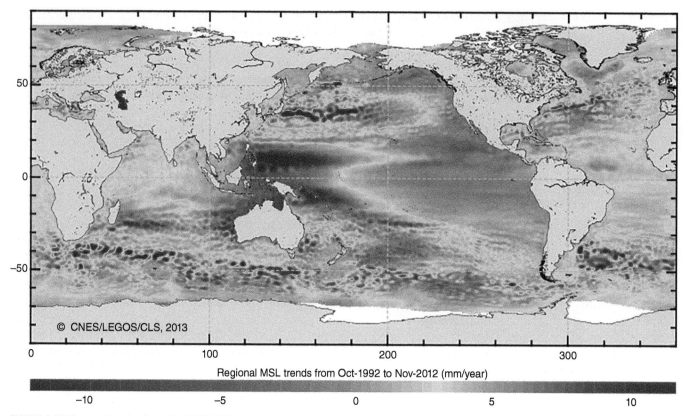

FIGURE 6.18 Regional sea-level trends, 1992–2012. For over two decades, a region of rapidly rising sea level persisted in the western tropical Pacific. This pattern reflected a sustained La Niña–like condition of pronounced trade winds.

Source: http://www.soest.hawaii.edu/coasts/sealevel/MSL_Map_MERGED_Global_IB_RWT_NoGIA_Adjust.png

The second type of aerosol that may have a significant effect on climate is desert dust. Wind blows large amounts of dust from the deserts of North Africa and Asia high into the atmosphere. The dust is composed of minerals, which absorb sunlight as well as scatter it. By absorbing sunlight, the dust warms the layer of the atmosphere where they reside. Researchers now think that this warming inhibits the formation of storm clouds and suppresses rainfall, thus causing desert expansion.

The third type is anthropogenic aerosols consisting of solid tiny sulfate particles or liquid droplets of sulfur dioxide, similar to the sulfur aerosols produced by volcanic eruptions. The largest amount of anthropogenic aerosols comes from burning coal and oil, mostly in the northern hemisphere where industrial activity is centered. Sulfur aerosols, whether produced by volcanic eruption, anthropogenic emission, or some other process, will reflect sunlight and also make clouds brighter and cause them to reflect more sunlight. The impact of sulfur aerosols is to reduce the amount of sunlight reaching Earth's surface, causing cooling.

Impact of Aerosols on Hadley Circulation A common trait among CMIP3 and CMIP5 model projections is that global warming will result in a more general average state

of the climate that is El Niño-like,[93] stimulating a simultaneous increase in tropical cyclone activity in the Pacific.[94] However, against the projections of these climate models, a persistent La Niña-like state emerged over the past two decades in the Pacific. These windy conditions pushed so much water into the western Pacific warm pool that it rapidly raised sea level, producing some of the fastest rates of sea level rise on Earth[95] (**Figure 6.18**). In fact, this is consistent with studies of ENSO variability over the past 1000 years.[96] Researchers found that during the medieval climate anomaly, megadroughts that occurred in western North America might have been the result of stronger or more frequent La Niña than El Niño events. In **Chapter 7**, we will see that drought is once again visiting the western United States.

[93] Cai et al. (2015)

[94] Chand, S.S., et al. (2017) Projected increase in El Niño-driven tropical cyclone frequency in the Pacific, *Nature Climate Change* 7, 123-127, DOI: 10.1038/nclimate3181

[95] Merrifield, M. A. (2011) A shift in western tropical Pacific sea level trends during the 1990s. *Journal of Climate*, 24, 4126-4138, doi:10.1175/2011JCLI3932.1.

[96] Khider, D., et al. (2011) Assessing El Niño Southern Oscillation Variability during the Past Millennium, *Paleoceanography* 26, PA3222, doi:10.1029/2011PA002139.

However, the perseverance of strong trade winds, and intensification of the Hadley circulation, has been perplexing as it is not consistent with the implications of model findings which suggest that these winds should be weakening as global surface temperature increases.[97] One study found that changes in Hadley circulation were related to the north–south temperature gradient[98] (known as the "meridional temperature gradient") from the equator toward the midlatitudes. If the meridional temperature gradient was somehow strengthening, the reason why was unclear.

Researchers attributed the sea-level pattern to accelerated and sustained trade winds associated with a decadal-scale La Niña–like climate and assigned it to long-term climate processes such as the Interdecadal Pacific Oscillation[99] (IPO) or the Pacific Decadal Oscillation[100] (PDO).[101] It was cited as a cause of the "global warming hiatus," where the rate of global mean temperature rise slowed considerably in the decade and a half after the 1998 very strong El Niño, despite continued greenhouse gas emissions.[102] The correspondence of the global surface temperature record and the invigorated state of tropical Pacific winds, suggested that the Pacific was "back in the driver's seat"[103] in terms of controlling global temperature change and was

burying heat in the enhanced western Pacific warm pool; heat that would otherwise be in the air and driving continued atmospheric warming.

A new analysis of the situation emerged in 2016 based on CMIP5 modeling.[104] Looking at the cooling effect of anthropogenic aerosols originating in China and southeast Asia, researchers discovered that rather than being caused by natural Pacific climate variability such as the PDO or IPO, intensification of the Pacific Hadley circulation could be accounted for by industrialization in China. The study suggested that variations in the Pacific Ocean trade wind system were actually triggered by changing aerosol emissions from human activity—particularly by increases in China's burning of fossil fuels. Further, the modeling results indicated that the slowdown in surface warming could have been predicted and that future cuts to aerosol emissions could prompt rapid warming in coming years.

Although the PDO can vary naturally, the modeling showed that increased anthropogenic aerosol emissions could be responsible for the extended negative phase, which has persisted since the 1998 El Niño. As we discussed earlier, aerosols are tiny particles and droplets in the atmosphere that tend to have a cooling effect on Earth's climate. They scatter incoming sunlight and stimulate clouds to form, preventing the Sun's energy from reaching Earth's surface. There are natural sources of aerosols such as volcanic eruptions, and dimethyl sulfide, a sulfur compound released by phytoplankton in the ocean. When released into the atmosphere, dimethyl sulfide encourages clouds to form, helping to keep Earth cool by reflecting incoming sunlight back out to space. It also directly scatters incoming sunlight. Hence, there is a link between tiny algae in the ocean and cloud formation.[105] However, humans also churn out aerosols via vehicle exhausts and burning fossil fuels and wood (**Figure 6.19**).

Strong industrialization in China over the period 1998 to 2012 led to pronounced aerosol emissions from coal burning and automobile exhaust. So strong, in fact, that air quality in urban areas such as Beijing became a serious health concern for Chinese authorities. The aerosols scattered sunlight and caused local cooling over China and across the north Pacific. Aerosol reductions elsewhere caused warming over North America and Europe.

Cooling in the Pacific strengthened the temperature difference between the tropics and midlatitudes. This intensified the Hadley circulation and increased the trade winds, the tell-tale sign of La Niña, and the La Niña–like character of a negative phase of the PDO. Using CMIP5 models,

[97] D'Agostino et al. (2017)

[98] Seo et al. (2014)

[99] Meehl, G.A., et al. (2016) Contribution of the interdecadal Pacific oscillation to twentieth-century global surface temperature trends. *Nature Climate Change* 6, 1005–1008, doi:10.1038/nclimate3107

[100] Trenberth, K.E. (2015) Has there been a hiatus? *Science*, 14 Aug, Vol. 349, Issue 6249, pp. 691-692 DOI:10.1126/science.aac9225

[101] We studied the Pacific Decadal Oscillation in Chapter 5. The PDO is a dominant year-round pattern of sea surface temperature variability that oscillates between two typical states: positive (also known as the "warm" phase) and negative ("cool"), which each affect global weather in different ways. During a negative PDO phase, the trade winds strengthen, driving heat into the western Pacific warm pool which brings cooler water to the surface in the eastern Pacific. This has a cooling effect on global surface temperatures. During a positive PDO, weak trade winds have the opposite effect, transferring heat from the oceans to the atmosphere and causing surface temperatures to rise more quickly. The negative PDO has been described as having an effect on climate that is "La Niña-like" and the positive PDO as "El Niño-like". The Interdecadal Pacific Oscillation (IPO) is a long-term oscillation of the Pacific Ocean. It lasts much longer than the El Niño Southern Oscillation, nominally in the range of decades. The IPO can affect the strength and frequency of El Niño and La Niña, and is similar, and nearly equivalent, to the Pacific Decadal Oscillation (PDO). The PDO, however is a predictor of the impact of climate oscillation in the northern Pacific, whereas the IPO includes the Southern Hemisphere as well. See discussion at http://climatology.co.uk/interdecadal-pacific-oscillation#.WdzGNkyZOV4

[102] Kosaka, Y., and Xie, S.-P. (2013) Recent global-warming hiatus tied to equatorial Pacific surface cooling, *Nature* 501, 403-407, 19 September, doi:10.1038/nature12534 see also Meehl, G.A., et al. (2013) Externally Forced and Internally Generated Decadal Climate Variability Associated with the Interdecadal Pacific Oscillation, *Journal of Climate*, American Meteorological Society Journals Online, https://doi.org/10.1175/JCLI-D-12-00548.1

[103] Clement, A., and DiNezio, P. (2014) The tropical Pacific Ocean – Back in the driver's seat? *Science*, 28 Feb, Vol. 343, Issue 6174, pp. 976-978, DOI: 10.1126/science.1248115

[104] Smith, D.M., et al. (2016) Role of volcanic and anthropogenic aerosols in the recent global surface warming slowdown, *Nature Climate Change* 6, 936-940, 03 March, doi:10.1038/nclimate3058

[105] Researchers are concerned that growing ocean acidification will make the ocean less hospitable for phytoplankton, leading to declining emissions of dimethyl sulfide. If lower sulfur emissions lead to less reflective clouds, that would mean more sunlight reaching Earth's surface, and more warming.

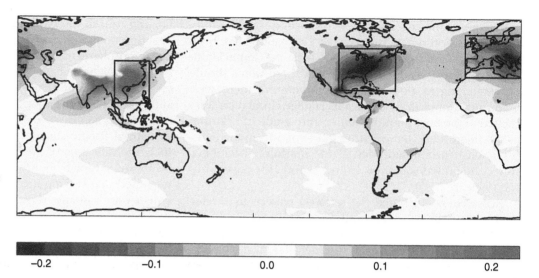

FIGURE 6.19 Trend in the concentration of aerosols between 1998 and 2012. This is measured in "optical depth," which doesn't have units. Aerosols decreased across North America, the Atlantic Ocean, and Europe, causing warming, but increased in Asia and the Pacific, causing cooling.

Source: https://www.nature.com/nclimate/journal/v6/n10/full/nclimate3058.html

| –0.2 | –0.1 | 0.0 | 0.1 | 0.2 |

researchers found that they could recreate the negative PDO—and the global surface warming slowdown—by simulating these changes in human aerosol emissions.

One particularly interesting element of this research is the implication that future reductions in aerosol emissions from China to improve air quality could promote a positive phase of the PDO and a period of increased trends in global surface temperatures.[106] Sure enough, China has embarked on a **coal abatement** program as part of a national pivot toward clean energy.[107] They have cancelled dozens of planned coal-fired power plants and coal use dropped by 1.28 percent in 2014, the first time that coal use has declined in China this century. It fell further in 2016, down by 4.7 percent over 2015 levels, while overall energy consumption declined as well.

Remarkably, at the same time, that China was reducing its aerosol production by moving away from coal and other fossil fuels, Pacific sea level changed from its previous, decades-long mode of La Nina–like intensification, to a more El Niño–like pattern of falling sea level in the western tropical Pacific and rising sea level in the east (**Figure 6.20**; this figure is also discussed in **Chapter 5**). Were these events related, or just coincidental? According to the CMIP5 modeling by Smith et al.,[108] this is exactly the scenario their simulations of aerosol impacts would predict.

This research proposes that aerosol emissions from burning coal and oil in Asia are responsible for a chain of linked phenomena that have major global implications. This linkage involves the following sequence:

1. Regional cooling over SE Asia caused by anthropogenic aerosols from an expanding Chinese economy over recent decades;

2. Spread of emissions and associated cooling across the Pacific by the midlatitude westerlies;

3. Accelerated Hadley circulation and associated trade winds due to increased meridional temperature gradient;

4. Rapid sea-level rise in the western tropical Pacific, and concomitant sea level fall in the eastern Pacific;

5. Increased heat storage in the tropical Pacific Ocean associated with the expansion of the west Pacific warm pool;

6. A hiatus in the rise of the global mean surface temperature.

Rather than traditional climate processes such as the IPO or the PDO, is it fossil fuel burning by the expanding Chinese economy that has been controlling global climate? Do the IPO and PDO even exist as natural processes? Or, if they are natural, to what extent are they modulated by anthropogenic aerosols?

A related, and corroborating, research result was published at about the same time that the aerosol story emerged. Researchers found[109] that the "tilt" of the tropical Pacific Ocean surface, what they refer to as **dynamic sea level**, can be used to predict changes in global mean surface temperature. This must be related to the heat storage capacity of the west Pacific warm pool. That is, when the tropical Pacific sea surface is tilted up to the west, it is a sign of enhanced heat storage in the west Pacific warm pool signaled by a high regional sea level. This is associated with decreased global mean surface temperature and is consistent with a La Niña–like condition. When the ocean is tilted down to the west, it is a sign of increased global mean surface temperature and is consistent with an El Niño–like condition where the trade winds are reduced or absent.

Using CMIP5 models, researchers calculated that over the period 1998–2012, global surface warming was

[106] Aerosol emissions: https://www.carbonbrief.org/aerosol-emissions-key-to-the-surface-warming-slowdown-study-says

[107] China: https://cleantechnica.com/2017/08/16/china-halts-construction-150-gw-new-coal-power-plants/

[108] Smith et al. (2016)

[109] Peyser, C. E., et al. (2016) Pacific sea level rise patterns and global surface temperature variability, *Geophys. Res. Lett.*, 43, 8662–8669,

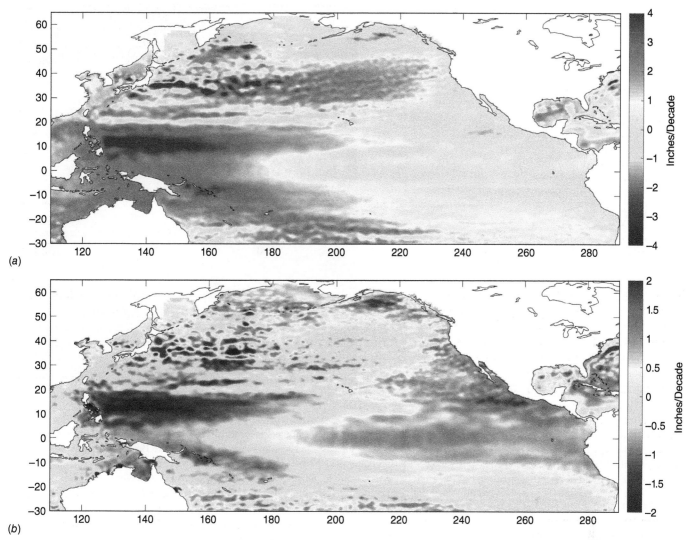

FIGURE 6.20 (a) The pattern of Pacific sea level 1993–2011 reflected the characteristic La Nina-like, negative phase of the Pacific Decadal Oscillation. However, (b) the pattern from 2011-2015 was distinctly different. Sea level shifted from rising to falling in the western Pacific warm pool and from falling to rising along the coasts of North and South America.[111] This figure is also discussed in Chapter 5 (note: scale in inches/decade).

Source: B. Hamlington, Old Dominion University

suppressed by 0.16 ± 0.06 degrees Celsius (0.29 ± 0.11 degrees Fahrenheit) by this effect. In contrast, during the strong El Niño of 1997/1998, the Pacific contributed 0.29 ± 0.10 degrees Celsius (0.52 ± 0.18 degrees Fahrenheit) to global surface temperature. On the basis of the relationship between dynamic sea level and global surface temperature, they predicted that by the end of 2016, global surface temperature would increase by 0.21 ± 0.07 degrees Celsius (0.38 ± 0.13 degrees Fahrenheit) over 2014. These projections have since been borne out. According to NASA global surface temperature records,[110] the annual mean temperature for 2016 was 0.26 degrees Celsius (0.47 degrees Fahrenheit), greater than 2014.

As this story of anthropogenic aerosol emissions continues to play out, it will be interesting to watch the Hadley circulation, the Pacific trade winds, sea-level variability, and global temperature over the next decade. Were the record setting years of 2014, 2015, and 2016 a result of the Chinese coal abatement program? Will continued coal abatement in China lead to more record-setting warm years? Will Hadley circulation and trade winds settle into a weaker mode that is more reflective of the influence of global warming? Or will they strengthen again, telling us that we do not yet fully understand the nature of variability in the climate system?

[110] NASA Goddard Institute for Space Studies maintains a global temperature database at this site: https://data.giss.nasa.gov/gistemp/tabledata_v3/GLB.Ts+dSST.txt

[111] Hamlington, B. D., et al. (2016) An Ongoing Shift in Pacific Ocean Sea Level, *CCPO Publications*. 205. http://digitalcommons.odu.edu/ccpo_pubs/205

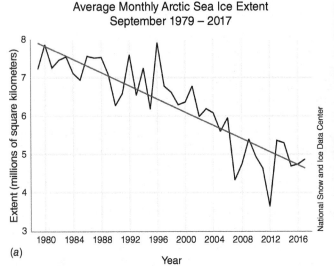

Average Monthly Arctic Sea Ice Extent
September 1979 – 2017

(a)

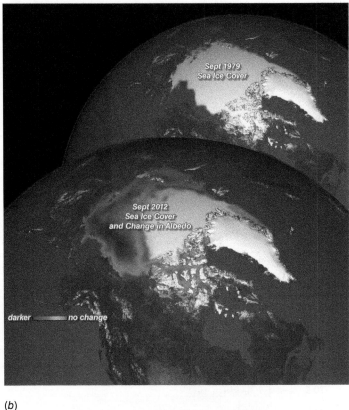

(b)

FIGURE 6.21 Arctic sea ice has been declining for almost 40 years according to the National Snow and Ice Data Center.[115] (a) Arctic sea ice extent typically reaches its low point each year in September. The rate of decline in 2017 was 13.2% per decade relative to the 1981 to 2010 average. For comparison, the decline rate was calculated at 13.7% after the 2013 minimum and 13.4% in 2016. Although sea ice shows significant year-to-year variability, the overall trend of decline remains strong. (b) Arctic ice reflects sunlight, helping to cool the planet. As this ice begins to melt, less sunlight gets reflected into space and more is absorbed into the oceans and land, raising the overall temperature and fueling further melting. This results in a positive feedback loop called the ice–albedo feedback, which causes the loss of the sea ice to be self-compounding. The more it disappears, the more likely it is to continue to disappear. (top) Average September Arctic sea ice from 1979; (bottom) Average September Arctic sea ice from 2012 with change in albedo overlaid.[116]

Source: (a) National Snow and Ice Data Center. (b) NASA/Goddard Space Flight Center Conceptual Image Lab: https://svs.gsfc.nasa.gov/4138

6.5.7 Ice-Albedo Feedback

Worldwide attention is paid to the annual summer retreat, and persistent long-term decline, of Arctic sea ice.[112] Scientists devote great effort to understanding the behavior of Arctic sea ice because as the summer extent of the ice pack decreases, white, sunlight-reflecting ice and snow is replaced by dark, heat-absorbing seawater (**Figure 6.21**). This switch constitutes a positive climate feedback that amplifies global warming called the ice–albedo feedback. In fact, it is this Arctic amplification of global warming that is viewed as the cause for the dramatic warming that has come to characterize the Arctic over the past two decades[113] and is thought to

be the driver of changes in the jet stream that are causing many types of extreme weather in the Northern Hemisphere[114] (a subject we discussed in **Chapter 3**).

In addition to the excess warming produced by the albedo switch,[117] researchers also worry that continued decline of Arctic sea ice will reach a tipping point[118] where so much heat-absorbing water has been exposed that the ice decline[119] becomes self-amplified and unstoppable.

[112] National Snow and Ice Data Center: https://nsidc.org/arcticseaicenews/

[113] Ghatak, D., et al. (2010) On the Emergence of an Arctic Amplification Signal in Terrestrial Arctic Snow Extent, *Journal of Geophysical Research* 115: D24105, doi:10.1029/2010JD014007. Perovich, D., et al. (2011) Solar Partitioning in a Changing Arctic Sea-Ice Cover, *Annals of Glaciology* 52, no. 57: 192–196.

[114] Overland, J.E., et al. (2016) Nonlinear response of mid-latitude weather to the changing Arctic. *Nature Climate Change*, 6 (11): 992, DOI:10.1038/nclimate3121

[115] NSIDC: http://nsidc.org/

[116] Pistone, K., et al. (2014) Observational determination of albedo decrease caused by vanishing Arctic sea ice, *PNAS* 111 (9) 3322-3326; doi:10.1073/pnas.1318201111

[117] VIDEO: NASA, Older Arctic sea ice disappearing. https://www.youtube.com/watch?v=Vj1G9gqhkYA

[118] Duarte, C.M., et al. (2012) Abrupt Climate Change in the Arctic, *Nature Climate Change* 2: 60–63.

[119] Still hope for Arctic Sea Ice: http://www.sciencedaily.com/releases/2011/02/110204092149.htm

The tipping-point[120] idea goes like this: With less sea ice, more sunlight is absorbed by the dark open water of the Arctic Ocean. Warm ocean water leads to additional sea ice melting, thus producing more open water, which absorbs even more heat and melts even more ice, and so forth until this becomes a self-fulfilling process that cannot be stopped. Researchers have long thought that this feedback loop can in principle become self-operating and independent of prevailing climate conditions.

The sea-ice story is focused on summer pack ice. Winter in the Arctic is dark and cold, and there is little worry that ice would not continue to form each winter, at least for several decades. But as global warming and Arctic amplification advance, the winter ice is increasingly characterized by thin annually forming ice that readily melts the following summer, rather than the thicker multiyear ice that seemingly is more stable. Eventually, researchers fear that the Arctic will become ice free[121] in the summer months and that the summer ice-free months will expand into early spring and late fall.[122] By opening the Arctic Ocean to resource exploitation such as oil drilling, fishing, sea floor dredging, and others, this ice-free condition can put at risk fragile arctic ecosystems.

Surprisingly, studies[123] indicate that the oldest and thickest Arctic sea ice is disappearing at a faster rate than the younger and thinner ice at the edges of the Arctic Ocean's floating ice cap. Researchers have documented that the average thickness of the Arctic sea ice cover is declining because it is rapidly losing its thick component, the multiyear ice. At the same time, the surface temperature in the Arctic is going up, which results in a shorter ice-forming season. It would take a persistent cold spell for most multiyear sea ice and other ice types to grow thick enough in the winter to survive the summer melt season and reverse the trend.

There are global implications for the rapid and geologically unusual warming that is happening today.[124] Because of the ice–albedo feedback, a future small increase in temperature could lead to larger warming over time, making the polar regions the most sensitive areas to climate change on Earth. The ice–albedo feedback has the potential to turn a small climate change into a big climate change. Why does this matter? There are three reasons:

- Human populations in the Arctic depend on their ecosystems for their food and other basic needs. Rapid ecosystem change threatens the very survival of these human communities. Without options, they become dependent on government support, displaced, and bereft of their traditional culture and identity.
- Ecosystem destruction anywhere on the planet reduces the diversity, interconnectedness, and complexity of living communities and thereby weakens the whole biological kingdom.
- Melting Greenland ice leads to global sea-level rise, and excessive meltwater may be capable of slowing the thermohaline circulation.

Greenland (**Figure 6.22**) is the largest island in the world, and it is covered in ice left over from the last ice age, which peaked about 20,000 to 30,000 years ago, covering approximately 27 percent of the world's land area with ice (compared with 10.4 percent today). Although Greenland ice has largely resisted the warm temperatures of the Holocene Epoch (from 10,000 years ago to present), it is succumbing now to the additional heat produced by anthropogenic global warming.[125]

Since 1979, scientists have tracked the extent of summer melting of the Greenland Ice Sheet. In 2007, the extent of melting broke the record set in 2005 by 10 percent, making it at the time, the largest season of melting ever recorded. That record was broken in the 2010 melt season, in which melting started earlier, ended later, and peaked with more melting than any previous melt season.[126]

Until recently, calving glaciers that carry ice from the interior into the sea accounted for about as much mass loss as surface melting and shifts in snowfall. But the balance tipped dramatically between 2011 and 2014, when satellite data and modeling suggested that 70 percent of the annual 269 billion tons of snow and ice shed by Greenland was lost through surface melt, not calving. The accelerating surface melt has doubled Greenland's contribution to global sea-level rise since 1992–2011, to 0.74 millimeters per year (0.03 inches per year). Scientists expressed alarm that melting was occurring so much faster than expected.[127]

The Greenland ice sheet is not melting simply because of high temperatures. Although the Arctic is warming twice as fast as the rest of the world, the melting appears to be aided by microbes and algae[128] that grow on the wet surface of the glacier. These produce dark pigments that increase the absorption of solar energy. Soot and dust that blow from lower latitudes and darken the ice are also playing a role, as are changes in weather patterns that increasingly steer warm, moist air over the vulnerable ice. Satellite data

[120] Wassmann, P. and Lenton, T. (2012) Arctic Tipping Points in an Earth System Perspective, *AMBIO* 41, no 1: 1–9.

[121] Wang, M. and Overland, J. (2009) A sea ice free summer Arctic within 30 years? *Geophysical Research Letters* 36: L07502, doi: 10.1029/2009GL037820.

[122] Markus, T., et al. (2009) Recent Changes in Arctic Sea Ice Melt Onset, Freezeup, and Melt Season Length, *Journal of Geophysical Research* 114: C12024, doi:10.1029/2009JC005436. Ghatak et al. (2010). Perovich (2012).

[123] Hall, D., et al. (2012) A Satellite-Derived Climate-Quality Data Record of the Clear-Sky Surface Temperature of the Greenland Ice Sheet, *Journal of Climate*, doi: 10.1175/JCLI-D-11-00365.1; http://ntrs.nasa.gov/search.jsp?R=20120009049

[124] Axford, Y., et al. (2009) Recent Changes in a Remote Arctic Lake are Unique within the Past 200,000 Years. *PNAS*, 106, no. 44: 18443–18446; doi: 10.1073/pnas.0907094106.

[125] van den Broeke, M., et al. (2009) Partitioning Recent Greenland Mass Loss, *Science* 326, no. 5955: 984–986, doi: 10.1126/science.1178176.

[126] Tedesco, M., et al. (2011) The Role of Albedo and Accumulation in the 2010 Melting Record in Greenland, *Environmental Research Letters* 6: 014005, doi: 10.1088/1748-9326/6/1/014005

[127] Science Magazine, The Great Greenland Meltdown http://www.sciencemag.org/news/2017/02/great-greenland-meltdown

[128] Stibal, M. (2017) Algae drive enhanced darkening of bare ice on the Greenland ice sheet. *Geophysical Research Letters*, 44. https://doi.org/10.1002/2017GL075958

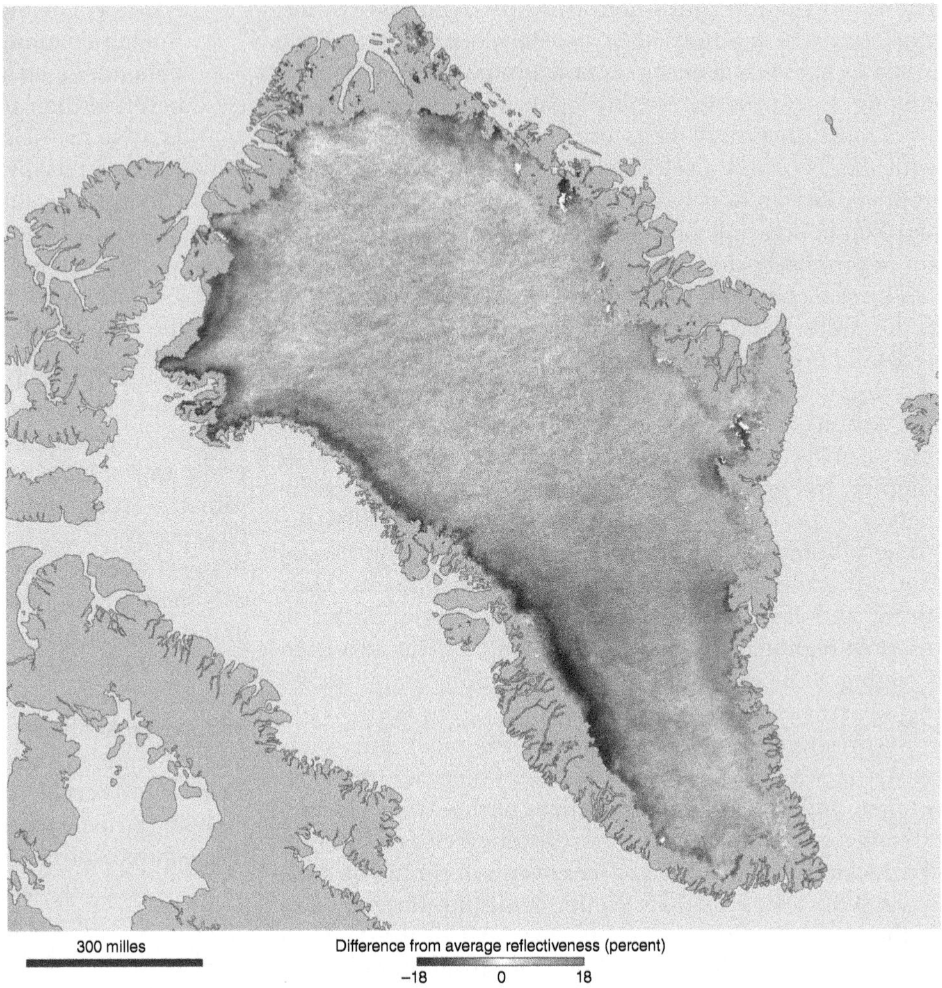

FIGURE 6.22 Map of changes in the percentage of light reflected by the Greenland ice sheet in summer (June, July, August). Virtually the entire surface has grown darker owing to surface melting, dust and soot on the surface, temperature-driven changes in the size and shape of snow grains, and the growth of dark microbes and algae on the wet surface of the glacier. Previously, the bright surface of the ice reflected more than half of the sunlight that fell on it. This helped keep the ice sheet stable, as less absorbed sunlight meant less heating and melting. However, in the past decade, satellites have observed a decrease in Greenland's reflectiveness. This darker surface now absorbs more sunlight, which accelerates melting.

Source: R. Lindsey "Greenland Ice Sheet Getting Darker," NOAA Climate Watch Magazine, 2011, https://www.climate.gov/news-features/understanding-climate/greenland-ice-sheet-getting-darker

300 milles

Difference from average reflectiveness (percent)

−18   0   18

show that the margins of the ice sheet have darkened by as much as 5 percent per decade since 2001.

An additional reason has been identified for the Greenland melt-down, there are fewer clouds to block the Sun. Scientists[129] have discovered a significant decrease in summer cloud cover during the past 2 decades, and this has accelerated melt on the surface of the Greenland ice sheet. The findings show that less cloud cover and more summer sunshine allow increased solar radiation to reach the surface providing more energy for melting. In fact, clouds have decreased so much that a 1 percent reduction in summer cloud cover is equivalent to 27 gigatons of extra ice melt on the Greenland ice sheet—roughly equivalent to the annual domestic water supply of the United States or 180 million times the weight of a blue whale. Since 1995, researchers found that Greenland has lost a total of about 4,000 gigatons of ice, which has become the biggest single contributor to the rise in global sea levels. This process explains about two-thirds of Greenland's melting signal in recent decades.

6.5.8 Clouds

It is apparent to anyone who has been outside that clouds can exert influence over the climate.[130] A sunny moment can change to a cool one as a cloud passes overhead. Clouds interact with solar radiation and reflect incoming sunlight in significant amounts, causing the albedo (reflectivity) of the entire Earth to be about twice what it would be in the absence of clouds.[131] Clouds also absorb the long-wave (infrared) radiation emitted by Earth's surface, similar to the effects of atmospheric greenhouse gases. But clouds do not cause climate change; they are a feedback to climate change caused by humans.[132] The question is, "Are clouds a positive feedback or a negative feedback to anthropogenic global warming?"

[129] Hofer, S., et al. (2017) Decreasing cloud cover drives the recent mass loss on the Greenland Ice Sheet. *Science Advances*, June, DOI: 10.1126/sciadv.1700584

[130] National Science Foundation, Clouds: The Wild Card of Climate Change, http://www.nsf.gov/news/special_reports/clouds/downloads.jsp

[131] Ramanathan, V., et al. (1989) Cloud Radiative Forcing and Climate: Results from the Earth Radiation Budget Experiment, *Science* 24, no. 4887: 57-63.

[132] Dessler, A.E. (2011) Cloud variations and the Earth's energy budget. *Geophysical Research Letters* 38: L19701, doi: 10.1029/2011GL049236

FIGURE 6.23 Earth is a cloudier place than many people realize. Low dense clouds reflect sunlight, a negative feedback, but high-altitude clouds trap heat coming off Earth's surface, a positive feedback. How will these opposite effects change in a warmer world, and what will be the net effect of clouds on future climate change? Research suggests that clouds tend to produce a positive feedback to global warming, amplifying the effects of greenhouse gases.

Source: NASA Visible Earth, "The Blue Marble," https://visibleearth.nasa.gov/view.php?id=57735

Getting the balance of cooling and warming effects right, and attributing these effects accurately to various cloud types at different altitudes, has been troubling for climate models.[133] Typical modeling experiments consist of a researcher running several general circulation model scenarios and finding that they do not agree on how clouds of various types respond to a warming atmosphere. Another example is to compare observations of clouds (by satellite, for instance) to model predictions and identify failures of the models to depict true cloud conditions.

In a warmer world, will clouds (**Figure 6.23**) provide a positive or negative feedback? That is, will there be fewer clouds or more, at what elevations, and how will this affect the balance of cooling and warming caused by clouds?[134] Fine-tuning answers to these questions is still the target of active research; however, scientists are increasingly concluding that clouds are not the cause of surface temperature changes, they are instead a feedback in response to those temperature changes because the radiative impact of clouds accounts for little of observed temperature variations.[135]

In the past, state-of-the-art climate models have disagreed on how clouds will respond to warming. Clouds have both warming and cooling effects. Clouds are the result of microscopic water droplets, or ice crystals, coalescing like grains of sand. Yet they cover 75 percent of Earth's surface. This huge range of scales, and the dependence of clouds on microscopic processes, makes them difficult to accurately model. Climate models are fundamentally large-scale tools making it hard for them to get the very small-scale processes right. Researchers have had to develop parameterizations[136] of clouds to simulate their role in the climate system. Nonetheless, clouds are very important in regulating the climate. They block a lot of sunlight, and they also trap a lot of heat radiating up from Earth's surface.

Low-level dense clouds tend to reflect sunlight, thus playing a cooling role; high-altitude clouds tend to trap heat, providing amplification to warming caused by other processes. Some models predict that low-level cloud cover will increase in a warmer climate, reflecting more sunlight, and limiting the level of global warming (a negative feedback). Other models predict less cloudiness, thus amplifying global warming (a positive feedback). The way clouds change with warming is of huge importance to global warming predictions. This is the main reason

[133] Kay, J.E., et al. (2012) Exposing Global Cloud Biases in the Community Atmosphere Model (CAM) Using Satellite Observations and their Corresponding Instrument Simulators, *Journal of Climate* 25, no. 4, http://journals.ametsoc.org/doi/abs/10.1175/JCLI-D-11-00469.1

[134] VIDEO: What role do clouds play in climate change? https://www.youtube.com/watch?v=BxFDJBrdOh0

[135] Dessler (2011)

[136] See discussion of parameterizations in climate models earlier in this chapter.

for the differences in warming produced by different climate models.[137]

All of the IPCC climate models[138] reduce low- and middle-altitude cloud cover with warming, a positive feedback. However, there is published research pointing to a negative feedback attributed to clouds. One study[139] of seasonal changes in the tropics using satellite data observed a decrease in net radiation (cooling) during the rainy season; this was related to a decrease in ice formation in the atmosphere. Another study[140] used detailed climate modeling to study the behavior of clouds above a warming ocean. Researchers found that low-level clouds thickened (reflecting more sunlight) as the ocean warmed, providing a natural cooling effect in response to the warming.

But there are a number of papers concluding that clouds amplify warming.[141] In one study,[142] researchers examined measurements from the Clouds and Earth's Radiant Energy System (CERES[143]) instrument onboard NASA's Terra satellite to calculate the amount of energy trapped by clouds as the climate varied over the last decade. The study concluded that warming due to increases in greenhouse gases will cause clouds to trap more heat, which will lead to additional warming, meaning clouds trap more heat, which in turn leads to even more warming—a positive feedback.

Another study[144] focused on a region of the atmosphere over the eastern Pacific Ocean and adjacent land. The clouds here are known to influence present climate, yet most models do poorly in representing them. The model developed by the authors performed well and simulated key features of the modern cloud field, including the response of clouds to El Niño. The improved model was then turned to focus on a warmer climate at the end of the century. The result? The model projected thinner and fewer clouds, and these trends were more pronounced than in other models. The study authors concluded that if their results prove to be representative of the real global climate, then climate is actually more sensitive to greenhouse gases than current global models predict, and even the highest warming predictions for the future would underestimate the real change we could see.[145]

Yet another study[146] examined the change in cloudiness that could occur as storm tracks shift poleward with continued warming. In the first study to document that storm tracks have indeed shifted poleward, researchers also found a related reduction in cloudiness and an increase in the net flux of radiation at the top of Earth's atmosphere in storm track regions. These observations point to a positive feedback: Poleward migration of storms produces a reduction in cloudiness that leads to amplified warming.

Groundbreaking studies indicate that clouds are moving in two important ways: toward the poles[147] and to higher altitudes in the atmosphere.[148] Climate change is predicted to expand atmospheric circulation patterns toward the poles[149] and clouds are following the same trajectory. The movement of clouds toward the poles produces a positive feedback effect that amplifies global warming. The reason is that low clouds that block sunlight and have a cooling effect are going to be most effective in the tropics where the greatest heating from the Sun is located. If these clouds are shifting toward the poles, they are going to be less effective at blocking sunlight thus producing less cooling. This enhances global warming.

It has also been observed that the height of the highest cloud tops is moving to higher altitudes in the atmosphere. Because the air is warming, they are essentially remaining at the same temperature. Therefore, even as the planet heats up, high clouds don't. They are not losing more heat to space, yet they are trapping more heat from the warming Earth below. This also enhances global warming. A notable aspect of this research is that both of these trends are predicted by models. Thus, it appears that novel models associated with CMIP5 are successfully reproducing observed cloud behavior and that this behavior constitutes a positive feedback to global warming.[150]

So, what does all this discussion about clouds mean? Scientists are still working to nail down the complexities of clouds but so far two things are clear: (1) the behavior of clouds that have been observed so far is acting as a positive

[137] Trenberth, K.E. and Fasullo, J.T. (2009) Global Warming Due to Increasing Absorbed Solar Radiation, *Geophysical Research Letters* 36; L07706, doi:10.1029/2009GL037527.

[138] Boucher, O., et al. (2013) Clouds and Aerosols. In: *Climate Change 2013: The Physical Science Basis. Contribution of Working Group I to the Fifth Assessment Report of the Intergovernmental Panel on Climate Change* [Stocker, T.F., et al. (eds.)]. Cambridge University Press, Cambridge, United Kingdom and New York, NY, USA.

[139] Spencer, R., et al. (2007) Cloud and Radiation Budget Changes Associated with Tropical Intraseasonal Oscillations, *Geophysical Research Letters* 34; L15707, doi:10.1029/2007GL029698.

[140] Caldwell, P. and Bretherton, C.S. (2009) Response of a Subtropical Stratocumulus-Capped Mixed Layer to Climate and Aerosol Changes. *Journal of Climate* 22: 20–38.

[141] Clement, A.C., et al. (2009) Observational and Model Evidence for Positive Low-Level Cloud Feedback, *Science* 325, no. 5939: 460–464.

[142] Dessler, A.E. (2010) A Determination of the Cloud Feedback from Climate Variations over the Past Decade, *Science* 330, no. 6010: 1523–1527, doi: 10.1126/science.1192546

[143] CERES homepage, http://ceres.larc.nasa.gov/

[144] Lauer, A., et al. (2010) The Impact of Global Warming on Marine Boundary Layer Clouds over the Eastern Pacific—A Regional Model Study, *Journal of Climate* 23: 5844–5863, doi: 10.1175/2010JCLI3666.1.

[145] Dr. Kevin Hamilton http://www.sciencedaily.com/releases/2010/11/101122172010.htm

[146] Bender, F., et al. (2011) Changes in Extratropical Storm Track Cloudiness 1983–2008: Observational Support for a Poleward Shift, *Climate Dynamics*, doi:10.1007/s0038-011-1065-6.

[147] Marvel, K., et al. (2015) External influences on modeled and observed cloud trends, *Journal of Climate*, v. 28, 4820-4840, DOI: 10.1175/JCLI-D-14-00734.1

[148] Norris, J.R., et al. (2016) Evidence for climate change in the satellite cloud record. *Nature*, DOI: 10.1038/nature.2016.20230

[149] Kate Marvel, Interview with Yale360: http://e360.yale.edu/features/investigating-the-enigma-of-clouds-and-climate-change

[150] Scientists find evidence for climate change in satellite cloud record: https://www.sciencedaily.com/releases/2016/07/160711151319.htm

feedback to global warming, and (2) there is no observational evidence that clouds will substantially slow down global warming. A range of instrumentation (satellites, weather balloons, aircraft) is being used by researchers to study a range of cloud types (ice clouds, low-level clouds, high-level clouds, tropical clouds, midlatitude clouds[151]). The intense work to document cloud processes with direct observations and the constant effort to improve modeling capabilities promise to keep alive the field of clouds and climate change for quite some time to come.[152]

6.5.9 Solar Radiation

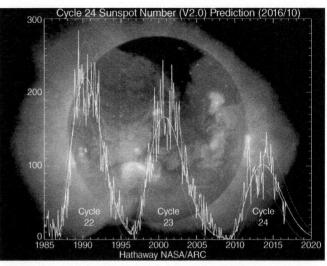

FIGURE 6.24 In July 2011, researchers at NASA's Marshall Space Flight Center[157] predicted that the solar maximum of cycle 24 would peak in June 2013 with a relatively low total solar irradiance (or TSI). As predicted, cycle 24 had the lowest TSI in several decades. In fact, the declining TSI over the past 30 years is evidence that global warming is not caused by the Sun. Solar cycles are numbered beginning with the first confirmed cycle 1755–1766; they average about 10.66 years in length, but cycles as short as 9 years and as long as 14 years have been observed.[158]

Source: Solar Cycle Prediction https://solarscience.msfc.nasa.gov/predict.shtml

In **Chapters 2** and **4**, we established that the Sun is not responsible for recent climate change.[153, 154] Satellites have not detected any increase in solar radiation over the past 35 years, a period when global mean temperature has dramatically risen. Had the Sun been responsible for global warming, the entire atmosphere would warm, not just the troposphere, as has been observed; in fact, the stratosphere has cooled over the same period. This is because greenhouse gases are trapping heat in the lower atmosphere. Indeed, none of the following would be true[155] if the Sun were the cause of global warming:

* Warming has been greater at the poles than at the Equator.
* Warming has been the same rate at night, as during the day.
* Warming has been greatest in the winter, not in the summer.

In each of these cases, the opposite would be true had the Sun caused the warming.

However, as we discussed in **Chapter 2**, solar output does vary through time (**Figure 6.24**), and it is critical that global climate models continue to account for the role that the Sun plays in Earth's climate. Total solar irradiance (TSI)[156]

varies in what is known as the *solar cycle*, the rise and fall (over approximately 11 years) of the number of sunspots on the Sun's surface. Sunspots are dark cool regions, but along the edge of a sunspot solar activity is high and when there are a high number of sunspots, TSI is at a maximum. During the 11-year solar cycle, the total energy given off by the Sun varies by about 0.1 percent. The solar cycle also causes a sizeable change in the ultraviolet (UV) radiation produced by the Sun, where most of the impacts are located in the stratosphere (above ~10 kilometers or 6.2 miles).

If TSI only varies by 0.1 percent and UV radiation affects mainly the stratosphere, are Earth's weather and climate unaffected by the solar cycle? NASA scientists tested[159] this question by simulating 1600 years of varying UV and TSI in climate models. They found[160] that the solar cycle can account for 15 to 20 percent of rainfall in certain areas. For instance, a solar maximum favors increased precipitation north of the equator (the South Asian monsoon) and decreased precipitation near the equator and at northern midlatitudes. Complex changes in UV and TSI drive these patterns; increased UV radiation leads to a rise in stratospheric ozone, which warms the tropics and (because of

[151] Cloud research: http://www.sciencedaily.com/releases/2011/03/110315142526.htm

[152] VIDEO: Can clouds buy us more time to solve climate change? https://www.ted.com/talks/kate_marvel_can_clouds_buy_us_more_time_to_solve_climate_change

[153] Lockwood, M. (2009) Solar Change and Climate: An Update in the Light of the Current Exceptional Solar Minimum, *Proceedings of the Royal Society A* 2 December, doi 10.1098/rspa.2009.0519. Lean, J. (2010) Cycles and Trends in Solar Irradiance and Climate, *Wiley Interdisciplinary Reviews: Climate Change* 1, January/February: 111–122.

[154] VIDEO: Solar-Schmolar. http://www.youtube.com/watch?v=_Sf_UIQYc20

[155] Menne, M., et al. (2010) On the Reliability of the U.S. Surface Temperature Record, *Journal of Geophysical Research – Atmospheres* 115: D11108, doi:10.1029/2009JD013094. Parker, D.D. (2006) A Demonstration that Large-Scale Warming is not Urban, *Journal of Climate* 19, no. 12: 2882–2895.

[156] Kopp, G. and Lean, J.L. (2011) A new, Lower Value of Total Solar Irradiance: Evidence and Climate Significance, *Geophysical Research Letters*, 38: L01706

[157] NASA: http://solarscience.msfc.nasa.gov/predict.shtml

[158] Solar Cycle *Wikipedia*, http://en.wikipedia.org/wiki/Solar_cycle

[159] Rind, D., et al. (2008) Exploring The Stratospheric/Tropospheric Response To Solar Forcing, *Journal of Geophysical Research* 113: D24103, doi:10.1029/2008JD010114.

[160] NASA: http://www.giss.nasa.gov/research/briefs/rind_03/

various interactions between the stratosphere and the troposphere) shifts the zone of Hadley circulation to the north, accounting for regional shifts in climate. Increased TSI during the solar cycle causes a rise in sea-surface temperature where cloudiness is low (Northern Hemisphere subtropics), an effect that also favors reduced rainfall near the equator and in the northern midlatitudes.

Remember, the solar cycle influence on these processes is relatively minor, on the order of 15 to 20 percent. This influence is likely to change as rising greenhouse gases cause their own changes in climate; stratospheric cooling, increased sea surface temperatures, expanding tropics, changing winds, new patterns of precipitation, and shifts in Hadley circulation have all been attributed to global warming. How these balance with solar cycle influences adds significant complexity to the challenge of modeling global climate.

6.6 Concluding Thoughts

By now it is obvious to you that Earth's climate system is very complex. There are many oceanographic, atmospheric, solar, and terrestrial processes (all with some degree of uncertainty) that need to be individually depicted, and their interrelationships portrayed, in the form of mathematical calculations. These include frequent but unpredictable processes such as ENSO (located in the Pacific Basin but with global impacts), quasi-predictable processes such as the solar cycle (but each cycle is not identical, and there is uncertainty regarding long-term trends in solar strength), uncertainties related to clouds,[161] random and unpredictable explosive volcanism, ice–albedo feedback that is not fully understood, anthropogenic sulfur aerosols, that change with the waxing and waning of industrial activities, and others. These processes all interact with one another over different lengths of time and on different geographic scales and produce various types of feedbacks.

The complexity and variability of the climate system make it all the more remarkable that climate models are able to get it right so much of the time. How do we know when they get it right? Because they can reproduce, with amazing fidelity, a range of complex historical observations (refer back to Figure 6.9).

But climate models are not perfect. Perhaps their greatest weakness is their difficulty in projecting climate at high resolution for individual regions and localities, a problem compounded by the need for computers of extraordinary power. They also have trouble projecting future

precipitation trends that are significantly different than natural variability. However, when their simulations of historical climate are compared to observations, even early generations of models show remarkable conformity with the data.[162]

In one analysis,[163] climate models published since 1973 were compared to surface temperature observations. In general, they have been quite skillful in reproducing observed warming. While some projected temperatures that were too low, some too high, they all show outcomes reasonably close to what has actually occurred, especially when discrepancies between predicted and actual CO_2 concentrations and other random and unpredictable (e.g., volcanic eruptions) climate forcings are taken into account.

Comparing models with observations requires comparing apples to apples. The most often used projections from climate models are global surface air temperatures. However, observational datasets come from surface air temperatures over land and sea surface temperatures of ocean water.

To account for these differences, researchers create blended model fields. These fields include sea surface temperatures (ocean) and surface air temperatures (land), in order to match what is actually measured in the observations. Blended fields show slightly less warming than global surface air temperatures, as models depict the air over the ocean warming faster than sea surface temperatures. Global surface air temperatures in CMIP5 models have warmed about 16 percent faster than observations since 1970. About 40 percent of this difference is due to air temperatures over the ocean warming faster than sea surface temperatures in the models. Blended model fields only show warming 9 percent faster than observations and thus are a better match and represent more of an apples-to-apples comparison.

Figure 6.25 compares a number of observational datasets to the 95 percent confidence interval of CMIP5 models using the RCP4.5 scenario. The comparison reveals that the models have clear skill in matching historical temperature patterns. While there are a few periods where models and observations diverge—around 1910 and the early 2000s, for example—current temperatures for 2015, 2016, and the first half of 2017 are pretty close to the model average.[164] The "warming hiatus," the slowdown in global warming in the early 2000s, is a decade where models and observations disagree. However, as we discussed, incorporating sulfur aerosols from SE Asia[165] in CMIP5 modeling

[161] Andrews, T., et al. (2012) Forcing, Feedbacks and Climate Sensitivity in CMIP5 Coupled Atmosphere-Ocean Climate Models, *Geophysical Research Letters* 39: L09712, doi:10.1029/2012GL051607.

[162] CarbonBrief: https://www.carbonbrief.org/analysis-how-well-have-climate-models-projected-global-warming
[163] CarbonBrief: https://www.carbonbrief.org/analysis-how-well-have-climate-models-projected-global-warming
[164] CarbonBrief: https://www.carbonbrief.org/factcheck-climate-models-have-not-exaggerated-global-warming
[165] Smith et al. (2016)

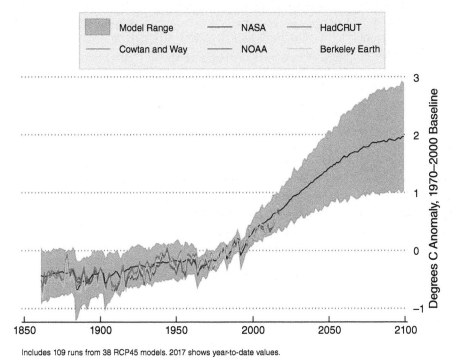

Climate Models and Observations, 1861–2100

Includes 109 runs from 38 RCP45 models. 2017 shows year-to-date values.

FIGURE 6.25 CMIP5 model depiction of RCP4.5 is shown compared to temperature observations from several research groups. The black line represents the average of all the climate models while the gray band is the 95 percent confidence interval. Observations of global temperature shown from NASA,[166] NOAA,[167] HadCRUT,[168] Cowtan and Way,[169] and Berkeley Earth[170] from 1861 to 2100.

Source: https://www.carbonbrief.org/factcheck-climate-models-have-not-exaggerated-global-warming

neatly reproduces the slowdown and reveals that modeling is indeed capable of tracking detailed temperature observations provided that biases and assumptions are closely scrutinized.

The close match between projected and observed warming since 1970 suggests that estimates of future warming may prove similarly accurate.[171] Climate models are effective tools that grow more powerful with continued research and technological improvement. Model projections warn us that global warming will likely cause serious problems in the future including increased drought, sea-level rise, frequency and intensity of heat waves, heavy rainfall events, intensity of tropical cyclones, sea-ice reductions, and other dangerous physical impacts. Models improve our understanding of the risks posed by climate change, and aid the development of appropriate socioeconomic responses.

Animations and Videos

1. National Center of Atmospheric Research, Recipe for a Better Climate Model http://www.outube.com/watch?v=TLvCCHNCEdc&NR=1

2. National Center for Atmospheric Research Meso-to Planetary Scale Processes in a Global Ultra-High-Resolution Climate Model https://www.youtube.com/watch?v=4794mgJLTbU

3. NASA Center for Climate Simulation: Data Supporting Science, Supercomputing the Climate, http://www.nasa.gov/topics/earth/features/climate-sim-center.html

4. CMIP5: 21st Century Temperature and Precipitation Scenarios - RCP 8.5: 936 ppm https://vimeo.com/70674503

5. Vox: How Climate Change Makes Hurricanes Worse https://www.youtube.com/watch?v=_0TCrGtTEQM

6. Yale Climate Connections Humans experimenting with Climate's playing nice https://www.yaleclimateconnections.org/2017/10/humans-experimenting-with-climates-playing-nice/

7. El Niño What is it? https://www.youtube.com/watch?v=WPA-KpldDVc

8. NASA's Goddard Space Flight Center, Monsoons: Wet, Dry, Repeat... https://svs.gsfc.nasa.gov/12255

[166] NASA https://data.giss.nasa.gov/gistemp/
[167] NOAA https://www.ncdc.noaa.gov/monitoring-references/faq/anomalies.php
[168] HadCRUT https://www.metoffice.gov.uk/hadobs/hadcrut4/
[169] Cowtan and Way http://www-users.york.ac.uk/~kdc3/papers/coverage2013/series.html
[170] Berkeley Earth http://berkeleyearth.org/data/
[171] VIDEO: NASA, Global temperature model (1885-2100). https://www.youtube.com/watch?v=tBithxUmPiA

9. NASA Goddard Institute for Space Science, Older Arctic Sea Ice Disappearing https://www.youtube.com/watch?v=Vj1G9gqhkYA

10. What Role to Clouds Play in Climate Change? from Huffpost Talk Nerdy To Me https://www.youtube.com/watch?v=BxFDJBrdOh0

11. TED Talk by Climate Scientist Kate Marvel Can clouds buy us more time to solve climate change? https://www.ted.com/talks/kate_marvel_can_clouds_buy_us_more_time_to_solve_climate_change

12. Climate Denial Crock of the Week: Solar Schmolar, http://www.youtube.com/watch?v=_Sf_UIQYc20

13. NASA Global Temperature Model (1885–2100) https://www.youtube.com/watch?v=tBithxUmPiA

Comprehension Questions

1. Why are climate models important? How are they useful?

2. How are climate observations used in climate models?

3. Describe why scientists are working to increase the resolution of climate model projections.

4. What are some of the impacts of El Niño?

5. What are the implications of declining sea ice in the Arctic?

6. What role do clouds play in climate?

7. How do aerosols affect climate?

8. What is the solar cycle? How does it influence climate?

9. Identify some negative impacts from global warming that are projected by climate models.

10. Describe precipitation changes by the end of the century as projected by climate models.

Thinking Critically

1. Pick three areas experiencing the worst warming by the end of the century. Describe the major impacts to human civilizations and to natural ecosystems in each area.

2. How would you prepare for climate change in your region?

3. Why are clouds difficult to incorporate into climate models?

4. What is the evidence that climate models are skillful?

5. Describe how modern society might prepare for and adapt to precipitation changes caused by global warming.

6. There is evidence that the solar cycle and TSI will decline over the decade 2011–2021. What additional information would you need to anticipate the impact of this on global climate? Speculate about the potential impacts of a return of TSI to previous levels over the decade 2021–2031 if greenhouse gas production does not decrease over the same period.

7. Consider the various representative concentration pathways used by the IPCC in AR5. Which do you consider most likely to occur over the next decade? Over the next half-century? Why?

8. Study the IPCC AR5 climate impacts projected for the end of the century. Describe the most significant threats to human health.

9. Among most scientists, it is a foregone conclusion that significant climate change is unavoidable, especially because greenhouse gases continue to accumulate. Pick one area where global warming will affect your life and describe steps you would take to adapt to the change.

10. You have just stepped into an elevator with the mayor of your town; he is a climate skeptic. You have a captive audience for the next 30 seconds. Convince him that climate change is real and that he needs to incorporate this in his leadership.

Activities

1. View the National Center of Atmospheric Research video "Recipe for a Better Climate Model," http://www.youtube.com/watch?v=TLvCCHNCEdc&NR=1, and answer the following questions.

   (a) Why are climate models important? How are they useful?
   (b) How are climate observations used in climate models?
   (c) What is AIRS and why is it useful?
   (d) Why is working with people in developing countries important?

2. Study how El Niño works in this video: "Brian Slocum Explains El Niño," http://www.youtube.com/watch?v=uySu7Zv2cbU, answer the following questions.

   (a) What is an El Niño?
   (b) What are some of the impacts of El Niño?
   (c) What happens in the Pacific Ocean during an El Niño?
   (d) How does El Niño affect hurricane formation in the Atlantic Ocean (and Gulf of Mexico)?
   (e) How do winds change in the Southern Hemisphere when an El Niño occurs?

3. View the NASA Global Temperature Model (1885–2100) https://www.youtube.com/watch?v=tBithxUmPiA, and answer the following questions.

   (a) What trends do you see in the historical projections?
   (b) What trends do you see in the future projections?
   (c) What parts of the globe warm the fastest?
   (d) What significant impacts over land are likely to develop in the future based on this projection?

4. Study the video "Solar Schmolar," http://www.youtube.com/watch?v=_Sf_UIQYc20, and answer the following questions.

   (a) Is the Sun responsible for observed global warming?
   (b) What are the global warming patterns we observe and are they consistent with solar warming?
   (c) What is TSI and what is its relationship to global warming?
   (d) When comparing the Earth surface temperature to the solar sunspot cycle, are they related?

5. TED Talk by Climate Scientist Kate Marvel Can clouds buy us more time to solve climate change? https://www.ted.com/talks/kate_marvel_can_clouds_buy_us_more_time_to_solve_climate_change, and answer the following questions.

   (a) Describe the role that clouds play in the greenhouse effect.
   (b) What trends have been observed in clouds that may change the role they play in climate change?
   (c) Will clouds slow global warming?
   (d) Why do climate models have difficulty in representing clouds?
   (e) What answer does the speaker provide to the question that is the title of her talk?

Key Terms

Anthropogenic aerosols

Boundary conditions

Calibrated uncertainty language

Central-Pacific El Niño

Climate model

CMIP

Coal abatement

Control runs

Coupled general circulation model

Desert dust

Dynamical downscaling

Dynamic sea level

Exposure

Findings

General circulation models

Hazards

Ice-albedo feedback

Model ensemble

Monsoon

Multimodel mean

Nested modeling

Parameterization

RCPs

Representative Concentration Pathways

Risk

Spatial resolution

Statistical downscaling

Sulfur aerosols

Temporal resolution

Uncertainty

Upwelling

Variability

Volcanic aerosols

Vulnerability

CHAPTER 7

Warming Impacts

LEARNING OBJECTIVE

According to the **U.S. Global Change Research Program** (USGCRP),[1] which reports to the U.S. Congress and the President on the effects of global climate change, global annually averaged surface air temperature has increased by about 1.0°C (1.8°F) over the last 115 years (1901–2016). This period is now the warmest in the history of modern civilization. Recent years have seen record-breaking, climate-related weather extremes, and the warmest years on record for the globe. These trends are expected to continue over the course of this century and beyond.

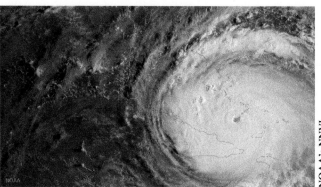

NOAA's NNVL

On August 30, 2017, Hurricane Irma formed in the tropical Atlantic. It reached peak intensity on September 6 with maximum sustained winds of 82.7 m/s (185 mi/hour). This wind speed lasted for 37 hours, the longest on record. Irma caused widespread and catastrophic damage throughout its long lifetime, particularly in parts of the northeastern Caribbean and the Florida Keys. Complete devastation was reported in the northern Leeward Islands and Virgin Islands. In the Florida Keys, 25% of buildings were destroyed while 65% were significantly damaged. The hurricane brought heavy rains, strong winds, and record storm surge to most of Florida. Irma was responsible for at least 134 deaths, including 3 in Puerto Rico and 90 on the U.S. mainland.[2]

Source: NOAA; https://www.ncdc.noaa.gov/sotc/tropical-cyclones/201709

7.1 Chapter Summary

Climate change impacts to human communities include the following: stresses to water resources, threats to human health, shifting demand on energy supply; disruptions to transportation and agriculture, and increased vulnerability of society and ecosystems to future climate change. In the United States, extreme weather events have increased in number and magnitude and are likely to do so in the future. Severe heat waves and record-setting temperatures are occurring with greater frequency.[3]

Among other impacts are the spread of diseases into areas where they have not been historically prevalent, retreat of tundra and arctic ecosystems, increased drought and flooding, rapidly multiplying wildfire, sea-level rise, decreased snow pack, retreating glaciers, changes in the timing of seasons, multiple and severe ecological impacts, and others.

In this chapter, you will learn that

1. The USGCRP coordinates and integrates federal research on changes in the global environment and their implications for society. The program reports to Congress on the state of climate change globally and in North America. Key findings of the Fourth

---

[1] U.S. Global Change Research Program: globalchange.gov
[2] NOAA National Centers for Environmental Information: https://www.ncdc.noaa.gov/sotc/tropical-cyclones/201709.

[3] VIDEO: The Climate 25. https://vimeo.com/123438759

National Climate Assessment[4] (NCA4) include the following:

a. The global, long-term, and unambiguous warming trend has continued during recent years.

b. Annual average temperature over the contiguous United States has increased by 1.0°C (1.8°F) for the period 1901–2016 and is projected to continue to rise.

c. There have been marked changes in temperature extremes across the contiguous United States. The number of high temperature records set in the past two decades far exceeds the number of low temperature records.

d. Heavy precipitation events in most parts of the United States have increased in both intensity and frequency since 1901.

e. Extreme temperatures in the contiguous United States are projected to increase even more than average temperatures.

f. Future decreases in surface soil moisture from human activities over most of the United States are likely as the climate warms under the higher scenarios.

g. The world's oceans have absorbed about 93% of the excess heat caused by greenhouse gas warming since the mid-20th century, making them warmer and altering global and regional climate feedbacks.

h. Global mean sea level has risen by about 16–21 cm (7–8 in) since 1900, with about 7 cm (3 in) occurring since 1993.

i. The world's oceans are currently absorbing more than one-quarter of the CO_2 emitted to the atmosphere annually from human activities, making them more acidic, with potential detrimental impacts to marine ecosystems.

j. Annual average near-surface air temperatures across Alaska and the Arctic have increased over the last 50 years at a rate more than twice as fast as the global average temperature.

k. Since the early 1980s, annual average arctic sea ice has decreased in extent 3.5% to 4.1% per decade, has become thinner by between 1.3 and 2.3 m (4.3 to 7.5 ft), and is melting at least 15 more days each year. September sea ice extent has decreased between 10.7% and 15.9% per decade.

l. The observed increase in global carbon emissions over the past 15–20 years has been consistent with higher scenarios (e.g., RCP8.5). In 2014 and 2015, emission growth rates slowed as economic growth became less carbon intensive. Even if this slowing trend continues, however, it is not yet at a rate that would limit the increase in the global average temperature to well below 2°C (3.6°F) above preindustrial levels

m. Choices made today will determine the magnitude of climate change risks beyond the next few decades.

n. Unanticipated and difficult or impossible-to-manage changes in the climate system are possible throughout the next century as critical thresholds are crossed and/or multiple climate-related extreme events occur simultaneously.

2. In addition to the quadrennial NCA4, many other reports emerge every year that come to similar conclusions. In this chapter, we highlight some of these.

7.2 Air Temperature and Precipitation Patterns in the United States

The effects of global warming can be identified in every sector of the U.S. economy and every region of the continent. Effects are found in human society and health; ecosystems, water resources, energy, and transportation infrastructure; and agriculture. Many of these effects are distributed regionally. For instance, the south and Southwest have plunged into severe drought, leading to an increase in wildfires, decreased drinking water, and growing heat stress in cities and within ecosystems.

The New England, Midwest, and southern Canadian regions have seen a sharp increase in flooding owing to higher levels of rainfall and more violent rainstorms. Nationwide,[5] extreme summer temperatures are already occurring more frequently, and heat waves will become normal by mid-century if the world continues on a business-as-usual schedule of emitting greenhouse gases.[6]

---

[4] Wuebbles, D.J., et al. (2017) Executive Summary. In: *Climate Science Special Report: Fourth National Climate Assessment, Volume* I [Wuebbles, D.J., et al. (eds.)]. U.S. Global Change Research Program, Washington, DC, USA, pp. 12–34, doi: 10.7930/J0DJ5CTG

[5] Duffy, P. and Tebaldi, C. (2012) Increasing Prevalence of Extreme Summer Temperatures in the U.S., *Climatic Change* 111, no. 2: 487, doi: 10.1007/s10584-012-0396-6

[6] The International Energy Agency sees four large-scale shifts occurring in the global energy system: (1) rapid deployment and falling costs of clean energy, (2) growing electrification of energy, (3) shift to a more services-oriented economy and a cleaner energy mix in China, and (4) resilience of shale gas and tight oil in the United States. Oil demand continues to grow to 2040, but at a steadily decreasing pace. Natural gas use rises by 45% to 2040; with more limited room to expand in the power sector, industrial demand for energy becomes the largest area for growth. The outlook for nuclear power has dimmed, but China continues to lead a gradual rise in output, overtaking the U.S. by 2030 to become the largest producer of nuclear-based electricity. Renewables capture two-thirds of global investment in power plants to 2040 as they become, for many countries, the least-cost source of new electricity generation.

Overall, there is a new intensity and destructiveness to weather events.[7] Across the country, air temperature has increased and ecosystems that have evolved under the regular timing of seasonal patterns are experiencing increased disease and infestation. Coastal communities have sustained losses due to shoreline erosion and the landfall of storms. Unusually intense snowstorms have hit the east coast, winters across North America have become warmer, climate is establishing new records for warmth,[8] there are unusually high numbers of tornadoes in the Midwest, and summers throughout the continent are characterized by record-breaking heat waves.

7.2.1 Air Temperature

NCA4 reports that scientists have concluded that these trends are consistent with the expected influence of rising air temperature created by global warming.[9] There is 40 percent more carbon dioxide in the atmosphere now than there was only 19 years ago. As we saw in earlier chapters, this greenhouse gas is effective at trapping heat in the

troposphere and even under natural conditions is responsible for Earth's habitable climate.

Researchers report that global surface temperatures in 2014–2016 were the warmest ever observed.[10] In the 28-month span between May 2014 and August 2016, 24 monthly global temperature records were broken,[11] and 14 of the 15 largest all-time monthly temperature departures were set during 2015 and 2016.[12] Annual average temperature over the contiguous United States has increased by 1.0 degree Celsius (1.8 degrees Fahrenheit) for the period 1901–2016 and is projected to continue to rise (**Figure 7.1**).

According to NCA4,[14] more than 95 percent of the land surface of the contiguous U.S. saw an increase in annual average temperature. In contrast, the only cooling observed was on small and somewhat dispersed parts of the Southeast and southern Great Plains.

From a seasonal perspective, warming was greatest and most widespread in winter, with increases of over 0.8 degree Celsius (1.5 degrees Fahrenheit) in most areas. In summer, warming was less extensive and located mainly along the east coast and in the western third of the nation. Summer

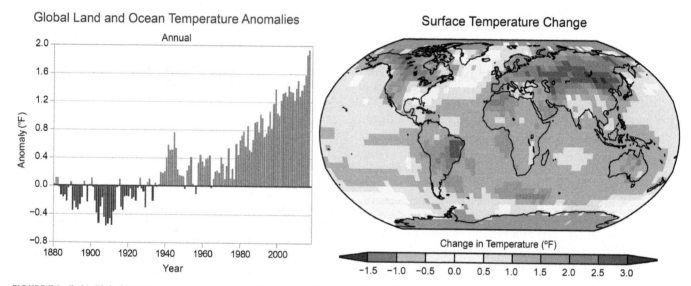

FIGURE 7.1 (left) Global annual average temperature has increased by more than 0.7°C (1.2°F) for the period 1986–2016 relative to 1901–1960. (right) Surface temperature change (in °F) for the period 1986–2016 relative to 1901–1960. Gray indicates missing data.[13]

Source: https://science2017.globalchange.gov/chapter/executive-summary/

[7] Herring, S.C., et al. Eds. (2017) Explaining Extreme Events of 2016 from a Climate Perspective, *Bulletin American Meteorological Society*, 98 (12), S1–S157.

[8] NASA, http://www.earthobservatory.nasa.gov/IOTD/view.php?id=77465

[9] Guirguis, K., et al. (2011) Recent Warm and Cold Daily Winter Temperature Extremes in the Northern Hemisphere, *Geophysical Research Letters* 38: L17701, doi: 10.1029/2011GL048762

[10] NASA, NOAA Data Show 2016 Warmest Year on Record Globally: https://www.nasa.gov/press-release/nasa-noaa-data-show-2016-warmest-year-on-record-globally

[11] How unusual is 2016's record-temperature? NOAA, https://www.climate.gov/news-features/blogs/beyond-data/how-unusual-2016s-record-temperature-three-peat-and-will-hot-streak

[12] NOAA https://www.climate.gov/news-features/blogs/beyond-data/how-unusual-2016s-record-temperature-three-peat-and-will-hot-streak

[13] Figure 1, Executive Summary, 4th National Climate Assessment: https://science2017.globalchange.gov/chapter/executive-summary/

[14] Wuebbles et al. (2017)

Annual Temperature

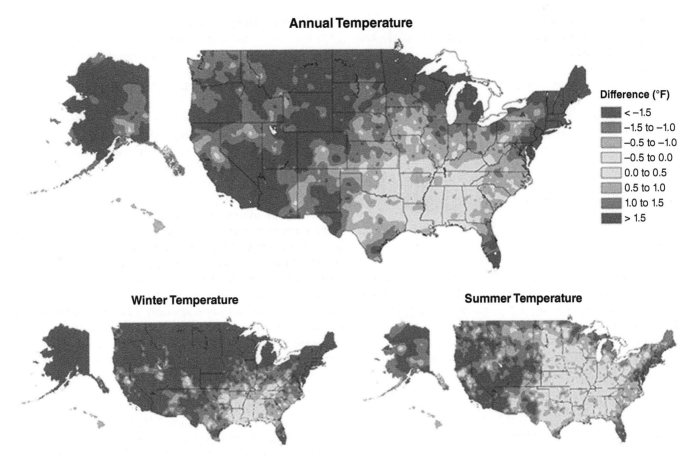

FIGURE 7.2 Observed changes in annual, winter, and summer temperature (in °F). Changes are the difference between the average for present-day (1986–2016) and the average for the first half of the last century (1901–1960 for the contiguous United States, 1925–1960 for Alaska and Hawai'i).

Source: https://science2017.globalchange.gov/chapter/6/

cooling was evident in parts of the Southeast, Midwest, and Great Plains.

The largest changes in net warming were in the western states, where average temperature increased by more than 0.8 degree Celsius (1.5 degrees Fahrenheit) in Alaska, the Northwest, the Southwest, and in the northern Great Plains. The Southeast region had the least warming, resulting from a combination of natural variations and human influences.

In most regions, the average minimum temperature increased at a slightly higher rate than the average maximum temperature. The Midwest had the largest discrepancy between these two measurements, and the Southwest and Northwest had the smallest. This differential rate of warming resulted in a continuing decrease in the daily temperature range that is consistent with other parts of the globe.

Annual average sea surface temperature also increased along all U.S. coastlines, though changes were generally smaller than over land owing to the higher heat capacity of water. Increases were largest in Alaska (greater than 0.6 degree Celsius [1.0 degree Fahrenheit]) while increases were smallest (less than 0.3 degree Celsius [0.5 degree Fahrenheit]) in coastal areas of the Southeast (**Figure 7.2**).

Extreme Temperatures There have been marked changes in temperature extremes across the contiguous United States. The number of high temperature records set in the past two decades far exceeds the number of low temperature records. Temperature and precipitation extremes can affect water quality and availability, agricultural productivity, human health, vital infrastructure, iconic ecosystems and species, and the likelihood of disasters.

Some extremes have already become more frequent, intense, or of longer duration, and many extremes are expected to continue to increase or worsen, presenting substantial challenges for built, agricultural, and natural systems. Some storm types such as hurricanes, tornadoes, and winter storms are also exhibiting changes that have been linked to climate change, although the current state of the science does not yet permit detailed understanding (**Figure 7.3**).

Overall, there is a new intensity and destructiveness to weather events.[15] Across the country, air temperature has

[15] Rahmstorf, S. and Coumou, D. (2011) Increase of Extreme Events in a Warming World, *PNAS* 108, no. 44: 17905–17909, www.pnas.org/cgi/doi/10.1073/pnas.1101766108.

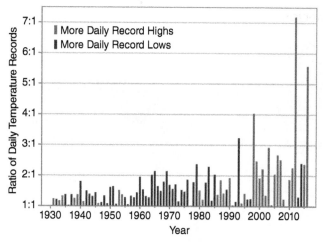

FIGURE 7.3 Observed changes in the occurrence of record-setting daily temperatures in the contiguous United States. Light grey, or red bars indicate a year with more daily record highs than daily record lows, while dark, or blue bars indicate a year with more record lows than highs. The height of the bar indicates the ratio of record highs to lows (light grey or red) or of record lows to highs (dark, or blue). For example, a ratio of 2:1 for a blue (dark) bar means that there were twice as many record daily lows as daily record highs that year.

Source: https://science2017.globalchange.gov/chapter/executive-summary/

increased, and ecosystems that have evolved under the regular timing of seasonal patterns are experiencing increased disease and infestation. Coastal communities have sustained losses due to shoreline erosion and the landfall of storms. Unusually intense snowstorms have hit the east coast, winters across North America have become warmer, climate is establishing new records of warmth, there are unusually high numbers of tornadoes in the Midwest, and summers throughout the continent are characterized by record-breaking heat waves.

Projected Temperature Model projections of annual average air temperature nation-wide (contiguous states only) indicate increases of about 1.4 degrees Celsius (2.5 degrees Fahrenheit) for the period 2021–2050 relative to 1976–2005 in all RCP scenarios. Hence, recent record-setting years may be common over the next few decades. Much larger air temperature increases are projected by the end of the century: 1.6 degrees to 4.1 degrees Celsius (2.8 degrees to 7.3 degrees Fahrenheit) in the lower (RCP4.5) scenario and 3.2 degrees to 6.6 degrees Celsius (5.8 degrees to 11.9 degrees Fahrenheit) in the higher scenario (RCP8.5).

Extreme temperatures in the contiguous states are projected to increase even more than average temperatures. The temperatures of extremely cold days and extremely warm days are both expected to increase. Cold waves are projected to become less intense while heat waves will become more intense. The number of days below freezing is projected to decline while the number above 32 degrees Celsius (90 degrees Fahrenheit) will rise.

7.2.2 Precipitation

According to the Executive Summary of the Fourth National Climate Assessment,[16] precipitation in most parts of the United States has increased in both intensity and frequency since 1901. However, there are important regional differences. The largest increases have occurred in the Northeastern region. Annual precipitation has decreased in much of the west, Southwest, and Southeast and increased in most of the northern and southern plains, Midwest, and Northeast. A national average increase of 4 percent in annual precipitation since 1901 is mostly a result of large increases in the fall season.

Precipitation changes differ across the seasons in the contiguous states. Fall precipitation has increased the most by an average 10 percent averaged across the country. In much of the northern Great Plains, the Southeast, and the Northeast, precipitation increased by 15 percent. Nation-wide, winter increases average only 2 percent, with most of the western region and parts of the Southeastern region experiencing drying.

Spring and summer have experienced increases of about 3.5 percent. In spring, the northern half of the contiguous U.S. has become wetter, and the southern half has become drier. In summer, there is a mixture of increases and decreases across the country. Alaska shows little change in annual precipitation; however, in all seasons, central Alaska shows declines and the Southeast coastal area shows increases. Hawai'i shows a decline of more than 15 percent in annual precipitation (**Figure 7.4**).

Extreme Rain The frequency and intensity of heavy precipitation events has increased across the United States (**Figure 7.5**) and are projected to continue to increase over the 21st century. In the north, including Alaska, more precipitation is projected in the winter and spring, and in parts of the Southwest, less precipitation is projected for the winter and spring.

Snowfall In the United States, snowfall has declined. Northern Hemisphere spring snow cover, North America maximum snow depth, snow water equivalent in the western United States, and extreme snowfall years in the southern and western United States have all declined. However, extreme snowfall years in parts of the northern states have increased. There has been a trend toward earlier snowmelt and a decrease in snowstorm frequency on the southern margins of snowy areas.

Winter storm tracks have shifted northward since 1950 over the Northern Hemisphere. Potential linkages between the frequency and intensity of severe winter storms in the United States and accelerated warming in the Arctic have been postulated (**Chapter 3**), but they are complex

---

[16] Wuebbles et al. (2017)

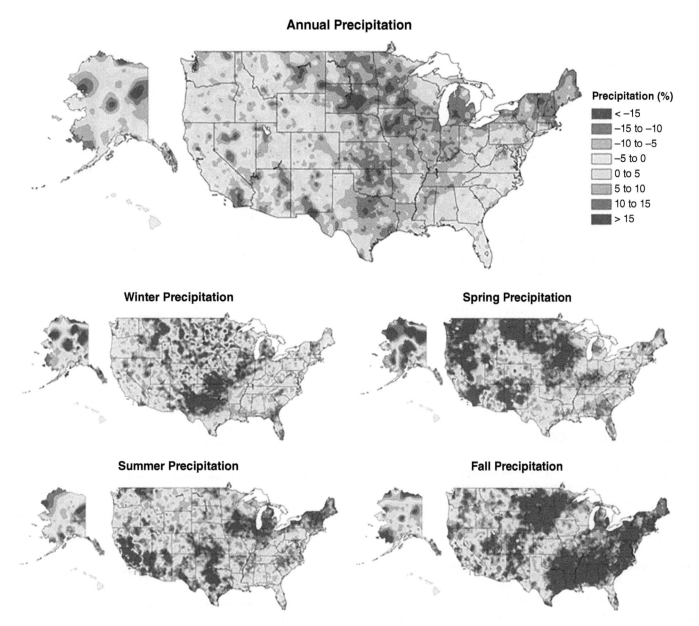

FIGURE 7.4 Annual and seasonal changes in precipitation over the United States. Changes are the average for present-day (1986–2015) minus the average for the first half of the last century (1901–1960 for the contiguous United States, 1925–1960 for Alaska and Hawai'i) divided by the average for the first half of the century.

Source: https://science2017.globalchange.gov/chapter/7/

and need further research to increase confidence in this connection.

Projections CMIP5 model projections for precipitation in the United States indicate both increases and decreases in precipitation by region. In winter and spring, the northern part of the country is projected to become wetter as global warming continues. In the early to middle parts of this century, this will likely be manifested as increases in snowfall.

By the latter half of the century, as temperature continues to increase, it will be too warm to snow in many current snow-producing locations, and precipitation will mostly be rainfall. In the Southwestern region, precipitation will decrease in the spring but the changes are only a little larger than natural variations. Many other regions of the country will not experience significant changes in average precipitation. This is also the case over most of the country in the summer and fall.

7.3 Climate Impacts to Planning Sectors

When planning for the future, decision makers typically consider seven planning sectors. These are water resources, energy supply and use, transportation, agriculture, ecosystems,

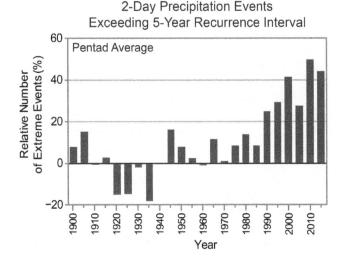

FIGURE 7.5 Number of 2-day extreme precipitation events exceeding the 5-year recurrence interval in the contiguous U.S., expressed as a percentage difference from 1901 to 1960 mean. The annual values are averaged over 5-year periods.

Source: https://science2017.globalchange.gov/chapter/7/

 human health, and society. Climate change affects each of these.[17]

7.3.1 Water Resources

The water cycle is powered by solar energy. When the troposphere becomes warmer—as is happening now—the process is accelerated. Evaporation rates increase, which increases the amount of moisture circulating in the atmosphere, leading to an increase in the frequency of intense rainfall and snowfall events, mainly over land areas. More precipitation falls as rain rather than snow, which causes the peak discharge of streams in the spring season to arrive earlier; the danger of summer drought increases; and there is less freshwater available in summer and fall for agriculture, drinking, and other human uses, when demand is highest. In addition, more precipitation comes in the form of heavier rains and snow storms rather than light events.[18]

Over the past 50 years, the heaviest 1 percent of rain events increased by nearly 20 percent.[19] In the United States, increases have been greatest in the Northeast region. For every 0.55 degree Celsius (1 degree Fahrenheit) rise in

temperature, the capacity of the atmosphere to hold water vapor increases by about 4 percent. In addition, with the expansion of the tropics and other climate zones, changes in atmospheric circulation shift storm tracks northward.

As a result, dry areas experiencing a decrease in storminess can become drier and wet areas experiencing an increase in storminess can become wetter. Because of this, the arid Southwest and south regions of the United States are projected to experience longer and more-severe droughts from the combination of increased evaporation and reductions in precipitation.[20] Precipitation and runoff are likely to increase in the Northeast and Midwest in winter and spring.

Regional changes in water resources can be summarized as follows:

1. Coastal Alaska, Yukon River Basin, Coastal British Columbia
   - Increased spring flood risks[21]
   - Retreat and disappearance of glaciers, leading to impacts on stream discharge and associated aquatic ecology
   - Flooding of coastal wetlands by sea-level rise
   - Changes in estuary salinity and ecology
2. Hawai'i[22]
   - Stronger El Niño and La Niña more frequent
   - Increase in tropical cyclones due to shift in tracks toward poles[23]
   - Average daily wind speeds slowly declining in Honolulu and Hilo
   - Consecutive wet days and consecutive dry days both increasing[24]
   - Heavy rainfall events and droughts more common: increasing runoff, erosion, flooding, and water shortages
   - Overall decline in rainfall over the past 30 years, with widely varying precipitation patterns on each island[25]
   - Windward sides of the major islands becoming cloudier and wetter
   - Dry leeward sides generally fewer clouds and less rainfall

[17] VIDEO: Most accurate models predict highest climate warming. https://climatecrocks.com/2017/12/18/most-accurate-models-predict-highest-climate-warming/

[18] VIDEO: How will climate change affect it?—The Water Cycle. https://www.youtube.com/watch?v=fI5b5bwpdVE

[19] Gutowski, W.J., et al. (2008) Causes of Observed Changes in Extremes and Projections of Future Changes. In Karl, T.R., et al. (eds.) *Weather and Climate Extremes in a Changing Climate: Regions of Focus: North America, Hawai'i, Caribbean, and U.S. Pacific Islands.* Synthesis and Assessment Product 3.3 (Washington, DC, U.S. Climate Change Science Program), pp. 81–116.

[20] Christensen, J.H., et al. (2007) Regional Climate Projections. In Solomon, S., et al. (eds) *Contribution of Working Group I to the Fourth Assessment Report (AR4) of the Intergovernmental Panel on Climate Change* (Cambridge, U.K., Cambridge University Press) pp. 847–940.

[21] Loukas, A. and Quick, M. (1999) The Effect of Climate Change on Floods in British Columbia, *Nordic Hydrology* 30: 231–256.

[22] Marra, J.J., and Kruk, M.C. (2017) State of Environmental Conditions in Hawai'i and the U.S. Affiliated Pacific Islands under a Changing Climate: https://coralreefwatch.noaa.gov/satellite/publications/state_of_the_environment_2017_hawaii-usapi_noaa-nesdis-ncei_oct2017.pdf

[23] Murakami, H., et al. (2013) Projected increase in tropical cyclones near Hawai'i. *Nature Climate Change*, v. 3, August, pp. 749–754.

[24] Kruk, M. C., et al. (2015), On the state of the knowledge of rainfall extremes in the western and northern Pacific basin, *Int. J. Climatol.*, 35(3), 321–336.

[25] Frazier, A.G. and Giambelluca, T.W. (2017) Spatial trend analysis of Hawaiian rainfall from 1920 to 2012. *Int. J. Climatol*, 37: 2522–2531, DOI: 10.1002/joc.4862

- Stream flow declined over past century, consistent with observed decreases in rainfall
- Declining groundwater levels.

3. Pacific Coast States: Alaska, Washington, Oregon, California
   - More winter rainfall and less snowfall
   - Earlier spring peak in stream runoff
   - Increased fall and winter flooding
   - Decreased summer water supply,[26] which is likely to produce changes in estuary and stream ecologies and negatively affect the availability of water for irrigation

4. Washington, Oregon, Idaho, Montana, Wyoming
   - Increasing drought and wildfire

5. California, Nevada, Utah, Arizona, New Mexico, parts of Oklahoma, and Colorado
   - Decreased runoff

6. Rocky Mountain Region[27]
   - Rise in the snow line, switch from snowfall to rainfall earlier in spring and late winter, earlier snowmelt
   - More frequent rain or snow
   - Earlier peaks in stream discharge and reductions in summer stream discharge and summer soil moisture
   - Likely decrease in surface runoff by mid-century
   - Rising stream temperatures[28] with impacts on stream ecology and species composition

7. Southwest[29]
   - Changes in snowpack and runoff leading to declines in groundwater recharge (freshwater used for drinking and irrigation)[30]
   - Increased stream water temperatures that will change aquatic ecosystems
   - Higher frequency of intense precipitation events and risk of flash floods

8. Midwest and Canadian Prairies[31]
   - Annual stream flow could increase in some areas and decrease in others, but a decrease in summer stream discharge is expected

- Severe drought[32] and heat waves are likely to become more frequent
- Semiarid zones may become drier

9. Arctic and Sub-Arctic Coasts of Alaska and Canada[33]
   - Arctic sea ice is retreating and thinning with each year, resulting in a 1- to 3-month extension of the annual ice-free seasons
   - Where possible, ecosystems will shift to more northerly and higher elevation conditions as they give way to ecosystems migrating from the south
   - Decreased ice cover produces a positive-feedback effect of reduced sunlight reflection, and increased ground heating could lead to acceleration of snow and ice cover loss

10. Great Lakes Region[34]
    - Precipitation increases, with lake level declines due to reduced snow pack melt and resulting runoff
    - Reduced ice cover, or some years with no ice cover
    - Changes in zooplankton and phytoplankton biomass
    - Northward migration of fish species and loss of cold water species
    - Reduced runoff[35] leading to reduced hydro power production and shallow shipping channels

11. Northeast and East[36]
    - Decreased snow cover and large reductions in stream flow
    - Rising sea level and decreasing sea ice could produce accelerated coastal erosion, saline intrusion into coastal aquifers, and changes in the magnitude and timing of ice freeze-up and break-up on lakes and cold coastal regions, which can affect spring flooding,[37] eliminate bog ecosystems, and shift the distribution and migration patterns of fish species

12. Southeast, Gulf, and mid-Atlantic regions
    - Heavily populated coastal floodplains at risk from flooding from extreme precipitation and hurricanes

[26] Melack, J.M., et al. (1997) Effects of Climate Change on Inland Waters of the Pacific Coastal Mountains and Western Great Basin of North America, *Hydrological Processes* 11: 971–992. See also Hamlet, A.F. and Lettenmaier, D.P. (1999) Effects of Climate Change on Hydrology and Water Resources in the Columbia River Basin, *Journal of the American Water Resources Association* 35, no. 6: 1597–1624.

[27] Williams, M.W., et al. (1996) Changes in Climate and Hydrochemical Responses in a High-Elevation Catchment in the Rocky Mountains, USA, *Limnology and Oceanography* 41, no. 5: 939–946.

[28] Hauer, F.R., et al. (1997) Assessment of Climate Change and Freshwater Ecosystems of the Rocky Mountains, USA and Canada, *Hydrological Processes* 11: 903–924.

[29] Hurd, B., et al. (1999) Relative Regional Vulnerability of Water Resources to Climate Change, *Journal of the American Water Resources Association* 35, no. 6: 1399–1410.

[30] U.S. Environmental Protection Agency (1998) *Climate Change and Arizona*. Publication EPA 236-F-98-007c (Washington, D.C., Environmental Protection Agency.

[31] Woodhouse, C.A. and Overpeck, J.T. (1998) 2000 Years of Drought Variability in the Central United States, *American Meteorological Society Bulletin* 79: 2693–2714.

[32] Wolock, D.M. and McCabe, G.J. (1999) Estimates of Runoff Using Water-Balance and Atmospheric General Circulation Models, *Journal of the American Water Resources Association* 35, no. 6: 1341–1350.

[33] Maxwell, B. (1997) *Responding to Global Climate Change in Canada's Arctic*, Vol. II of the Canada Country Study: Climate Impacts and Adaptation (Downsview, Ontario: Environment Canada).

[34] Hofmann, N.L., et al. (1998) *Climate Change and Variability: Impacts on Canadian Water*, Vol. II of the Canada Country Study: Climate Impacts and Adaptation (Downsview, Ontario: Environment Canada). 1–120.

[35] Chao, P. (1999) Great Lakes Water Resources: Climate Change Impact Analysis with Transient GCM Scenarios, *Journal of the American Water Resources Association* 35, no. 6: 1499–1508.

[36] Hare, F.K., et al. (1997) Climatic Variation over the Saint John Basin: An Examination of Regional Behavior, *Climate Change Digest*, CCD. 97-02, Atmospheric Environment Service, Toronto.

[37] Moore, M.V., et al. (1997) Potential Effects of Climate Change on Freshwater Ecosystems of the New England/Mid-Atlantic Region, *Hydrological Processes* 11: 925–947.

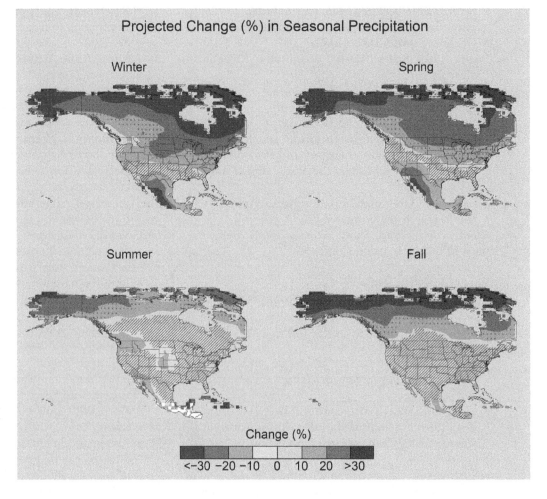

FIGURE 7.6 Projected change (%) in total seasonal precipitation from CMIP5 simulations for 2070–2099. The values are weighted multimodel means and expressed as the percent change relative to the 1976–2005 average. These are results for the higher scenario RCP8.5. Dots indicate that changes are large compared to natural variations. Hatching indicates that changes are small compared to natural variations. Blank regions are where projections are inconclusive.

Source: https://science2017.global change.gov/chapter/7/

- Lower base flow to streams, larger peak flows, and longer droughts possible, as well as precipitation increase and changes in runoff and stream discharge
- Prolonged drought of growing severity likely in the Southeast,[38] even while heavy rainfall increases threat of flooding
- Gulf of Mexico hypoxic zone and other impacts to coastal systems likely to continue to grow as a result of polluted runoff,[39] sea-level rise, accelerated coastal erosion, and saltwater intrusion to low-lying coastal plain aquifers
- Changes to estuarine and wetland ecosystems, biotic processes, and species distribution also possible.

Are U.S. states preparing for these changes in water resources? The answer is "some." One study[40] found that more than one in three counties in the United States could face a "high" or "extreme" risk of water shortages due to climate change by the middle of the 21st century. Every region of the United States is potentially vulnerable to adverse water-related impacts from climate change.[41]

Some states are taking action by reducing the greenhouse gas pollution that contributes to climate change and by planning for projected climate change-related impacts. However, many states are not. Nonetheless, the effects of climate change on the nation's water resources are already being observed. Warmer temperatures are causing changes to the water cycle that include: Changes in precipitation patterns and intensity; Increases in evaporation; Changes in runoff and soil moisture; Changes in the occurrence of drought; Widespread melting of snow and ice; Loss of lake and river ice, and; Rising water temperatures. **Figure 7.6**).

[38] Mulholland, P.J., et al. (1997) Effects of Climate Change on Freshwater Ecosystems of the Southeastern United States and the Gulf Coast of Mexico, *Hydrological Processes* 11: 949–970.

[39] Cruise, J.F., et al. (1999) Assessment of Impacts of Climate Change on Water Quality in the Southeastern United States, *Journal of the American Water Resources Association* 35, no. 6: 1539–1550.

[40] Roy, S., et al. (2012) Projecting Water Withdrawal and Supply for Future Decades in the U.S. under Climate Change Scenarios, *Environmental Science & Technology* 46, no. 5: 2545–2556, 120210153558000, doi: 10.1021/es2030774

[41] National Resources Defense Council (2012) Ready or Not: An Evaluation of State Climate and Water Preparedness Planning: https://assets.nrdc.org/sites/default/files/Water-Readiness-full-report.pdf?_ga=2.264404810.419588606.1527480352-1224335704.1522129701

7.3.2 Transportation

Transportation is a huge daily activity in North America. People and materials are moved by vast fleets of cars, trains, airplanes, trucks, and ships. The great majority of passenger travel occurs by automobile for shorter distances and by airplane for longer distances. In descending order, most types of cargo travel by railroad, truck, pipeline, or boat; air shipping is typically used for perishables and premium express shipments. According to the U.S. Department of Transportation, employment in the national transportation and material movement industry accounts for approximately 7.4 percent of all employment and more than $1 out of every $10 produced in the U.S. gross domestic product.

Climate change poses definite threats to transportation activities. According to a 2008 study by the National Research Council (NRC),[42] five categories of climate change are of particular concern: 1) Increases in very hot days and heat waves, 2) Decreases in Arctic temperatures, 3) Rising sea levels, 4) Increases in intense precipitation events, and 5) Increases in hurricane intensity. These changes in the environment will have significant effects on transportation, affecting the way professionals design and maintain the system of roads, airports, harbors, bridges, rail lines, and other elements that keep transportation moving. Decisions made today will affect how well the transportation system adapts to climate change in the future.[43]

Sea-Level Rise and Storm Surge According to NCA4, **global mean sea level** (GMSL) has risen by about 16–21 centimeters (7–8 inches) since 1900. Relative to the year 2000, GMSL is very likely to rise by 9–18 centimeters (0.3–0.6 feet) by 2030, 15–38 centimeters (0.5–1.2 feet) by 2050, and 30–130 centimeters (1.0–4.3 feet) by 2100. They report that future greenhouse gas emissions have little effect on projected GMSL rise in the first half of the century, but emissions scenarios do significantly affect projections of sea level for the second half of the century.

Emerging science regarding Antarctic ice sheet stability suggests that, if greenhouse gas emissions remain high, a GMSL rise exceeding 2.4 meters (8 feet) by 2100 is physically possible, although the probability of such an extreme outcome is unknown. Regardless of emissions, it is extremely likely that GMSL rise will continue beyond 2100.

Owing to changes in Earth's gravitational field resulting from melting of alpine glaciers, and the Greenland and Antarctic ice sheets, sea-level rise will vary from place to place. This is known as **relative sea-level rise** (RSL). Because of this effect, most of the U.S. coastline will see

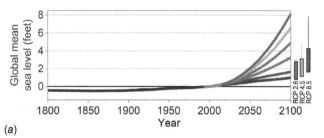

(a)

Projected Relative Sea Level Change for 2100 under the Intermediate Scenario

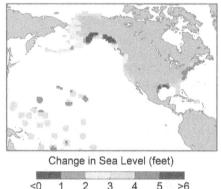

Change in Sea Level (feet)

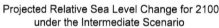

(b)   <0 1 2 3 4 5 >6

FIGURE 7.7 (a) Global mean sea level (GMSL) rise from 1800 to 2100 showing six scenarios. Also shown as the very likely ranges in 2100 for different RCPs (boxes), and lines augmenting the very likely ranges with new understanding of ice sheet stability. (b) Relative sea-level rise (ft) in 2100 projected for the intermediate scenario (1 m [3.3 ft] GMSL rise by 2100).[44]

Source: https://science2017.globalchange.gov/chapter/12/

RSL that is greater than the global average (**Figure 7.7**). In intermediate and low GMSL scenarios, RSL is likely to be less than the global average in much of the Pacific Northwest and Alaska. Almost all U.S. coastlines experience more than GMSL rise in response to Antarctic ice loss, and thus would be particularly affected under extreme GMSL rise scenarios involving substantial Antarctic mass loss.

Historically, the growth of communities has been tied to the transportation advantages of ports and harbors. But coastal areas are repeatedly assaulted by high waves and winds, storms, and tsunamis, and as sea level rises, the potential for these hazards to do even worse damage increases. In one study,[45] it was calculated that today's 100-year floods could instead occur every decade or two because of the effects of sea-level rise.

Along the Gulf Coast area, an estimated 3,862 kilometers (2,400 miles) of major roadway and 395 kilometers (246 miles) of freight rail lines are at risk of marine inundation within the next 50 to 100 years owing to a combination of

[42] Transportation Research Board (2008) *Special Report 290: Potential Impacts of Climate Change on U.S. Transportation*, Transportation Research Board, Division on Earth and Life Studies (Washington, D.C., National Research Council of the National Academies).
[43] Karl, T.R., et al. (2009) Global Climate Change Impacts in the United States, http://www.globalchange.gov/publications/reports/scientific-assessments/us-impacts

[44] Sweet, W.V., et al. (2017) Sea Level Rise. In Climate Special Report: Fourth National Climate Assessment, Volume I [Wuebbles et al. (2017)]
[45] Lin, N., et al. (2012) Physically Based Assessment of Hurricane Surge Threat under Climate Change, *Nature Climate Change* 2: 462–467, doi: 10.1038/nclimate1389

sea-level rise and land subsidence.[46] Because the Gulf Coast transportation network is interdependent and relies on minor roads and other low-lying infrastructure, the risks of service disruptions are likely to be even greater.

Sea-level rise causing marine and groundwater inundation of roads, railroads, airports, seaports, and pipelines would potentially affect commercial transportation activity valued in the hundreds of billions of dollars annually. The Transportation Research Board study[47] concluded that six of the nation's top 10 freight systems may be threatened by sea-level rise. Seven of the 10 largest ports are located on the Gulf Coast. The region is also home to the U.S. oil and gas industry, with its offshore drilling platforms, refineries, and pipelines. Roughly two-thirds of all U.S. oil imports are transported through the region.

Global climate change is very likely to change the nature of storms, especially hurricanes, and their impact on the coast. Some studies have identified an increase in storm frequency for some areas and a decrease in others.[48] Stronger and larger storms are likely due to increased air temperature and higher water vapor content. When coupled with sea-level rise, stronger storms will result in far-reaching and damaging effects (greater storm surge coupled with more flooding due to increased rainfall). An estimated 96,560 kilometers (60,000 miles) of coastal highway are already exposed to periodic flooding from coastal storms and high waves.[49]

Rainfall Intensity Although total precipitation has increased by only 5 percent, the heaviest 1 percent of events increased by 20 percent.[50] Intense precipitation can cause severe damage to transportation assets. For instance, the Great Flood of 1993 caused catastrophic flooding along 805 kilometers (500 miles) of the Mississippi and Missouri river system, paralyzing rail, truck, and marine traffic and affecting a fourth of all U.S. freight.

During the June 2008 Midwest flood, the second "100-year" flood in 15 years, dozens of levees in Iowa, Illinois, and Missouri were breached or overtopped, and the runoff inundated huge populated areas. Although highway and rail bridges largely survived, access and approach roads and rail lines were under water, and rail, roadway, and marine

transport was shut down for weeks. Events like these are likely to occur more often in a warming world.

If more precipitation falls as rain rather than snow in winter and spring, the increased runoff raises the risk of landslides, mudflows, stream floods, and rock falls, which can prompt road closures, more road repair, and reconstruction of rail lines and roadways. More-frequent heavy precipitation also causes increases in weather-related accidents, traffic and rail delays, and disruptions in a network already challenged by increasing congestion.[51] Local governments also need to anticipate and budget for the impact of increased flooding on evacuation routes, construction activities, urban infrastructure, and congested traffic locations (e.g., commuter choke points).

Extreme Heat As the world warms, the frequency and duration of days characterized by extreme heat will increase. Extreme heat, especially when 32.2 degrees Celsius (90 degrees Fahrenheit) and above for sustained periods (**Figure 7.8**), affects the transportation sector in several costly and potentially dangerous ways: Asphalt softens and can develop ruts from heavy traffic, affecting the safe operation of cars and trucks; railroad tracks can warp and deform, leading to speed restrictions and derailment in the worst cases; transportation vehicles of all types can overheat; and tires can deteriorate, leading to concerns about safe operation and raising maintenance costs.[52]

Extreme heat also raises the possibility of health and safety problems for highway workers, construction crews, and vehicle operators. The U.S. Occupational Safety and Health Administration states that concern over heat stress for moderate to heavy work begins at about 26.6 degrees Celsius (80 degrees Fahrenheit) and varies from place to place depending on humidity levels, urban heat island (the tendency of cities to be hotter than surrounding countryside) effects, and winds.[53]

Drought Drought accompanies rising air temperatures. Drought develops with decreasing precipitation and increasing evaporation which together create pronounced dry conditions. In a warmer world, the impact of increasing drought is likely to be severe in certain areas. In North America, even in those parts where total annual precipitation might not decrease, the frequency of rainfall and snowfall events are projected to drop.[54]

[46] Kafalenos, R.S., et al. (2008) What are the Implications of Climate Change and Variability for Gulf Coast Transportation? In Savonis, M.J., et al. (eds.), *Impacts of Climate Change and Variability on Transportation Systems and Infrastructure: Gulf Coast Study, Phase I.* Synthesis and Assessment Product 4.7. (Washington, DC, U.S. Department of Transportation) pp. 4–1 to 4F-27.

[47] Transportation Research Board (2008)

[48] Li, T., et al. (2010) Global Warming Shifts Pacific Tropical Cyclone Location, *Geophysical Research Letters* 37: L21804, doi: 10.1029/2010GL045124

[49] Transportation Research Board, 2008

[50] Kunkel, K.E., et al. (2008) Observed Changes in Weather and Climate Extremes. *Weather and Climate Extremes in a Changing Climate. Regions of Focus: North America, Hawai'i, Caribbean, and U.S. Pacific Islands. A Report by the U.S. Climate Change Science Program and the Subcommittee on Global Change Research,* Karl, T.R., et al. (eds).

[51] Potter, J.R., et al. (2008) Executive Summary. In Savonis et al. (eds.)

[52] C. B. Field, et al. (2007) North America. In M.L. Parry, et al. (eds.), Climate Change 2007: Impacts, Adaptation and Vulnerability. Contribution of Working Group II to the Fourth Assessment Report of the Intergovernmental Panel on Climate Change (Cambridge, U.K., Cambridge University Press), pp. 617–652.

[53] Occupational Safety and Health Administration (2008) Heat stress, in OSHA Technical Manual, Section III: Chap 4. Washington, DC., Occupational Safety and Health Administration, http://www.osha.gov/dts/osta/otm/otm_iii/otm_iii_4.html

[54] Gutowski et al. (2008)

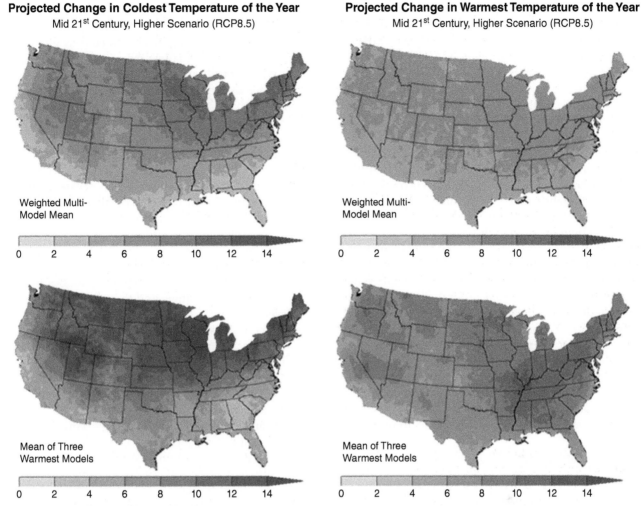

Projected Change in Coldest Temperature of the Year
Mid 21ˢᵗ Century, Higher Scenario (RCP8.5)

Projected Change in Warmest Temperature of the Year
Mid 21ˢᵗ Century, Higher Scenario (RCP8.5)

FIGURE 7.8 Projected changes in the coldest and warmest daily temperatures (°F) of the year in the contiguous U.S. Changes are the difference between the average for mid-century (2036–2065) and the average for recent past (1976–2005) under the higher scenario (RCP8.5). Maps in the top row depict the weighted multimodel mean, whereas maps on the bottom row depict the mean of the three warmest models (i.e., the models with the largest temperature increase). Maps are derived from 32 climate model projections that were statistically downscaled. Increases are statistically significant in all areas (i.e., more than 50% of the models show a statistically significant change, and more than 67% agree on the sign of the change).

Source: https://science2017.globalchange.gov/chapter/6/

Drought causes significant problems for transportation activities. For example, because of drought, wildfires are projected to grow in duration, frequency, and intensity, especially in the Southwest. These catastrophic events threaten communities and infrastructure and cause road and rail closures in affected areas.

Increased susceptibility to wildfires during droughts threatens roads and other transportation infrastructure directly, or it causes road closures because of fire threat or reduced visibility, such as has occurred in Texas, Oklahoma, New Mexico, Florida, California, and other states. Areas deforested by wildfires are also at increased susceptibility to mudslides. River transport is seriously affected by drought, with reductions in the routes available, shipping season, and cargo-carrying capacity.

Storminess Future projections[55] of tropical cyclones indicate that greenhouse warming will cause the globally averaged intensity of tropical cyclones to shift toward stronger storms, with intensity increases of 2 percent to 11 percent by 2100. Studies also project the globally averaged frequency of tropical cyclones to decrease by 6 percent to 34 percent.

Modeling studies project substantial increases in the frequency of the most intense cyclones and increases in the precipitation rate on the order of 20 percent within 100 kilometers (62 miles) of the storm center. As a result, the transportation sector would experience increased precipitation impacts, wind impacts, and storm surge impacts.

[55] Knutson, T., et al. (2010) Tropical Cyclones and Climate Change, *Nature Geoscience* 3 (2010): 157–163, doi: 10.1038/ngeo779

Stronger hurricanes have longer periods of intense precipitation, and the amount of rainfall is expected to be higher. Higher wind speeds lead to greater damage, and damage increases exponentially with wind speed.[56] Higher wind speeds and low air pressures in a storm produce higher storm surges and waves. These increases in transportation vulnerability require new methods of planning for future impacts, because statistics based on historical patterns are less likely to provide accurate projections of future conditions.[57]

Storms have costly results: a higher probability of infrastructure failures, such as damaged decking on bridges, washed away roads and rail lines, debris left on roads and rail lines, emergency evacuations, damage to signs and lighting fixtures, and reduction in the useful life of highways and rail lines exposed to flooding. On the Gulf Coast, more than one-third of the railways are likely to flood when hit by a storm surge of 5.5 meters (18 feet).[58]

Shipping is especially vulnerable to major storms. Freighters have to be diverted around storms, and their sailing schedules are delayed. Work on offshore drilling platforms, coastal pumping facilities, sewage-treatment plants, and other marine-dependent activities comes to a halt, and costly evacuations are needed. Infrastructure associated with these activities is heavily damaged by high winds, waves, and storm surge. Harbor infrastructure, such as cranes, docks, and other terminal facilities, damaged during severe storms costs billions of dollars to replace.

7.3.3 Ecosystems

An ecosystem is an interdependent system of plants, animals, and microorganisms. The natural resources that humans depend on are largely made possible by the healthy state of ecosystems around the world. The air we breathe, clean water, lumber, food, and even our safety from a number of natural hazards (such as landslides, flooding, and others) are due in part or in whole to a healthy planet made up of healthy ecosystems.

Although greenhouse gas emissions have only warmed the surface 1 degree Celsius (1.8 degrees Fahrenheit), impacts to environments and ecosystems are already severe and widespread, in part because they magnify existing stress that human communities have placed on natural systems. One study[59] found that across 65 percent of the terrestrial surface, land use and related pressures have caused biotic intactness to decline beyond 10 percent, considered a "safe" planetary boundary. Changes have been most pronounced in grassland biomes and biodiversity hotspots.

Health The key to a healthy ecosystem is the interdependence of its components, and if one or more of these components is negatively affected by global warming, the entire system is less robust and less resistant to stress. Without the support of the other organisms within their own ecosystem, life forms would not survive, much less thrive.

Changing Conditions High-altitude and high-latitude ecosystems across the world have already been negatively affected by changes in climate. The Intergovernmental Panel on Climate Change (IPCC) reviewed studies of biological systems and concluded[60] that 20 percent to 30 percent of species assessed may be at risk for extinction from climate change impacts in this century if global mean temperatures exceed 2 to 3 degrees Celsius (3.6 to 5.4 degrees Fahrenheit) relative to preindustrial levels. The USGCRP[61] concludes that ecosystem processes have been affected by climate change, and they document the following developments:

- Large-scale shifts have occurred in the ranges of species and the timing of the seasons and animal migration.
- Fires, insect pests, disease pathogens, and invasive weed species have increased.
- Deserts and semiarid lands are likely to become hotter and drier, feeding a self-reinforcing cycle of invasive plants, fire, and erosion.
- Coastal and near-shore ecosystems are already under multiple stresses. Climate change and ocean acidification will exacerbate these stresses.
- Arctic sea ice ecosystems are already being adversely affected by the loss of summer sea ice, and rapidly rising temperatures and further changes are expected.
- The habitats of some mountain species and coldwater fish, such as salmon and trout, are very likely to contract in response to warming.
- Some of the benefits that ecosystems provide to society will be threatened by climate change, and others will be enhanced.

Ecosystems are very sensitive to changing temperatures, shifts in precipitation, variations in seasonal timing, and other processes normally associated with climate change. These shifts in established patterns have a strong influence on the processes that control growth and development in ecosystems. Higher temperatures generally speed up plant growth, rates of decomposition, and how rapidly nutrients are cycled; however, factors, such as extreme temperatures, lack of water, soil desiccation, the spread of hardy weeds, and others also influence these rates.

[56] Landsea, C.W. (1993) A Climatology of Intense (or Major) Atlantic Hurricanes, *Monthly Weather Review* 121, no. 6: 1710–1713.

[57] Hay, J.E., et al. (2005) Climate Proofing: A Risk-based Approach to Adaptation, (Manila, Asian Development Bank), http://www.adb.org/Documents/Reports/Climate-Proofing/default.asp

[58] Kafalenos et al. (2008)

[59] Newbold, T., et al. (2016) Has land use pushed terrestrial biodiversity beyond the planetary boundary? A Global Assessment, *Science*, 15 July, v. 353, Is. 6296, 288–291, DOI:10.1126/science.aaf2201.

[60] Parry, M.L., et al. (eds.), Climate Change 2007: Impacts, Adaptation and Vulnerability. Contribution of Working Group II to the Fourth Assessment Report of the Intergovernmental Panel on Climate Change (Cambridge, U.K., Cambridge University Press).

[61] Karl et al. (2009)

1960–1990   2070–2100

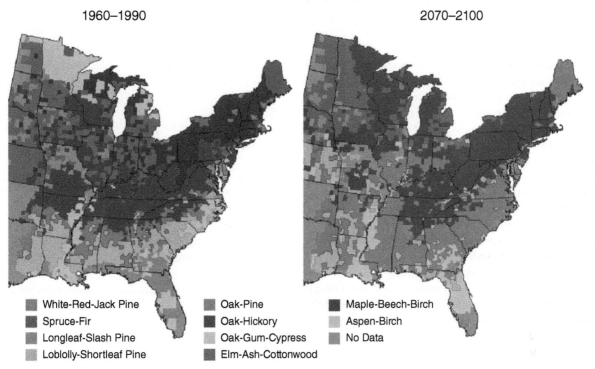

White-Red-Jack Pine   Oak-Pine   Maple-Beech-Birch
Spruce-Fir   Oak-Hickory   Aspen-Birch
Longleaf-Slash Pine   Oak-Gum-Cypress   No Data
Loblolly-Shortleaf Pine   Elm-Ash-Cottonwood

FIGURE 7.9 These maps show current and projected future forest types. Major changes are projected for many regions in the United States. For example, in the northeast, under a mid-range warming scenario, the currently dominant maple–beech–birch forest type is projected to be completely displaced by other forest types in a warmer future.

Source: USGCRP, https://downloads.globalchange.gov/usimpacts/allimages/7-Ecosystems/7-Hi%20res/7-Ecosystems-pg-81.jpg

Researchers[62] have observed that spring now arrives an average of 10 days to two weeks earlier than it did 20 years ago, and the growing season is lengthening over much of North America. Migratory bird species are returning earlier.

Species Shifts Northeastern birds that winter in the south now arrive back in the northeast an average of 13 days earlier than they did during the first half of the 20th century. Birds wintering in South America arrive back in the northeast an average of four days earlier. The range boundaries of species have shifted poleward, with a mean velocity of 6 kilometers (3.7 miles) per decade, as well as upward in elevation.[63]

Measurements indicate that forest growth has risen over the past several decades owing to young forests reaching maturity sooner, more carbon dioxide in the atmosphere, longer growing seasons, and increased deposition of nitrogen from the atmosphere.[64]

Climate change is also causing the geographic range of species to shift northward and upward in elevation. Trees, flowers, birds and insects face the arrival of spring at ever-earlier dates compared to the past 30 years, providing some of the clearest evidence that nature is responding to climate change. The timing of life-cycle events such as blooming, migration, and insect emergence has changed unevenly, however, prompting concern that further warming could disrupt interactions between species, such as feeding and pollination.[65] For example, the ranges of many butterfly species have expanded northward, contracted at the southern edge, and shifted to higher elevations as warming has continued.

In the future, forest tree species are expected to shift their ranges northward and upslope in response to climate change (**Figure 7.9**). Some common forest types, such as oak and hickory, are projected to expand; others, such as maple, beech, and birch, are projected to contract. Still others, such as spruce and fir, are likely to disappear from the United States altogether.[66] Although some forests might derive near-term benefits from an extended growing season, longer periods of hot weather could stress trees and make them more susceptible to wildfires, insect damage,

[62] Ryan, M.G., et al. (2009) Land Resources. In P. Backlund, et al. (eds.), The Effects of Climate Change on Agriculture, Land Resources, Water Resources, and Biodiversity in the United States, Synthesis and Assessment Product 4.3 (Washington, DC, U.S. Department of Agriculture), pp. 75–120.

[63] Parmesan, C and Yohe, G. (2003) A Globally Coherent Fingerprint of Climate Change Impacts Across Natural Systems, *Nature* 421, 37–42, doi: 10.1038/nature01286

[64] Ryan et al. (2009)

[65] Wilson, R. and Roy, D. (2011) Ecology: Butterflies Reset the Calendar, *Nature Climate Change* 1: 101–102, doi: 10.1038/nclimate1087

[66] Ryan et al. (2009)

and disease. Climate change has likely already increased the size and number of forest fires, insect outbreaks, and tree deaths, particularly in Alaska and the West.

Wildfire Global warming changes air temperature, precipitation patterns, and soil moisture. When it comes to precipitation changes, a general rule of thumb used by scientists is that dry areas become dryer, and wet areas become wetter. In fact, this pattern has already been documented by researchers.[67] In forested areas, where these changes combine to create dry conditions, the risk of wildfires grows.

Higher spring and summer temperatures, earlier spring snow-melt, and greater rates of evaporation cause soils to be drier for longer. The prospect of drought increases and creates a longer wildfire season, particularly in the western United States. Hot, dry conditions raise the likelihood that, once fires are started by lightning strikes or human error, they will be more intense and long-burning.[68]

The global temperature is increasing and the climate is changing due to the greenhouse gas emissions we have already produced. By engaging in mitigation efforts—creating buffer zones between human habitation and susceptible forests, and meeting home and city fire-safety standards—and by taking steps to reduce our impact on the climate, we can help to reduce the costs inflicted by wildfires. Ultimately, however, the only true solution is to reduce greenhouse gas emissions that lead to further climate change. Because reducing global warming is likely to take more than a century of effort, wildfire will continue to be a problem.

The incidence of large forest fires in the western United States and Alaska has increased since the early 1980s and is projected to further increase in those regions as the climate warms.[69] In the western United States, wildfires are occurring nearly four times more often, burning more than six times the land area, and lasting almost five times as long compared to the period 1970–1986 (**Figure 7.10**).[70]

Wildfire costs include the risk to human life, health impacts from poor air quality related to smoke and aerosols, property damage, state and federal dollars spent in fighting a fire, and damage to local ecosystems and hydrology. In most cases, these costs are very high and are very likely to increase because the occurrence of wildfire in the United States is rising faster than the ability of local authorities to suppress them.

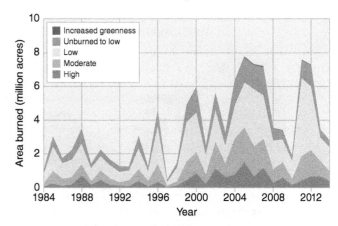

FIGURE 7.10 Distribution of acreage burned by large wildfires, based on the level of damage caused to the landscape–a measure of wildfire severity. Large wildfires are defined as fires with an area larger than 1,000 acres in the western United States and 500 acres in the eastern United States.

Source: https://www.epa.gov/climate-indicators/climate-change-indicators-wildfires

The economic costs of wildfires can be crippling. Between 2000 and 2009, the property damage from wildfires averaged $665 million per year. In addition, wildfires cost states and the federal government millions in fire-suppression management; the U.S. Forest Service's yearly fire-suppression costs have exceeded $1 billion dollars at least twice since FY 2000.

Climate variability, human activities such as land-use (clearing, development, mining) and fire exclusion, and climate change all influence the likelihood of wildfires. However, many of the areas that have seen these increases—such as Yosemite National Park and the Northern Rockies—are protected from or relatively unaffected by human land-use and behaviors. This suggests that climate change is a major factor driving the increase in wildfires.

Wildfire seasons (seasons with higher wildfire potential) in the United States are projected to lengthen, with the Southwest region lengthening from 7 months to 12 months (all year). Additionally, wildfires themselves are likely to become more severe. Moist, forested areas are the most likely to face greater threats from wildfires as conditions grow drier and hotter. However, some dry grassland areas may grow less at risk because they become so barren from wildfire stress that they lack even the fuel to feed a fire.

Additionally, a parallel aspect of wildfires—especially in the semi-arid southwest United States—is that when the rains do come, mountain forest wildfires increase flash flood risk in lower-lying areas in the days and months following the fire due to loss of vegetation and the inability of burned soil to absorb moisture. For example, residents in Arizona living along usually dry stream beds have lost property and life following such tragic and unexpected post wildfire flash floods.

[67] Skliris, N., et al. (2016) Global Water Cycle Amplifying at Less than the Clausius-Clapeyron Rate. *Scientific Reports* 6, 38752. doi: 10.1038/srep38752

[68] Union of Concerned Scientists https://www.ucsusa.org/global-warming/science-and-impacts/impacts/global-warming-and-wildfire.html#.WkVNVSOZOV4

[69] Wehner, M.F., et al. (2017) Droughts, floods, and wildfires. In: Climate Science Special Report: Fourth National Climate Assessment, Volume I [Wuebbles, D.J., et al. (eds.)]. U.S. Global Change Research Program, Washington, DC, USA, pp. 231–256, doi: 10.7930/J0CJ8BNN

[70] Westerling, H.G., et al. (2006) Warming and Earlier Spring Increase Western U.S. Forest Wildfire Activity, *Science*, 18 Aug, 940–943.

Marine Systems Global warming is changing marine ecosystems as well. Oceanic plankton and various species of marine fish are shifting northward into cooler water.[71] The timing of plankton blooms is changing, and coral reefs are experiencing stress related to warmer waters. In 2014–2016, the Caribbean basin experienced high water temperatures that resulted in widespread coral bleaching, with some sites seeing 90 percent of the coral bleached.

Coral bleaching occurs when symbiotic algae that provide coral polyps with food leave the polyp as a result of stress, such as high-water temperatures. Coral might not immediately die, and may recover if the microscopic algae return, but if the algae do not return the coral usually die within a period of months. Some corals begin to recover when water temperatures decrease, but often disease will appear, striking the previously bleached and weakened coral. Growth of fleshy, invasive algae on the bleached coral is another problem that can happen beginning immediately following bleaching, preventing any chance of recovery.

Severe Effects A number of studies[72] indicate that without major changes in the emission of greenhouse gases within the next decade, severe ecosystem effects are likely by the end of the century. If allowed to occur, the following conditions may have a negative impact on the quality of human life, health, and happiness:

- High temperature rise, especially over land—some 5 to 6°C (10°F) over much of the United States;
- Dust Bowl conditions over the U.S. Southwest and many other heavily populated regions around the globe;
- Sea-level rise of around 32 cm (1 ft) by 2050, then 80 cm to 1.8 m (2.8 to 6 ft) (or more) by 2100, rising some 15 to 30 cm (6 to 12 in) (or more) each decade thereafter;
- Species loss on land and sea—perhaps 50% or more of all biodiversity;
- More extreme weather;
- Loss of food security—the increasingly difficult task of feeding 8 billion, then 9 billion, and then 10 billion people in a world in an ever-worsening climate;
- Myriad direct health effects; and
- Unanticipated effects known as "unknown unknowns."

7.3.4 Agriculture

Agriculture is a key element in the development of human civilization. Prior to the Industrial Revolution, most humans labored for the production of food, animal, and plant goods and fuels from agriculture. Today one-third of the world's workers are still employed in agriculture; however, despite the size of the workforce, agricultural production accounts for less than 5 percent of the gross world product.

The major products of agriculture can be broadly categorized as food (e.g., cereals, vegetables, fruits, and meat), fibers (e.g., cotton, wool, hemp, silk, and flax), fuels (e.g., various biofuels, methane, ethanol, and biodiesel), and raw materials (e.g., lumber, bamboo, and plant resins). Food production, among the most important direct activities of agriculture, is an increasingly global concern, especially because of tensions arising from human population growth, global warming impacts to soil and water availability, and international trade.

Impacts on Crops Crops respond to changing climate based on the interrelationship of three factors: rising temperatures, changing water resources, and increasing carbon dioxide. Warming air temperatures and increased carbon dioxide generally cause plants that are below their optimum temperature to grow faster, thus producing benefits in the form of increased yields (yield is a measure of agricultural output); however, for some plants, such as cereal crops, faster growth means there is less time for the grain itself to grow and mature, and instead the stalk grows faster. This leads to lower yield. Many weeds, insects, and pathogens also respond positively to increased warmth and CO_2 levels.

Humans get 75 percent of their food—either directly, or indirectly as meat—from four crops: maize (corn), wheat, rice and soybeans. The world's rising population—likely to reach 10.1 billion before the end of the century—has access to food thanks to increasing yields of these four crops. But our preference for burning fossil fuels is cutting crop production and nutritional value.[73] For instance, cereals and grasses will have reduced zinc and iron levels under the elevated CO_2 conditions predicted for the middle of this century.[74]

Climate change may actually benefit some plants by lengthening growing seasons and increasing CO_2 available for photosynthesis. Yet, other effects of global warming, such as increased pests, droughts, and flooding, will be less benign. In a warmer world, yields of wheat are expected to decrease by 6 percent, rice by 3.2 percent, maize by 7.4 percent, and soybean by 3.1 percent.[75]

Research suggests that extra CO_2 improved yields for these crops by roughly 3 percent over the past 30 years. Unfortunately, in the case of wheat and maize, that wasn't enough to overcome losses caused by warm temperatures.

[71] Janetos, A., et al. (2010) Biodiversity. In P. Backlund, et al. (eds.), *The Effects of Climate Change on Agriculture, Land Resources, Water Resources, and Biodiversity in the United States, Synthesis and Assessment Product 4.3* (Washington, DC, U.S. Department of Agriculture), pp. 151–181.

[72] Sokolov, A.P.. et al. (2009) Forest, Probabilistic Forecast for Twenty-First-Century Climate Based on Uncertainties in Emissions (Without Policy) and Climate Parameters, *Journal of Climate* 22: 5175–5204, doi: 10.1175/2009JCLI2863.

[73] Lobell, D.B., et al. (2011) Climate Trends and Global Crop Production Since 1980. *Science* (05 May) 1204531, DOI:10.1126/science.1204531

[74] Myers, S.S., et al. (2014) Increasing CO_2 threatens human nutrition, *Nature*, 510, 139–142, doi: 10.1038/nature13179.

[75] Zhao, C., et al. (2017) Temperature increase reduces global yields of major crops in four independent estimates *PNAS* 114 (35) 9326–9331; doi:10.1073/pnas.1701762114

Temperature problems are overriding CO_2 benefits. Climate-related losses contribute as much as 18.9 percent to the average price of a given crop. Climate change is not disastrous to global food security, but it is a multibillion-dollar-per-year effect.

Loss of crop yield and nutritional value are bad news. But, a more insidious food problem is lurking. Research shows that rising CO_2 deteriorates the quality of the plants we eat.[76] Every leaf and every blade of grass on Earth makes more sugar as CO_2 levels rise. Mathematical biologist Irakli Loladze has said "We are witnessing the greatest injection of carbohydrates into the biosphere in human history—[an] injection that dilutes other nutrients in our food supply."[77]

Across nearly 130 varieties of plants and more than 15,000 samples collected from experiments over the past three decades, the overall concentration of minerals such as calcium, magnesium, potassium, zinc, and iron had dropped by 8% on average. The ratio of carbohydrates to minerals is going up. Plants that humans rely on for basic nutrition are becoming junk food.

Some non cereal crops show positive responses to elevated carbon dioxide and low levels of warming, but higher levels of warming negatively affect the growth of food plants. This is because soil moisture is decreased, competition with invasive weeds increases costs, and the combined effect of these competing factors leads to diminished yields. Analysis of crop responses suggests that even a moderate increase in temperature will decrease yields of corn, wheat, sorghum, bean, rice, cotton, and peanut crops in the United States.

As a result of global warming, extreme weather events such as heavy downpours and droughts, extreme temperature days, and an early end to winter are growing in frequency.[78] These reduce crop yields because excesses or deficits of water have negative impacts on plant growth.[79] Rain and snowfall have become less frequent but more intense, a pattern that is projected to continue across the United States.[80]

Excessive rainfall delays spring planting, which jeopardizes profits related to early season production of high-value crops, such as melon, sweet corn, and tomatoes. When flooding causes fields to become unusable during the growing season, crop losses occur owing to low oxygen levels in the soil, increased susceptibility to root diseases, and increased soil compaction due to the use of heavy farm equipment on wet soils.

For instance, in spring 2008, heavy rains caused the Mississippi River to rise to about 2.1 meters (7 feet) above flood stage. Hundreds of thousands of acres of cropland were inundated just as farmers were preparing to harvest wheat and plant corn, soybeans, and cotton, with net losses estimated at around $8 billion.[81] Some farmers were put out of business and others required many years to recover. The flooding caused severe erosion in some areas and also caused an increase in runoff and leaching of agricultural chemicals into surface water and groundwater.

Weeds, insect pests, and various types of crop and animal diseases benefit from warming, and weeds also benefit from higher carbon dioxide concentration. Some historically aggressive weeds have been confined to the South because they cannot cross certain winter temperature thresholds. For instance, the kudzu vine has invaded 2.5 million acres of the Southeast and is a carrier of the fungal disease soybean rust, which represents a major and expanding threat to U.S. soybean production.

Sixty-four percent of the southern soybean crop is lost each year to weeds, whereas only 22 percent is lost on farms to the north. As winter warming increases (**Figure 7.11**), these weeds will find a foothold on northern farmland. Stress on crop plants will increase, requiring more attention to pest and weed control. As pesticide and herbicide use increases, so will costs and consumer prices.

Effects on Farm Animals High temperature and high humidity stress animals, too. Milk production declines in dairy operations, it takes longer for cows in meat operations to reach their target weight, the conception rate in cattle falls, and swine growth rates decline due to heat. Swine, beef, and milk production are all projected to decline in a warmer world.

Cool night air allows animals stressed by heat to recover. However, global warming causes nights to increase in temperature as rapidly as the days thus there is no relief at night. Heat from heat waves does not lift at night and livestock, unable to recover, have died. (Individual states have reported losses of 5,000 head of cattle in a single heat wave.) Warmer winter temperatures, the early arrival of spring, and summer heat also increase the presence of parasites and disease pathogens. The cost of new housing facilities, treatments, food types, medicines, and other animal care logistics necessary to cope with these new stresses is passed on to the consumer.

A warming world poses challenges to crop and livestock production. Because some plant species respond positively to warmer conditions and higher CO_2 concentrations,

[76] Loladze, I. (2014) Hidden shift of the ionome of plants exposed to elevated CO2 depletes minerals at the base of human nutrition, *eLife* doi:10.7554/eLife.02245

[77] Evich, H.B. (2017) The Great Nutrient Collapse. Politico, Sept. 13, http://www.politico.com/agenda/story/2017/09/13/food-nutrients-carbon-dioxide-000511

[78] Medvigy, D. and Beaulieu, C. (2011) Trends in Daily Solar Radiation and Precipitation Coefficients of Variation Since 1984, *Journal of Climate* 25: 1330–1339, doi: 10.1175/2011JCLI4115.1

[79] Kahn, B. (2012) Innovative Farmers Look to Climate Forecasts for an Edge, NOAA Climate Services, *Climate Watch Magazine* http://www.climatewatch.noaa.gov/article/2012/innovative-farmers-look-to-climate-forecasts-for-an-edge

[80] Kunkel et al. (2008)

[81] National Climatic Data Center, "Climate of 2008: Midwestern U.S. Flood Overview," http://www.ncdc.noaa.gov/oa/climate/research/2008/flood08.html

Projected Changes in Key Climate Variables Affecting Agricultural Productivity

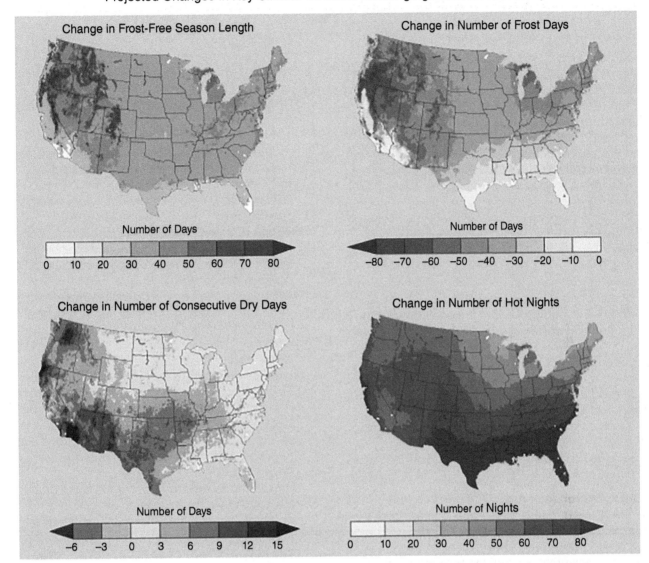

FIGURE 7.11 Projected changes in key climate variables affecting agricultural productivity for the end of the century (2070–2099) compared to 1971–2000. Changes in climate parameters critical to agriculture show lengthening of the frost-free or growing season and reductions in the number of frost days (days with minimum temperatures below freezing), under an emissions scenario that assumes continued increases in heat-trapping gases. On the frost-free map, white areas are projected to experience no freezes for 2070–2099, and gray areas are projected to experience more than 10 frost-free years during the same period. In the lower left graph, consecutive dry days are defined as the annual maximum number of consecutive days with less than 0.01 in of precipitation. In the lower right graph, hot nights are defined as nights with a minimum temperature higher than 98% of the minimum temperatures between 1971 and 2000.

Source: USGCRP, http://www.globalchange.gov/resources/gallery?func=viewcategory&catid=1

agriculture may be one of the sectors most adaptable to climate change. Even so, increased heat, pests, water stress, diseases, flooding, and weather extremes will require crop and livestock production to adapt.

7.3.5 Society

Population Growth Human population growth and vulnerability to climate impacts are inextricably linked. In late October 2011, the world population reached 7 billion; in just over 10 years, more than one billion people have been added to the human race, the same number gained over the entire 19th century. Although there is evidence that the rate of population growth will slow, it is far from stabilizing: UN predictions indicate that by mid-century, the total number will have reached somewhere between 8 and 10.5 billion.

The consequences of expanding human population look severe for a world with increasingly stressed energy and food supplies. Most of the two to three billion people born between now and 2050 will live in the cities and towns

Box 7.1 | Spotlight on Climate Change

Surprises in the Climate System

Scientific studies have revealed that Earth's climate system is a complex mosaic of physical processes that interact in complicated ways across a range of time and space scales. Negative feedbacks, also known as **self-stabilizing cycles**, can dampen changes. For instance, warming of the Southern Ocean might lead to more evaporation and thus more snowfall on Antarctica, stabilizing sea-level rise even as the ocean grows warmer. Positive feedbacks, also known as **self-reinforcing cycles**, can magnify changes. For instance, melting Arctic sea ice causes Arctic amplification that may lead to an unstable jet stream that produces more frequent and deeper heat waves in the Arctic, melting more sea ice.

Some components of the climate system, such as sea ice or ocean pH, may have thresholds beyond which positive feedbacks drive the component into a fundamentally different state. State shifts may have high consequences, could occur rapidly, and may be unstoppable. These are called *tipping points*. Tipping points may exceed projections by climate models, and thus would occur as surprises. Scientists are worried about tipping points because they hold the potential for devastating consequences during the coming century.[82]

Here is a list of some of the tipping points that scientists worry about.[83]

Disappearance of Arctic Summer Sea Ice—As Arctic sea ice disappears, dark ocean water absorbs heat and releases dissolved methane and carbon dioxide, reinforcing Arctic amplification. Some scientists fear that this self-reinforcing cycle has already passed a tipping point, and Arctic summers will be ice-free before mid-century.

Melting Greenland Ice Sheet—Loss of Arctic sea ice might produce rapid Arctic warming. Rapid warming may push the Greenland Ice Sheet to melt faster than expected, with very dangerous consequences for coastal communities.

Disintegration of the West Antarctic Ice Sheet—Warming ocean water is causing floating ice shelves to retreat into water-filled valleys, a self-reinforcing cycle. There is evidence that this is already an unstoppable process.

Collapse of Coral Reefs—Ocean warming and acidification may lead to annual coral bleaching by mid-century. Research predicts that most of our remaining coral systems will collapse even before a global temperature rise of 2°C (3.6°F).

Dieback of the Amazon Rainforest—Deforestation, lengthening of the dry season, and increased summer temperatures each place stress on rainfall in the Amazon. The dieback of the Amazon Rainforest ultimately depends on regional land-use management, and on how magnification of both La Niña and El Niño will influence future precipitation patterns.

Dieback of Boreal Forests—Increased water and heat stress, wildfires, and permafrost melt could lead to a decrease in boreal forest cover by up to half of its current size.

Weakening of the Marine Carbon Pump—Marine plankton store carbon by photosynthesis and shell building. As oceans become warmer, more acidic, and anoxic, this natural carbon storage could be threatened.

Disruption of Ocean Circulation Patterns—Atlantic Meridional Overturning Circulation has slowed by 20%. This threatens the collapse of Atlantic Deep-Water Formation. Some models suggest that this will cause heat to build-up in the tropics and south Atlantic leading to super storms and accelerated Antarctic melting.

Atmospheric Super-rotation—As the band around the equator warms, intense upward rising warm air adds energy to the winds aloft. If they speed up and become faster than Earth's rotation, they can trap water vapor. This super-rotation state can confine water vapor within this band and reduce cloud formation elsewhere. Loss of sunlight reflectance from clouds can increase climate response to greenhouse gases.

Release of Marine Methane Hydrates—Frozen methane on the ocean floor is threatened by warming ocean temperatures. If disrupted, the released methane could cause a rapid spike in air temperature.

Ocean Anoxia—Warming of the ocean surface reduces seawater mixing with the atmosphere. Release of fertilizers into major rivers causes localized eutrophic conditions. Together, these two effects cause regions of the ocean to be depleted in oxygen.

---

[82] Some content from NCA4 Chapter 15 https://science2017.globalchange.gov/chapter/15/

[83] Some content from http://blogs.edf.org/climate411/2017/11/01/everything-you-need-to-know-about-climate-tipping-points/

Stronger El Niño and La Niña States—Ninety percent of the extra heat trapped on Earth's surface by greenhouse gases is absorbed by the oceans. One consequence of this oceanic heat uptake is a gradual transition to more intense and permanent El Nino/Southern Oscillation (ENSO) conditions, with implications including extensive drought throughout Southeast Asia, South America, and beyond.

Permafrost Melting—As global temperatures rise and the high latitudes experience amplified warming, melting permafrost gradually releases carbon dioxide and methane into the atmosphere and creates a feedback for even more warming.

Tundra Transition to Boreal Forest—Much like the conversion of the Amazon Rainforest and boreal forests to other biomes, tundra environments may transition into forests as temperatures increase.

Extreme Warm/Dry and Warm/Wet Conditions—Analysis of billion-dollar disasters since 1980 reveals that the costliest climate and weather events are related to temperature and precipitation extremes (**Figure 7.12**). Already hot summers have become more frequent and droughts more intense.[84]

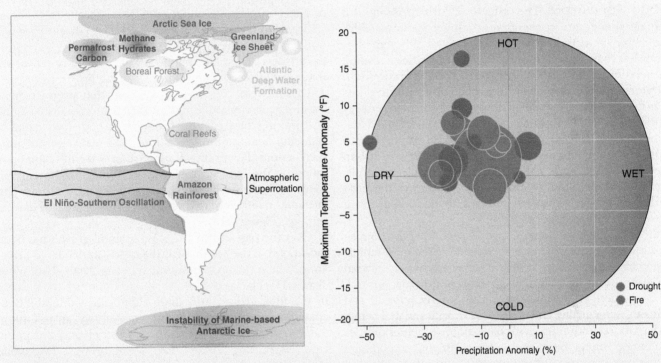

FIGURE 7.12   Potential climate tipping elements affecting the Americas. Right—Wildfire and drought events from the NOAA Billion Dollar Weather Events list (1980–2016), and associated temperature and precipitation anomalies. Dot size scales with the magnitude of impact, as reflected by the cost of the event. These high-impact events occur preferentially under hot, dry conditions.

Source: https://science2017.globalchange.gov/chapter/15/

of low-income countries in Africa and Asia, where fertility rates continue to be high. What makes this especially worrisome is that the areas likely to experience most growth are also those likely to be most affected by climate change and least able to cope with the extra demand on resources. At present, 13 percent of the world's population lives in at-risk coastal areas, and 75 percent of those people are located in Asia. Just over 50 percent of the world's inhabitants now live in urban areas; this number will rise to almost 69 percent by 2050.[85]

Globally, one billion people are now suffering from food shortages and water scarcity, a figure that could triple within 40 years. Yet, the planet is already feeling the strain of its seven billion human inhabitants, documented in lost species, rivers that have run dry, air that is polluted with chemicals, and sprawling development that has degraded ecosystems. If those being born between now and 2050 continue on the

[84] Hao, Z., et al. (2013) Changes in Concurrent Monthly Precipitation and Temperature Extremes, *Environmental Research Letters* 8, 034014, doi: 10.1088/1748-9326/8/3/034014

[85] A Scary Statistic, *Nature Climate Change* 1 (2011), doi: 10.1038/nclimate1255

development path of the present population, they will experience these effects to an even greater extent. They could also experience a lesser-appreciated outcome of the relationship of extreme climate and overcrowding: conflict.

Researchers[86] have found evidence of a link between civil unrest and higher temperatures; one result showed that 21 percent of 234 conflicts during the period 1950 to 2004 were probably set off as a result of the high temperatures associated with El Niño events. Much of the civil conflict that arises globally is also sparked by resource scarcity and access issues. As the climate warms and populations grow, regions that are hotspots for both are more likely to become conflict zones. This in turn will lead to more asylum seekers migrating to stable nations such as the United States.[87]

Urban Populations The effects of urbanization and climate change are converging in dangerous ways. Cities are major contributors to climate change: although they cover less than 2 percent of Earth's surface, cities consume 78 percent of the world's energy and produce more than 60 percent of all carbon dioxide and significant amounts of other greenhouse gas emissions, mainly through energy generation, vehicles, industry, and biomass use. At the same time, cities and towns are heavily vulnerable to climate change. Hundreds of millions of people in urban areas across the world will be affected by rising sea levels, increased precipitation, inland floods, more frequent and stronger cyclones and storms, and periods of more extreme heat and cold.

In the United States, more than 80 percent of the population resides in urban areas. They are among the most rapidly changing environments on Earth[88] and are host to myriad social problems, including neighborhood degradation, traffic congestion, crime, unemployment, poverty, poor air and drinking water quality, and inequities in health and well-being.

Urban communities also have unique vulnerabilities to climate change because they are analogous to complex ecosystems consisting of multifaceted and interconnected regional and national economies and infrastructure. The growth in size and complexity compound the impact of increased heat, water shortages, and extreme weather events. The negative influence of these stressors is intensified by the aging infrastructure, buildings, and populations that abound in cities; however, urban settings also present opportunities for adaptation through technology, infrastructure, planning, and design.[89]

Because cities absorb, produce, and retain more heat than the surrounding countryside, they alter local climates through the **urban heat island** effect. This process has raised average urban air temperatures by 1.1 to 2.7 degrees Celsius (2 to 5 degrees Fahrenheit) more than surrounding areas over the past century and by up to 11 degrees Celsius (20 degrees Fahrenheit) more at night.[90] These temperature increases, on top of warmer air induced by global warming, affect the health, comfort, energy costs, air quality, water quality and availability, and even violent crime rate in urban areas.[91]

Sea-level rise, storm surge, and increased hurricane intensity, projected to grow worse in future decades, are all looming threats for coastal cities. New Orleans, Miami, and New York are particularly at risk, and they would have difficulty coping with the sea-level rise projected by the end of the century under a higher emissions scenario (1 meter [3.3 feet] or more of higher sea level).[92] Analyses of population centers in the U.S. northeast indicate that the potential impacts of climate change are likely to be negative, but that policy changes can reduce vulnerability.[93]

Urban areas concentrate activities that produce heat-trapping emissions and thus afford advantages in managing and limiting these gasses.[94] Cities have a large role to play in reducing heat-trapping emissions, and many are pursuing such actions. For example, more than 900 cities have committed to the U.S. Mayors' Climate Protection Agreement to advance emissions-reduction goals by making transportation more efficient and relieving stress on nearby natural settings (**Figure 7.13**).[95]

Over the past century, U.S. population growth has been most rapid in the South, near the coasts, and in large urban areas. The four most populous states in 2000—California, Texas, Florida, and New York—account for 38 percent of the total growth and share significant vulnerability to coastal storms, severe drought, sea-level rise, air pollution, water shortages, and urban heat island effects. But population is shifting toward the Mountain West (Montana, Idaho, Wyoming, Nevada, Utah, Colorado, Arizona, and New Mexico), a region projected to experience a population increase of 65 percent from 2000 to 2030 and representing one-third of all U.S. population growth.

[86] Hsian, S.M., et al. (2011) Civil Conflicts are Associated with Climate Change, *Nature* 476, 438–441, doi: 10.1038/nature10311

[87] Refugees. http://blogs.ei.columbia.edu/2017/12/21/hotter-temperatures-will-accelerate-migration-europe-says-study/

[88] van Kamp, I., et al. (2003) Urban Environmental Quality and Human Well-Being: Towards a Conceptual Framework and Demarcation of Concepts: A Literature Study, *Landscape and Urban Planning* 65 (1–2): 5–18.

[89] Wilbanks, T.J., et al. (2008) Effects of Global Change on Human Settlements. In Gamble, J.L. (ed.), *Analyses of the Effects of Global Change on Human Health and Welfare and Human Systems, Synthesis and Assessment* Product 4.6 (Washington, DC, U.S. Environmental Protection Agency), pp. 89–109.

[90] Grimmond, S. (2007) Urbanization and Global Environmental Change: Local Effects of Urban Warming, *Geographical Journal* 173, no. 1: 83–88.

[91] Anderson, C.A. (2001) Heat and Violence, *Current Directions in Psychological Science* 10, no. 1: 33–38.

[92] Vermeer, M. and Rahmstorf, S. (2009) Global Sea Level Linked to Global Temperature, *PNAS*, 106, no. 51: 21527–21532, http://www.pnas.org/content/106/51/21527.full

[93] Rosenzweig, C. and Solecki, W. (eds.), Climate Change and a Global City: The Potential Consequences of Climate Variability and Change—Metro East Coast (New York, Columbia Earth Institute, 2001).

[94] Gamble, J.L., et al. (2008) Introduction. In, *Analyses of the Effects of Global Change on Human Health and Welfare and Human Systems, Synthesis and Assessment*, Product 4.6 (Washington, DC, U.S. Environmental Protection Agency), pp. 13–37.

[95] United States Conference of Mayors, "U.S. Conference of Mayors Climate Protection Agreement," as endorsed by the 73rd Annual U.S. Conference of Mayors meeting, Chicago, 2005. http://usmayors.org/climateprotection/agreement.htm

FIGURE 7.13 There are many steps that cities can take to be more sustainable. The term "ecocity" is applied to cities inhabited by people who are dedicated toward minimizing energy, water, food, waste, release of heat, air pollution (e.g., CO_2, methane), and water pollution. For instance, green roofs alter the surface energy balance and can help reduce the urban heat island effect. Incorporating green roofs in a design will help with air quality, climate, and water runoff. This green roof is on Chicago City Hall.

Source: https://commons.wikimedia.org/wiki/File:Chicago_City_Hall_green_roof_edit.jpg

Wikimedia Commons

Simultaneously, populations of southern coastal areas on the Atlantic and on the Gulf of Mexico are projected to continue to grow. As a result, more Americans will be living in the areas most vulnerable to the effects of climate change including flooding, decreased water resources, shifts in weather patterns toward more extreme events, and increased sea-level rise and storm surge.

Heat waves and poor air quality are projected to increase in a warmer world. Atmospheric conditions that produce heat waves are often accompanied by stagnant air and poor air quality. The simultaneous occurrence of these factors plus drought negatively affects quality of life, especially in cities. Poor air quality resulting from the lack of rainfall, high temperatures, and stagnant conditions during a heat wave can lead to a rise in unhealthy air quality days throughout large parts of the country. Climate change is projected to increase the likelihood of such episodes.[96]

7.3.6 Energy Supply and Use

Energy is at the heart of the global warming challenge. The production and use of energy, and the resultant greenhouse gas emissions, are the primary cause of global warming. Climate change, in turn, will affect our production and use of energy.

U.S. Emissions The largest source of greenhouse gas emissions from human activities in the United States is from burning fossil fuels for electricity, heat, and transportation. Since 1990, U.S. greenhouse gas emissions have increased by about 4 percent. From year to year, emissions can rise and fall due to changes in the economy, the price of fuel, and other factors. In 2015, U.S. greenhouse gas emissions decreased compared to 2014 levels. This decrease was largely driven by a decrease in emissions from fossil fuel combustion, which was a result of multiple factors including substitution from coal to natural gas consumption in the electric power sector; warmer winter conditions that reduced demand for heating fuel in the residential and commercial sectors; and a slight decrease in electricity demand. From 2015 to 2016, and again in 2017, U.S. emissions continued to decline. U.S. carbon dioxide emissions from energy sources hit a 25-year low in 2017 as the power sector burned less carbon-intensive coal and more low-cost natural gas (**Figure 7.14**).[97]

As global warming continues, in most American cities, the demand for air conditioning will grow while the demand for space heating will decrease. Studies[98] find that the demand for cooling energy increases from 5 percent to 20 percent per 1 degree Celsius (1.8 degrees Fahrenheit) of warming. Simultaneously, the demand for heating energy drops by 3 percent to 15 percent. These are only partially offsetting trends, and the net change is an increase in demand.

[96] Leung, L.R., and Gustafson Jr., W.I. (2005) Potential Regional Climate Change and Implications to U.S. Air Quality, *Geophysical Research Letters* 32: L16711, doi: 10.1029/2005GL022911

[97] U.S. Carbon Emissions. https://www.reuters.com/article/usa-natgas-eia-steo/update-1-u-s-carbon-emissions-seen-at-25-year-low-in-2017-idUSL1N1J311B

[98] Bull, S.R., et al. (2007) Effects of Climate Change on Energy Production and Distribution in the United States. In Wilbanks, T.J., et al. (eds.), *Effects of Climate Change on Energy Production and Use in the United States. Synthesis and Assessment,* Product 4.5 (Washington, DC, U.S. Climate Change Science Program), pp. 45–80.

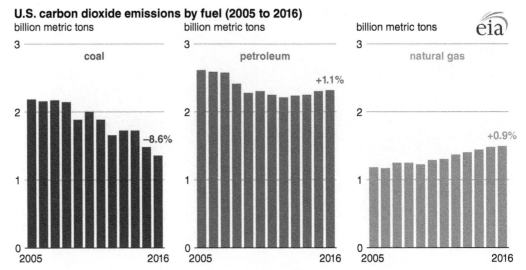

U.S. carbon dioxide emissions by fuel (2005 to 2016)

FIGURE 7.14 U.S. energy-related carbon dioxide (CO_2) emissions in 2016 totaled 5,170 million metric tons (MMT), 1.7% below their 2015 levels, after dropping 2.7% between 2014 and 2015. These recent decreases are consistent with a decade-long trend, with energy-related CO_2 emissions 14% below the 2005 level in 2016.

Source: https://www.eia.gov/todayinenergy/detail.php?id=30712

Cooling a building is primarily powered by electricity. This can be supplied by renewable energy sources such as hydropower, solar and wind power, geothermal energy, and traditional carbon-based power sources. Heating is supplied primarily by natural gas and fuel oil. Because approximately 30 percent[99] of the nation's electricity is currently generated from coal, these factors together have the potential to increase total national carbon dioxide emissions. However, improved energy efficiency, development of non-carbon energy sources, and/or carbon capture and storage technologies can combine to reduce emissions.

Climate change also places stress on the energy production network of human communities. Generation of electricity in thermal power plants (coal, nuclear, gas, or oil) is water intensive. Power plants rank only slightly behind irrigation in terms of freshwater withdrawals in the United States. There is a high likelihood that water shortages will limit power plant electricity production in many regions. By 2025, water limitations on electricity production in thermal power plants are projected for Arizona, Utah, Texas, Louisiana, Georgia, Alabama, Florida, California, Oregon, and Washington state.

Extreme Events A warmer climate is characterized by more-extreme weather events such as windstorms, ice storms, floods, tornadoes, and hail.[100] As a result, the transmission systems of electric utilities could experience a higher rate of failure. Development of new energy facilities could be restricted by siting concerns related to sea-level rise, exposure to extreme events, and increased costs resulting from a need to provide greater protection from extreme events.

Power plant operations can be affected by extreme heat waves. For example, intake water that is normally used to cool power plants becomes so warm during extreme heat events that it endangers power plant operations. High demand for cooling can overwhelm electricity production, causing blackouts. In the summer heat wave of 2006, for example, electric power transformers failed in several areas (including St. Louis, Missouri, and Queens, New York) as a result of high temperatures, causing interruptions of electric power supply. During the record-setting heat wave of 2012, a rolling windstorm called a "derecho"[101] passed from west to east across the continental United States and brought down power lines causing black-outs in hundreds of communities just when the demand for cooling was greatest.

Changes in the pattern of extreme events have been noted by the USGCRP[102]:

- Heat waves have become more frequent and intense, especially in the west.
- Cold waves have become less frequent and intense across the nation.
- Droughts in the Southwest and heat waves everywhere are projected to become more intense and cold waves less intense everywhere.
- Heavy downpours are increasing nationally, especially over the last three to five decades. Largest increases are in the Midwest and northeast. Increases in the frequency and intensity of extreme precipitation events are projected for all U.S. regions.
- The intensity, frequency, and duration of North Atlantic hurricanes, as well as the frequency of the strongest (Category 4 and 5) hurricanes, have all increased since the early 1980s.
- Hurricane-associated storm intensity and rainfall rates are projected to increase as the climate continues to warm.
- Winter storms have increased in frequency and intensity since the 1950s, and their tracks have shifted northward over the United States.

---

[99] Energy Information Administration. https://www.eia.gov/tools/faqs/faq.php?id=427&t=3

[100] USGCRP Extreme Events https://www.globalchange.gov/explore/extreme-events

[101] NOAA "Historic Derecho of June 29, 2012," http://www.erh.noaa.gov/rnk/events/2012/Jun29_derecho/summary.php

[102] USGCRP Extreme Event https://www.globalchange.gov/explore/extreme-events

Population Change, 2015–2016

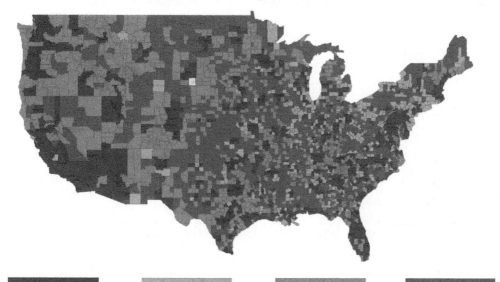

Metro counties that gained population

Metro counties that lost population

Rural counties that gained population

Rural counties that lost population

FIGURE 7.15 This map shows the percentage change in county population between 2015 and 2016 and illustrates increases in places that require air conditioning (rural and metro counties that gained population). The map shows growth in coastal metropolitan regions and decline in rural counties. This trend has persisted for over a decade. Counties in the vicinity of South Florida, Atlanta, Los Angeles, Phoenix, Las Vegas, Denver, Dallas, and Houston all had very large increases.

Source: http://www.dailyyonder.com/rural-population-drops-fifth-straight-year/2017/04/14/18850/

- Other trends in severe storms, including the intensity and frequency of tornadoes, hail, and damaging thunderstorm winds, are uncertain and are being studied intensively.

If climate change leads to increased cloudiness, solar energy production could be reduced. Wind energy production would be reduced if wind speeds increase above or fall below the acceptable operating range of the technology. Changes in growing conditions could affect biomass production, a transportation and power plant fuel source that is rising in importance.

7.3.7 Shifting U.S. Populations

Demographic trends in the United States are increasing energy use. The population is shifting to the south and the Southwest (**Figure 7.15**), where air conditioning use is high. There is an increase in the square footage built per person and increased electrical needs in residential and commercial buildings, and air conditioning is being implemented in new places and by persons in income levels who had not previously embraced it.

As changes in precipitation take place in various regions of the country, the hydropower industry may be affected positively or negatively. Increases in hurricane intensity, frequency, and location will likely cause disruptions to oil and gas operations in the Gulf of Mexico (such as occurred in 2005 with Hurricane Katrina, 2008 with Hurricane Ike, and in 2017 with hurricanes Harvey, Irma, and Maria[103]). Public concerns about

global warming will alter perceptions and valuations of energy technology alternatives. These effects will play a growing role in energy policies in the United States.

7.3.8 Human Health

Human health[104] could suffer impacts (**Figure 7.16**) from climate change related to heat stress, extreme weather events and flooding, waterborne diseases, poor air quality, and diseases transmitted by insects and rodents. There are direct health effects from ailments caused or exacerbated by air pollution and airborne allergens and many climate-sensitive infectious diseases. In general, warming is likely to make it more challenging to meet air quality standards necessary to protect public health. For instance, rising temperature and carbon dioxide concentration increase pollen production and prolong the pollen season in a number of plants with highly allergenic pollen, presenting a health risk.

As temperatures rise and heat waves occur with greater frequency and intensity, the population of senior citizens (currently 12 percent and projected to be 21 percent by 2050; more than 86 million people), those with diabetes, and those with heart disease, is increasing. This population is vulnerable to the stresses associated with heat waves. Heat is already the leading cause of weather-related deaths in the United States. More than 3,400 deaths between 1999 and 2003 were reported as resulting from exposure to excessive heat and humidity.[105] A study of climate change impacts in

---

[103] VIDEO: The Weather Channel, 2017 Atlantic Hurricane Season: 17 Moments We'll Never Forget https://weather.com/storms/hurricane/news/2017-11-11-moments-hurricane-season-atlantic-irma-maria-harvey

[104] World Health Organization, *Constitution of the World Health Organization—Basic Documents*, 45th ed., Supplement, October 2006, http://www.who.int/governance/eb/who_constitution_en.pdf

[105] Zanobetti, A. and Schwartz, J. (2008) Temperature and Mortality in Nine US Cities, *Epidemiology* 19, 4: 563–570.

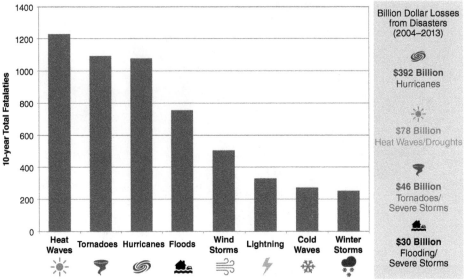

Estimated Deaths and Billion Dollar Losses
from Extreme Events in the U.S., 2004–2013

FIGURE 7.16 10-year estimates of fatalities related to extreme events from 2004 to 2013, as well as estimated economic damages from 58 weather and climate disaster events with losses exceeding $1 billion. These statistics are indicative of the human and economic costs of extreme weather events over this time period. Climate change will alter the frequency, intensity, and geographic distribution of some of these extremes, which has consequences for exposure to health risks from extreme events. Trends and future projections for some extremes, including tornadoes, lightning, and wind storms are still uncertain.

Source: U.S. Global Change Research Program.

California[106] projects that by the 2090s, annual heat-related deaths in Los Angeles would increase by two to three times under a lower-emissions scenario and by five to seven times under a higher-emissions scenario.

Poor air quality, especially in cities, is a serious concern across North America. Half of all Americans, 158 million people, live in counties where air pollution fails to meet national health standards. Breathing ozone results in short-term decreases in lung function and damages the lining the lungs. A warmer climate is projected to accelerate troposphere ozone formation and increase the frequency and duration of stagnant air masses that allow pollution to accumulate, which will exacerbate health symptoms. Under constant pollutant emissions, by the middle of this century, red-ozone-alert days (when the air is unhealthy for everyone) in the 50 largest cities in the eastern United States are projected to increase by 68 percent as a result of warming alone.[107]

Heavy downpours have increased in recent decades and are projected to increase further as the world continues to warm and the amount of water vapor increases in the atmosphere. This can lead to increased incidence of waterborne diseases due to pathogens such as *Cryptosporidium* and *Giardia*. Downpours can trigger sewage overflows, contaminating drinking water and endangering beachgoers. The consequences will be particularly severe in the roughly 770 U.S. cities and towns, including New York, Chicago, Washington, Milwaukee, and Philadelphia, that have combined sewer systems, an older design that carries storm water and sewage in the same pipes. During heavy rains, these systems often cannot handle the volume, and raw sewage spills into

lakes or waterways, including drinking-water supplies and places where people swim.

- Some diseases transmitted by food, water, and insects are likely to increase.
- Cases of food poisoning due to *Salmonella* and other bacteria peak within one to six weeks of the highest-reported ambient temperatures.
- Cases of waterborne *Cryptosporidium* and *Giardia* increase following heavy downpours. These parasites can be transmitted in drinking water and through recreational water use.
- Climate change affects the life cycle and distribution of the mosquitoes, ticks, and rodents that carry West Nile virus, equine encephalitis, Lyme disease, and Hantavirus.
- Heavy rain and flooding can contaminate certain food crops with feces from nearby livestock or wild animals, increasing the likelihood of foodborne disease associated with fresh produce.
- *Vibrio* species (shellfish poisoning) accounts for 20% of the illnesses and 95% of the deaths associated with eating infected shellfish. The U.S. infection rate increased 41% from 1996 to 2006 concurrent with rising temperatures.
- As temperatures rise, tick populations that carry Rocky Mountain spotted fever are projected to expand northward to new regions.
- The introduction of disease-causing agents from other regions of the world is a threat.

Communities have the capacity to adapt to climate change, but during extreme weather and climate events, actual practices have not always protected people and property. Vulnerability to extreme events is variable (**Figure 7.17**), with disadvantaged groups experiencing more disruption

[106] Hayhoe, K., et al. (2004) Emissions Pathways, Climate Change, and Impacts on California, *PNAS*, 101, no. 34: 12422–12427.
[107] Bell, M., et al. (2007) Climate Change, Ambient Ozone, and Health in 50 U.S. Cities, *Climatic Change* 82, no. 1–2: 61–76.

Elements of Vulnerability to Climate Change

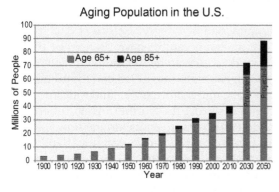

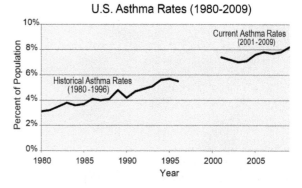

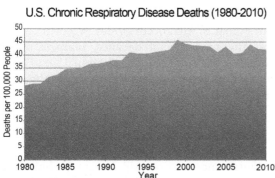

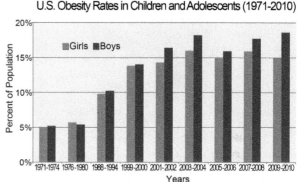

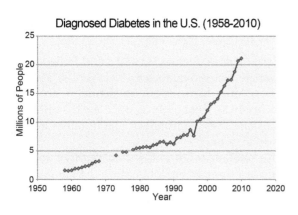

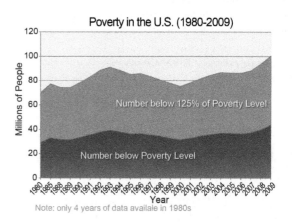

FIGURE 7.17 A variety of factors can increase the vulnerability of specific demographic groups to health effects due to climate change. For example, older adults are more vulnerable to heat stress because their bodies are less able to regulate their temperature. Overall population growth is projected to continue through at least 2050, with older adults comprising a growing proportion of the population. Similarly, there are an increasing number of people who are obese and/or have diabetes, heart disease, or asthma, which makes them more vulnerable to a range of climate-related health impacts. The poor are less able to afford the kinds of measures that can protect them from and treat them for various health impacts.

Source: U.S. Climate Resilience Toolkit; NOAA

to their lives than other groups. Adaptation tends to be reactive, unevenly distributed, and focused on coping rather than on preventing problems.[108]

---

[108] VIDEO: Climate Change is Making Us Sick, Top U.S. Doctors Say https://www.usatoday.com/story/news/health/2017/03/15/climate-change-making-us-sick-top-us-doctors-say/99218946/

7.4 Climate Impacts to Geographic Regions

As global warming continues, the USGCRP provides American citizens and decision makers with periodic updates on the climate-related changes observed in the United States, its coastal waters, and globally. As discussed earlier in this and previous chapters, these climate changes include

Change in Maximum Number of Consecutive Dry Days

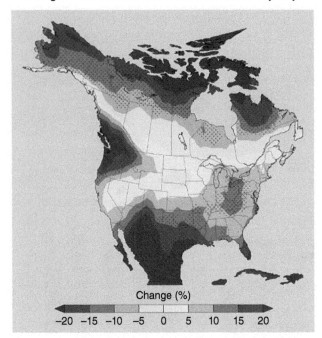

Change (%)

-20 -15 -10 -5   0   5   10   15   20

FIGURE 7.18 Change in the number of consecutive dry days (days receiving less than 1 mm [0.04 in] of precipitation) at the end of this century (2070–2099) relative to the end of last century (1971–2000) under the higher scenario, RCP 8.5. Dots indicate areas where changes are consistent among at least 80% of the 25 models used in this analysis.

Source: https://www.globalchange.gov/browse/multimedia/change-maximum-number-consecutive-dry-days

increases in heavy downpours, rising temperature and sea level, rapidly retreating glaciers, thawing permafrost, lengthened growing seasons, changes in consecutive dry days (**Figure 7.18**), lengthened ice-free seasons in the ocean and on lakes and rivers, earlier snowmelt, expanding drought, and shifts in stream characteristics. Nearly all of these changes are projected to increase. An important contribution of USGCRP assessments is regional descriptions of climate-related impacts.

7.4.1 Northeast Region

The northeast region of the United States includes Maine, Vermont, New Hampshire, Massachusetts, Connecticut, New York, Pennsylvania, Rhode Island, New Jersey, Delaware, Maryland, West Virginia, and Washington, D.C. The climate in this area has changed in noticeable ways: more frequent days with temperatures above 32 degrees Celsius (90 degrees Fahrenheit); a longer growing season; increased heavy precipitation; less winter precipitation falling as snow and more as rain; reduced snowpack; earlier breakup of winter ice on lakes and rivers; earlier spring snowmelt resulting in earlier peak river flows; and rising sea surface temperatures and sea

level.[109] All of these measured changes are consistent with the rise of atmospheric temperature.

Since 1970, the yearly average temperature in the northeast has increased by 1 degree Celsius (1.8 degrees Fahrenheit). Winter temperatures have risen twice this much.[110] In one study,[111] researchers examined daily wintertime temperature extremes since 1948 and found that the warm extremes were much more severe and widespread than the cold extremes during the Northern Hemisphere winters of 2009–2010 (which featured an extreme snowfall episode on the East Coast dubbed "Snowmaggedon") and 2010–2011. Moreover, while the extreme cold was mostly attributable to a natural climate cycle, the extreme warmth was not. Overall, by late this century, under the higher-emission scenario, residents of New Hampshire could experience a summer climate similar to what occurs today in North Carolina.

Over the next several decades, temperatures in the northeast are likely to rise to an additional 1.4 to 2.2 degrees Celsius (2.5 to 4 degrees Fahrenheit) in winter and 0.8 to 2.0 degrees Celsius (1.5 to 3.5 degrees Fahrenheit) in summer.[112] It is projected that

- Winters will be shorter, with fewer cold days and more precipitation;
- Winter snow season will be cut in half across the northern states and will be reduced to a week or two in southern parts of the region;
- Cities that today experience few days above 37.8°C (100°F) will average 20 such days per summer;
- Certain cities, such as Hartford and Philadelphia, will average nearly 30 days over 37.8°C (100°F);
- Short, one- to three-month droughts are projected to occur as frequently as once each summer in the Catskill and Adirondack Mountains and across New England;
- Hot summer conditions will arrive three weeks earlier and last three weeks longer into the fall;
- Sea level will rise more than the global average because of localized land subsidence in this area and changes in North Atlantic circulation.

7.4.2 Southeast and Caribbean Region

The Southeast region includes Virginia, Kentucky, Tennessee, North Carolina, South Carolina, Georgia, Florida, Louisiana, Alabama, Mississippi, coastal Texas, Arkansas,

[109] USGCRP, (2009) *Global Climate Change Impacts in the United States,* http://www.globalchange.gov/publications/reports/scientific-assessments/us-impacts

[110] Hayhoe, K., et al. (2007) Past and Future Changes in Climate and Hydrological Indicators in the U.S. Northeast, *Climate Dynamics* 28, no. 4: 381–407

[111] Guirguis, K., et al. (2011) Recent Warm and Cold Daily Winter Temperature Extremes in the Northern Hemisphere, *Geophysical Research Letters* 38: L17701, doi: 10.1029/2011GL048762

[112] Hayhoe et al. (2007)

Trends in Water Availability

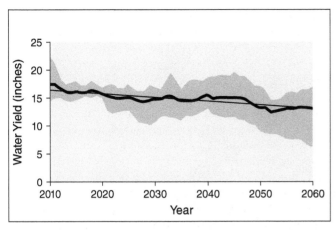 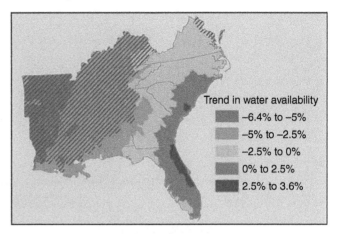

FIGURE 7.19 Left: Projected trend in Southeast-wide annual water yield (equivalent to water availability) due to climate change. The green (grey) area represents the range in predicted water yield from four climate model projections. Right: Spatial pattern of change in water yield for 2010–2060 (decadal trend relative to 2010). The hatched areas are those where the predicted negative trend in water availability associated with the range of climate scenarios is statistically significant (with 95% confidence). As shown on the map, the western part of the Southeast region is expected to see the largest reductions in water availability.

Source: https://nca2014.globalchange.gov/report/regions/Southeast

and Caribbean islands of Puerto Rico and the U.S. Virgin Islands. Compared with the rest of the nation, the Southeast is warm and wet, with mild winters and high humidity. Over most of the past century, the average temperature of the region did not change significantly. However, since 1970, the annual average temperature has risen to about 1.1 degrees Celsius (2 degrees Fahrenheit), and the greatest increase in temperature has occurred in the winter months.

The number of freezing days has declined by 4 to 7 days per year since the mid-1970s, and the average autumn precipitation has increased by 30 percent over the 20th century. Regions experiencing moderate to severe drought in the spring and summer have increased by over 10 percent since the 1970s. Even in the autumn months, when precipitation tended to increase in most of the region, the extent of drought increased by 9 percent. Higher temperatures lead to more evaporation of moisture from soils and water loss from plants; hence, the frequency, duration, and intensity of droughts are likely to continue to increase.

Water resources in the Southeast are abundant and support heavily populated urban areas, rural communities, unique ecosystems, and economies based on agriculture, energy, and tourism. However, the region also experiences extensive droughts. While change in projected precipitation for this region has high uncertainty, there is still a reasonable expectation that there will be reduced water availability due to the increased evaporative losses resulting from rising temperatures alone (**Figure 7.19**).

Over the 21st century, the number of very hot days is likely to rise at a greater rate than the average temperature. If greenhouse gas emissions are kept to a low level, average temperatures could rise about 2.5 degrees Celsius (4.5 degrees Fahrenheit) by the 2080s, but higher emissions

could result in 5 degrees Celsius (9 degrees Fahrenheit) of average warming; the increase in summer may be as much as 5.8 degrees Celsius (10.5 degrees Fahrenheit).

The coastal area of the Southeast is home to thousands of communities that have built in low-lying areas on the shores of the Atlantic ocean and Gulf of Mexico. This region, more than any other the United States, is prone to the deadly impacts of hurricanes, which pack winds capable of demolishing buildings and storm surges that rises as much as 4.5 to 6 meters (15 to 20 feet) in the streets of coastal towns and cities.

The destructive potential of Atlantic hurricanes has increased since 1970, correlated with an increase in sea-surface temperature. Notably, researchers have failed to establish a relationship between rising sea-surface temperature and the frequency of land-falling hurricanes.[113] However, in IPCC AR5, researchers concluded that the intensity of Atlantic hurricanes is likely to increase during this century, with higher peak wind speeds, rainfall intensity, and storm surge height and strength. Rising sea-surface temperatures are thought to be one reason for these increases. Even absent an increase in hurricane frequency, coastal inundation and shoreline erosion will increase as sea-level rise accelerates, which is one of the most certain and costly consequences of a warming climate.

[113] Hoyos, C.D., et al. (2006) Deconvolution of the Factors Contributing to the Increase in Global Hurricane Intensity, *Science* 312, no. 577, 94–97. Mann, M.E. and Emanuel, K.A. (2006) Atlantic Hurricane Trends Linked to Climate Change, *Eos* 87, no. 24, 244. Trenberth, K.E. and Shea, D.J. (2006) Atlantic Hurricanes and Natural Variability in 2005, *Geophysical Research Letters* 33: L12704, doi: 10.1029/2006GL026894. Webster, P.J., et al. (2005) Changes in Tropical Cyclone Number, Duration, and Intensity in a Warming Environment, *Science* 309, no. 5742: 1844–1846.

Projected Change in Number of Days Over 95°F

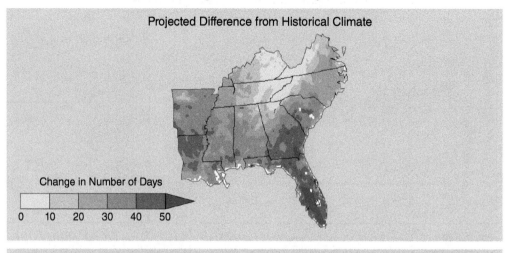

Projected Difference from Historical Climate

Change in Number of Days

0 10 20 30 40 50

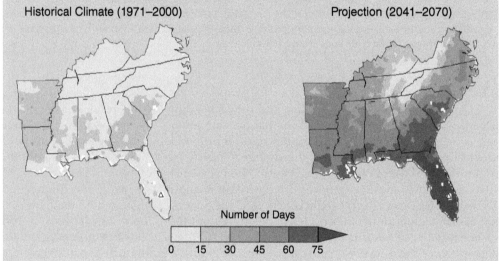

Historical Climate (1971–2000)

Projection (2041–2070)

Number of Days

0 15 30 45 60 75

FIGURE 7.20 Projected average number of days per year with maximum temperatures above 95°F for 2041–2070 compared to 1971–2000, assuming emissions continue to grow. Patterns are similar, but less pronounced, assuming a reduced emissions scenario.

Source: https://nca2014.globalchange. gov/report/regions/Southeast

Studies indicate that warming could cause the globally averaged intensity of tropical cyclones to shift toward stronger storms, with intensity increases of 2 percent to 11 percent by 2100. Studies project decreases in the globally averaged frequency of tropical cyclones by 6 percent to 34 percent. Balanced against this, research indicates substantial increases in the frequency of the most intense cyclones and increases of the order of 20 percent in the precipitation rate within 100 kilometers (62 miles) of the storm center.

Each year, the number of days with peak temperature over 32 degrees Celsius (90 degrees Fahrenheit) is expected to rise significantly, especially under a higher-emissions scenario. By the end of the century, global circulation models indicate that North Florida could have more than 165 days (nearly six months) per year over 32 degrees Celsius (90 degrees Fahrenheit), which is a significant increase from roughly 60 days in the 1960s and 1970s (**Figure 7.20**). The increase in very hot days to nearly half the days in the year could have consequences for human health, drought, and wildfires.

7.4.3 Midwest Region

The Midwest states include Michigan, Ohio, Illinois, Indiana, Missouri, Wisconsin, Minnesota, and Iowa. Located far from the climate-moderating effect of the ocean, the air temperature in the Midwest is subject to large seasonal swings. Hot, humid summers alternate with cold winters. However, in recent decades, the average annual temperature has increased, and the largest increase has been in wintertime.[114] Despite strong year-to-year variations, the length of the frost-free or growing season has extended by more than one week, mainly as a result of earlier dates for the last spring frost.

The major global warming issues for this region revolve around increases in both heat and flooding. Summer heat waves in cities can lead to health problems[115] as well as

[114] Wuebbles, D.J. and Hayhoe, K. (2004) Climate Change Projections for the United States Midwest, *Mitigation and Adaptation Strategies for Global Change* 9, no. 4: 335–363.
[115] Sheridan, S.C., et al. (2008) Trends in Heat-Related Mortality in the United States, 1975–2004, *Natural Hazards* 50, no. 1: 145–160.

Projected Mid-Century Temperature Changes in the Midwest

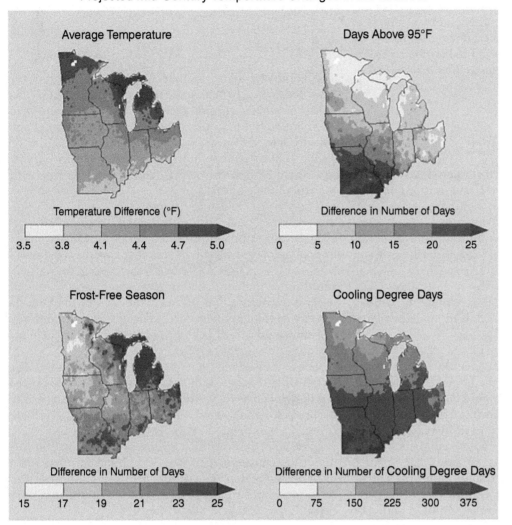

FIGURE 7.21 Projected increase in annual average temperatures by mid-century (2041–2070) as compared to the 1971–2000 period. Maps also show annual projected increases in the number of the hottest days (days over 95°F), longer frost-free seasons, and an increase in cooling degree days, defined as the number of degrees that a day's average temperature is above 65°F, which generally leads to an increase in energy use for air conditioning. Projections are from global climate models that assume rising emissions of heat-trapping gases.

Source: https://www.globalchange.gov/browse/multimedia/projected-mid-century-temperature-changes-midwest

placing increased energy demand on public services. There may be reduced air quality, increases in insect and water-borne diseases, more heavy downpours, and increased evaporation in summer. This could produce more periods of both flooding and water deficits.

A longer growing season provides the potential for increased crop yields in this important agricultural district, but growth in heat waves, floods, droughts, and insects and weeds migrating in from the south present mounting challenges to managing crops, livestock, and forests.

Scientists have also observed changes in rainfall in the Midwest; heavy downpours are now twice as frequent as they were a century ago, and both summer and winter precipitation have been above average for the last three decades, the wettest period in a century.[116] The Midwest has experienced both increasing extreme events and long-term trends: two record-breaking floods in the past two decades,

a decrease in lake ice (including on the Great Lakes), and increased frequency of large heat waves since the 1980s, which have been more frequent than any time in the past century, other than the Dust Bowl years of the 1930s.[117]

Models predict that Midwest summers could feel progressively more like summers currently experienced in states to the south and west (**Figure 7.21**).[118] By mid-century and toward the end of the century, Midwest states are projected to get considerably warmer and have less summer precipitation. Heat waves that are more frequent, more severe, and longer lasting are anticipated. The frequency of hot days and the length of the heat-wave season both may be more than twice as great under a higher-emissions scenario compared to a lower-emissions scenario.

In 1995, a heat wave hit the city of Chicago and resulted in more than 700 deaths. Events of this nature are expected to become more common. Under a low-emissions scenario, a heat wave equivalent to the 1995 event is projected to

---

[116] NOAA National Climatic Data Center, Climate of 2008: Midwestern U.S. Flood Overview, http://www.ncdc.noaa.gov/oa/climate/research/2008/flood08.html

[117] Kunkel et al. (2008)
[118] Wuebbles and Hayhoe (2004)

occur every other year in Chicago by the end of the century; under a high scenario, there would be approximately three such heat waves per year. Even more severe heat waves, such as the one that claimed tens of thousands of lives in Europe in 2003, are projected to become more frequent in a warmer world, occurring as often as every other year in the Midwest by the end of this century under a high-emissions scenario.[119]

7.4.4 Great Plains Region

The Great Plains states include North and South Dakota, Nebraska, Kansas, Oklahoma, Texas, and portions of Colorado, Wyoming, and Montana. Major global warming issues in this area include increases in temperature, evaporation, extreme weather events, and drought frequency and magnitude. These trends are likely to lead to declining water resources with impacts on agriculture, ranching, and natural lands as well as on key habitats such as playa lakes, prairie potholes, and other wetland ecosystems. Human population shifts toward cities will lead to heat-related problems as well.[120]

Climate has changed in the past few decades in the Great Plains. Average temperatures have increased the most in the northern states, and the largest increases have occurred in the winter. Relatively cold days are becoming less frequent and relatively hot days more frequent.[121] Temperatures are projected to continue to increase over the 21st century, with larger changes expected under scenarios of higher greenhouse gas emissions (**Figure 7.22**).

Summer changes are projected to be larger than those in winter in the southern and central Great Plains.[122] There has also been an increase in rainfall, with the greatest increases in states to the Southeast. However, with continued global warming, conditions are anticipated to become wetter in the north and drier in the south. Changes in long-term climate will include more-frequent extreme events such as heat waves, droughts, and heavy rainfall. These will affect many aspects of life in the Great Plains including threats to water resources, essential agricultural and ranching activities, unique

natural and protected areas, and the health and prosperity of inhabitants.[123]

7.4.5 Southwest Region

The Southwest states include California, Nevada, Utah, New Mexico, Arizona, and portions of Colorado and Texas. As global warming continues, the biggest problems that could develop in this region are related to water scarcity, drought, and heat. Studies indicate that much of the region is likely to have more than twice as many days per year above 32 degrees Celsius (90 degrees Fahrenheit) by the end of the century.

The prospect of future droughts becoming more severe as a result of global warming is a significant concern, especially because the Southwest continues to lead the nation in population growth. In an area that already wrestles with competing demands for scarce water resources, warming could force tradeoffs among rival water uses, potentially leading to conflict. Temperature increases throughout the century could amplify the frequency of drought and wildfire and could accentuate problems related to invasive species and shifts in agriculture.

The Southwest region is one of the most rapidly warming in the United States, with some areas significantly exceeding the global average. Temperature increases are driving declines in spring snowpack, and consequently river discharge in the region is down as well.[124] Model projections (**Figure 7.23**) indicate that strong warming will continue under low-emissions scenarios, with much-larger increases likely under higher scenarios. There will almost certainly be serious water supply shortages in the future along with expanding urban heat island effects.[125]

7.4.6 Northwest Region

The Northwest region includes the states Washington, Oregon, Idaho, and western Montana. Global warming has caused the average annual temperature to rise 0.8 degree Celsius (1.5 degrees Fahrenheit) throughout the region. Some areas experienced an increase of 2.2 degrees Celsius (4 degrees Fahrenheit) over the same period.

Models project increases of an additional 1.7 to 5.5 degrees Celsius (3 to 10 degrees Fahrenheit) this century, with higher-emissions scenarios resulting in warming

[119] Ebi, K.L., and Meehl, G.A. (2007) The Heat is On: Climate Change and Heat Waves in the Midwest. In *Regional Impacts of Climate Change: Four Case Studies in the United States*, Arlington, VA, Pew Center on Global Climate Change, 8–21. See: http://www. pewclimate.org/regional_impacts

[120] USGCRP (2009) *Global Climate Change Impacts in the United States*, http:// www.globalchange.gov/publications/reports/scientific-assessments/ us-impacts

[121] DeGaetano, A.T. and Allen, R.J. (2002) Trends in Twentieth-Century Temperature Extremes Across the United States, *Journal of Climate* 15, no. 22: 3188–3205.

[122] Christensen, J.H., et al. (2007) Regional Climate Projections. In Solomon, S., et al. (eds.), *Climate Change 2007: The Physical Basis*. Contribution of Working Group I to the Fourth Assessment Report of the Intergovernmental Panel on Climate Change. Cambridge, U.K., Cambridge University Press), pp. 847–940.

[123] Parton, W., et al. (2007) Long-Term Trends in Population, Farm Income, and Crop Production in the Great Plains, *Bioscience* 57, no. 9: 737–747.

[124] Barnett, T.P., et al. (2008) Human-Induced Changes in the Hydrology of the Western United States, *Science* 319, no. 5866: 1080–1083.

[125] Rauscher, S.A., et al. (2008) Future Changes in Snowmelt-Driven Runoff Timing over the Western United States, *Geophysical Research Letters* 35: L16703, doi: 10.1029/2008GL034424; see also Guhathakurta, S. and Gober, P. (2008) The Impact of the Phoenix Urban Heat Island on Residential Water Use, *Journal of the American Planning Association* 73, no. 3: 317–329.

Historic conditions

Mid-century

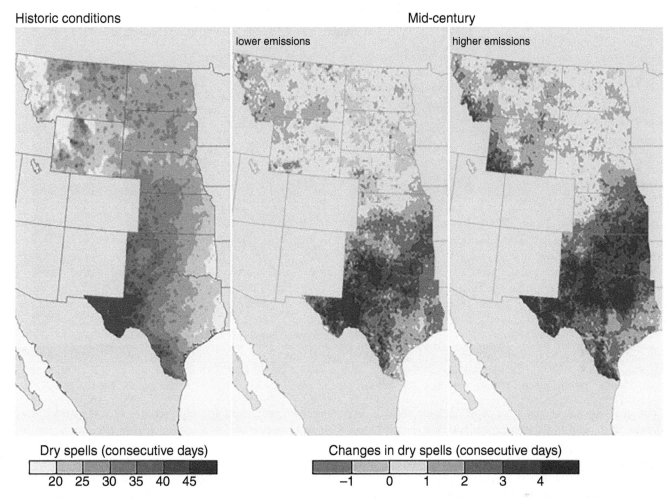

Dry spells (consecutive days)

20 25 30 35 40 45

Changes in dry spells (consecutive days)

−1 0 1 2 3 4

FIGURE 7.22 Historical and projected patterns in the number of consecutive dry days experienced in different parts of the Great Plains. The historical map shows the average annual maximum number of consecutive dry days during 1971–2000. The projected maps show changes in consecutive dry days for 2041–2070, compared to 1971–2000. One scenario assumes substantial reductions in emissions, and the other scenario assumes continued rising emissions.

Source: https://www.climate.gov/news-features/featured-images/longer-dry-spells-store-us-great-plains

at the upper end of this range. Warming is likely to bring increased winter precipitation and decreased summer precipitation, with related changes to streamflow, snowpack, forest ecosystems, wildfires, and other important aspects of life and ecology in the Northwest.[126]

The key issues related to global warming in the region include decreased spring snowpack (reducing summer stream flow and straining water resources), increased wildfires and insect outbreaks (**Figure 7.24**), shifting species composition in forest ecologies and impacts on the lumber industry, and stresses to salmon ecosystems with rising water temperatures and declining discharge. Sea-level rise and increased wave height along vulnerable

coastlines could result in accelerated coastal erosion and land loss.[127]

Snow that collects throughout the winter feeds streams and groundwater. These sources of freshwater sustain human communities, aquatic ecosystems, and forest environments. Human demands for water in the Northwest are intense. Seasonal snow pack provides water to meet growing demand from municipal and industrial uses, agricultural irrigation, hydropower production, navigation, recreation, and fish industries. As global warming raises temperatures in the Northwest, more precipitation could fall as rain rather than snow and contribute to earlier snowmelt.

[126] USGCRP (2009) *Global Climate Change Impacts in the United States,* http://www.globalchange.gov/publications/reports/scientific-assessments/us-impacts

[127] Petersen, A.W. (2007) *Anticipating Sea Level Rise Response in Puget Sound,* M.M.A. thesis, School of Marine Affairs, University of Washington, Seattle.

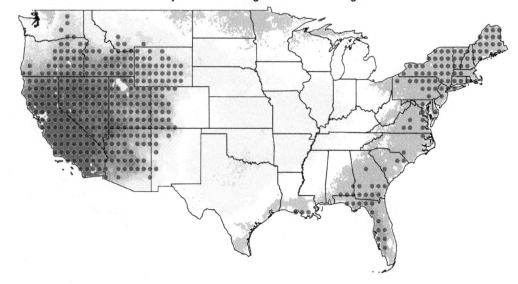

Drying the Southwest

Weather systems that bring rain are becoming more rare

FIGURE 7.23 Weather systems that typically bring moisture to the Southwestern United States are forming less often, resulting in a drier climate across the region. This map depicts the portion of overall changes in precipitation across the United States that can be attributed to these changes in weather system frequency. The gray dots represent areas where the results are statistically significant.

Source: https://www2.ucar.edu/atmosnews/news/19173/Southwest-dries-wet-weather-systems-become-more-rare

Percent change in precipitation per decade (1980–2010)

−6.5 −5.5 −4.5 −3.5 −2.5 −1.5 −0.5 0.5 1.5 2.5 3.5 4.5 5.5 6.5

The thickness and cover of April 1 snowpack, a key indicator of natural water storage available for the warm season, has declined substantially throughout the Northwest. For example, in the Cascade Mountains, the average snowpack declined approximately 25 percent over the past 40 to 70 years (mostly due to a 1.4 degrees Celsius [2.5 degrees Fahrenheit] increase in cool season temperatures).[128] It is likely that continued warming will contribute to further snowpack declines. The April 1 snowpack is projected to decline as much as 40 percent in the Cascades by mid-century.[129]

7.4.7 Alaska

Alaska has warmed twice as fast as the rest of the nation, bringing widespread impacts. Sea ice is rapidly receding and glaciers are shrinking. Thawing permafrost is leading to more wildfire, and affecting infrastructure and wildlife habitat. Rising ocean temperatures and acidification will alter valuable marine fisheries.[130]

Arctic temperatures have reached their warmest level of any decade in at least 2,000 years. To determine this, researchers[131] used geologic records and computer simulations that provide new evidence that the Arctic would be cooling if not for greenhouse gas emissions that are overpowering natural climate patterns. Part of this recent trend has been shown to originate with the positive climate feedback relating to the loss of Arctic sea ice. Sea ice melting (**Figure 7.25**) is changing the albedo (sunlight reflectivity) of the high north and causing dark ocean water to absorb solar radiation, whereas previously the white icy surface reflected the radiation back to space.[132]

Global warming is hitting Alaska in profound ways. As in many high-latitude locations, warming has exceeded the global average, and in Alaska the rate of warming has been more than twice the rate in the rest of North America. The primary impacts of global warming have already been seen. These include the following: an increase in wildfires and insect outbreaks; declining lakes and ponds resulting from drying; longer summers and higher temperatures causing drier conditions even in the absence of strong trends in precipitation; thawing permafrost that damages roads, pipelines, airports, water and sewer systems, and other infrastructure designed for colder conditions; coastal erosion that increases the risk to fishing villages, coastal towns; and growing storm vulnerability.

Alaska's annual average temperature has increased 1.9 degrees Celsius (3.4 degrees Fahrenheit), and winters have warmed by 3.5 degrees Celsius (6.3 degrees Fahrenheit). Warming is reducing sea ice, bringing an earlier

[128] Mote, P.W. (2006) Climate-Driven Variability and Trends in Mountain Snowpack in Western North America, *Journal of Climate* 19, no. 23, 6209–6220.

[129] Payne, J.T., et al. (2004) Mitigating the Effects of Climate Change on the Water Resources of the Columbia River Basin, *Climatic Change* 62, no. 1–3, 233–256.

[130] VIDEO: Climate Change Intensifies Winter Extremes. https://climate-crocks.com/2017/12/10/climate-change-intensifies-winter-extremes/

[131] Kaufman, D., et al. (2009) Recent Warming Reverses Long-Term Arctic Cooling," *Science*, 325, 1236–1239.

[132] Screen, J. and Simmonds, I. (2010) The Central Role of Diminishing Sea Ice in Recent Arctic Temperature Amplification, *Nature* 464, 1334–1337, doi: 10.1038/nature09051

Insects and Fire in Northwest Forests

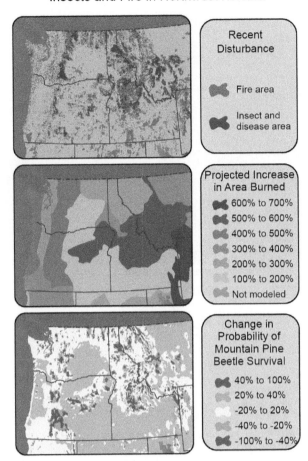

FIGURE 7.24 Top—Insects and fire have cumulatively affected large areas of the Northwest and are projected to be the dominant drivers of forest change in the near future. Map shows areas historically burned (1984 to 2008) or affected by insects or disease (1997 to 2008). Middle—Map indicates the increases in area burned that would result from the regional temperature and precipitation changes associated with 1.2°C (2.2°F) global warming across areas that share broad climatic and vegetation characteristics. Bottom—Projected changes in the probability of climatic suitability for mountain pine beetles for the period 2001 to 2030 (relative to 1961 to 1990).

Source: https://nca2014.globalchange.gov/report/regions/northwest

spring snowmelt, melting permafrost, eroding coastlines,[133] and causing the retreat of glaciers throughout the state. These changes are consistent with model predictions that warming will exceed the pace of the rest of the nation, especially in winter. Sea ice reductions also alter the timing and location of plankton blooms, which is expected to drive important shifts in marine species such as pollock and other commercial fish stocks.[134]

7.4.8 Hawai'i and Pacific Islands

Island communities in the Pacific are isolated, trade-dependent, and ocean-oriented cultures that are especially vulnerable to climate change. In Hawai'i, air and ocean temperatures are rising, rainfall is decreasing, sea level is rising, the ocean is acidifying, and winds are shifting both direction and speed. These trends signal decreased water resources, increased coastal erosion (**Figure 7.26**) and marine inundation, and increased economic expense. There are two types of Pacific Islands: high volcanic islands that generate rain from the orographic process, and low atoll reef islands that rely on convective rainfall. In either case, the resulting water is the lifeblood of island communities, and with rising air temperatures this precious resource is growing more scarce.

Marine and coastal ecosystems of the islands are particularly vulnerable to the impacts of climate change.[135] Sea-level rise, increasing water temperatures, rising storm intensity, coastal inundation and flooding from extreme events, beach erosion, ocean acidification, increased incidences of coral disease, and increased invasions by non-native species are among the threats that endanger the ecosystems that provide safety, sustenance, economic viability, and cultural and traditional values to island communities.

Reefs are under stress owing to rising water temperatures and acidification. Many fringing reefs are already stressed from a history of overfishing and polluted runoff from nearby watersheds. Changing ocean conditions further threaten these reefs, as does population growth along the desirable shorelines of islands, where development often leads to declining water quality related to sewage and other pollutants.

7.4.9 Coastal Regions

It has been estimated that approximately 3.9 million people in the United States[136] and more than 145 million people worldwide[137] live within 1 meter (3.3 feet) of modern sea level and thus risk losing their land and property under most scenarios of global warming by the end of the century (**Figure 7.27**). The resulting disruption threatens the economy and social well-being of many more. This realization

[133] Jones, B.M., et al. (2009) Increase in the Rate and Uniformity of Coastline Erosion in Arctic Alaska, *Geophysical Research Letters*, 36, L03503, doi: 10.1029/2008GL036205

[134] Grebmeier, J.M., et al. (2006) A Major Ecosystem Shift in the Northern Bering Sea, *Science*, 311, no. 5766, 1461–1464

[135] Gillespie, R.G., et al. (2008) Biodiversity Dynamics in Isolated Island Communities: Interaction Between Natural and Human-Mediated Processes, *Molecular Ecology* 17, no. 1, 45–57.

[136] Tebaldi, C., et al. (2012) Modeling Sea Level Rise Impacts on Storm Surges along US Coasts, *Environmental Research Letters* 7, 014032, doi: 10.1088/1748-9326/7/1/014032; and Strauss, B., et al. (2012) Tidally Adjusted Estimates of Topographic Vulnerability to Sea-Level Rise and Flooding for the Contiguous United States, *Environmental Research Letters* 7, 014033, doi: 10.1088/1748-9326/7/1/014033

[137] Anthoff, D., et al. (2006) *Global and Regional Exposure to Large Rises in Sea-Level: A Sensitivity Analysis*, Working Paper 96. Norwich, Tyndall Centre for Climate Change Research, http://www.tyndall.ac.uk/biblio/working-papers?biblio_year=2006

ARCTIC HAD WARMEST YEAR ON RECORD

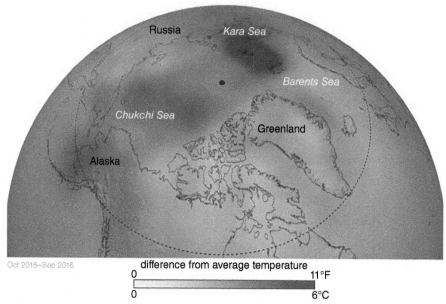

Oct 2015–Sep 2016

difference from average temperature

0   11°F

0   6°C

ARCTIC IS WARMING TWICE AS FAST AS THE GLOBAL AVERAGE

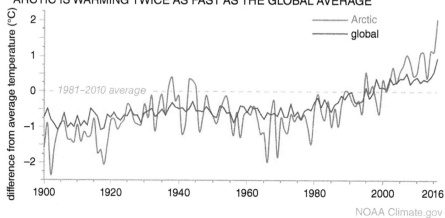

FIGURE 7.25 Temperatures across the Arctic from October 2015 to September 2016 compared to the 1981–2010 average. (graph) Yearly temperatures since 1900 compared to the 1981–2010 average for the Arctic, and the globe.

Source: https://www.climate.gov/news-features/ understanding-climate/noaas-2016-arctic-report-card-visual-highlights

is driving some coastal communities to consider various ways to adapt to sea-level rise, including the development of guidance in the form of new government policies, engineering solutions, and other strategies to accommodate rising waters and its attendant problems. However, making the transition to an adapted community that has successfully reduced vulnerability to sea-level rise impacts is only beginning.

A study[138] of lands that are vulnerable to sea-level rise reveals that almost 60 percent of the land below 1 meter (3.3 feet) along the U.S. Atlantic coast is expected to be developed and thus not able to accommodate the inland migration of wetlands, beaches, estuarine zones, and other tidal ecosystems. Less than 10 percent of the land below 1 meter has been set aside for conservation. Development not only threatens the migration path of tidal ecosystems

but also entails population growth on the world's riskiest lands.

It has been estimated that about one-third of all Americans live in counties that border ocean coasts,[139] and coastal and ocean activities contribute more than $1 trillion to the nation's gross domestic product. The ecosystems of the coast and the 322 kilometers (200 miles) wide Exclusive Economic Zone hold rich biodiversity and provide invaluable services.[140]

However, over the past 50 years, population growth in the coastal zone outpaced the ability of resource managers and community leaders to ensure the sustainability of coastal environments and resources. Fish stocks have been severely diminished by overfishing, large dead zones in

[138] Titus, J.G., et al. (2009) State and Local Governments Plan for Development of Most Land Vulnerable to Rising Sea Level along the U.S. Atlantic Coast, *Environmental Research Letters* 4, 044008.

[139] Crowell, M., et al. (2007) How Many People Live in Coastal Areas?" *Journal of Coastal Research* 23, no. 5: iii–vi.

[140] U.S. Commission on Ocean Policy, *An Ocean Blueprint for the 21st Century* (Washington, D.C., U.S. Commission on Ocean Policy, 2004) http://www.oceancommission.gov/documents/full_color_rpt/welcome.html

FIGURE 7.26 Unusual coastal erosion at Sunset Beach, Oahu. Seventy percent of beaches on the islands of Kauai, Oahu, and Maui are in a state of chronic erosion.[141]

Source: Mike Foley.

Michael J. H. Foley

FIGURE 7.27 Surging population growth in the coastal zone exposes more people to the dangers of geologic hazards, such as storms, hurricanes, tsunamis, and others, than in any other geologic environment. The world's coasts are home to fragile ecosystems, beautiful vistas, pristine waters, and major growing cities, all coexisting in a narrow and restricted space. Expanding communities compete for more space at the expense of extraordinary wild lands. There are problems with coastal erosion, waste disposal, a dependency on imported food and water, and rising sea level.

iStockphoto

coastal waters are depleted of oxygen because of excess nitrogen runoff, toxic algae blooms are growing in frequency and geographic diversity, seawall construction results in beach loss, and coral reefs are in decline in some areas from human causes. About half of the nation's coastal wetlands have been lost, and most of this loss has occurred during the past 50 years.[142]

[141] Fletcher, C.H., et al. (2012) *National Assessment of Shoreline Change: Historical Shoreline Change in the Hawaiian Islands*, U.S. Geological Survey Open-File Report 2011–1051, 55 p.
[142] USGCRP (2009) *Global Climate Change Impacts in the United States*, http://www.globalchange.gov/publications/reports/scientific-assessments/us-impacts

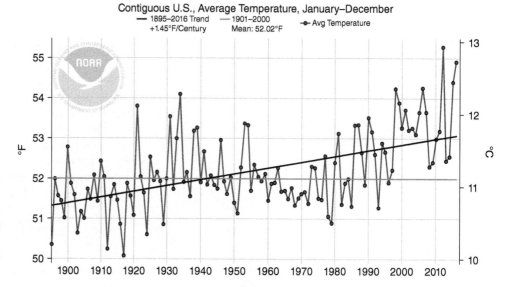

Contiguous U.S., Average Temperature, January–December

FIGURE 7.28 Annual average temperature for the contiguous United States from 1895 to 2016. Individual years are shown as black dots, connected by a purple (dark grey) line. The linear warming trend is 1.45°F/century.

Source: https://www.climate.gov/news-features/blogs/beyond-data/mapping-us-climate-trends

Global warming places new stresses on this situation. Rising sea level is eroding shorelines, drowning wetlands, and threatening communities on the coast.[143] The potential of Atlantic tropical storms and hurricanes to cause damage has grown since 1970 because more people have moved onto and built along the nation's coastlines and because rising Atlantic sea surface temperatures are fueling increased hurricane rainfall and wind speeds.[144] Studies[145] reveal that because of sea-level rise, the odds of flooding by catastrophic "100-year" floods (floods expected only once per century) will double for most coastal cities by 2030.

Over the past 50 years, coastal water temperatures have risen by about 1.1 degrees Celsius (2 degrees Fahrenheit) in several regions, and the distribution of marine species has shifted. Where rainfall has increased on land, greater river runoff pollutes coastal waters with nitrogen and phosphorous, sediments, partially treated human sewage, and other contaminants that are carried from farm fields and polluted streets.

Among other stressors, coral reefs are affected by the mixture of atmospheric carbon dioxide with seawater, which lowers the pH of seawater, causing ocean acidification. This threatens corals, mollusks, plankton, and other marine organisms that form their shells and skeletons from calcium carbonate, which is not as stable in the new seawater chemistry. Ocean acidification threatens the ability of these organisms to secrete the calcium carbonate materials they need to live.[146] All of these forces converge and interact at the coasts, making these areas particularly sensitive to the impacts of climate change.

7.4.10 Overall

The United States is already feeling the impacts of climate change in the form of rising temperatures (**Figure 7.28**), increasing extreme events, and accumulating negative effects on agriculture, ecosystems, and human communities. In coming decades, most states are likely to see increasing hot days, intense rainfall, and other forms of climate change.

Here are some of the impacts that are currently visible throughout the United States and will continue to affect these regions:

Northeast. Heat waves, heavy downpours, and sea-level rise pose growing challenges to many aspects of life in the northeast. Infrastructure, agriculture, fisheries, and ecosystems will be increasingly compromised. Many states and cities are beginning to incorporate climate change into their planning.

Northwest. Changes in the timing of streamflow reduce water supplies for competing demands. Sea-level rise, erosion, inundation, risks to infrastructure, and increasing ocean acidity pose major threats. Increasing wildfire, insect outbreaks, and tree diseases are causing widespread tree die-off.

Southeast. Sea-level rise poses widespread and continuing threats to the region's economy and environment. Extreme heat will affect health, energy, agriculture, and more. Decreased water availability will have economic and environmental impacts.

[143] Williams, S.J., et al. (2009) Sea-Level Rise and its Effects on the Coast, In Titus, T.G., et al. *Coastal Elevations and Sensitivity to Sea-level Rise: A Focus on the Mid-Atlantic Region*, Synthesis and Assessment Product 4.1. (Washington, DC, U.S. Environmental Protection Agency), pp. 11–24.

[144] Kunkel et al. (2008)

[145] Strauss et al. (2012)

[146] Orr, J.C., et al. (2005) Anthropogenic Ocean Acidification over the Twenty-First Century and its Impact on Calcifying Organisms, *Nature* 437, no. 7059: 681–686.

Midwest. Extreme heat, heavy downpours, and flooding will affect infrastructure, health, agriculture, forestry, transportation, air and water quality, and more. Climate change will also exacerbate a range of risks to the Great Lakes.

Southwest. Increased heat, drought, and insect outbreaks, all linked to climate change, have increased wildfires. Declining water supplies, reduced agricultural yields, health impacts in cities due to heat, and flooding and erosion in coastal areas are additional concerns.

7.5 Conclusion

The IPCC is planning its next assessment report in 2020. This will be an important study that sets the tone for climate research over the following decade. Congress continues to investigate the science of global warming and it is hoped that as public understanding of climate change grows and improves, the United States will soon adopt a national carbon-control law that strongly contributes to mitigating the problem of greenhouse gas emissions and thereby improves the prospects for avoiding the very worst aspects of long-term climate change.

Animations and Video

1. The Climate 25. https://vimeo.com/123438759
2. The Most Accurate Models Predict the Highest Climate Warming. https://climatecrocks.com/2017/12/18/most-accurate-models-predict-highest-climate-warming/
3. How will climate change affect it? The water cycle. https://www.youtube.com/watch?v=fI5b5bwpdVE
4. The Weather Channel, 2017 Atlantic Hurricane Season, 17 Moments We'll Never Forget https://weather.com/storms/hurricane/news/2017-11-11-moments-hurricane-season-atlantic-irma-maria-harvey
5. Climate Change is Making Us Sick, Top U.S. Doctors Say. https://www.usatoday.com/story/news/health/2017/03/15/climate-change-making-us-sick-top-us-doctors-say/99218946/
6. Climate Change Intensifies Winter Extremes. https://climatecrocks.com/2017/12/10/climate-change-intensifies-winter-extremes/

Comprehension Questions

1. Describe the role of the USGCRP.
2. How much has the U.S. average temperature risen over the past 50 years?
3. Describe the U.S. average temperature increase that might occur under the low-emissions scenario and under the high-emissions scenario by the end of the century.
4. What is the general rule among climatologists with regard to precipitation changes due to global warming?
5. Describe how global warming has changed precipitation in the United States.
6. Describe some ways climate change could affect the water resources and transportation sectors of the U.S. economy.
7. Describe some ways climate change could affect human health and energy supply and use in the United States.
8. Describe some ways climate change could affect ecosystems and agriculture in the United States.
9. How has the summer growing season changed in the Great Plains region?
10. Describe the impact of global warming where U.S. population growth has been the greatest in recent decades.

Thinking Critically

1. There are regions in the United States where annual precipitation has increased but there has been an increase in seasonal drought. Describe why this is a concern.
2. Rising air temperature is causing changes in the length of seasons. Describe how this can affect water resources.
3. Is global warming causing more or less extreme weather? Explain your answer and describe the type of extreme weather you are referring to.
4. Transportation operations are affected by global warming in several ways. Describe these.
5. How will changing air temperature affect demand for electricity? In what areas will this be most noticeable?
6. You are mayor of a town in New England. What effects can you expect from climate change and what should you do to prepare the town for these effects?
7. What special risks and vulnerabilities do coastal communities have in the face of climate change?
8. How is climate change affecting rain and what are the positive and negative impacts to local communities?
9. If the USGCRP has documented such wide-ranging impacts to climate change, why has the U.S. Congress failed to take significant action on the problem?
10. The U.S. Southwest and Southeast have rapidly growing populations. Describe the special risks they face from climate change.

Activities

1. Visit the USGCRP website http://www.globalchange.gov/ and answer the following questions.
   (a) What types of reports have they produced other than the Fourth National Climate Assessment?
   (b) What is adaptation science? Why is it important?
   (c) Visit the agencies that are coordinated under the USGCRP and describe the climate activities and concerns of six of them.
   (d) How has the current executive branch of the U.S. modified steps to mitigate and adapt to climate change that were achieved in the previous administration?
2. Watch the video "Climate Denial Crock of the Week: Bad, Badder BEST" http://www.youtube.com/watch?v=tciQts-8Cxo and answer the following questions.
   (a) Describe the BEST study and what its conclusions were.
   (b) What is the urban heat island effect and what has the BEST study concluded about the effect?
   (c) Go to the Web and see if you can find any blogs of climate denialists and describe their reaction to the BEST study.
   (d) Are climate denialists driven by facts or by some other motivation?
3. Watch the video "Lone Star State of Drought" http://www. youtube.com/watch?v=0VMpes8EyIw and answer the following questions.
   (a) Describe the 2011 Texas drought and its effects.
   (b) How does the information in this video compare to the description of climate change in the Great Plains states?
   (c) Are extreme events expected in a warming climate? What types of events are likely?
   (d) Describe the relationship between drought in Texas and the Pacific pattern known as ENSO.
   (e) How will the ratio of record warm days to record cold days change by mid-century compared to the current pattern?

Key Terms

Global mean sea level

Relative sea-level rise

Self-stabilizing cycles

Self-reinforcing cycles

Urban heat island

United States Global Change Research Program (USGCRP)

CHAPTER 8

Dangerous Climate

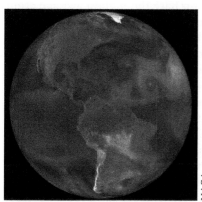 

Global warming is increasing the number of places that are susceptible to wildfires. Fires are a powerful and widespread phenomenon that now touch 30% of the land surface. NASA's Terra and Aqua satellites detect fires as small as 250 m (820 ft) across and measure carbon aerosols within the smoke they produce. Black carbon (soot) impacts air quality and human health and contributes to climate change. Using the satellite data, the NASA GEOS-5/GOCART atmosphere model simulates the global circulation of these aerosols in these two scenes.[1]

Source: https://svs.gsfc.nasa.gov/11192

LEARNING OBJECTIVE

As the atmosphere continues to warm, researchers continue to document the consequences; the polar regions are melting, the oceans are losing dissolved O_2, heating, and acidifying, extreme weather is increasing, ecosystems are unraveling, and the world is poised to change in ever more dangerous ways. Efforts to abandon the carbon-based energy of the past and move strongly toward cleaner, renewable energy are not happening as quickly as needed to ensure future generations inherit a livable planet.

8.1 Chapter Summary

Weather extremes continue to set new records. To some scientists, this is a signal that the impacts of climate change are appearing more rapidly than anticipated. This raises the possibility that some climate models are underestimating projections of future impacts resulting from global warming. **Equilibrium climate sensitivity (ECS)** is the global mean surface temperature resulting from a doubling of preindustrial CO_2 content (estimated to be 280 ppm). Researchers debate the value for ECS, and there is evidence that this value should be higher than assumed by many models.

Damaging consequences of climate change that have already emerged make it clear that future costs stemming from additional warming will be significant. Aware of this, and buoyed by strong market forces and subsidies, many

cities, states, and corporations have pivoted toward renewable energy and now constitute a global movement. While this trend is encouraging, it is nonetheless "too little, too late" to stop warming at the 1.5 degrees Celsius (2.7 degrees Fahrenheit) target set by he United Nations Framework Convention on Climate Change. It may also prove too little to halt warming at the alternate target, 2 degrees Celsius (3.6 degrees Fahrenheit).

On the basis of peer-reviewed literature, it can be argued that warming beyond 2 degrees Celsius risks the socio-economic framework that forms the foundation for modern humanity. Pervasive heat waves, massive storms, crop failure, dramatic sea-level rise, and water scarcity could increase to the point that large parts of the world become unlivable. The resulting tide of climate refugees would strain geopolitical relationships, giving rise to radical authoritarian nationalism, characterized by dictatorial power. The early signs of this future are already in play.

Recognizing that these disastrous conditions are already emerging, acting immediately and effectively to prevent their further development, and convincing a still skeptical

---

[1] VIDEO: NASA Planet on Fire, computer simulation of global fires and aerosols https://svs.gsfc.nasa.gov/11192

U.S. public that it is in their own best interest to accept the science of climate change, must become the mission of an emerging class of young leaders in order to maintain any hope that the world will continue to support a thriving human population.

In this chapter, you will learn that:

- The warming caused by human activities ranges from 0.170 to 0.175°C/decade (0.31 to 0.32°F/decade).
- 2016 had the largest annual increase in atmospheric CO_2 on record.
- NOAA has declared 2017 the most extreme weather year in U.S. history.
- Day-to-day weather has grown increasingly erratic and extreme, with severe, record-breaking events increasing in frequency and magnitude.
- The Arctic has settled into a new normal. It is warmer, greener, less icy, and suffering extreme ecosystem damage.
- 2012 saw the worst heat wave on record and the most severe drought conditions to hit the U.S. since the Dust Bowl era.
- If global warming persists as expected, it is estimated that almost one-third of all plant and animal species worldwide could become extinct.
- Sea surface temperature has increased by an average of 1°C (1.8°F) in the past 100 years, and the acidity of the ocean surface has increased tenfold.
- An increase of 3.1°C (5.6°F) in global average surface temperatures seems most likely as a result of doubling the CO_2 concentration above preindustrial levels.
- A global movement to adopt clean, renewable energy is moving forward with great strength, but the demand for new energy by a growing economy means fossil fuel use is growing also.
- Buoyed by government subsidies, the prospect of job growth, and profits for investors, advances in renewable technology are making strong progress.
- The annual UN Gap Report reveals that pledges by nations to reduce their emissions add up to no more than one-third of what is needed to keep global temperature below 2°C (3.6°F).
- In a world that has warmed 2°C:
  - Temperature will exceed 40°C (104°F) every year in many areas that are heavily populated.
  - Many croplands will experience debilitating drought that drives up food prices, famine, and political unrest.
  - Sea-level rise will exceed 1 m (3.3 ft), disrupting many of the world's major cities.
  - Superstorms will develop as heat accumulates in the tropics.
  - Infectious disease, malnutrition, and hunger will intensify and expand to new communities.

- Climate refugees, displaced by unlivable conditions, will catalyze political turmoil as they seek new homes.
- To avoid risking the socioeconomic fabric of modern humanity, carbon emissions must decline to zero by mid-century.[2]

8.2 Extreme Weather

Since 1980, damage from billion-dollar disaster events in the United States has been dominated by tropical cyclones.[3] Tropical cyclones are the most expensive ($850.5 billion, inflation-adjusted) and also have the highest average event cost ($22.4 billion per event). These numbers have risen with the inclusion of 2017 Hurricanes Harvey, Irma, and Maria, each of which caused massive damage. Drought ($236.6 billion), severe storms ($206.1 billion), and inland flooding ($119.9 billion) are also responsible for considerable damage over the past 4 decades.[4]

Severe storms are the most frequent type of billion-dollar disaster event (91), while the average event cost is the lowest ($2.3 billion). Tropical cyclones and flooding represent the second and third most frequent event types (38 and 28), respectively. Tropical cyclones are responsible for the highest number of deaths (3,461), followed by drought/heatwave events (2,993) and severe storms (1,578). Overall, the United States has sustained 219 weather and climate disasters since 1980 where overall damages/costs reached or exceeded $1 billion (inflation adjusted to 2017). The total cost of these 219 events exceeds $1.5 trillion (**Figure 8.1**).

8.2.1 Event History

In 2011 the United States was pummeled by 16 extreme weather events,[5] each of which caused more than $1 billion in damage (inflation adjusted); in several states, the months of January to October were the wettest ever recorded. According to NOAA[6] scientists, 2011 was a record-breaking year for climate extremes, as much of the U.S. faced historic levels of heat, precipitation, flooding, and severe weather.

---

[2] VIDEO: NOAA Visualizations - Climate Predictions, https://www.gfdl.noaa.gov/visualizations-climate-prediction/

[3] NOAA Billion Dollar Weather and Climate Disasters: Overview https://www.ncdc.noaa.gov/billions/

[4] NOAA Billion-Dollar Weather and Climate Disasters: Summary Stats, https://www.ncdc.noaa.gov/billions/summary-stats

[5] NOAA https://www.ncdc.noaa.gov/billions/events/US/1980-2017

[6] NOAA, 2011 a Year of Climate Extremes in the United States http://www.noaanews.noaa.gov/stories2012/20120119_global_stats.html

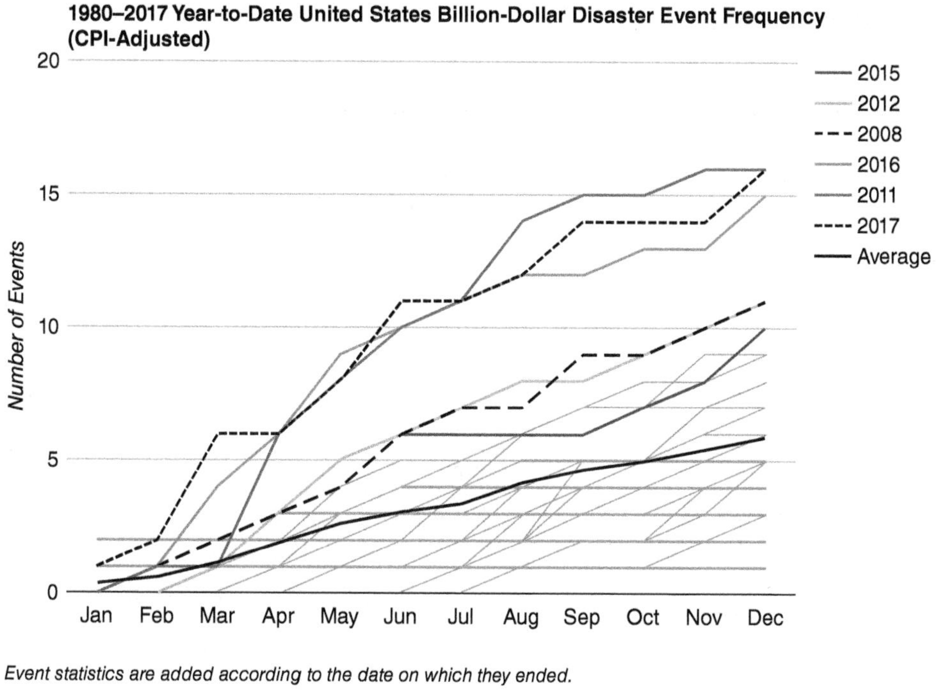

1980–2017 Year-to-Date United States Billion-Dollar Disaster Event Frequency (CPI-Adjusted)

Event statistics are added according to the date on which they ended.

FIGURE 8.1 1980–2017, $billion-dollar weather disasters in the United States (adjusted for inflation). In 37 years of record-keeping, the years 2011, 2016, and 2017 experienced a roughly 50% increase in disaster frequency compared to the full record.

Source: https://www.ncdc.noaa.gov/billions/

Japan also registered record rainfalls, and the Yangtze River basin in China suffered a record drought. Similar record-breaking events occurred also in previous years. In 2010, Western Russia experienced the hottest summer in centuries, and in Pakistan and Australia record-breaking amounts of rain fell. Europe had its hottest summer in at least half a millennium in 2003[7]; in 2002, Germany measured more rain in one day than ever before, followed by the worst flooding of the Elbe River for centuries.

In the spring and summer of 2012, one month after another set records for high temperatures as the Northern Hemisphere was enveloped in a heat wave that refused to end. March 2012 set over 1,000 record-high temperatures[8] across the United States. The hottest March in U.S. history, turned into an April that set a record for average global land temperature. Global temperatures in May were the second warmest since recordkeeping began in 1880. In the United States, June was 1.1 degrees Celsius (2 degrees Fahrenheit) above the 20th century average.[9] Temperatures late in the month broke or tied over 170 all-time records across North America. June also culminated the warmest 6-month and 12-month periods in national history.

The 2012 spring and summer heat brought drought to the nations agriculture from northern Florida to eastern Washington state. Conditions ranging from "abnormally dry" to "exceptional drought" prompted the U.S. Department of Agriculture to declare[10] more than 1,000 counties in 26 states as natural disaster areas. This nation-wide emergency established the largest natural-disaster area in U.S. history. Drought impacts across the central agriculture states resulted in widespread harvest failure for corn, sorghum, and soybean crops, among others. The summer heat wave also caused 123 direct deaths, but an estimate of the excess mortality due to heat stress is still unknown.

Since the 2012 heat waves, the United States has experienced a deadly and costly parade of billion-dollar weather disasters.

- 2012, 11 events, 1 drought, 7 severe storms, 2 tropical cyclones, 1 wildfire, $125.1 billion in losses, 377 deaths.
- 2013, 9 events, 1 drought, 2 floods, 6 severe storms, total cost $24.5 billion, 113 deaths.
- 2014, 8 events, 1 drought, 1 flood, 5 severe storms, 1 winter storm, $18.5 billion in losses, 53 deaths.
- 2015, 10 events, 1 drought, 2 floods, 5 severe storms, 1 wildfire, 1 winter storm, $23.2 billion in losses, 155 deaths.

[7] Extreme Weather of Last Decade Part of Larger Pattern Linked to Global Warming http://www.sciencedaily.com/releases/2012/03/120325173206.htm
[8] NASA, Historic Heat in North American turns Winter to Summer. http://earthobservatory.nasa.gov/IOTD/view.php?id=77465&src=eoa-iotd
[9] NOAA, June 2012 Brings More Record-Breaking Warmth to U.S. http://www.climatewatch.noaa.gov/image/2012/june-2012-brings-more-record-breaking-warmth-to-u-s

[10] USDA Announces Streamlined Disaster Designation Process with Lower Emergency Loan Rates and Greater CRP Flexibility in Disaster Areas, http://www.usda.gov/wps/portal/usda/usdahome?contentid=2012/07/0228.xml&navid=NEWS_RELEASE&navtype=RT&parentnav=LATEST_RELEASES&edeployment_action=retrievecontent

- 2016, 15 events, 1 drought, 4 floods, 8 severe storms, 1 tropical cyclone, 1 wildfire, $48.2 billion in losses, 138 deaths.
- 2017, 16 events, 1 drought, 2 floods, 1 winter freeze, 8 severe storms, 3 tropical cyclones, 1 wildfires, $306 billion in losses, 362 deaths.[11]

2017 As 2017 came to a close, NOAA officials announced that it had broken the record for most expensive year in U.S. history for weather and climate disasters—$306 billion dollars in losses.[12]

California was drenched in the wettest winter on record, ending years of drought. Then came California's most destructive and largest wildfire season ever. The Tubbs Fire in Northern California killed 22 people and damaged more than 5,600 structures. In Southern California, the Thomas Fire lasted 5 weeks, and was the biggest blaze in state history. It spanned Santa Barbara and Ventura Counties, scorched 281,893 acres, and destroyed 1,063 structures.

In early 2018, before the Thomas Fire was even fully contained, torrential rain triggered massive debris flows on the slopes above Montecito, a small coastal community perched above the ocean that had barely escaped destruction by the fire. Denuded by the blaze, barren hills released rivers of mud and rock that destroyed over 100 houses, tossed cars around like toys, and carried meter-wide boulders all the way to the coastline below. Experts describe[13] a "one-two punch of fire and debris flows." Not only does an intense wildfire consume the vegetation and shallow root system that holds soil in place; it dries the soil to the point that it can no longer soak up rain, like a shriveled sponge. Over 30 fatalities occurred and total losses spiraled into $hundreds of millions of dollars.

Hurricane Harvey broke a rainfall record for a single tropical storm with more than 4 ft of rain. Puerto Rico stayed mired in the longest blackout in U.S. history after Hurricane Maria struck in September. San Francisco reported its hottest temperature ever, 41.1 degrees Celsius (106 degrees Fahrenheit), while other parts of the country set records for high-temperature streaks. Fourteen places across Oklahoma, Missouri, and Arkansas reported record-high water levels during floods in April and May. Requests for federal disaster aid jumped tenfold compared to 2016, with 4.7 million people registering with the Federal Emergency Management Agency. By the end of 2017, the U.S. had experienced 16 separate billion-dollar weather and climate disasters, tying 2011 for the most in one year (**Figure 8.2**).

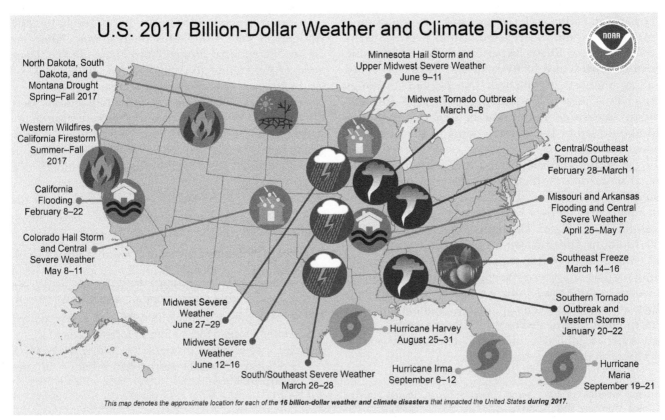

FIGURE 8.2 Map of 2017 weather disasters. In 2017, there were 16 weather and climate disaster events with losses exceeding $1 billion each across the United States. Overall, these events totaled over $306 billion in losses. Since 1980, the annual average is 5.8 events, and the annual average for the most recent 5 years (2012–2017) is 11.6 events.

Source: https://www.ncdc.noaa.gov/billions/

---

[11] NOAA Billion-Dollar Weather and Climate Disasters: Table of Events https://www.ncdc.noaa.gov/billions/events/US/1980-2017

[12] NOAA https://www.ncdc.noaa.gov/billions/

[13] Los Angeles Times http://www.latimes.com/local/lanow/la-me-mud-flows-science-montecito-20180112-story.html

Are these weather extremes coincidental, or are they the product of global warming? Scientists investigating[14] this question have concluded that a clear link connects extreme rainfall, heat waves, and other types of severe weather to human-caused global warming, a link that is supported by elementary physical principles, statistical trends, and computer simulations.[15]

8.2.2 Attribution

As the climate system has warmed, the frequency, intensity, and duration of some extreme weather events have changed.[16] Some of these changes are well understood and can be simulated using climate models. Surface warming is expected to increase the likelihood of extremely hot days and nights and to increase evaporation. This can intensify droughts, and the increased water vapor can increase the frequency of heavy rainfall and snowfall events.

Detecting changes in climate that are the result of anthropogenic global warming is known as **attribution**. Attributing general patterns of changing temperature or precipitation (such as described above) to anthropogenic global warming has become widely accepted by researchers. However, attributing individual weather events to anthropogenic global warming has been more difficult and the number of examples has been limited (but is growing).

Global Warming Attribution In what is now a classic example of attribution, researchers compared a century of temperature observations to CMIP model simulations of the same time period. In one simulation, only natural factors are used; in another, natural factors plus anthropogenic greenhouse gas emissions are used. Figure 6.9 (**Chapter 6**) reveals that when only natural factors are used, model simulations do not match the observations. But when natural factors plus anthropogenic greenhouse gas emissions are used, model simulations closely match the observations.[17]

Another example of attribution is illustrated in **Figure 8.3**. In the top panel, the figure shows global mean surface temperature observations since the late 1800s. In lower panels, it shows the climate system components that produced the trend and variability of that temperature history: changes in solar radiation, temporary cooling from volcanic eruptions,

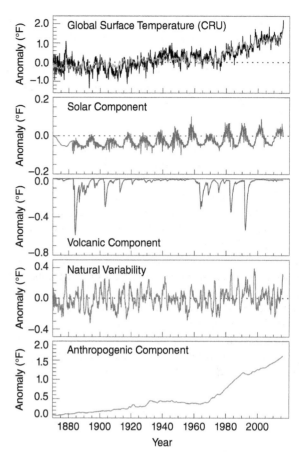

FIGURE 8.3 Estimates of the contributions of several factors to global mean temperature change since 1870. The top panel shows observed global temperature anomalies (°F). The lower four panels show: (1) solar variability, (2) volcanic eruptions, (3) temperature variability related to the El Niño/Southern Oscillation, and (4) temperature resulting from anthropogenic greenhouse gas emissions (including a cooling component from anthropogenic aerosols).

Source: https://science2017.globalchange.gov/chapter/3/

variability related to the El Niño–Southern Oscillation, and anomalous warming resulting from anthropogenic greenhouse gas emissions.

Analysis, based on the approach in Figure 8.3, revealed to researchers[18] that anthropogenic greenhouse gas emissions substantially contribute to the rise in global mean temperature since the late 1800s.

Attribution studies are important because they can help determine whether human influence on climate variables (e.g., temperature) can be distinguished from natural variability. Studies can help evaluate whether model simulations are consistent with observed trends or other changes in the climate system. Attribution studies can also inform decision-making on climate policy and adaptation.

[14] Coumou, D. and Rahmstorf, S. (2011) A Decade of Weather Extremes, *Nature Climate Change*, doi: 10.1038/NCLIMATE1452

[15] Stott, P. (2016) How Climate Change Affects Extreme Weather Events, *Science* 352: 1517–1518, doi:10.1126/science.aaf7271

[16] National Academies of Sciences, Engineering, and Medicine (2016) *Attribution of Extreme Weather Events in the Context of Climate Change.* Washington, DC: The National Academies Press, https://doi.org/10.17226/21852.

[17] Stocker, T.F., et al. (2013) Technical Summary. In *Climate Change 2013: The Physical Science Basis. Contribution of Working Group I to the Fifth Assessment Report of the Intergovernmental Panel on Climate Change* [Stocker, T.F., et al. (eds.)]. Cambridge University Press, Cambridge, United Kingdom and New York, NY, USA, p. 82-85.

[18] Canty, T., et al. (2013) An Empirical Model of Global Climate—Part 1: A Critical Evaluation of Volcanic Cooling, *Atmospheric Chemistry and Physics* 13: 3997–4031, doi:10.5194/acp-13-3997-2013

For example, a study[19] by the Australian government's Pacific Climate Change Science Program reported that future weather and climate in the region will be characterized by more-intense tropical cyclones, more-frequent deluges, and a greater proportion of hot days and warm nights. Already, people living on Pacific Islands are experiencing changes in their climate, such as higher temperatures, shifts in rainfall patterns, changing frequencies of extreme events, and rising sea levels. These changes are affecting peoples' lives and livelihoods as well as important industries such as agriculture and tourism.

Extreme rainfall events among Pacific islands that currently occur once every 20 years on average are projected to occur four times per year on average by 2055 and seven times per year on average by 2090 under a high-emissions scenario. By 2030, the projected regional warming for the South Pacific is around 0.5 to 1.0 degrees Celsius (0.9 to 1.8 degrees Fahrenheit), regardless of the emissions scenario. This type of information can underpin local decisions about water and food resources, building codes appropriate for adapting to climate change, transportation and infrastructure planning, and other key aspects of planning resilient communities.

Single Event Attribution For years, most scientists have been hesitant to connect single weather events, such as powerful rain storms or hot nights, to climate change. When queried about the connection between global warming and extreme weather, scientists usually begged off with statements such as, "It's too early to say for sure that weather is changing," or "Climate and weather are different and you cannot predict one with the other." That attitude is changing, however, as new studies emerge that link extreme weather events to climate change.[20]

When an extreme weather event occurs, the question is often asked: was this event caused by climate change? However, a more appropriate question is: has climate change altered the odds of occurrence of this extreme event?

Extreme Rain Rainfall, both the amount and the rate, represents one of the strongest signals of climate change. As the atmosphere warms, its ability to hold water (its water vapor content) is expected to increase exponentially with temperature. Warmer air increases the evaporation rate of water, and for every degree Celsius increase in temperature, a given volume of air can hold 7 percent more water. Thus, the odds of an extreme rainfall event occurring have increased.

Although single weather extremes have always occurred and are related to localized processes, they are now unfolding against a background of a warmer atmosphere and amplified water cycle that can turn extreme weather into

a record-breaking event. Studies[21] identify a link between the increase in atmospheric water-holding capacity and increases in heavy precipitation. Thus far, atmospheric water content is increasing in agreement with theoretical expectations,[22] leading researchers to conclude that global warming plays an important role in heavy precipitation **Figure 8.4**.[23]

Because of the limited availability of daily observations, however, early studies examined only model results.[24] However, in early 2011, a study[25] emerged that connected observations of extreme precipitation events with human-induced climate change. Researchers used observational data of precipitation events over two-thirds of Northern Hemisphere land areas and showed that increases in extreme precipitation were consistent with the increase in greenhouse gases and that models[26] may be underestimating observed increases in heavy precipitation with warming.

In the United States, 2017 started off with torrential rainfall in California, marking the wettest winter in a century. Parched after years of drought, the rainfall officially brought the dry spell to an end as floods inundated hundreds of homes, landslides buried roads, and high-water levels threatened to burst dams. Flooding across Missouri and Arkansas in the spring also claimed 20 lives and carried a $1.7 billion price tag.

In 2017, hurricanes Harvey, Irma, and Maria all made landfall as powerful Category 4 storms with winds exceeding 209 kilometers per hour (130 miles per hour). Harvey in particular dumped a truly staggering amount of rain over Houston. The estimated 90.8 trillion liters (24 trillion gallons) that fell was so heavy it actually depressed the land surface more than 1.27 centimeters (0.5 inches) in some spots.[27] Attribution studies indicate that human-induced climate change likely increased Harvey's total rainfall by at least 19 percent, with a best estimate of 37 percent.[28] Another study concluded that climate change increased flooding by around 15 percent.[29]

[19] The Pacific Climate Change Science Program (2011) Climate Change in the Pacific: Scientific Assessment and New Research, www.cawcr.gov.au/projects/PCCSP/publications.html

[20] Trenberth, K.E., et al. (2015) Attribution of climate extreme events, *Nature Climate Change* 5: 725–730, doi:10.1038/nclimate2657

[21] Min, S.-K., et al. (2011) Human Contribution to More Intense Precipitation Extremes, *Nature* 470: 378–381, doi: 10.1038/nature09763

[22] Santer, B.D., et al. (2007) Identification of Human-Induced Changes in Atmospheric Moisture Content, *PNAS*, 104: 15248–15253.

[23] Trenberth, K.E., et al. (2003) The Changing Character of Precipitation, *Bulletin of the American Meteorological Society* 84: 1205–1217.

[24] Min, S.-K., et al. (2009) Signal Detectability in Extreme Precipitation Changes Assessed from Twentieth Century Climate Simulations, *Climate Dynamics* 32: 95–111.

[25] Min et al. (2011)

[26] Allan, R.P. and Soden, B.J. (2008) Atmospheric Warming and the Amplification of Precipitation Extremes, *Science* 321: 1481–1484.

[27] BBC News http://www.bbc.com/news/science-environment-42347510

[28] Risser, M.D. and Wehner, M.F. (2017) Attributable Human-induced Changes in the Likelihood and Magnitude of the Observed Extreme Precipitation During Hurricane Harvey, *Geophysical Research Letters* 44, https://doi.org/10.1002/2017GL075888

[29] Oldenborgh, G.J., et al. (2017) Attribution of Extreme Rainfall from Hurricane Harvey, August, 2017, *Environmental Research Letters* 12: 124009, https://doi.org/10.1088/1748-9326/aa9ef2

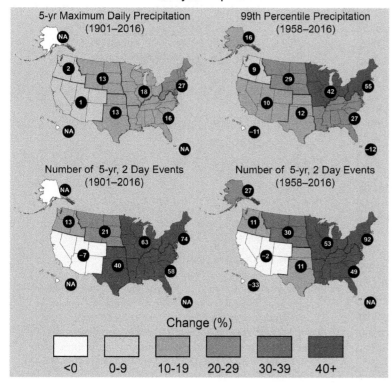

Observed Change in Heavy Precipitation

FIGURE 8.4 Maps showing change in extreme precipitation by region. Upper left, maximum daily precipitation in consecutive 5-year blocks, Upper right, amount of precipitation falling in daily events that exceed the 99th percentile of all nonzero precipitation days. Lower left, number of 2-day events with a precipitation total exceeding the largest 2-day amount that is expected to occur, on average, only once every 5 years, as calculated over 1901–2016. Lower right, number of 2-day events with a precipitation total exceeding the largest 2-day amount that is expected to occur, on average, only once every 5 years, as calculated over 1958–2016. The numerical value is the percent change over the entire period, either 1901–2016 or 1958–2016.

Source: https://science2017.globalchange.gov/chapter/7/

NOAA experts report that extreme weather events have grown more frequent in the United States since 1980. According to Tom Karl, Director of the National Climatic Data Center, part of that shift is due to climate change. "Extremes of precipitation are generally increasing because the planet is actually warming and more water is evaporating from the oceans," he said. "This extra water vapor in the atmosphere then enables rain and snow events to become more extensive and intense than they might otherwise be."[30]

Increasingly, published research[31] is drawing connections between weather and climate change. October and November 2000 marked the wettest autumn in England and Wales since recordkeeping began in 1766. Flooding damaged nearly 10,000 properties, severely disrupted services, and caused billions of dollars in losses. To assess the role of atmospheric warming in these events, researchers used thousands of climate-model simulations of autumn 2000 weather under current conditions and under conditions as they might have been had global warming not occurred. Their data show that in nine out of 10 cases, 20th-century anthropogenic greenhouse gas emissions increased the risk of floods occurring in England and Wales in autumn 2000

by more than 20 percent, and in two out of three cases by more than 90 percent.

U.S. Climate Change Science Program

The U.S. Climate Change Science Program (CCSP) studied the issue of extreme weather under global warming.[32] They found that extreme events have significant impacts on our economy and environment and are among the most serious challenges associated with climate change. Over the past three decades, North America has experienced an increase in unusually hot days and nights, a decrease in unusually cold days and nights, and a reduction in frost days.[33]

According to the CCSP study, heavy rain events and severe droughts have become more frequent and more intense.[34] Hurricanes have increased in power and frequency, although the annual number of North American

---

[30] NOAA Makes It Official: 2011 Among Most Extreme Weather Years in History.

[31] Pall, P., et al. (2011) Anthropogenic Greenhouse Gas Contribution to Flood Risk in England and Wales in Autumn 2000, *Nature* 470, no. 7334: 382, doi: 10.1038/nature09762

[32] U.S. Climate Change Science Program (2008) *Weather and Climate Extremes in a Changing Climate. Regions of Focus: North America, Hawaii, Caribbean, and U.S. Pacific Islands.* A Report by the U.S. Climate Change Science Program and the Subcommittee on Global Change Research. (Washington, D.C., Department of Commerce, NOAA's National Climatic Data Center).

[33] Peterson, T.C., et al. (2008) Changes in North American Extremes Derived from Daily Weather Data, *Journal of Geophysical Research* 113: D07113, doi: 10.1029/2007JD009453

[34] Easterling, D.R., et al. (2007) The Effects of Temperature and Precipitation Trends on U.S. Drought, *Geophysical Research Letters* 34: L20709, doi: 10.1029/2007GL031541

land-falling storms has not shown any trend.[35] Mid-latitude storm tracks are shifting northward, and the strongest storms are becoming even stronger. Throughout these patterns, the CCSP finds extreme weather events can be attributed to anthropogenic climate change. For instance, increased atmospheric water vapor due to warming is theoretically and statistically associated with the increase in heavy precipitation events.

The CCSP found several potential effects of global warming. Continued global warming will lead to future increases in the frequency and intensity of heat waves and heavy downpours. Droughts of greater severity and frequency are likely to occur across substantial areas of North America. There will be future increases in hurricane wind speeds, rainfall intensity, and storm surge levels. During winter, the strongest storms are likely to be more frequent and have stronger winds and more extreme wave heights.

Temperature The Fourth National Climate Assessment[36] reports that the likely range of human contribution to global mean temperature increase over the period 1951–2010 is 0.6 to 0.8 degree Celsius (1.1 to 1.4 degrees Fahrenheit) and the central estimate of the observed warming of 0.65 degree Celsius (1.2 degrees Fahrenheit) lies within this range. This translates to a likely human contribution of 93–123 percent of the observed 1951–2010 change. It is extremely likely that more than half of the global mean temperature increase since 1951 was caused by human influence on climate. The likely contributions of natural forcing and internal variability to global temperature change over that period are minor **Figure 8.5**.

In another study, researchers[38] examining daily wintertime temperature extremes since 1948 found that the warmest days were much more severe and widespread than the coldest days during the Northern Hemisphere winters of 2009–2010. Furthermore, whereas the extreme cold was mostly attributable to a natural climate cycle, the extreme warmth was not.

Researchers[39] also found that the number of unusually warm nights increased during the second half of the 20th century, the rate of increase was greatest in the most recent period, the increase could not be explained by natural climate variability alone, and at least part of the change was attributable to global warming. Unusually warm nights and daily winter temperature extremes: These are just some of the weather phenomena that are attributable to climate change.

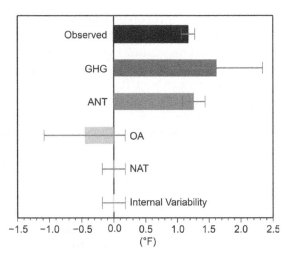

GHG - well-mixed greenhouse gases
ANT - all anthropogenic forcings combined
OA - other anthropogenic forcings
NAT - natural forcings

FIGURE 8.5 Observed global mean temperature trend (HadCRUT4, black bar) and attributable warming or cooling influences of anthropogenic and natural forcings over 1951–2010, with 5-95% error bars. Likely ranges (bar-whisker plots) and midpoint values for attributable forcings are from IPCC AR5. GHG refers to well-mixed greenhouse gases, OA to other anthropogenic forcings, NAT to natural forcings, and ANT to all anthropogenic forcings combined.[37]

Source: https://science2017.globalchange.gov/chapter/3/

Analysis of 2016 The American Meteorological Society (AMS) annually publishes special reports explaining extreme events of the prior year from a climate perspective. Issued in the second half of 2017,[40] the report is sixth in the series but first to find that some extreme events were only possible because of anthropogenic climate change. In the report, researchers from around the world analyzed 27 extreme weather events from 2016 and found that human-caused climate change was a "significant driver" for 21 of them. Six extreme weather events from 2016 were made more likely by global warming.

- Earth reached its highest temperature on record, beating marks set in 2015 and 2014. While that partly reflected the influence of El Niño, the record warmth was only possible because of anthropogenic warming.

[35] Knutson, T.R. and Tuleya, R.E. (2008) Tropical Cyclones and Climate Change: Revisiting Recent Studies at GFDL. In H. Diaz and R. Murnane (eds.), *Climate Extremes and Society*. Cambridge, U.K., Cambridge University Press, pp. 120–144.

[36] Knutson, T., et al. (2017) Detection and Attribution of Climate Change. In *Climate Science Special Report: Fourth National Climate Assessment*, Volume I [Wuebbles, D.J., et al. (eds.)]. U.S. Global Change Research Program, Washington, DC, USA, pp. 114–132, doi: 10.7930/J01834ND

[37] Bindoff, N. L., et al. (2013) Detection and Attribution of Climate Change: From Global to Regional. T. F. Stocker, et al. (eds.), Cambridge University Press, pp. 867–952.

[38] Guirguis, K., et al. (2011) Recent Warm and Cold Daily Winter Temperature Extremes in the Northern Hemisphere, *Geophysical Research Letters* 38: L17701, doi: 10.1029/2011GL048762.

[39] Morak, S., et al. (2011) Detectable Regional Changes in the Number of Warm Nights, *Geophysical Research Letters* 38: L17703, doi: 10.1029/2011GL048531.

[40] Herring, S.C., et al. (2017) (eds.) Explaining Extreme Events of 2016 from a Climate Perspective, *Bulletin of the American Meteorological Society* 99, no. 1: S1–S157.

- Two studies found that unusually high temperatures across Asia and the Arctic in 2016 would not have been possible without anthropogenic climate change.
- Greenhouse gas emissions likely increased the risk of bleaching on the Great Barrier Reef, by increasing thermal stress in the ocean; bleaching risks are likely to increase in the future.
- In southern Africa, global warming was mainly responsible for severe droughts (called **flash droughts**) and heat waves that triggered local food and water shortages that affected millions. These have tripled over the past 60 years.
- Wildfires burned about 8.9 million acres in North America, including a particularly destructive fire in Canada that forced mass evacuations and destroyed 2,400 homes. Anthropogenic climate change most likely played a supporting role by making drying of vegetation and wildfire risk in the region five times more likely during the summer months.
- Unprecedented warm water off the coast of Alaska allowed the spread of toxic algae blooms that killed seabirds by the thousands and forced local fisheries to close. Researchers concluded that it cannot be explained without anthropogenic climate warming, and that more such events were likely with further warming. These will result in a profound shift for people, eco systems, and species.

The AMS report concludes that while these results are novel, they were not unexpected. Scientists have been predicting that the influence of human-caused climate change would eventually push weather events beyond the bounds of natural variability. It was also predicted that we would first observe this phenomenon for heat events where the climate change influence is most pronounced.[41]

Extreme weather events are devastating in their impacts and affect nearly all regions of the globe.[42] Nearly twice as many record hot days as record cold days occur in the United States and Australia. The length of summer heat waves in Western Europe has almost doubled and the frequency of hot days has almost tripled. Extremely hot summers are now observed in over 10 percent of the global land area, compared with only about 0.1 to 0.2 percent for the period 1951–1980.

Daily Extremes

One study[43] found that day-to-day weather has grown increasingly erratic and extreme, with significant fluctuations in sunshine and rainfall affecting more than a third of the planet. Researchers reported that extremely sunny or cloudy days are more common today than they were in the early 1980s. Analysis of daily weather data revealed that swings from thunderstorms to dry days rose considerably since the late 1990s. These swings can have consequences for ecosystem stability and the control of pests and diseases as well as for industries such as agriculture and solar-energy production, all of which are vulnerable to inconsistent and extreme weather.

Probabilities

A special report[44] by the Intergovernmental Panel on Climate Change (IPCC) provides an analysis of extreme weather events, their attribution to climate change, and probabilities of future changes. Extreme weather events vary from year to year and place to place, but overall the number of events and the economic loses they cause have increased over time.

The probability that the frequency of heavy precipitation will increase in the 21st century over many regions is 66 to 100 percent, and it is virtually certain (99 to 100 percent probability) that increases in the frequency of warm daily temperature extremes and decreases in cold daily extremes will occur on a global scale throughout the 21st century. It is also very likely (90 to 100 percent probability) that heat waves will increase in length, frequency, and/or intensity over most land areas.

Projected precipitation and temperature changes imply changes in floods; however, because of limited evidence and because the causes of regional climate changes are complex, there is low confidence overall at the global scale regarding climate-driven changes in magnitude or frequency of river-related flooding.

It is likely (66 to 100 percent probability) that the average maximum wind speed of tropical cyclones will increase throughout the coming century, although possibly not in every ocean basin. It is also likely that overall there will be either a decrease or essentially no change in the number of tropical cyclones. It is very likely (90 to 100 percent probability) that average sea-level rise will contribute to upward trends in extreme coastal (high) water levels.

Overall, the IPCC special report finds that human activity related to global warming has driven increases in some extreme weather and climate events around the world in recent decades. Those events and other weather extremes will worsen in coming decades as greenhouse gases build.

Extreme weather is consistent with what we know is occurring as a result of climate change. For instance, on average, the United States is 1.1 degrees Celsius (2 degrees Fahrenheit) warmer than it was 40 years ago. Warmer air increases the odds of extreme precipitation[45] because the air holds more moisture and can release more of it during

[41] Horton, R.M., et al. (2016) A Review of Recent Advances in Research on Extreme Heat Events, *Current Climate Change Reports*, 2: 242–259, doi:10.1007/s40641-016-0042-x

[42] NAS (2016)

[43] Medvigy, D. and Beaulieu, C. (2011) Trends in Daily Solar Radiation and Precipitation Coefficients of Variation since 1984, *Journal of Climate* 25: 1330–1339, doi: 10.1175/2011JCLI4115.1

[44] IPCC (2012) Managing the Risks of Extreme Events and Disasters to Advance Climate Change Adaptation. A Special Report of Working Groups I and II of the Intergovernmental Panel on Climate Change [C. B. Field, et al. (eds.)]. Cambridge University Press, Cambridge, UK, and New York, NY, USA, 582 pp.

[45] Min (2009)

rainstorms and snowstorms. Heavy precipitation, both rain and snow, is happening more often[46] than it used to.

Heat-related extreme events are also on the rise around the globe, and global warming has significantly increased the odds of some specific events, including the killer European heat wave of 2003[47] and the Russian heat wave of 2010.[48] Even small increases in average temperatures raise the risk of heat waves, droughts, and wildfires.[49] Twice as many record highs have been set in the past decade as record lows in the United States.[50] By 2050, record highs could outpace record lows by 20 to one in the United States. By the end of the century, the ratio could jump to 100 to one if greenhouse gas emissions continue unabated.

Extending the Record of Extreme Weather In **Chapter 3**, we discussed the hypothesis that instabilities in the Northern Hemisphere Polar Jetstream lead to extreme weather. Large meanders to the north in the jetstream tend to pull tropical air masses into temperate and subArctic latitudes and produce heat waves. Meanders to the south pull down cold Arctic air producing freezing conditions. Storminess, too, is associated with these instabilities as low-pressure fronts develop ahead of meanders and produce stormy weather.

It has been proposed that these instabilities in the jet have increased as a result of global warming.[51] Some researchers, however, have criticized this hypothesis as not provably related to climate change because the record of Polar Jetstream position only extends as far as the era of satellite monitoring, the past 3 decades.

Using the fact that August temperatures in the British Isles and the Northeastern Mediterranean reflect the summer position of the North Atlantic portion of the Polar Jetstream, researchers used a 300-year record of tree rings to assess historical jet stream positions from 1725 to 1978. To bring the record up to present day, they relied on data from meteorological observations spanning 1979 to 2015. The part of an annual tree ring that forms late in the growing season is called latewood. The density of latewood in a tree ring reflects the August temperature that year.

Assessment of the centuries-long record revealed that after 1960, there were more years when the jetstream was in an extreme position. Increased fluctuations in the path of the North Atlantic jet stream since the 1960s coincide with more extreme weather events in Europe such as heat waves, droughts, wildfires, and flooding. When the North Atlantic Jet is in the extreme northern position, the British Isles and western Europe have a summer heat wave while southeastern Europe has heavy rains and flooding. When the jet is in the extreme southern position, the situation flips: Western Europe has heavy rains and flooding while southeastern Europe has extreme high temperatures, drought and wildfires. Researchers relate that heat waves, droughts and floods affect people, and on top of already increasing temperatures and global warming, they describe it as a "double whammy."[52]

8.3 Drought

The nations of the world face a growing threat of severe and prolonged drought in coming decades.[53] Warming temperatures associated with climate change will likely create increasingly dry soil conditions across much of the globe in the next 30 years, possibly reaching a scale in some regions by the end of the century that has rarely if ever been observed in modern times.[54]

As we learned in **Chapter 3**, the region around 30 degrees latitude is characterized by dry, sinking air associated with the Hadley Cell (part of global atmospheric circulation). Known as the subtropics, this great belt around the globe is characterized by deserts, few clouds, and little precipitation. Climate studies[55] indicate that as global warming continues, and the tropics expand,[56] the subtropics will expand as well, changing precipitation patterns around the world.

In Assessment Report 4, the IPCC projected that expanding drought will be associated with expansion of the subtropical belt, with high-latitude areas getting more precipitation. In their *Special Report on Managing the Risks of Extreme Events and Disasters to Advance Climate Change Adaptation*, the IPCC calculated a large drying trend over many Northern Hemisphere land areas since mid-1950 and an opposite trend in eastern North and South America.[57] They report that one study found that very dry land areas across the globe have more than doubled in extent since the 1970s, initially as a result of a short-term El Niño event and subsequently due to surface heating (air warming).[58]

[46] Alexander, L.V., et al. (2006) Global Observed Changes in Daily Climate Extremes of Temperature and Precipitation, *Journal of Geophysical Research—Atmospheres*, 111.D5: D05109, doi:10.1029/2005JD006290

[47] Christidis, N., et al. (2011) The Role of Human Activity in the Recent Warming of Extremely Warm Daytime Temperatures, *Journal of Climate*, doi: 10.1175/2011JCLI4150.1

[48] Rahmstorf and Coumou (2011)

[49] Romps, D.M., et al. (2014) Projected Increase in Lightning Strikes in the United States Due to Global Warming, *Science*, 14 Nov., 851–854.

[50] Meehl, G.A., et al. (2009) Relative Increase of Record High Maximum Temperatures Compared to Record Low Minimum Temperatures in the U.S., *Geophysical Research Letters* 36: L23701, doi: 10.1029/2009GL040736

[51] Francis, J.A. and Vavrus, S.J. (2012) Evidence Linking Arctic Amplification to Extreme Weather in Mid-latitudes, *Geophysical Research Letters* 39, L06801.

[52] Trouet, V., et al. (2018) Recent Enhanced High-summer North Atlantic Jet Variability Emerges from Three-century Context, *Nature Communications* 9, no. 1, doi: 10.1038/s41467-017-02699-3

[53] Dai, A. (2011) Drought under Global Warming: A Review, *Climate Change* 2: 45–65, doi: 10.1002/wcc.81

[54] Drought may threaten much of globe within decades, https://www2.ucar.edu/atmosnews/news/2904/climate-change-drought-may-threaten-much-globe-within-decades

[55] Dai (2011)

[56] Lu, J., et al. (2009) Cause of the Widening of the Tropical Belt Since 1958, *Geophysical Research Letters* 36: L03803, doi: 10.1029/2008GL036076

[57] IPCC (2012)

[58] CO2NOW.org: http://co2now.org/Know-the-Changing-Climate/Climate-Changes/ipcc-faq-changes-in-extreme-events.html

Drought severity has been increasing across the Mediterranean, southern Africa, and the eastern coast of Australia over the course of the 20th century. Semi-arid areas of Mexico, Brazil, southern Africa, and Australia have also encountered growing desertification as the world has warmed. Research[59] predicts these trends will accelerate and expand as warming continues. Significantly drier conditions will spread across 20 to 30 percent of the global land surface if global warming reaches 2 degrees Celsius (3.6 degrees Fahrenheit), a level of warming that could arrive as early as mid-century.[60]

Increased drought and wildfire will escalate as a result in Central America, Southern Europe, Southern Australia, parts of South East Asia, and Southern Africa, areas home to about one-fifth of the human population. Limiting warming to under 1.5 degrees Celsius (2.7 degrees Fahrenheit) would dramatically reduce the fraction of the Earth's surface that undergoes such changes.

8.3.1 Seasonal Patterns

Drought can also take on seasonal patterns. For instance, as climate warms, more precipitation will take the form of rain and less as snow, and snow that does accumulate during winter will melt faster and earlier in the spring. Planners and managers have long worried that the consequence of this pattern will be growing flood risk early in the year and decreasing discharge in the mid- and late summer, producing drought in the summer and fall growing season.

Researchers[61] have found this pattern verified in model studies that show faster and earlier snowpack melting due to rising air temperature, which in turn produces an increase in catastrophic events such as flooding and summer droughts. That this pattern is present today and is historically unusual has been confirmed by a reconstruction[62] of 800 years of snowpack size for the watersheds feeding the Colorado, Columbia, and Missouri rivers. Results show that snowpack in the northern Rocky Mountains has shrunk at an unusually rapid pace during the past 30 years. The research shows that recent declines are nearly unprecedented, owing to a combination of natural variability and human-induced atmospheric warming.

Drought has many impacts. For instance, in the U.S. Rocky Mountain region, a steady decline in the winter snowfall[63] has produced some important effects. As the snowpack at high elevations decreases, elk browse on plants that were previously inaccessible during the snow season. As a result, deciduous trees and associated songbirds in mountainous Arizona have decreased over the past two decades.[64] The increased browsing results in a trickle-down effect, such as lowering the quality of habitat for mountain songbirds.

8.3.2 Regional Drought

Dry periods are not unusual in history, and they have occurred many times over the past thousand years. North America, West Africa, and East Asia have experienced megadroughts triggered by irregular tropical sea-surface temperatures. La-Niña-like sea-surface temperature conditions lead to drought in North America, and El-Niño-like sea-surface temperatures affect the famous wet season known as the monsoon, causing drought in East Asia and westward.

Models project increased aridity in the 21st century over most of Africa, southern Europe, the Middle East, most of the Americas, Australia, and Southeast Asia. Research indicates this process is already under way and that drought has expanded and deepened over the 20th century.[65] Studies found that the percentage of Earth's land area afflicted by serious drought more than doubled from the 1970s to the early 2000s, and as a result, some of the world's major rivers are losing water,[66] threatening drinking water and crop irrigation in previously stable areas.

8.3.3 Drought Severity

Droughts are events associated with reduced precipitation, dry soils leading to crop failure, and imperiled drinking-water supplies. Drought is measured by the **Palmer Drought Severity Index (PDSI)**, which tracks precipitation and evaporation and compares them to historical patterns. Model studies[67] indicate that by the 2030s some regions could experience particularly severe drought, including much of the central and western United States; lands bordering the Mediterranean, Central America, and the Caribbean region; and portions of Europe and Asia. By the end of the century, drought could intensify and spread with continued warming. Many populated areas, including the United States, could reach unprecedented levels of drought severity (**Figure 8.6**).

[59] Park, C.-E., et al. (2018) Keeping Global Warming Within 1.5°C Constrains Emergence of Aridification, *Nature Climate Change*, doi: 10.1038/s41558-017-0034-4

[60] Reuters https://www.reuters.com/article/us-climatechange-draft/warming-set-to-breach-paris-accords-toughest-limit-by-mid-century-draft-idUSKBN1F02RH

[61] Molini, A., et al. (2011) Maximum Discharge from Snowmelt in a Changing Climate, *Geophysical Research Letters* 38: L05402, doi: 10.1029/2010GL046477.

[62] Pederson, G., et al. (2011) The Unusual Nature of Recent Snowpack Declines in the North American Cordillera, *Science* 333: 332–335.

[63] Pierce, D.W., et al. (2008) Attribution of Declining Western US Snowpack to Human Effects, *Journal of Climatology* 21: 6425–6444.

[64] Martin, T.E. and Maron, J.L. (2012) Climate Impacts on Bird and Plant Communities from Altered Animal–Plant Interactions, *Nature Climate Change* 2: 195–200, doi: 10.1038/nclimate1348

[65] Dai, A., et al. (2004) A Global Dataset of Palmer Drought Severity Index for 1870–2002: Relationship with Soil Moisture and Effects of Surface Warming, *Journal of Hydrometeorology* 5: 1117–1130.

[66] Drought May Threaten Much of Globe Within Decades. http://www2.ucar.edu/news/2904/climate-change-drought-may-threaten-much-globe-within-decades

[67] Dai (2011)

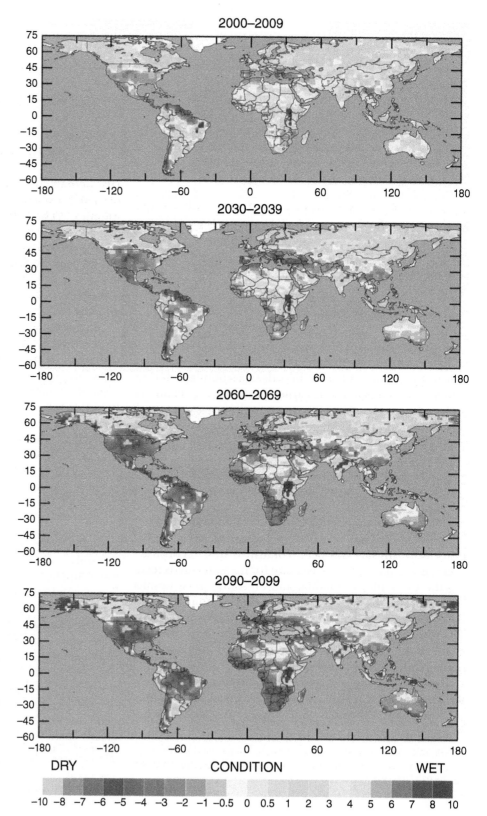

FIGURE 8.6 These four maps illustrate the potential for future drought worldwide, based on current projections of future greenhouse gas emissions. The maps use the Palmer Drought Severity Index, which assigns positive numbers when conditions are unusually wet for a particular region and negative numbers when conditions are unusually dry. A reading of –4 or below is considered extreme drought.

Source: https://www2.ucar.edu/atmosnews/news/2904/climate-change-drought-may-threaten-much-globe-within-decades

Box 8.1 | Spotlight on Climate Change

Syria and the Rise of European Demagogues

Severe drought is causing civil unrest, and political turmoil in the Middle East. Originating with, and amplified by climate change, it has contributed to a civil war in Syria and the rise of populism and nationalism throughout Europe.

Drought

Over 250,000 people have died and millions more have become refugees as a result of political turmoil arising from a severe drought in Syria that scientists describe is the worst in at least the past 900 years. Using tree rings, which provide reliable evidence of precipitation, NASA scientists concluded that the drought was larger than any seen in the region for centuries.[68] "We are starting to push the climate system outside of what it would normally do," said Benjamin Cook, a climate scientist at NASA's Goddard Institute for Space Studies.

The drought caused 85% of livestock to die and 75% of Syrian farms to fail, forcing up to 1.5 million Syrians to leave family homelands, and move to Syrian cities such as Homs, Aleppo, and Damascus. The addition of so many hungry and homeless families in Syria's cities created a stressful condition that ultimately led to the outbreak of civil war.[69]

In this way, climate change, in the guise of severe drought, multiplied existing weaknesses in the socio-economic framework of Syria. The corrupt government of President Bashar al-Assad and the lack of adequate civil services that resulted, had created a long simmering resentment among the population of the country. Climate change fueled the explosion of this resentment into a civil war and opened the door to religious extremism in the form of Al-Qaida and ISIS.[70]

Refugees

In 2016, from an estimated pre-war population of 22 million, the United Nations identified 13.5 million Syrians requiring humanitarian assistance. Of these, more than 6 million are internally displaced within Syria, and 5 million are refugees outside of Syria. Syrian refugees fled the violence through two primary corridors; north through Turkey and south, through North Africa. Both routes require Syrian families to navigate open ocean to access Europe.

Taking advantage of this opportunity, smugglers in Turkey and Libya offer families boat passage to European shores. The resulting situation is one of recent history's most horrific and shameful chapters.

In a typical situation, a family will pay thousands of U.S. dollars for transit on overcrowded, unstable, and unsanitary vessels. Typical boats are 9 m (30 ft) rubber inflatables, with a wooden floor, and an outboard engine. The one-person cost of transit on these boats ranges from $500 to $700 per adult. It is common practice for smugglers to take these boats out to international waters, 3 miles offshore, remove the engine, and set the boat adrift after telling the refugees that they will be rescued by either the Greek or Italian coast guard. Too often, however, overcrowded boats flip or sink and result in the loss of hundreds of lives.

Symbolic of this human tragedy, is the image of a drowned Syrian boy (**Figure 8.7**) washed up on a Turkey beach that brought worldwide attention to this problem.

Political Turmoil

In the European Union (E.U.), the refugee catastrophe is the worst humanitarian crisis since World War II. It has had a destabilizing effect on the political landscape. Faced with millions of Syrians in need of assistance, the nations of the E.U. are confronted with the choice of closing their borders to preserve national resources

FIGURE 8.7 Three-year-old Aylan Kurdi, from Syria, part of a group of 23 trying to reach the Greek island of Kos. They'd set out in two boats on the 13-mile Aegean journey, but the vessels capsized. Aylan's brother and mother also drowned.

Source: https://www.npr.org/sections/parallels/2015/09/03/437132793/photo-of-dead-3-year-old-syrian-refugee-breaks-hearts-around-the-world

[68] Cook, B.I., et al. (2016) Spatiotemporal Drought Variability in the Mediterranean over the last 900 Years, *Journal of Geophysical Research and Atmosphere* 121, no. 5: 2060–2074, doi:10.1002/2015JD023929.
[69] Obama, B. (2015) *Remarks by the president at the United States coast guard academy commencement, (20 May);* https://www.whitehouse.gov/the-press-office/2015/05/20/remarks-president-united-states-coast-guard-academy-commencement
[70] Cole, J. (2015) Did ISIL arise partly because of climate change? *The Nation,* (July 24); available at: http://www.thenation.com/article/did-isil-arise-partly-because-of-climate-change/

and culture, or offer aid typically at a high cost. In many cases, where refugees have been assimilated, local citizen groups reject the challenge of hosting the immigrants, and have turned to populism and nationalism.

This has given rise to a new class of politicians, who are riding to power on demagoguery. These actors play on the fears of their constituents by promising to protect them from the loss of jobs, affordable housing, and other imagined fears connected to the refugees. These politicians encourage the growth and spread of xenophobia across the E.U. For the first time since before World War II, ultraconservative politicians are finding a voice and widespread public support.

Two prominent examples characterize this situation: Brexit and the Austrian Presidency. Brexit, as it has come to be known, was a vote by British citizens to determine their continued membership in the E.U. To the surprise of the world, the British voted to leave the E.U. This is widely perceived as a reaction against the influx of Muslim, Syrian families. Catalyzed by acts of terrorism in London and other European locations, many voters chose to leave the E.U. because of fears for their safety and security. This is a case of spreading prejudice where English families associate typical Syrian families with terrorists.

The second example is the recent election of 31-year-old conservative Sebastian Kurz, Austria's new chancellor. Kurz rode a wave of anti-immigrant anxiety to become the world's youngest leader. He advocates a hard line on immigrants. He is against the distribution of social benefits for new comers. He is committed to shutting down the "Balkan" route that asylum seekers have taken toward Austria. He has promised to dramatically lower taxes. These actions have found widespread appeal to common voters, as they promote personal safety, national security, and keeping workers' paychecks in their pockets. Kurz's election is widely seen by political observers as a sign that the E.U. is turning to the right and will become characterized by more conservative politics.

Is the Syrian sequence of events a crystal ball revealing the future of humanity? Global warming will stress resources in every location. What nation, already stressed by climate change, population growth, and declining natural resources, can afford to open its doors to an influx of refugees that look different, act different, practice different religions, speak a different language, eat different food, and raise their children in different ways? Is the world fated to become a labyrinth of locked borders? As global warming intensifies, and more places in the world become unlivable, will a swelling population of climate refugees push global politics toward totalitarianism?

8.4 Wildfire

In 2015, for the first time on record, over 10 million acres were burned in one year in the United States. In 2017, a total of 58,000 wildfires burned more than 9.2 million acres, the second worst on record.

8.4.1 Influence of Climate Change

The role played by climate change in the rising occurrence of wildfires is at the same time obvious and subtle. Humans ignite the vast majority of fires, and land-use and management practices contribute to the problem. However, a warmer world sets the stage for fires to be larger, hotter, and more frequent. Therefore, the story of growing wildfire occurrence involves many factors and complexities.

For example, plants absorb carbon dioxide through tiny pores in their leaves called stomata. They also release water through these same pores every time they open. This process of respiration is fundamental to plant metabolism. However, the higher the air temperature, the more water they lose, causing plants to become drier as global warming progresses. Warmer weather also causes more water to evaporate from soil and dead plants littering the ground.

Thus, in a warmer world, forests are drier and it's easier for fires to start and spread.

Typically, fires are started by human negligence (or intent) and lightning strikes. Both are growing in number. Human communities are pushing out from urban centers in the form of ever-widening suburbs, highways, hiking trails, and recreation areas. Lightning strikes, too, may become more frequent. Generally, lightning occurs more frequently in hot weather than cold weather. One study[71] found that the number of lightning strikes in the United States could increase by about 12 percent for every degree Celsius of warming. Research[72] indicates that climate change has already played a significant role in making forests in the western U.S. drier and more likely to burn, nearly doubling the area affected by wildfires over the last three decades.

Rising temperatures don't just increase the chance that a fire will start at any given time—they also lengthen the total time throughout the year that conditions are right for wildfires. Research shows that spring is arriving earlier

---

[71] Romps, et al. (2014)
[72] Abatzoglou, J.T. and Williams, A.P. (2016) Impact of Anthropogenic Climate Change on Wildfire Across Western US Forests, *PNAS* 113, no. 42: 11770–11775.

in many places, based on the timing of snowmelt and the reemergence of vegetation from their dormant winter.[73]

In the southeastern U.S., spring is arriving up to 3 weeks earlier. Three-quarters of the U.S. national park system have seen[74] an advance in the timing of spring over the past century, and spring is coming an average 8 days earlier in the northern forests of North America and Eurasia.[75] An earlier spring means an earlier beginning and end to the period of winter snowmelt, earlier onset of late spring and summer drought, and an earlier start to the wildfire season.

Convergence of Factors Local conditions can amplify the likelihood of wildfires. For instance, in 2017, a unique combination of weather events and Pacific climate variability set the stage for the massive Thomas wildfire in Southern California. The fire scorched 281,893 acres, destroyed 1,063 structures, and killed 42 people, as key weather and climate factors drove the flames forward.

Following several years of debilitating drought, California as a whole experienced a wet winter in 2016/2017. Fast-growing grasses and shrubs that dominate the chaparral landscape around Los Angeles flourished in the moist conditions. A hot summer in 2017 was followed by a record-breaking October heat wave. This dried out the vegetation, turning it into perfect fuel. Typical season-ending rains failed to materialize, due in part to cooler Pacific Ocean temperatures under recurring La Niña conditions. This complex, self-reinforcing tapestry of weather and climate processes left vast areas of Southern California covered in desiccated tinder. When the hot, arid Santa Ana winds picked up with unusual speed, the likelihood of a fire starting, and spreading, increased dramatically. The heat, the vegetation, the rain (and lack thereof), and the wind all aligned at the end of the year to create a record-setting season of wildfire in Southern California.

Positive Feedback Every large fire turns forests into massive carbon polluters, releasing carbon dioxide that was locked up in soil and foliage and contributing to the rise of greenhouse gases and the expansion of climate change in a dangerous feedback loop. This is of particular concern for boreal forests in North American and Eurasia, which contain heavy deposits of carbon-rich peat. A large fire, capable of burning 1 to 2 meters (3.3 to 6.6 feet) deep into the soil, could release many thousands of years of carbon accumulation in one blast.

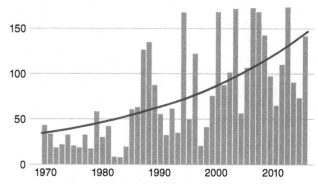

Large Wildfires Increasing Across the West
Number of fires larger than 1,000 acres per year on U.S. Forest Service land

Source: Climate Central analysis of U.S. Forest Service records

FIGURE 8.8 The number of large fires on U.S. Forest Service land is increasing dramatically. The area burned by these fires is also growing at an alarming rate.[77]

Source: http://www.climatecentral.org/news/western-wildfires-climate-change-20475

8.4.2 Large Fires

An analysis[76] of 45 years of U.S. Forest Service records from the western U.S. shows that the number of large fires on Forest Service land is increasing dramatically (**Figure 8.8**). Compared to the 1970s, today the average annual number of fires larger than 1,000 acres has more than tripled, the fire season is 105 days longer, and more than six times as many acres were burned in the 2010s than in the 1970s. It is approaching the point where the notion of a fire season will be made obsolete by the reality of year-round wildfires across the West.

Megafires Studies[78] of large, high intensity wildfires (megafires) suggest that high temperatures, dry conditions, and strong winds are key components that amplify fires into extreme wildfire events. Using satellite data from 2002 to 2013, researchers examined nearly 23,000 fires worldwide, identifying 144 large, high-intensity, extreme wildfire events where people died and homes were destroyed. With monthly world weather data from 2000 to 2014, the researchers modeled the likely changes in fire behavior from 2041 to 2070, predicting a 20 to 50 percent increase in the number of days when conditions are conducive to such fires.

By 2041, there will likely be 35 percent more large, catastrophic fires per decade. That translates to four extreme

[73] National Phenology Network (2017) Status of Spring, https://www.usanpn.org/data/spring

[74] Monahan, W.B., et al. (2016) Climate Change is Advancing Spring Onset Across the U.S. National Park System, *Ecosphere* 7, no. 10, http://dx.doi.org/10.1002/ecs2.1465

[75] Pulliainen, J., et al. (2017) Early Snowmelt Significantly Enhances Boreal Springtime Carbon Uptake, *PNAS* 114, no. 42, 11081–11086, doi: 10.1073/pnas.1707889114

[76] Kenward, A., et al. (2016) Western Wildfires, A Fiery Future, a report to Climate Central, http://assets.climatecentral.org/pdfs/westernwildfires-2016vfinal.pdf

[77] Climate Central, Climate change is tipping scales toward more wildfires http://www.climatecentral.org/news/western-wildfires-climate-change-20475

[78] Bowman, D., et al. (2017) Human Exposure and Sensitivity to Globally Extreme Wildfire Events, *Nature Ecology & Evolution*, 1, no.3: 0058, doi: 10.1038/s41559-016-0058

fire events for every three that occur now. Forests in the western United States, southeastern Australia, Europe and the eastern Mediterranean region that extends from Greece to Lebanon and Syria are among those areas at highest risk. Wind-driven fires accounted for nearly 35 percent of these catastrophic events, while severe drought was a factor in nearly 22 percent. Other extreme fire weather conditions, largely due to high temperatures and low humidity, accounted for slightly more than 20 percent of these costly fires.

8.4.3 Fire Season Cost

It was only the beginning of July, and 2017, unfortunately, was already unfolding as the new normal: an early start to the fire season, and millions of acres of forest and range burned or was ablaze as the summer just began to heat up. At least 60 large blazes were at work devouring parts of the West, adding to the 3.4 million acres already burned that year. As early as April, wildfires had scorched more than 2 million acres in the United States, nearly the average consumed in entire fire seasons during the 1980s. At least 20 new, large fires had ignited in the first week of July, forcing thousands of people from their homes.

Described by one historian as the "Pyrocene," global warming is creating a new epoch in the American west, where the summer season is changing from a time of growth and renewal, into a season of burning. Climate change[79] has nearly doubled the amount of forest burned in the United States western states since 1984. The length of the season, along with bigger, more intense fires, is taxing budgets.

The U.S. Forest Service, which is under the U.S. Department of Agriculture, dedicated half of its budget to fighting fires in 2015, exceeding 50 percent for the first time in its 112-year history.[80] Former Agriculture Secretary Tom Vilsack joked that it should be called the "Fire Service." The fire season used to be approximately 60 days in length, then 90 days, and now it is nearly year-round.

The pressure on the Forest Service to cope with bigger, more dangerous fires, has meant more funding going toward fighting blazes and not enough toward preventing them in the first place. In 1995, fire made up 16 percent of the Forest Service's annual budget compared to more than 50 percent in 2015. This shift in funds to provide firefighters, aircraft, and other assets necessary to protect lives, property, and natural resources from catastrophic wildfires, has meant a corresponding reduction elsewhere. All non-fire personnel have been reduced by 39 percent, and fewer and fewer funds and resources are available to support other agency work (such as programs that reduce the fire threat).

Extrapolating this trend to 2025 indicates the budget devoted to fire could exceed 67 percent, meaning that in just 10 years, two out of every three dollars the Forest Service gets from Congress as part of its appropriated budget will be spent on fire programs **Figure 8.9**.

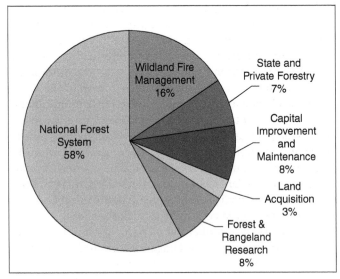
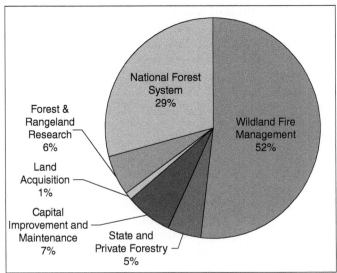

FIGURE 8.9 From 1995 (left) to 2015 (right), the Wildland Fire Management appropriation has more than tripled in its portion of the Forest Service budget from 16% to 52%.

Source: https://www.fs.fed.us/sites/default/files/2015-Fire-Budget-Report.pdf

[79] Abatzoglou and Williams (2016)

[80] U.S. Forest Service (2015) The Rising Cost of Wildfire Operations: Effects on the Forest Service's Non-Fire Work, Aug. 4 https://www.fs.fed.us/sites/default/files/2015-Fire-Budget-Report.pdf

8.4.4 Human Health

Since the 1970s, the number of large wildfires in the United States (10,000 acres or more) has increased fivefold. This is worrying because exposure to particles and gases associated with wildfire smoke may cause breathing and heart-related problems in human populations. To assess this issue, researchers measured air quality associated with wildfires between 2006 and 2013 using satellite information about the presence and spread of smoke plumes. Levels of fine particles and ozone were also analyzed using the U.S. Environmental Protection Agency's Air Quality System that monitors air pollution levels at different sites across the United States.

Researchers found that ozone, a gas associated with poor air quality and related to breathing problems, was 11.1 percent higher on days with large wildfire plumes than on clear days. Fine particle levels were also significantly higher than normal, occurring 33.1 percent more often on such days. A striking finding was that although plumes had occurred only on 6 to 7 percent of the days, these plumes accounted for 16 percent of unhealthy days due to small particles and 27 percent of unhealthy days due to ozone.[81]

8.5 Ecosystem Impacts

Climate is an important environmental influence on ecosystems.[82] Changing climate affects ecosystems in a variety of ways. For instance, warming may force species to migrate to higher latitudes or higher elevations where temperatures are more conducive to their survival. Similarly, as sea-level rises, saltwater intrusion into a freshwater system may force some key species to relocate or die, thus removing predators or prey that are critical in the existing food chain.

8.5.1 Variety of Impacts

Climate change not only affects ecosystems and species directly but it also interacts with other human stressors such as land development, pollution, and noise. Although some stressors cause only minor impacts when acting alone, their cumulative impact may lead to dramatic ecological changes. For instance, climate change may exacerbate the stress that land development places on fragile coastal areas. Additionally, recently logged forested areas may become vulnerable to erosion as climate change leads to increases in heavy rainstorms.

Plant Physiology In a study[83] based on decades of observations, researchers have documented how a broad group of plant species living in open conditions (rather than a controlled laboratory) have responded to rising temperatures. Data from historical records of 1,558 species of wild plants on four continents show that leafing and flowering advances, on average, five to six days per degree Celsius of warming.

The power of this finding is the global distribution of the database and the fact that it records plant behavior under real-life conditions of seasons, weather, predator–prey relationships, and other natural wild conditions. The results are statistically consistent across species and geographic datasets.

When compared to the usual method of understanding how plants react to warming (small-scale experiments of a few plants under laboratory conditions), these data show that previous estimates of plant response to global warming grossly under predict advances in flowering by eight and a half times and advances in leafing by four times. These results suggest that the way global warming experiments on plant health are currently conducted needs to be reevaluated, especially because data of this type are used to parameterize global climate models when predicting responses to global warming and changes in the carbon cycle.

Multiple Stressors Ecosystems are sensitive to the balance of multiple stressors, both natural and human related. Studies[84] show that natural decreases in biodiversity are as potentially damaging as the negative impacts resulting from climate change, pollution, and other major forms of environmental stress. Because natural stressors are ever present, the growth of negative impacts related to climate change and human population growth could cause increasing damage to ecosystems that are already stressed as a natural condition.

Researchers combined data from published accounts of how environmental factors affect two important ecosystem processes: plant growth and decomposition of dead plants by bacteria and fungi. They found that species losses of 1 to 20 percent have negligible effects on ecosystem plant growth; losses of 21 to 40 percent reduce plant growth by 5 to 10 percent, which is comparable to the impact of global warming and increased ultraviolet radiation due to stratospheric ozone loss; and losses of 41 to 60 percent equate with the effects of major damage such as ozone pollution, acid deposition on forests, and nutrient pollution. This research suggests that natural stressors to global biodiversity will be amplified by the growth of climate change.

[81] Larsen, A.E., et al. (2017) Impacts of fire smoke plumes on regional air quality, 2006–2013. *Journal of Exposure Science & Environmental Epidemiology*, doi: 10.1038/s41370-017-0013-x

[82] EPA Climate Impacts on Ecosystems https://19january2017snapshot.epa.gov/climate-impacts/climate-impacts-ecosystems_.html

[83] Wolkovich, E.M., et al. (2012) Warming Experiments Under predict Plant Phenological Responses to Climate Change, *Nature*, doi: 10.1038/nature11014

[84] Hooper, D.U., et al. (2012) A Global Synthesis Reveals Biodiversity Loss as a Major Driver of Ecosystem Change, *Nature*, doi: 10.1038/nature11118

Range Shifts Climate change causes many terrestrial species to shift to higher elevations and higher latitudes in order to maintain the same climate conditions that are optimal to their survival. But researchers[85] have found that species are responding to climate change up to three times faster than previously appreciated. Species have moved toward the poles at three times the rate previously accepted in the scientific literature, and they have moved to cooler, higher altitudes at twice the rate previously realized.

On average, species have moved to higher elevations at 12.2 meters (40 feet) per decade and, more dramatically, to higher latitudes at 17.6 kilometers (11 miles) per decade. Scientists estimate that these changes are equivalent to animals and plants shifting away from the equator at around 20 centimeters (8 inches) per hour, every hour of the day, and every day of the year. They estimate that this trend has been going on for the last 40 years and that it will continue for at least the rest of this century.

Growing Season Global warming is even changing the routine of America's home gardeners. On the back of seed packets bought by 80 million U.S. gardeners each year is a color-coded map[86] of plant hardiness zones. The map provides guidance on where various species of flowers, vegetables, and ornamental plants will have an optimal growing climate.

For the first time since 1990, the U.S. Department of Agriculture revised this official guide and shifted about half the continental United States approximately a half zone to the north. The new map reflects the fact that climate (and growing) zones have shifted strongly to the north as a consequence of changing climate. Nearly entire states, including Ohio, Nebraska, and Texas, have been updated to warmer zones. On average, the growing season in the contiguous U.S. has increased approximately 2 weeks compared to the average of 1895–2015 (**Figure 8.10**).

Extinction Changes in habitat quality cause changes in the distribution of food sources and place wildlife populations under stress. Species extinction and the degradation of ecosystems are proceeding rapidly, and the pace is accelerating. Biodiversity is declining rapidly throughout the world, and the challenges of conserving the world's species are made even larger in light of the negative effects of global climate change. The world is losing species at a rate that is 100 to 1000 times faster than the natural extinction rate.[87]

The ways humans use land, the ocean, and other natural resources affect the distribution and quality of plant and

FIGURE 8.10 The length of the growing season in the contiguous 48 states compared with a long-term average (1895–2015), where "growing season" is defined by a daily minimum temperature threshold of 5°C (41°F). For each year, the line represents the number of days shorter or longer than the long-term average. The line was smoothed using an 11-year moving average. Choosing a different long-term average for comparison would not change the shape of the data over time.

Source: https://science2017.globalchange.gov/chapter/10/

animal habitats. The area of undeveloped space for wildlife is continually declining under the pressure of a growing human population. Essential freshwater systems are affected by pollution, damming, and diversion of water for human use. No area of the ocean is untouched by human pollution in some form.[88] Climate change is driving aquatic and forest ecosystems toward the heads of their watersheds at the highest elevations, with little recourse thereafter as warming continues.

One group of researchers[89] have concluded that if global warming persists as expected, almost one-third of all flora and fauna species worldwide could become extinct and that by 2080 more than 80 percent of genetic diversity within species could disappear in certain groups of organisms.

Climate change can affect species in relation to their role in an ecosystem. Scientists hypothesize that species in rich, biodiverse ecosystems are exposed to heightened threats by the consequences of global warming, specifically extreme weather events. High winds, torrential downpours, and droughts have become more frequent; this increases the risk for species extinction in diverse ecosystems such as coral reefs and tropical rainforests.

In a rainforest or on a coral reef there are a wide variety of species of primary producers. Primary producers are organisms (such as green plants and algae) that produce biomass from inorganic compounds and thus provide a

[85] Chen, I.-C., et al. (2011) Rapid Range Shifts of Species Associated with High Levels of Climate Warming, *Science* 333, no. 6045: 1024, doi: 10.1126/science.1206432.

[86] Plant hardiness. http://planthardiness.ars.usda.gov/PHZMWeb/; Climate change comes to your backyard. http://wwwp.dailyclimate.org/tdc-newsroom/usda/climate-change-comes-to-your-backyard

[87] Biodiversity Crisis is Worse, Climate Change Experts Say. http://www.sciencedaily.com/releases/2012/01/120120010357.htm

[88] Global Map of Human Impacts to Marine Ecosystems, http://www.nceas.ucsb.edu/globalmarine

[89] Bálint, M., et al. (2011) Cryptic Biodiversity Loss Linked to Global Climate Change, *Nature Climate Change*, doi:10.1038/NCLIMATE1191.

foundation to the food web. Because they are competitors, relatively few individuals of the same species exist, exposing them to a greater risk of extinction should environmental conditions change, such as during and after an extreme weather event. This could result in a depletion of food sources for species (such as herbivores) that rely on primary producers. This extinction, in turn, affects a predator at the top of the food web. Biologists call this transformation a **cascading extinction**. Using models of this process, researchers[90] found that flora and fauna in these conditions are 100 to 1,000 times more likely to become extinct than normal.

Ecological Sensitivity As global warming continues, many plant and animal species face increasing competition for survival as well as significant species turnover as some species invade areas occupied by other species. NASA scientists have investigated[91] the influence of doubled CO_2 on **ecological sensitivity**, and their results show that changes accompanying higher carbon dioxide levels lead to increasing ecological change and stress in Earth's biosphere.

Researchers have found that water temperatures in many streams and rivers throughout the United States are increasing. Analysis[92] of historical records from 20 major U.S. streams and rivers reveals that annual mean water temperatures increased by 0.009 degree Celsius to 0.077 degree Celsius per year (0.02 degree Fahrenheit to 0.14 degree Fahrenheit per year). Long-term increases in stream water temperatures were correlated with increases in air temperatures, and rates of warming were most rapid in urbanized areas. Warming water can affect basic ecological processes, aquatic biodiversity, biological productivity, and the cycling of contaminants through the ecosystem.

Most of Earth's land that is not covered by ice or desert is projected to undergo at least a 30 percent change in plant cover—a change that will require humans and animals to adapt and often relocate. Other studies[93] have confirmed these results, finding that as species migrate at different rates to new ecosystems, conflict grows owing to competition for space and resources. The collision course that ensues leads to new ecosystems for which there is no historical analogue, and it increases stress and extinctions among the affected plant and animal community.

8.5.2 Marine Ecosystems

The impact of shifting climate on marine ecosystems has been widely observed.[94] When temperatures rise, plants and animals that need a cooler environment move to new regions. Land warms about three times faster than the ocean, but species do not necessarily move three times faster on land. If the land temperature becomes too hot, some species can move to higher elevations, where temperatures are cooler. That's not an easy option, however, for marine species that live at the surface of the ocean.

When the temperature of seawater rises, species such as fish will be able to move into deeper water to find the cooler environments they prefer. However, deeper water has reduced light levels, potentially changing aspects of metabolism and predator–prey relationships. Other species, such as marine plants or corals, are tied to specific characteristics of shallow water including light levels, water circulation, and oxygen content. These species have to move horizontally to find suitable habitats, and they could become trapped if there are no cooler places for them to go. Rising temperatures could leave some marine species with nowhere to go.

Sea Surface Temperature Ocean warming dominates the increase in energy stored in the climate system, accounting for more than 90 percent of the excess warmth accumulated between 1971 and 2010. More than 60 percent of this increase is stored in the upper 700 meters (2297 feet) of the ocean, and about 30 percent is stored in the ocean below 700 meters.[95]

Sea surface temperature has increased in the upper 75 meters (246 feet) of the ocean at a rate of 0.09 to 0.13 degree Celsius (0.162 to 0.234 degree Fahrenheit) per decade over the period 1971 to 2010.[96] Additionally, the acidity of the ocean surface has increased 10-fold. Corals cannot tolerate severely warming waters, however, and temperature stress causes a phenomenon known as *bleaching*, whereby corals expel the symbiotic algae that live in their tissues (**Figure 8.11**).

Anoxia In the past 50 years, there has been a fourfold increase in the amount of open ocean water with zero oxygen.[97] Along the coast, low-oxygen sites have increased more than 10-fold since 1950. *Dead zones* are locations where oxygen plummets so low that many animals suffocate and die. But the decline of oxygen has other effects, animal growth is stunted, reproduction is hindered, ecosystem boundaries

[90] Kaneryd, L., et al. (2012) Species-Rich Ecosystems are Vulnerable to Cascading Extinctions in an Increasingly Variable World, *Ecology and Evolution*, doi: 10.1002/ece3.218.

[91] Bergengren, J., et al. (2011) Ecological Sensitivity: A Biospheric View of Climate Change, *Climatic Change* 107, nos. 3–4, doi: 10.1007/s10584-011-0065-1

[92] Kaushal, S.S., et al. (2010) Rising Stream and River Temperatures in the United States, *Frontiers in Ecology and the Environment*: 461–466, http://dx.doi.org/10.1890/090037

[93] Urban, M.C., et al. (2012) On a Collision Course: Competition and Dispersal Differences Create No-Analogue Communities and Cause Extinctions during Climate Change, *Proceedings of the Royal Society B: Biological Sciences*, doi: 10.1098/rspb.2011.2367.

[94] Burrows, M.T., et al. (2011) The Pace of Shifting Climate in Marine and Terrestrial Ecosystems, *Science* 334, no. 6056: 652, doi: 10.1126/science.1210288.

[95] IPCC, AR5, WGI (2013) Summary for Policymakers

[96] IPCC, AR5, WGI (2013) Summary for Policymakers

[97] Breitburg, D., et al. (2018) Declining Oxygen in the Global Ocean and Coastal Waters, *Science*, doi: 10.1126/science.aam7240

Healthy - Dec 2014   Dying - Feb 2015   Dead - Aug 2015

Photographed by The Ocean Agency / XL Catlin Seaview Survey / Richard Vevers

FIGURE 8.11 The US National Oceanic and Atmospheric Administration (NOAA) announced the third global bleaching event in October 2015 and it has already become the longest event ever recorded, impacting some reefs in consecutive years. This image shows the same reef in American Samoa before, during, and after a coral bleaching event.

Source: The Ocean Agency / XL Catlin Seaview Survey / Richard Vevers. http://www.globalcoralbleaching.org

contract, and the prevalence of disease rises. Declining oxygen can trigger the release of dangerous compounds such as nitrous oxide, a greenhouse gas up to 300 times more powerful than CO_2. Toxic hydrogen sulfide is also released from the microbial breakdown of organic matter in the absence of oxygen gas. Overall, marine biodiversity in dead zones plummets to near zero.

In the open ocean, warming surface waters create stratification, making it harder for oxygen to reach the ocean interior. And basic physics tells us that as ocean water gets warmer, it holds less dissolved oxygen. In coastal waters, the culprit is largely excess nutrient pollution from human land use. Nutrients delivered to the coastal zone by polluted runoff, cause algal blooms. Massive populations of algae, fueled by nutrients, drain oxygen from the water as they die and decompose. In an unfortunate twist, animals also need more oxygen in warmer waters, even as it is disappearing.

Limiting the impact of anoxia requires strong effort in three directions:[98]

1. Address the causes: Nutrient pollution can be managed with better septic systems, limiting fertilizers in farming, reducing the likelihood of flooding in urbanized and intensely farmed watersheds, and keeping pollution out of the water. Limiting climate change requires cutting fossil fuel emissions by replacing coal and oil with renewable energy sources such as solar, wind, hydrodynamic and other forms of clean energy.

2. Protect vulnerable marine life: Create marine protected areas or no-catch zones in areas where animals go to escape low oxygen.

3. Improve monitoring of low-oxygen zones: Enhanced monitoring, especially in developing countries, and numerical models will help pinpoint which places are most at risk and determine the most effective solutions.

As proof that these approaches can be effective, researchers point to the ongoing recovery of Chesapeake Bay. Nitrogen pollution has dropped 24 percent since its peak because of improved sewage treatment, better farming practices and successful laws such as the Clean Air Act. The area of the Chesapeake with zero oxygen has almost disappeared.

Bleaching On the Great Barrier Reef of Australia, corals live within a temperature range of 23 to 25 degrees Celsius (73 to 77 degrees Fahrenheit).[99] This means that a change of more than 1 to 2 degrees Celsius (1.8 to 3.6 degrees Fahrenheit) can cause coral bleaching, resulting in the gradual decline of coral tissue. The expulsion of zooxanthellae during bleaching leads to a decline in chlorophyll and protein levels. If the temperature change lasts only a few days, a small portion of the zooxanthellae will be expelled from the polyps. But, if bleaching lasts for longer than five weeks, the coral is less likely to restore the population of zooxanthellae and will instead die of oxygen starvation, decreased energy, and overheating. Bleaching can also happen when there is a decrease in temperature, if the water becomes

---

[98] The Ocean is Losing Its Breath: Here's the Global Scope, https://www.sciencedaily.com/releases/2018/01/180104153511.htm

[99] Coral bleaching. https://21191603.weebly.com/causes-temperature.html

turbid with mud, and if temperature change (in any direction) is especially rapid.

In 1997 and 1998, an unusually strong El Niño event caused high sea-surface temperatures, which led to coral bleaching that was observed in almost all of the world's reefs during that record-setting year. An estimated 16 percent of the world's corals died in that strong bleaching event, at the time—an unprecedented occurrence.

Global bleaching occurred again in 2005, 2009, 2010, and 2014–2016. In 2015, for example, global sea surface temperature was 0.33 to 0.39 degree Celsius (0.6 to 0.7 degree Fahrenheit) higher than the 1981–2010 average, and 0.74 degree Celsius (1.33 degrees Fahrenheit) higher than the average for the 20th century.[100] The frequency and intensity of bleaching may be growing, and scientists have predicted that coral populations will be exposed to bleaching-level conditions annually by mid-century.[101]

Under natural conditions, stressful events that result in bleaching are relatively rare. This allows for recovery of a reef between events. However, an analysis[102] of 100 reefs globally over the period 1980 to 2016 found that the average interval between bleaching events is now less than half what it was early in the period. The median return time between pairs of severe bleaching events has diminished steadily since 1980 and is now only 6 years. A narrow recovery window does not allow for full reef recovery. Subsequent events are then likely to cause more damage than previously. Moreover, warming events such as El Niño are warmer than in past decades, as are general ocean conditions. Such changes are likely to make it more and more difficult for reefs to recover between stressful events.

In response to warming temperatures, apparently, reef-forming coral species along the coast of Japan have been shifting their range into cooler waters to the north since the 1930s at rates as high as 14 kilometers (8.7 miles) per year.[103] Many coral reefs also have the unfortunate circumstance of being located immediately adjacent to urbanized watersheds. Fifty-eight percent of the world's coral reefs are potentially threatened by human activity,[104] ranging from coastal development and destructive fishing practices to overexploitation of resources, marine pollution, and polluted runoff from inland deforestation and farming.

The list of reef stressors is long: eroded silt in muddy runoff, pollutants of various types that cause coral disease, overfishing of species that are important in cropping back invasive algae that compete with corals for seafloor space, excessive levels of nitrogen, phosphorus and other nutrients, direct human impact with anchors, explosive fishing methods, and other human impacts all add to the stress that threatens coral reefs around the world.

Coral reefs are also experiencing the effects of sea-level rise. In locations where turbid runoff from exposed watersheds has delivered mud to the coastal zone and muddy shorelines line the landward edges of coral reefs, scientists[105] fear that a rising sea will permit additional wave action across the reef flat to erode the muddy coast. Muddy coastal waters, released by higher wave energy, could stab reef ecosystems in the back, killing coral communities on the seaward reef edges that currently enjoy open ocean conditions of clean and appropriately cool water.

But not all reefs are experiencing these threats. Researchers[106] have documented increases in coral reef growth under rising seas in more than one locality.[107] The reason is simple: Over the past millennium or so, many reefs have grown upward to the limit of the water column. Wave energy, hot summer temperatures, and shallow water do not allow any further upward growth in many shallow reef flats of the world. By raising sea level, global warming offers the possibility of additional upward growth in some locations, thus stimulating new coral growth in waters that were previously too shallow. But this advantage can only be utilized by a reef system that is otherwise healthy and not exposed to the impacts that come with human coastal development.

Ocean Acidity The oceans have absorbed about one-third[108] of the carbon dioxide emitted by humans over the past two centuries. Increasing ocean acidification, brought on by dissolved carbon dioxide that mixes with seawater to form *carbonic acid*, makes it difficult for calcifying organisms (corals, mollusks, and many types of plankton[109]) to secrete the calcium carbonate they need for their skeletal components (a process called *calcification*) **Figure 8.12**.

Scientists have found[110] that carbon dioxide emissions in the last 100 to 200 years have already raised ocean acidity

[100] CarbonBrief https://www.carbonbrief.org/rising-seas-could-ease-coral-bleaching

[101] Heron, S.F., et al. (2016) Warming Trends and Bleaching Stress of the Worlds Coral Reefs 1985-2012, *Scientific Reports* 6, 38402, doi: 10.1038/srep38402.

[102] Hughes, T.P., et al. (2018) Spatial and Temporal Patterns of Mass Bleaching of Corals in the Anthropocene, *Science* 359, no. 6371: 80–83, doi: 10.1126/science.aan8048.

[103] Yamano, H., et al. (2011) Rapid Poleward Range Expansion of Tropical Reef Corals in Response to Rising Sea Surface Temperatures, *Geophysical Research Letters* 38: L04601, doi: 10.1029/2010GL046474.

[104] Bryant, D., et al. (1998) *Reefs at Risk: A Map-Based Indicator of Threats to the World's Coral Reefs.* (Washington, D.C., World Resources Institute).

[105] Field, M.E., et al. (2011) Rising Sea Level May Cause Decline of Fringing Coral Reefs, *Eos* 92: 273–280.

[106] Brown, B., et al. (2011) Increased Sea Level Promotes Coral Cover on Shallow Reef Flats in the Andaman Sea, Eastern Indian Ocean, *Coral Reefs* 30: 867–878.

[107] Scopelitis, J., et al. (2011) Coral Colonization of a Shallow Reef Flat in Response to Rising Sea Level: Quantification from 35 Years of Remote Sensing Data at Heron Island, Australia, *Coral Reefs* 30: 951–965.

[108] Coral reefs. http://www.sciencedaily.com/releases/2011/08/110803133517.htm

[109] Beaufort, L., et al. (2011) Sensitivity of Coccolithophores to Carbonate Chemistry and Ocean Acidification, *Nature* 476, no. 7358: 80, doi: 10.1038/nature10295.

[110] Friedrich, T., et al. (2012) Detecting Regional Anthropogenic Trends in Ocean Acidification against Natural Variability, *Nature Climate Change*, doi: 10.1038/NCLIMATE1372.

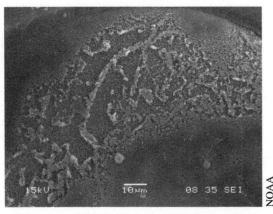

FIGURE 8.12 In 2014, researchers discovered[111] that the acidity of continental shelf waters off the U.S. West Coast is dissolving (left) the shells of tiny free-swimming marine snails, called *pteropods* (right), which provide food for pink salmon, mackerel and herring. Researchers estimate that in this region, the percentage of *pteropods* has doubled that are experiencing dissolving shells. Dissolution is due to ocean acidification that has doubled in the nearshore habitat since the preindustrial era and is on track to triple by 2050 when coastal waters become 70% more corrosive than in the preindustrial era due to human-caused ocean acidification.

Source: http://www.noaanews.noaa.gov/stories2014/20140430_oceanacidification.html

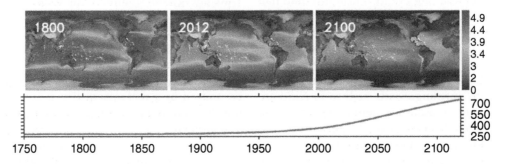

FIGURE 8.13 The upper panels show modeled surface seawater aragonite ($CaCO_3$) saturation for the years 1800, 2012, and 2100, respectively. Aragonite is a form of calcium carbonate that corals and other organisms use to build skeletons. As seawater becomes less saturated with aragonite (darker or reddish colors) it becomes more difficult for corals and other organisms to secrete skeletons; below zero, aragonite dissolves. White dots indicate present-day main coral reef locations. The lower panel shows atmospheric CO_2 concentration in parts per million, simulated for the years 1750 to 2100.

Source: University of Hawaii, International Pacific Research Center.

far beyond the range of natural variations. In some regions, the rate of change in ocean acidity since the Industrial Revolution is 100 times greater than the natural rate of change between the Last Glacial Maximum 20,000 years ago and preindustrial times.

When Earth started to warm 17,000 years ago, terminating the last glacial period, atmospheric CO_2 levels rose from 190 parts per million to 280 parts per million over 6,000 years, giving marine ecosystems ample time to adjust. Now, for a similar rise in CO_2 concentration to the present level above 400 parts per million, the adjustment time is reduced to only 100 to 200 years and might have decreased overall calcification rates by 15 percent.

On a global scale, pH conditions that support coral reefs are currently found in about 50 percent of the ocean, mostly in the tropics. By the end of the 21st century (**Figure 8.13**), this fraction is projected to be less than 5 percent. The Hawaiian Islands, which sit just on the northern edge of the tropics, will be one of the first to feel the impact.

Ocean acidification has other impacts. Acidification of seawater decreases the absorption of sound by up to 50 percent in the frequency range that is important to whales and other acoustic organisms.[112] Ship traffic,

[111] Bednaršek, N., et al. (2014) *Limacina helicina* shell dissolution as an indicator of declining habitat suitability owing to ocean acidification in the California Current Ecosystem, *Proc. R. Soc. B*; DOI: 10.1098/rspb.2014.0123

[112] Ilyina, T., et al. (2009) Future Ocean Increasingly Transparent to Low-Frequency Sound owing to Carbon Dioxide Emissions, *Nature Geoscience* 3: 18–22.

seismic testing, and industrial activities that were previously muted in the world's oceans will become more acute and potentially affect marine species. High levels of low-frequency sound have a number of behavioral and biological effects on marine life, including tissue damage, mass stranding of cetaceans, and temporary loss of hearing in dolphins.

Ocean acidification has also damaged a \$273 million per year oyster farming industry in the Pacific Northwest. A study[113] found that increased dissolved carbon dioxide levels in seawater resulted in more-corrosive ocean water and inhibited larval oysters from developing their shells. Because of this, larvae grew at a pace that prohibited cost-effective commercial production and contributed to a collapse of the oyster farming industry.

Ocean acidification is one of the consequences of CO_2 buildup that could have a great impact on the world's ocean ecology, which depends on the secretion of calcium carbonate by thousands of different species. As carbon dioxide emissions increase, it is anticipated that 450 parts per million CO_2 in the atmosphere will be reached before 2050.[114] Corals may already be far along a path to extinction as a result of increased ocean warmth, and ocean acidification may very well have driven the last nail into their coffin by that point. Such a catastrophe would not be confined to reefs but could be part of a domino-like sequence throughout the entire marine ecosystem.[115]

The loss of healthy coral reefs affects all the species that dwell there (such as turtles, mollusks, crabs, and fish) as well as the animals that depend on reef habitats as a food source (including sea birds, mammals, and humans). One quarter of all sea animals spend time in coral reef environments during their life cycle. There are economic impacts as well. Tourism and commercial fisheries generate billions of dollars in revenue annually. Biodiversity, food supplies, and economics could thus all be affected by the impacts of global warming and acidification on reefs.

8.6 Climate Sensitivity

How sensitive is climate to high levels of carbon dioxide? This issue is explored with estimates of a value researchers call the equilibrium climate sensitivity (ECS). ECS is the global mean near-surface temperature when it has equilibrated to atmospheric CO_2 concentrations that are double the preindustrial level of CO_2 (estimated to be 280 parts per million). Another way of putting it is, "How warm will it be when global climate is fully equilibrated to a CO_2 concentration of 560 ppm?"

Today the CO_2 level is above 410 parts per million and rising an average 2.5 parts per million per year (since 2010, the annual rise has ranged between 1.69 to 2.94 parts per million).[116] When will the level of CO_2 reach 560 parts per million? This depends on the level of continued greenhouse gas emissions, a major subject of IPCC modeling. Continued emissions at present rates, the "business as usual" scenario, would lead to doubled CO_2 levels sometime in the second half of the century.

True ECS requires that the climate system come to a steady-state equilibrium[117] around a CO_2 level of 560 parts per million. In reality, this would require centuries as fully integrating the temperature of the entire ocean with the temperature of the atmosphere requires a long time. Based on CMIP5 modeling, IPCC-AR5 reports that doubling preindustrial carbon dioxide levels will likely cause global average surface temperature to rise between 1.5 and 4.5 degrees Celsius (2.7 to 8.1 degrees Fahrenheit) compared to preindustrial temperatures.

To understand how sensitive the climate is to carbon dioxide on the shorter time frame that will require immediate response from the human communities (a century or less), scientists study the **transient climate response (TCR)**. Studies of TCR assume that carbon dioxide continues to increase at roughly the rate it has been and then analyze model simulations of the climate within a decade or two of when 560 parts per million is reached. The question addressed is "How much warming will occur by the time CO_2 reaches 560 ppm?" On this shorter time scale, estimates suggest the planet will warm between 1 and 2.5 degrees Celsius (2 to 4.5 degrees Fahrenheit).

As described by scientists at NOAA,[118] the difference between transient response and equilibrium sensitivity comes from the fact that some parts of the Earth system, mountain glaciers, sea ice, precipitation, and others, react within years or a few decades to a warming or cooling climate. Others, including ice sheets, permafrost, and especially the deep ocean, respond slowly, taking centuries to evolve into equilibrium with a change in climate.

[113] Barton, A., et al. (2012) The Pacific Oyster, *Crassostrea gigas,* Shows Negative Correlation to Naturally Elevated Carbon Dioxide Levels: Implications for Near-Term Ocean Acidification Effects, *Limnology and Oceanography* 57, no. 3, 698: doi:10.4319/lo.2012.57.3.0698.

[114] Zoological Society of London (2009) Coral Reefs Exposed to Imminent Destruction From Climate Change, http://www.sciencedaily.com/releases/2009/07/090706141006.htm

[115] Ullah, H., et al. (2018) Climate change could drive marine food web collapse through altered trophic flows and cyanobacterial proliferation. *PLOS Biology*; 16 (1): e2003446 DOI: 10.1371/journal.pbio.2003446

[116] NOAA, Earth System Research Laboratory, Global Monitoring Division, http://www.esrl.noaa.gov/gmd/ccgg/trends/global.html#global_data

[117] A "steady state equilibrium" simply means that other than natural variability, climate stops changing and settles into a state that fully reflects the influence of 560 ppm. It has been estimated that this would likely take a few centuries to over one thousand years to achieve because feedbacks need time to playout (such as ice to melt) and the ocean needs time to achieve a new temperature that is stable throughout the entire water column.

[118] How much will Earth warm if carbon dioxide doubles pre-industrial levels? Lindsey, R. (2014) https://www.climate.gov/news-features/climate-qa/how-much-will-earth-warm-if-carbon-dioxide-doubles-pre-industrial-levels

8.6.1 Estimating Climate Sensitivity

Researchers use various methods to estimate ECS and TCR, and they all have advantages and disadvantages.[119] Typical methods include reconstructing past temperature changes[120] that accompanied previous shifts in CO_2 concentration. These are estimated from geologic information (climate proxies; **Chapter 4**). Other methods of estimating ECS and TCR use model simulations to estimate[121] sensitivity, and calculating sensitivity from measurements of modern climate change.[122]

AR4 and AR5 The IPCC[123] Fourth Assessment Report (AR4) concluded that ECS is. . .

> "... likely to be in the range of 2 to 4.5°C (3.6 to 8.1°F), with a best estimate of about 3°C (5.4°F), and is very unlikely to be less than 1.5°C (2.7°F). Values substantially higher than 4.5°C (8.1°F) cannot be excluded, but agreement of models with observations is not as good for those values."

The problem with this wide range of estimates is that identifying the true effect of limiting fossil-fuel burning becomes highly uncertain, and it is difficult to rule out large temperature increases as a result of greenhouse gas emissions. Improved estimates of climate sensitivity would encourage governments to set emission targets with better-understood consequences.

Instead of narrowing estimates of climate sensitivity, the Fifth Assessment Report[124] widened the range by reducing the lower end from 2 degrees Celsius (3.6 degrees Fahrenheit) to 1.5 degrees Celsius (2.7 degrees Fahrenheit).

> "Based on the combined evidence from observed climate change including the observed 20th century warming, climate models, feedback analysis and paleoclimate ... ECS is likely in the range 1.5°C to 4.5°C with high confidence. ECS is positive, extremely unlikely less than 1°C (high confidence), and very unlikely greater than 6°C (medium confidence)."

Paleoclimate The problem with paleoclimate reconstructions is twofold: (1) Today's rapid climate changes and complex feedbacks might represent completely unique conditions that will not be accurately represented by studies of paleoclimate history and (2) the high degree of uncertainty that accompanies the use of climate proxies (dating inaccuracies, chemical changes in proxies when they are buried in the lithosphere, and low precision in characterizing climate).

Nevertheless, despite these criticisms, the period over which climate shifted from the last glacial maximum 23,000 to 19,000 years ago to warmer, preindustrial conditions (pre-19th century) has been used[125] to estimate climate sensitivity. This period is a potentially valuable episode for characterizing climate sensitivity for several reasons: Earth's climate changed relatively rapidly (in a geologic sense) from a glacial state to modern interglacial conditions, it was relatively recent in geologic history (better-resolved proxies), and the last glacial maximum has been robustly characterized by a number of independent, globally distributed climate proxies.

One set of investigators[126] used detailed paleoclimate proxy data to reconstruct the climate of the last glacial maximum and ran a series of global climate model simulations over the same period, with each simulation using a different value to represent climate sensitivity. By comparing the modeling results to the paleoclimate reconstruction, it was possible to identify the climate sensitivity value that most closely predicted the true paleoclimate ECS.

The study found that an increase of 3.1 degrees Celsius (5.6 degrees Fahrenheit) in global average surface temperatures seems most likely as a result of doubling the CO_2 concentration above preindustrial levels. The range of most-probable temperatures varied from 2 to 4.7 degrees Celsius (3.6 to 8.46 degrees Fahrenheit). Furthermore, the model simulations suggest that it would be difficult to exceed 4.7 degrees Celsius (8.46 degrees Fahrenheit) under doubled carbon dioxide levels.

Slow Mode Climate Change The role of carbon dioxide emissions in driving global warming is well established. However, temperature sensitivity to CO_2 changes as interpreted from paleoclimate proxy data, and temperature sensitivity to CO_2 changes recorded by historical observations over the past century, do not agree. Paleoclimate interpretations suggest that temperature is more sensitive to CO_2 changes than has been observed over the past century. That is, it should have warmed more by now than it has.

Because of the uncertainty that characterizes proxy data, this difference has been interpreted as a weakness in paleoclimate reconstructions. Global climate models are

[119] For a detailed discussion of climate sensitivity, see: http://www.skepticalscience.com/detailed-look-at-climate-sensitivity.html

[120] Zeebe, R.E., et al. (2009) Carbon Dioxide Forcing Alone Insufficient to Explain Palaeocene-Eocene Thermal Maximum Warming, *Nature Geoscience* 2: 576–580.

[121] Allen, M.R., et al. (2009) Warming Caused by Cumulative Carbon Emissions Towards the Trillionth Tonne, *Nature*, doi: 10.1038/nature08019.

[122] Forster, P.M. and Gregory, J.M. (2006) The Climate Sensitivity and its Components Diagnosed from Earth Radiation Budget Data, *Journal of Climate* 19, no. 1: 39–52, doi: 10.1175/JCLI3611.1

[123] IPCC, "Climate Sensitivity and Feedbacks." In Pachauri, R.K. and Reisinger, A. (eds.) *Climate Change 2007: Synthesis Report.* Contribution of Working Groups I, II and III to the Fourth Assessment Report of the Intergovernmental Panel on Climate Change. (Geneva, Intergovernmental Panel on Climate Change, 2007). http://www.ipcc.ch/publications_and_data/ar4/syr/en/mains2-3.html

[124] Stocker et al. (2013)

[125] Schmittner, A., et al. (2011) Climate Sensitivity Estimated from Temperature Reconstructions of the Last Glacial Maximum, *Science* 334, no. 6061: 1385–1388, doi: 10.1126/science.1203513

[126] Schmittner et al. (2011)

calibrated to paleoclimate reconstructions. Might this indicate that models have a bias and are overly sensitive to CO_2 changes? By extension, is the ultimate amount of warming Earth will experience at a given concentration of greenhouse gases going to be less than projected by models? If models are overestimating future climate change, perhaps global warming is not such a big deal.

Researchers[127] examined this problem by deconstructing the details of model projections. They found that significant levels of warming take a century or more to become fully developed. For instance, an important mode of warming associated with the Southern Ocean and the Eastern Equatorial Pacific will require more than 100 years before it is fully completed. Importantly, such slowly evolving changes to climate, which are termed **slow modes**, ultimately account for the greatest portion of warming found in climate models. These are missed when looking at the relatively brief span of historical observations. That is, a significant amount of warming still has yet to materialize from current levels of carbon dioxide.

Slow modes display stronger amplifying feedbacks and ultimately contribute 28 to 68 percent of equilibrium warming, yet they comprise only 1 to 7 percent of current warming. When accounted for, slow modes of feedback align the temperature range of climate sensitivity from historical observations with the range reported by climate models today, a range implying a higher sensitivity of climate to carbon dioxide emissions.

Reconciling Sensitivity[128] Global climate models predict a range of warming from unrestricted greenhouse gas emissions. Temperature projections under high emission scenarios vary from more than 10 degrees Celsius (18 degrees Fahrenheit) over coming centuries to only one-third of this amount. At the high end, temperatures are reached that have not been seen since the Eocene, an epoch when sea levels were 70 meters (230 feet) higher and tropical reptiles roamed the Arctic. Among these projections, which future are we to believe, and to plan for?

Two studies[129] that emerged after IPCC-AR5 indicate that this range of predictions is not realistic. These studies estimate climate sensitivity that is less than the high-end projections of CMIP5 models, implying that future warming is overestimated by models. On the basis of calculations that include the period of "climate hiatus," they resolve a climate sensitivity in the range 1.0 to 4.0 degrees Celsius

(1.8 to 7.2 degrees Fahrenheit), with a best estimate at around 2.0 degrees Celsius (3.6 degrees Fahrenheit). The studies conclude that CMIP5 models, with an ECS of 2.0 to 5.6 degrees Celsius (3.6 to 10.1 degrees Fahrenheit), are therefore too sensitive and project future temperatures that are too high.

However, subsequent analysis[130] reveals that comparing historical observations with climate model projections is an exercise in comparing apples to oranges. Climate sensitivity is defined as the "global mean near-surface air temperature change that will eventually result from doubling atmospheric CO_2." When examined closely, the specific elements of this definition reveal some interesting discrepancies. Richardson et al. (2016) examined the terms "global mean," "near-surface air temperature," "doubling of atmospheric CO_2," and "eventually." They learned that these specific statements need clear definition and are the keys to understanding the disagreement between modeled and observed climate.

The term "global mean" is the area-weighted average of temperatures from all over the globe, and "near-surface air temperature" refers to air that is a couple of meters (6 to 8 feet) above Earth's surface, whether over land, ocean or ice. These values can be calculated from model output, but it is a formidable challenge to acquire them as direct observations. Historical temperature records are sparse in polar regions, including the Arctic where the rate of warming is the highest worldwide. Observations come from diverse sources such as ships and weather stations. Ship-based measurements are actually taken in the water, which warms at a slower rate than the air. Accounting for just this difference leads to a 9 percent increase in global warming estimates. Accounting for incomplete geographic coverage requires a separate 15 percent adjustment.

These two effects mean that global near-surface air warming estimates should be revised upward by 24 percent in total. Consequently, observation-based estimates of climate sensitivity must also be revised upward by the same amount, resolving much of the mismatch with modeled values.

Additionally, the definition of climate sensitivity calls for "doubling of atmospheric CO_2." But global warming is produced by a variety of climate forcing agents such as methane, sunlight-scattering aerosols, changing land use (the shift from forests to farms), and others. A study[131] in 2016 showed that these non-CO_2 forcings have distinct effects on global warming and call for an upward revision of 30 percent to observational estimates of climate sensitivity.

The last term, "eventually," recognizes the full time needed for Earth to respond to climate forcing—the

[127] Proistosescu, C. and Huybers, P.J. (2017) Slow Climate Mode Reconciles Historical and Model-based Estimates of Climate Sensitivity, *Science Advances*, doi: 10.1126/sciadv.1602821

[128] Armour, K.C. (2016) Projection and Prediction: Climate Sensitivity on the Rise, *Nature Climate Change*, N&V, 6 October, 896–897.

[129] Otto A., et al. (2013) Energy Budget Constraints on Climate Response, *Nature Geoscience* 6: 415–416, doi:10.1038/ngeo1836. See also Lewis, N. and Curry, J.A. (2015) The Implications for Climate Sensitivity of AR5 Forcing and Heat Uptake Estimates, *Climate Dynamics* 45, no. 3–4: 1009–1023, https://doi.org/10.1007/s00382-014-2342-y

[130] Richardson, M., et al. (2016) Reconciled Climate Response Estimates from Climate Models and the Energy Budget of Earth, *Nature Climate Change*, 6, 931–935, doi:10.1038/nclimate3066

[131] Marvel, K., et al. (2016) Implications for Climate Sensitivity from the Response to Individual Forcings, *Nature Climate Change* 6: 386–389, doi:10.1038/nclimate2888

difference between ECS and TCR. Because of the large heat capacity of the oceans, this can be several centuries to over a thousand years, and todays observational record captures only the earliest phase of global warming. That is, estimates of ECS based on observations today make the implicit assumption that the same relationships will apply in the future. Several studies call this into question,[132] and require yet another upward revision, of about 25 percent, to observation based estimates.

Taken together, these findings indicate that observation-based estimates of climate sensitivity may be substantially higher than previously reported, aligning them more closely with projections by climate models and raising the possibility of a very warm future.

Climate Sensitivity Ranges As we have discussed, estimating climate sensitivity is an active area of research. The published results of diverse approaches to the problem, all of which analyze how sensitive the mean near-surface temperature has been under various conditions, are illustrated in **Figure 8.14**. These include the following: the modern instrumental period, the mean-climate state, global-climate modeling, paleoclimate patterns, and combinations of evidence including perturbations by volcanic eruptions, and expert judgment. Using expert assessment based on multiple lines of evidence, the authors of IPCC-AR5[133] conclude that ECS *likely* (66 to 100 percent probability) falls in the range 1.5 to 4.5 degrees Celsius (2.7 to 8.1 degrees Fahrenheit) and is *extremely unlikely* (0 to 5 percent probability) to be less than 1 degree Celsius (1.8 degrees Fahrenheit), and *very unlikely* (0 to 10 percent probability) to be greater than 6 degrees Celsius (10.8 degrees Fahrenheit).

The IPCC-AR5 also provides an estimate of the range of TCR (**Figure 8.15**). TCR is informative for stakeholders who are interested in knowing the short-term impacts of climate change in order to respond with appropriate and immediate adaptation and mitigation measures. The authors of AR5 conclude that TCR is *likely* (66 to 100 percent probability) in the range of 1 to 2.5 degrees Celsius (1.8 to 4.5 degrees Fahrenheit). This is an expert-assessed range, supported by several different and partly independent lines of evidence, each based on multiple studies, models and data sets. The estimate is close to the 5 to 95 percent probability range of CMIP5, 1.2 to 2.4 degrees Celsius (2.2 to 4.3 degrees Fahrenheit), and is assessed as *extremely unlikely* (0 to 5 percent probability) to be greater than 3 degrees Celsius (5.4 degrees Fahrenheit).

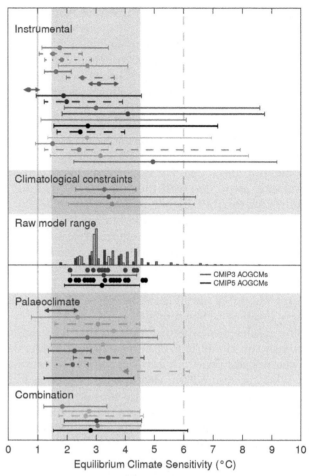

FIGURE 8.14 Temperature ranges of ECS from the published literature as illustrated by IPCC-AR5.[134] The gray shaded band marks the likely range, 1.5 to 4.5°C (2.7 to 8.1°F). The gray solid line shows the position of 1°C (1.8°F), and the gray dashed line shows the position of 6°C (10.8°F). AOGCMs stand for Atmosphere Ocean Global Circulation Models. For more detail, see IPCC-AR5, Technical Summary for Working Group I, Thematic Focus Element box, TFE.6, p. 82–85.[135]

Source: http://www.ipcc.ch/report/graphics/index.php?t=Assessment%20 Reports&r=AR5%20-%20WG1&f=Technical%20Summary

The assessed ranges of ECS and TCR are largely consistent with observed warming, the estimated forcing by greenhouse gas emissions and other human activities, and the projected future warming. No best estimate for ECS is given in AR5 because of a lack of agreement among authors. Climate models with ECS values in the upper part of the likely range show very good agreement with the present theoretical understanding of fundamental climate processes. Estimates of ECS derived from observed climate change tend to best-fit values in the lower part of the likely range. In estimates based on observed warming, the most likely value

[132] Winton, M., et al. (2010) Importance of Heat Uptake Efficacy to Transient Climate Change, *Journal of Climate*, https://doi.org/10.1175/2009JCLI3139.1. See also Armour, K.C., et al. (2013) Time-varying Climate Sensitivity from Regional Feedbacks, *Journal of Climate*, https://doi.org/10.1175/JCLI-D-12-00544.1. See also Gregory, J.M. and Andrews, T. (2016) Variation in Climate Sensitivity and Feedback Parameters During the Historical Period, *Geophysical Research Letters* 43, no. 8: 3911–3920, http://dx.doi.org/10.1002/2016GL068406
[133] Stocker et al. (2013)

[134] Stocker et al. (2013)
[135] See report at http://www.ipcc.ch/pdf/assessment-report/ar5/wg1/WG1AR5_TS_FINAL.pdf

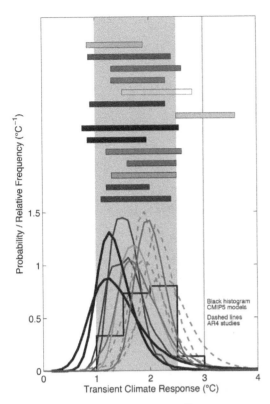

FIGURE 8.15 Temperature ranges of transient climate response from the published literature as illustrated by IPCC-AR5.[136] Shown are the 5% to 95% probabilities for the same studies illustrated in Figure 8.14 and from CMIP5 models (black histogram). Gray dashed lines are results from AR4. The gray shaded band marks the likely 1 to 2.5°C (1.8 to 4.5°F) range. The grey solid line shows the position of 3°C (5.4°F). For more detail, see IPCC-AR5, Technical Summary for Working Group I, Thematic Focus Element box, TFE.6, p. 82–85.[137]

Source: https://www.ipcc.ch/report/graphics/index.php?t=Assessment%20 Reports&r=AR5%20-%20WG1&f=Technical%20Summary

is sensitive to observational and model uncertainties, internal climate variability, and to assumptions about the prior distribution of ECS.

Best-Fit Models Dozens of models participate in the CMIP5 program. In the IPCC fifth assessment report, output from these models was used to define ranges of probability, and the "multimodel mean." Results in AR5, and in papers since, average the results of over 30 separate model outputs to report statistically combined projections. The problem with this methodology is that it assumes that model output will be normally distributed—models that project too much warming will be balanced by models that project to little warming, and by averaging the results of both, a reasonable estimate of ECS is produced. However, this may be a false assumption.

In late 2017, researchers published[138] an analysis that identified a subset of CMIP5 model outputs based on specific criteria: ability to best simulate observations of today's climate. The results show that models that best fit this criterion (called "constrained models") forecast greater amounts of warming later this century. These findings suggest that the models used by the Intergovernmental Panel on Climate Change, on average, may be underestimating future warming.

For example, raw CMIP5 results for a business-as-usual scenario (RPC8.5) indicate that we can expect global temperatures to increase anywhere in the range of 3.2 to 5.9 degrees Celsius (5.8 and 10.6 degrees Fahrenheit) over preindustrial levels by the end of the century—a difference of about a factor of two between the most- and least-severe projections. However, the constrained models tend to be the ones that project the most global warming over the remainder of the 21st century. The study eliminates the lower end of projected temperatures, finding that the most likely warming is about 0.5 degree Celsius (0.9 degree Fahrenheit) greater than what the raw results suggest **Figure 8.16**.

One outcome of this research is that it suggests that it doesn't make sense to dismiss the most-severe global warming projections simply because they fail to align with the multi-model mean. On the contrary, these results show that model shortcomings constitute a basis for dismissing the least-severe projections.[139]

The uncertainty in the range of future warming is mostly due to differences in how models simulate changes in clouds with global warming. Some models project increased cooling caused by clouds reflecting the Sun's energy back to space. Other models suggest that this cooling effect might decrease. The constrained models are the ones that simulate a reduction in future cloud cooling, and thus these are the models that predict the greatest future warming.

Of interest to this discussion, mean ECS value from the constrained models is 3.7 degrees Celsius (6.7 degrees Fahrenheit), with a likely range from 3 to 4.2 degrees Celsius (5.4 to 7.6 degrees Fahrenheit). Although IPCC AR5 did not report a mean ECS value, given their reported range of 1.5 to 4.5 degrees Celsius, the mean is 3.0 degrees Celsius. Thus, the constrained models indicate an ECS considerably higher than the AR5 mean.

One outcome of the constrained model analysis is that if emissions follow a commonly used business-as-usual scenario, there is a 93 percent chance that global warming will exceed 4 degrees Celsius (7.2 degrees Fahrenheit) by the

[136] Stocker et al. (2013)

[137] IPCC-AR5. http://www.ipcc.ch/pdf/assessment-report/ar5/wg1/ WG1AR5_TS_FINAL.pdf

[138] Brown, P.T. and Caldeira, K. (2017) Greater Future Global Warming Inferred from Earth's Recent Energy Budget, *Nature* 552, no. 7683: 45, doi: 10.1038/nature24672

[139] VIDEO: Greater Future Warming Inferred from Earth's recent Energy budget, https://patricktbrown.org/2017/11/29/greater-future-global-warming-inferred-from-earths-recent-energy-budget/

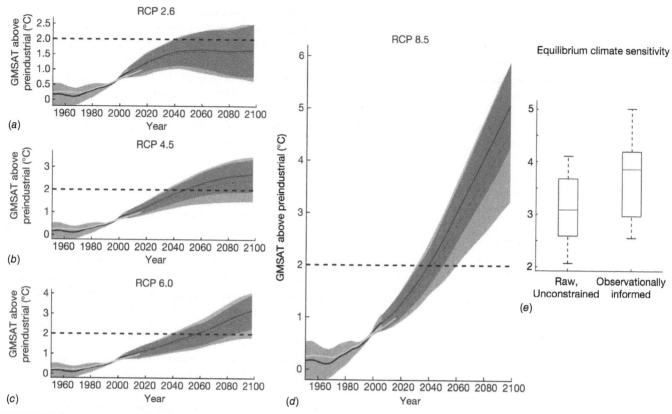

FIGURE 8.16 Comparison of all CMIP5 model output (dashed line and light shading) to constrained model output (solid line, dark shading) for global mean surface air temperature (GMSAT). Yellow (white) line is Berkeley Earth Surface Temperature dataset.[140] Each of the 4 primary RCP scenarios are shown (a–d). Right panel (e) shows box and whisker plots for raw and constrained distributions of ECS to a doubling of CO_2. The whiskers span the entire distribution range; the box spans the 25th to the 75th percentiles and the line indicates the median.

Source: https://www.nature.com/articles/nature24672/figures/2

end of this century. Previous studies (IPCC-AR5) had put this likelihood at 62 percent. World temperatures could rise 15 percent more than expected this century, obliging governments to make deeper cuts in greenhouse gas emissions to limit global warming. The results add to a broadening collection of research indicating that models that simulate today's climate best, tend to be the models that project the most global warming over the remainder of the 21st century.[141]

On the Other Hand In early 2018, an analysis[142] emerged that rules out the high end estimates of ECS. If accurate, this improves the chances of limiting warming to levels

sought by the *Paris Accord* (no more than 2 degrees Celsius; 3.6 degrees Fahrenheit).

In the study, researchers moved their focus away from the long-term warming trend of the past century, and instead analyzed the year-to-year fluctuations of temperature. They examined how the temperature of one year is statistically related to the temperature of the following year. This led to an equation that generalized the year-to-year relationship. They used this equation in 22 climate models to simulate historical temperature trends and found that it is a good predictor of changing surface temperature.

With the addition of their new equation, they ran projections of global climate under doubled future atmospheric CO_2. They found a 66 percent likelihood of ECS being between 2.2 and 3.4 degrees Celsius (4 and 6.1 degrees Fahrenheit), with a central estimate of 2.8 degrees Celsius (5 degrees Fahrenheit) and less than a 1 percent chance of it being greater than 4.5 degrees Celsius (8.1 degrees Fahrenheit).

[140] Rohde, R.A., et al. (2013) A New Estimate of the Average Earth Surface Land Temperature Spanning 1753 to 2011, *Geoinformation Geostatistics* 1: 1–7.

[141] Zhou, C., et al. (2015) The Relationship Between Interannual and Long-term Cloud Feedbacks, *Geophysical Research Letters* 42: 10463–10469. See also Tian, B. (2015) Spread of Model Climate Sensitivity Linked to Double-Intertropical Convergence Zone Bias, *Geophysical Research Letters* 42: 4133–4141. See also Myers, T.A. and Norris, J.R. (2016) Reducing the Uncertainty in Subtropical Cloud Feedback, *Geophysical Research Letters* 43: 2144–2148.

[142] Cox, P.M., et al. (2018) Emergent Constraint on Equilibrium Climate Sensitivity from Global Temperature Variability, *Nature* 553, no. 7688: 319, doi:10.1038/nature25450

8.7 Carbon Trends and Implications

The body of research that defines climate sensitivity is thorough, and creative. Broadly speaking, when atmospheric carbon dioxide concentration rises to 560 ppm, studies reviewed above indicate a transient climate response of around 2 degrees Celsius (3.6 degrees Fahrenheit) and an equilibrium climate sensitivity of 3 to 3.5 degrees Celsius (5.4 degrees Fahrenheit) or higher for both values if applying the findings of Brown and Caldeira (2017).

This level of warming will likely lead to a number of dramatic negative effects on human societies and the ecosystem, including drought, dangerous weather, accelerated sea-level rise, water and food stress, and widespread environmental damage. These in turn, are likely to make broad regions of planet Earth unlivable for human populations leading to waves of climate refugees, increased violence, and political turmoil.[143] These issues are explored later in this chapter. Studies show that the world is already committed to further warming, even if all emissions were to stop now.[144] However, as we will review later, emissions are projected to continue rising.

To provide some context to the question of how high global mean temperatures could rise this century and what the consequences may be, researchers at NASA[145] compared the present climate to the paleoclimate record of the Eemian. The most recent period of time marked by interglacial conditions similar to our own (**Chapter 4**), the Eemian began 130,000 years ago and lasted for about 15,000 years.[146] Sea surface temperatures were like those of today.[147] Currently, Earth's global-mean temperature is warming at approximately 0.17 degree Celsius (0.3 degree Fahrenheit) per decade,[148] and our studies of the ECS suggest additional warming will reach or exceed 3 degrees Celsius (5.4 degrees Fahrenheit) when atmospheric carbon dioxide concentration rises to 560 ppm. Thus, global mean temperature is on track to far exceed Eemian conditions, potentially in this century.

This is worrying because during the Eemian, sea levels were as much as 8 meters (26 feet) higher than present[149]— a prescription for disaster should the ocean respond the same way today. In the words of Jim Hansen of NASA, "We don't have a substantial cushion between today's climate and dangerous warming. Earth is poised to experience strong amplifying feedbacks in response to moderate additional global warming."[150]

8.7.1 Paris Accord

Under the auspices of the United Nations Framework Convention on Climate Change,[151] the world's governments have come to an agreement, known as the Paris Accord (**Chapter 1**), with regard to lowering future greenhouse gas emissions in order to limit the amount of damage to the planet. By the summer of 2018, 176 of the 197 parties (nations) have ratified, or legally adopted, the accord. The Paris Accord's central aim is to strengthen the global response to the threat of climate change by keeping global temperature rise this century well below 2 degrees Celsius (3.6 degrees Fahrenheit) above preindustrial levels and to pursue efforts to limit the temperature increase even further to 1.5 degrees Celsius (2.7 degrees Fahrenheit). These aspirations, and the modeling indications for TCR and ECS, obviously do not line-up on the same level of warming, indicating that if we are to achieve the Paris goals, greenhouse gas emissions cannot be allowed to approach 560 ppm.

Climate-change negotiations to assess progress on these goals and to identify emerging issues occur every year. Known as COP's (Conference of Parties), these are events where international talks focus on limiting global production of carbon dioxide and other heat-trapping gases. Often, there is disagreement between nations on the levels of sacrifice that should be apportioned among various economies. Some developing countries argue that most of the world's greenhouse gas emissions are produced by a few industrial nations whose quality of life generally has benefited from their rapid growth fueled by fossil energy, and they ask why developing countries should not aspire to the same benefits. Under guidance by the U.N., countries are working to rectify these differences of opinion, but disagreements persist.[152]

[143] Wallace-Wells, D. (2017) The Uninhabitable Earth, *New York Magazine*, http://nymag.com/daily/intelligencer/2017/07/climate-change-earth-too-hot-for-humans.html

[144] Armour, K. and Roe, G. (2011) Climate Commitment in an Uncertain World, *Geophysical Research Letters* 38, no. 1, doi:10.1029/2010GL045850

[145] Secrets from the Past Point to Rapid Climate Change in the Future, http://climate.nasa.gov/news/index.cfm?FuseAction=ShowNews&NewsID=649

[146] Rovere, A., et al. (2016) The Analysis of Last Interglacial (MIS5e) Relative Sea-level Indicators: Reconstructing Sea-level in a Warmer World, *Earth-Science Reviews* 159: 404–427, https://doi.org/10.1016/j.earscirev.2016.06.006

[147] Hoffman, J.S., et al. (2017) Regional and Global Sea-surface Temperatures During the Last Interglaciation, *Science* 355, no. 6322: 276, doi: 10.1126/science.aai8464

[148] Foster, G. and Rahmstorf, S. (2011) Global Temperature Evolution 1979–2010, *Environmental Research Letters* 6: 044022, doi: 10.1088/1748-9326/6/4/044022. See also Dahlmann, L. (2017) *Climate Change: Global Temperature*. https://www.climate.gov/news-features/understanding-climate/climate-change-global-temperature

[149] Dutton, A. and Lambeck, K. (2012) Ice Volume and Sea Level During the Last Interglacial, *Science* 337, no. 6091: 216–219, doi: 10.1126/science.1205749

[150] Hansen, J., et al. (2008) Target Atmospheric CO_2: Where Should Humanity Aim? *Open Atmospheric Science Journal* 2: 217–231, doi: 10.2174/1874282300802010217.

[151] United Nations Framework Convention on Climate Change, http://unfccc.int/2860.php

[152] Oroschakoff, K. (2018) Hot dispute over money threatens climate deal, *Politico*, updated 5/12/18: https://www.politico.eu/article/climate-change-goals-stalled-by-money-dispute-paris-agreement-katowice-summit/

Even though the Paris Accord sets significant emission thresholds, greenhouse gas production cannot stop on a dime. Thus, even the strongest practical steps at limiting warming are going to take some time to be implemented and additional time before a response from the climate system is witnessed.

8.7.2 The Carbon Budget

Climate change resulting from anthropogenic CO_2 emissions, and the associated risks, is essentially irreversible on human timescales. This is because 15 to 40 percent of CO_2 emitted until 2100 will remain in the atmosphere longer than 1,000 years.[153] Heat slowly released from the ocean into the atmosphere, even after anthropogenic emissions have ended, will further prolong warming.

Consequently, efforts to reduce anthropogenic emissions will have only modest temperature effects in the near term (the next one to two decades). But rapid reductions to a decarbonized world economy are necessary by mid-century in order to avoid some of the more dangerous aspects of global warming that will occur in the second half of this century. Near term climate is likely to be dominated by climate variability and extreme weather events, set upon a backdrop of long-term warming resulting from emissions over the past century.

Nonetheless, limiting the emission of short-lived climate pollutants such as methane, certain hydrofluorocarbons, ozone, and black carbon, can achieve more rapid climate benefits.[154] For instance, one study[155] found that heat waves would already be more severe by the 2030s if we exercised no mitigation versus moderate mitigation. Some of these short-lived substances exert stronger heat-trapping capacity than carbon dioxide so that the benefits are relatively enhanced and achieved more rapidly.

Limiting, eventually ending, emissions of strong heat-trapping gases that have short atmospheric residency times is important to control impacts over the next 2 to 3 decades. Stabilizing global temperature in the second half of this century and beyond requires that CO_2 emissions decline to zero.[156] Thus, an estimated range of cumulative CO_2 emissions can be calculated, known as a **carbon budget**. The key sources of uncertainty for this budget include the climate sensitivity, the response of the carbon cycle including feedbacks (e.g., permafrost thaw), the amount of past CO_2 emissions, and the influence of past and future non-CO_2 gases.

Since 1870, approximately 560 billion tons of carbon, in the form of CO_2, have been released to the atmosphere.[157] Studies[158] suggest that to stand a 66 percent chance of limiting warming to 2 degrees Celsius (3.6 degrees Fahrenheit) per the Paris Accord, no more than a total of 1,000 billon tons of carbon dioxide can be released to the atmosphere. However, this estimate ignores the effects of non-CO_2 heat-trapping releases such as methane and others. Considering these reduces the allowable CO_2 budget to 790 billion tons of carbon.[159] In light of past emissions, approximately 230 billion tons of carbon dioxide remain in the CO_2 budget to halt warming.

As illustrated by the Fourth National Climate Assessment (**Figure 8.17**), in order to stop warming at 2 degrees Celsius (3.6 degrees Fahrenheit) as called for in the Paris Accord, future emissions that follow a lower scenario consistent with RCP4.5, have until 2037 to cease emissions. Under the higher scenario, RCP8.5, this occurs by 2033. Unfortunately, limiting the global average temperature increase to the more stringent Paris goal, 1.5 degrees Celsius (2.7 degrees Fahrenheit), regardless of pathway, is already highly unlikely as the allowable carbon emissions are depleted by 2019 (Table 8.1).

8.7.3 1.5 Degree Celsius Very Unlikely

In early 2018, the unpublished draft of an IPCC report on stabilizing temperature at 1.5 degrees Celsius (2.7 degrees Fahrenheit) was leaked to the world press.[160] The report contained dire statements depicting a low likelihood of limiting global warming to 1.5 degrees Celsius. As reported by media outlets, among the report findings are the following:

- Curbing warming at 1.5°C would help limit heat extremes, droughts and floods, more migration of people and risks of conflict, compared to higher rates of warming.
- Limiting global warming to 1.5°C might not be enough to protect many coral reefs, already suffering

[153] Joos, F., et al. (2013) Carbon dioxide and Climate Impulse Response Functions for the Computation of Greenhouse Gas Metrics: A Multi-model Analysis, *Atmospheric Chemistry and Physics* 13, 2793–2825, doi:10.5194/acp-13-2793-2013

[154] Zaelke, D. and Borgford-Parnell, N. (2015) The Importance of Phasing Down Hydrofluorocarbons and Other Short-lived Climate Pollutants, *Jour. of Env. Studies and Sciences* 5: 169–175, doi:10.1007/s13412-014-0215-7.

[155] Tebaldi, C. and Wehner, M.F. (2016) Benefits of Mitigation for Future Heat Extremes Under RCP4.5 Compared to RCP8.5, *Climatic Change*, First Online: 1–13, doi:10.1007/s10584-016-1605-5.

[156] DeAngelo, B., et al. (2017) Perspectives on Climate Change Mitigation. In Climate Science Special Report: Fourth National Climate Assessment, Volume I [Wuebbles, D.J., et al. (eds.)]. U.S. Global Change Research Program, Washington, DC, USA, pp. 393-410, doi: 10.7930/J0M32SZG

[157] Le Quéré, C., et al. (2016) Global Carbon Budget, *Earth Sys. Sci.* 8: 605–649, doi:10.5194/essd-8-605-2016.

[158] Allen, M.R., et al. (2009) Warming Caused by Cumulative Carbon Emissions Towards the Trillionth Tonne, *Nature* 458: 1163–1166, doi:10.1038/nature08019.

[159] DeAngelo et al. (2017)

[160] Climate Change Draft Report. https://www.reuters.com/article/us-climatechange-draft/warming-set-to-breach-paris-accords-toughest-limit-by-mid-century-draft-idUSKBN1F02RH

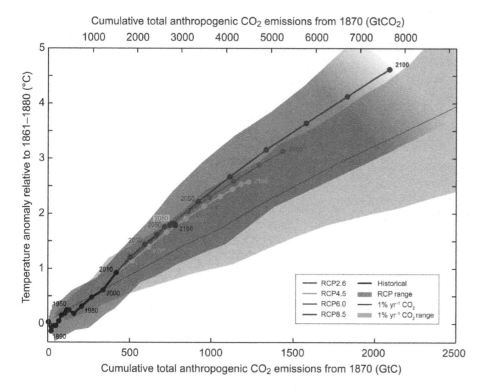

Cumulative total anthropogenic CO₂ emissions from 1870 (GtCO₂)

Cumulative total anthropogenic CO₂ emissions from 1870 (GtC)

FIGURE 8.17 Global mean temperature change caused by total (since preindustrial) carbon dioxide emissions for a number of scenarios. Time increases along each pathway. The dark colored field illustrates the multi-model spread over the four RCP scenarios and fades with the decreasing number of available models. The end of each line indicates that pathway leads to no further temperature increase. For a CO_2 increase of 1% per year, the multi-model mean and range simulated by CMIP5 models is shown by the thin black line and grey area. For a specific amount of cumulative CO_2 emissions, the 1% per year CO_2 simulation exhibits lower warming than those driven by RCPs. This is because the RCP scenarios include additional non-CO_2 forcings, and the 1% simulation does not. Temperature values are given relative to the 1861–1880 base period, emissions relative to 1870.[161]

Source: IPCC-AR5, SPM.10 https://www.ipcc.ch/pdf/assessment-report/ar5/wg1/WG1AR5_SPM_FINAL.pdf

---

TABLE 8.1

[162]Dates when cumulative carbon emissions (GtC) since 1870 commit warming to 2°C (3.6°F) and 1.5°C (2.7°F), when accounting for the addition of non-CO₂ forcings

2°C	66% = 790 GtC	50% = 820 GtC	33% = 900 GtC
RCP4.5	2037	2040	2047
RCP8.5	2033	2035	2040
1.5°C	66% = 593 GtC	50% = 615 GtC	33% = 675 GtC
RCP4.5	2019	2021	2027
RCP8.5	2019	2021	2025

GtC = billions of tons of carbon in the form of carbon dioxide

from higher ocean temperatures, and ice stored in Greenland and West Antarctica whose melt is raising sea levels.

- Average surface temperatures are about 1°C above preindustrial times, and average temperatures are on track to reach 1.5°C by the 2040s.
- There is no model that projects a 66-percent-or-better chance of holding global warming below 1.5°C.

- Limiting global warming to 1.5°C by 2100 will involve removal of carbon dioxide from the atmosphere (known as "negative emissions").
  - Removing carbon from the atmosphere could mean planting vast forests, which soak up carbon dioxide as they grow, or building power plants that burn wood or other plant matter and then capturing and burying the carbon dioxide they release.
  - However, these steps might not be feasible because so-called "energy crops" could divert land from food crops.
- There were no historic precedents for the scale of changes required in energy use, to shift from fossil fuels to renewable energies, and in reforms ranging from agriculture to industry to stay below the 1.5°C limit.
- Renewable energies such as solar and wind power would have to become the dominant form of primary energy by 2050 to achieve the 1.5°C goal. Coal would be phased out rapidly in most 1.5°C pathways. Removing carbon from the air is nonetheless required even in this case.
- Governments could let temperatures exceed 1.5°C if they found a way of turning down the global thermostat later in the century.[163]

---

[161] Stocker et al. (2013)
[162] DeAngelo et al. (2017)

[163] VIDEO: Is a Warmer World a Better World? https://www.youtube.com/watch?v=1SAqdG3gJH0&app=desktop

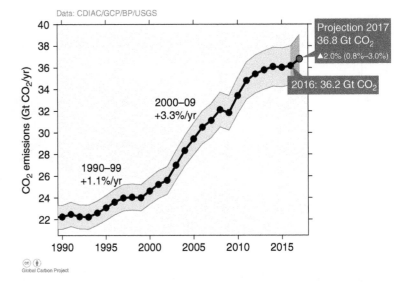

FIGURE 8.18 Emissions from fossil fuel use and industry. Over the three-year period 2014–2016, global CO_2 emissions from fossil fuels remained relatively flat. This changed in 2017 as emissions from fossil fuel use and industry grew by around 2%. Overall global carbon dioxide emissions from all sources combined (fossil fuels, industry, and land use change) rose to 41.5 +/-4.4 billion tons of carbon dioxide in 2017.

Source: https://www.icos-cp.eu/GCP/2017

8.7.4 Carbon Trends

It is already too late to stop significant warming from occurring, but as the world watches ongoing climate negotiations, the question in many scientists' minds is, "Can we act in time to avoid the most dangerous aspects of climate change?"

In 2015, global CO_2 accumulation in the atmosphere rose 3.03 parts per million (ppm), the largest annual increase on record. In 2016, CO_2 concentrations rose another 2.98 ppm above the 2015 level. Anthropogenic release of CO_2 during these record setting years was boosted by environmental emissions of CO_2 related to a strong El Niño, a positive feedback effect that had not previously been observed with such strength.[164] By 2017, the rate of carbon dioxide accumulation retreated to a more typical 2.13 ppm; El Niño had ended and a moderate La Niña was taking shape for 2018.

Rising carbon dioxide concentrations over these years were the result of environmental emissions coupled with anthropogenic emissions. The amount of CO_2 released into the atmosphere from burning fossil fuels, gas flaring, and cement production held steady over the period 2014–2016, neither increasing nor decreasing significantly. Nonetheless, the release of an average 40 billion tons of anthropogenic carbon dioxide each year represents an enormous amount of heat-trapping pollution entering the air annually. In 2017, global carbon dioxide emissions from all sources combined (fossil fuels, industry, and land use change) rose to 41.5 +/-4.4 billion tons of CO_2. Analysis by the Global Carbon Project,[165] reported by media outlets,[166] found that

global fossil fuel emissions grew by 0.7 percent in 2014, held steady in 2015, rose 0.2 percent in 2016, and unexpectedly jumped approximately 2 percent in 2017. After three years of essentially stable emissions, the large rise was startling to the world's carbon watchers (**Figure 8.18**).

Much of the slowdown in the growth of global emissions in recent years has been driven by a combination of reductions in the United States and China, as well as relatively little growth in emissions in other countries. This changed in 2017 with a sizeable increase in Chinese emissions. India's emissions increased a bit more slowly in 2017 than in the past few years, while the EU's emissions have remained relatively flat since 2014 and did not noticeably change in 2017. The growth in emissions from 2016 to 2017 also more than doubled in the rest of the world.[167]

A report[168] by the Word Resources Institute issued in late 2017 concluded that an encouraging trend is emerging: the number of countries that have reached peak greenhouse gas emissions or have a commitment that implies a peak in emissions in the future, grows from 19 countries in 1990 to 57 countries in 2030. They found the following:

- The number of countries that have already peaked their emissions grew from 19 in 1990, to 33 in 2000, to 49 in 2010.
- By 2020, the number of countries that have already peaked or have a commitment that implies an emissions peak grows to 53 and by 2030 grows to 57.
- The percentage of global emissions covered by countries that already peaked was 21% by 1990, 18% by 2000, and 36% by 2010.
- The percentage of global emissions covered by countries that have already peaked or have a commitment

[164] NASA. https://www.nasa.gov/press-release/nasa-pinpoints-cause-of-earth-s-recent-record-carbon-dioxide-spike

[165] Global Carbon Project, http://www.globalcarbonproject.org/about/index.htm

[166] Carbon Brief (2017) https://www.carbonbrief.org/analysis-global-co2-emissions-set-to-rise-2-percent-in-2017-following-three-year-plateau

[167] CarbonBrief (2017)

[168] Levin, K. and Rich, D. (2017) Turning Points: Trends in Countries' Reaching Peak Greenhouse Gas Emissions over Time. http://www.wri.org/ publication/turning-points.

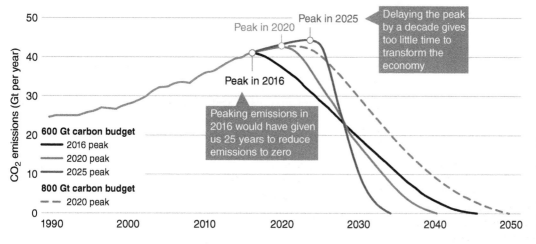

FIGURE 8.19 Implications of a delay in global emissions peaking on future decarbonization rates necessary to stay below 2°C.[170]

Source: http://www.wri.org/sites/default/files/turning-points-trends-countries-reaching-peak-greenhouse-gas-emissions-over-time.pdf

that implies an emissions peak by 2020 increases to 40% and, by 2030, to 60%.

- The large majority of developed countries as well as some developing countries have already peaked.
- We are witnessing a turning point with several developing countries taking on emissions reduction commitments that imply emissions peaks by 2020 or 2030.
- Nevertheless, the number of countries peaking and the emissions level at which they are peaking is insufficient to meet the Paris Agreement's temperature goals to limit warming to well below 2°C.

This news is encouraging and worrying at the same time. Global emissions are still not projected to peak in time to have a 66 percent chance of limiting warming to 2 degrees Celsius (3.6 degrees Fahrenheit). To achieve this, emissions need to peak by 2020 at the latest.[169] Unfortunately, projections show that even with national commitments to reduce emissions as promised in the Paris Accord, emissions are expected to continue to increase between 2020 and 2030. The world is not on track for a global emissions peak in 2020 (**Figure 8.19**).

According to the annual "Gap Report" issued by the United Nations, even meeting the pledges made as part of the Paris Accord will not limit warming to 2 degrees Celsius (3.6 degrees Fahrenheit). The promises to reduce greenhouse gas emissions that are embodied in the Paris treaty, add up to no more than one-third of the reductions needed to avoid the 2 degrees Celsius limit.[171] This "emissions gap" is possible to close, according to the U.N., but countries need to take action quickly. Focused action on just six key sectors will achieve a large portion of the necessary reductions: solar energy, wind energy, efficient appliances, efficient passenger cars, reforestation, and stopping deforestation. The report finds that strong actions in these six sectors sum up a potential savings of 18.5 billion tons of carbon dioxide in 2030, making up more than half of the needed reductions. Equally important, all these measures can be realized at modest cost and are predominantly achievable through proven policies.[172]

Falling Short According to a study[173] of economic, emissions and population trends, limiting warming to 2 degrees Celsius (3.6 degrees Fahrenheit) is highly unlikely. Researchers calculate a 95 percent chance that global temperatures will increase by more than 2 degrees Celsius, and a less than 1 percent chance they will not exceed 1.5 degrees Celsius (2.7 degrees Fahrenheit). The team looked at statistical data from 1960 to 2010 and found that temperatures over the next 80 years will likely increase between 2 to 4.9 degrees Celsius (3.6 to 8.8 degrees Fahrenheit), with a central estimate of 3.2 degrees Celsius (5.8 degrees Fahrenheit). There is a 90 percent chance that global temperatures will fall somewhere in the middle of the range.

Rather than look at how greenhouse gases will influence temperature, the new research analyzed the past 50 years of trends in world population, per capita gross domestic product and carbon intensity, which is the amount of carbon dioxide emitted for each dollar of economic activity. After building a statistical model covering a range of emissions scenarios, the researchers found that carbon intensity will be a crucial factor in future warming.

Technological advances are expected to cut global carbon intensity by 90 percent over the course of the century, with sharp declines in China and India. However, this will not be steep enough to avoid breaching the 2 degrees

[169] Figueres, C., et al. (2017) Three Years to Safeguard Our Climate, *Nature* 546: 593–595. doi:10.1038/546593a. https://www.nature.com/news/three-years- to-safeguard-our-climate-1.22201.

[170] After Figueres et al. (2017), from Levin and Rich (2017)

[171] UNEP (2017) The Emissions Gap Report, https://wedocs.unep.org/bitstream/handle/20.500.11822/22070/EGR_2017.pdf

[172] CarbonBrief (2017). https://www.carbonbrief.org/unep-six-crucial-actions-help-close-worlds-emissions-gap

[173] Raftery, A.E., et al. (2017) Less Than 2°C Warming by 2100 Unlikely, *Nature Climate Change*, doi: 10.1038/nclimate3352

Celsius limit.[174] The human population is projected to grow to 11 billion people by end of the century, but this will have a relatively small impact upon temperature as much of this growth will take place in sub-Saharan Africa, which is a minor contributor of greenhouse gas emissions. A breakthrough technology could dramatically change the outlook, but major advances of the past 50 years, such as the computer, robotics, hybrid cars, the Internet and electronic fuel injection, have improved carbon efficiency steadily at around 2 percent each year, rather than in huge jumps.

Mitigation In the IPCC 5th Assessment Report, fully 87 percent of the scenarios that limit warming to less than 2 degrees Celsius (3.6 degrees Fahrenheit) rely on removing carbon dioxide from the air (known as "negative emissions") in the second half of the century. Of even more concern, in all 56 scenarios without negative emissions, global emissions peak around 2010,[175] an assumption that was violated even before they went to press.

IPCC-AR5 projects that most carbon removal would come from an approach known as Bioenergy with Carbon Capture and Storage (BECCS) that would be deployed late in the 21st Century. BECCS removes atmospheric carbon with "energy crops" that are burned to produce electricity and whose emissions are captured and stored in geologic reservoirs in Earth's crust. However, BECCS is largely untested technology, requires large land areas (potentially equivalent to all todays cropland) that put at risk the security of food systems, and leads to negative environmental impacts.

In a comment published in the journal Nature Climate Change,[176] researchers argue that a larger range of scenarios is needed to provide guidance for policymakers seeking to limit warming. Many currently used emissions pathways assume that we can slowly decrease fossil fuel emissions today and make up for it later with heavy use of carbon removal technologies such as BECCS. However, this assumes that future generations will pick up the burden of negative emissions which is neither a realistic assumption nor is it morally acceptable. Additionally, the prospect of poorly understood, but potentially powerful climate feedbacks, such as methane release from thawing permafrost, puts the entire foundation of this approach at risk.

The authors describe 4 mitigation pathways:

1. Major reliance on future CO_2 removal, the current standard of many existing scenarios for limiting warming.
2. Rapid decarbonization starting immediately and halving every decade.
3. Earlier implementation of CO_2 removal technologies and phasing out by the end of the century.

4. Consistent implementation of CO_2 removal from now until the end of the century.

Notably, in all these pathways, current commitments under the Paris Accord are not sufficient to limit warming as desired. Negative emission technologies may be an asset, but they may also be an economic burden if not deployed with care. Many policymakers may not be aware that some mitigation scenarios are not realistic. For instance, scenarios that require converting massive land areas currently used for food production into negative emission farms, is not a reasonable alternative.

A simple **Global Carbon Law** has been defined that maps out a pathway to reach the Paris targets without turning to negative emissions. For a 50 percent chance of limiting warming to 1.5 degrees Celsius (2.7 degrees Fahrenheit) by 2100 and a >66 percent probability of meeting the 2 degree Celsius (3.6 degree Fahrenheit) target, global CO_2 emissions must peak no later than 2020. Thereafter, gross emissions must decline from ~40 billion tons of carbon dioxide per year in 2020, to ~24 by 2030, ~14 by 2040, and ~5 by 2050. By moderately increasing these targets in the form of cleanly halving emissions every decade, risks can be further reduced. This pathway defines a global carbon law, halving emissions every decade, that will limit cumulative total CO_2 emissions from 2017 until the end of the century to ~700 $GtCO_2$.

8.7.5 Human Vulnerability

An examination[177] of urban policies at the National Center for Atmospheric Research (NCAR) found that even though billions of urban dwellers are vulnerable to heat waves, sea-level rise, and other changes associated with warming temperatures, cities worldwide are failing to take the necessary steps to protect residents from the likely impacts of climate change. Not only are most cities failing to reduce emissions of carbon dioxide and other greenhouse gases, they are falling short in preparing residents for the likely impacts of climate change.

The study further noted that more than half the world's population lives in cities, where construction patterns are often dense, housing substandard, and access to reliable drinking water, roads, and basic services poor—all conditions that magnify the potential for humanitarian disaster.

Potential threats associated with climate change include storm surges, which can inundate coastal areas; development of steep hillsides and floodplains both of which are vulnerable to intense rains; and prolonged hot weather, which can heat heavily paved cities more than surrounding areas, exacerbate existing levels of air pollution, and cause widespread health problems. The study also identified factors that keep city leaders from making climate resilience a higher priority: Fast-growing cities are overwhelmed with

[174] Temperatures Rising: Achieving the global temperature goals laid out in the Paris Climate Agreement is unlikely, according to research: https://www.sciencedaily.com/releases/2017/08/170804082229.htm

[175] Anderson, K. (2015) Duality in Climate Science, *Nature Geoscience* 8: 898–900, doi:10.1038/ngeo2559

[176] Obersteiner, M., et al. (2018) How to spend a dwindling greenhouse gas budget, *Nature Climate Change* 8, no. 1: 7 doi: 10.1038/s41558-017-0045-1

[177] Hardoy, J. and Lankao, P.R. (2011) Latin American Cities and Climate Change: Challenges and Options to Mitigation and Adaptation Responses, *Current Opinion in Env. Sustainability*, doi: 10.1016/j.cosust.2011.01.004

other needs, city leaders are often pressured to choose economic growth over the need for health and safety standards, and climate projections are rarely fine-scale enough to predict impacts on individual cities.

Another study[178] combined climate change data with a global census of nearly 97 percent of the world's population to project human vulnerability to climate change by mid-century. The study concluded that populations in low-latitude tropical regions, such as central South America, the Arabian Peninsula, and much of Africa, may be most vulnerable to climate change.

Those communities already experience extremely hot and arid conditions that make agriculture challenging. Even a small temperature increase would have serious consequences on their ability to sustain a growing population. Communities in high-latitude temperate zones are already limited by cooler conditions, however. As such, researchers expect climate change will have less of an impact on people living in these areas.

Violence Climate change in stressed low-latitude nations can lead to war. In a first-of-its-kind study,[179] researchers examined the influence of El Niño, which every few years raises temperatures and cuts rainfall across broad swaths of tropical and subtropical regions. It was found that the onset of El Niño, used in the study as a proxy for longer-scale warming, doubles the risk of civil wars across 90 tropical countries and might account for one-fifth of worldwide conflicts during the past half century.

Study authors did not investigate why climate feeds conflict; however, they point out that a community characterized by poverty has underlying tensions, and it may be that warming delivers the final blow to peaceful solutions to persistent problems related to basic survival. For instance,

when crops fail, or water dries up, or other fundamental resources grow scarce, people may take up a gun simply to make a living. In fact, social scientists have shown in the past that individuals can become more aggressive when temperatures rise, but whether this behavior applies to whole societies is still speculative.

Food Impacts A study[180] found that global wheat production since 1980 was 5.5 percent lower than it would have been had climate remained stable and that global corn production was lower by almost 4 percent. In the United States, Canada, and Northern Mexico, a very slight regional cooling trend over the study period resulted in no significant production impacts. Outside of North America, most major agricultural countries experienced some decline in wheat and corn yields related to the rise in global temperature.

Although crop yields in most countries are still going up because of improvements in technology, fertilization, and other factors, they are not rising as fast as they would be without warming. Russia, India, and France experienced the greatest drop in wheat production, and China and Brazil experienced the largest losses in corn production.

Water Cycle Impacts Most evaporation and precipitation take place over the oceans, and as the atmosphere warms the rate of these processes accelerates. A study in the spring of 2012 revealed just how much the water cycle has sped up as a result of global warming.[181] Using 50 years of ocean surface salinity data (1.7 million measurements), scientists documented how the salinity of the ocean surface has changed as a result of changes in evaporation and precipitation. A map of their results (**Figure 8.20**) reveals that, as expected, wet areas are getting wetter, and dry areas are getting drier; high-latitude and equatorial parts of the

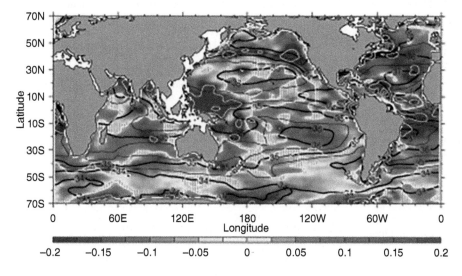

FIGURE 8.20 Absolute surface salinity change over the period 1950–2000. Rainfall and evaporation changes are making the oceans less salty in vast regions and saltier elsewhere. Research shows that while the surface warmed 0.5°C, the water cycle has sped up roughly 4%, twice as fast as predicted by most climate models. These results also indicate that in general, wet areas got wetter and dry areas got drier.[182]

Source: http://www.cmar.csiro.au/oceanchange/salinity.php

[178] Samson, J., et al. (2011) Geographic Disparities and Moral Hazards in the Predicted Impacts of Climate Change on Human Populations, *Global Ecology and Biogeography*, doi: 10.1111/j.1466-8238.2010.00632.x

[179] Hsiang, S.M., et al. (2011) Civil Conflicts are Associated with the Global Climate, *Nature* 476, no. 7361: 438, doi: 10.1038/nature10311.

[180] Lobell, D.B., et al. (2011) Climate Trends and Global Crop Production Since 1980, *Science*, doi: 10.1126/science.1204531

[181] Durack, P.J., et al. (2012) Ocean Salinities Reveal Strong Global Water Cycle Intensification During 1950 to 2000, *Science* 336, no. 6080: 455, doi: 10.1126/science.1212222.

[182] Durack et al. (2012)

oceans, where there is greater precipitation than average, became less salty; and mid-latitude areas (the central regions of ocean basins), where evaporation dominates, became saltier.

These results indicate that the water cycle has sped up roughly 4 percent while the surface warmed 0.5 degree Celsius (0.9 degree Fahrenheit), roughly twice as fast as predicted by most climate models. The study authors conclude that if the world warms 2 to 3 degrees Celsius (3.6 to 5.4 degrees Fahrenheit) by the end of the century, the water cycle will accelerate 16 to 24 percent. An amplified water cycle such as this would fuel violent storms in wet areas from tornadoes to tropical cyclones and produce severe and frequent flooding, and in dry areas it could mean long and intense droughts.

8.8 Dangerous Climate

Does climate change threaten the socio-economic framework of modern society? As radical and extreme as this sounds, it has been an ever-growing question in the minds of those who study the science behind global warming.[183] Two realizations raise concern:

1. The world's nations have only committed one-third of the greenhouse gas reductions necessary to stop warming at 2 degrees Celsius (3.6 degrees Fahrenheit),[184] and

2. After three years in which anthropogenic carbon dioxide release did not significantly increase, emissions surged by 2 percent in 2017.[185]

When carbon dioxide is released by human fossil fuel use and land management,[186] about half of it stays in the atmosphere over long periods of time (centuries or more[187]), the other half is absorbed by seawater (causing ocean acidification) and used by plants. The resulting global warming is changing the climate leading to more drought,[188] hotter and more frequent heat waves,[189] increased tropical cyclone intensity and rainfall,[190] warmer oceans that damage coral reefs and other parts of the marine ecosystem,[191] larger and more frequent wildfires,[192] more extreme forms of rainfall and flooding,[193] glaciers worldwide retreating,[194, 195] global sea-level rise,[196] global ecosystem damage,[197, 198] reduced food nutrition,[199] human health problems,[200] and others.

8.8.1 Political Turmoil

In some parts of the world, atmospheric warming makes freshwater scarce, farming impossible, summer temperatures deadly, and weather disasters larger and more frequent. These events make certain parts of the planet unlivable, driving people to leave their homelands to find new homes where life is not so difficult.

But the human population has been growing strongly over the past half century and now exceeds 7.5 billion people.[201] There are not very many places left on Earth that can support large and sudden increases to existing communities.

If climate change is making parts of the planet unlivable, who can afford to accept displaced populations? Most nations are already challenged with expensive problems and don't have the resources to help large groups of immigrants. Even the United States, the most prosperous nation in human history, is trillions of dollars in debt. And we have elected a President who has promised to build a wall to keep out immigrants. Where are climate immigrants to go? This is a recipe for violence, which rises with the air temperature.[202]

[183] Wallace-Wells (2017)

[184] UNEP (2017) *The Emissions Gap Report* 2017

[185] CarbonBrief (2017) https://www.carbonbrief.org/analysis-global-co2-emissions-set-to-rise-2-percent-in-2017-following-three-year-plateau

[186] Global Carbon Project (2017) http://www.globalcarbonproject.org/carbonbudget/index.htm

[187] Solomon, S., et al. (2009) Irreversible Climate Change Due to Carbon dioxide Emissions, *PNAS*, 1: 1704–1709, doi: 10.1073/pnas.-9128211-6

[188] Dai (2011)

[189] Mora, C., et al. (2017) Global Risk of Deadly Heat, *Nature Climate Change*, doi: 10.1038/NCLIMATE3322

[190] Knutson, T.R., et al. (2015) Global Projections of Intense Tropical Cyclone Activity for the Late Twenty-First Century from Dynamical Downscaling of CMIP5/RCP4.5 Scenarios, *Journal of Climate* 28, no. 18: 7203–7224.

[191] Ramírez, F., et al. (2017) Climate Impacts on Global Hot Spots of Marine Biodiversity, *Science Advances* 3, no. 2: e1601198, doi: 10.1126/sciadv.1601198

[192] Abatzoglou et al. (2016)

[193] Lehmann, J., et al. (2015) Increased Record-breaking Precipitation Events Under Global Warming, *Climatic Change*, doi: 10.1007/s10584-015-1434-y

[194] Zemp, M., et al. (2015) Historically Unprecedented Global Glacier Decline in the Early 21st Century, *Journal of Glaciology* 61, no. 228: 745, doi: 10.3189/2015JoG15J017

[195] Rignot, E., et al. (2011) Acceleration of the Contribution of the Greenland and Antarctic Ice Sheets to Sea Level Rise, *Geophysical Research Letters* 38: L05503, doi: 10.1029/2011GL046583

[196] Dangendorf, S., et al. (2017) Reassessment of 20th Century Global Mean Sea Level Rise, *PNAS*, doi: 10.1073/pnas.1616007114

[197] Wiens, J.J. (2016) Climate-related Local Extinctions are Already Widespread Among Plant and Animal Species, *PLOS Biology*, 14, no. 12: e2001104, doi: 10.1371/journal.pbio.2001104

[198] Ceballos, G., et al. (2017) Biological Annihilation via the Ongoing Sixth Mass Extinction Signaled by Vertebrate Population Losses and Declines, *PNAS*, 114, no. 30: E6089-E6096; doi:10.1073/pnas.1704949114

[199] Myers, S., et al. (2014) Increased CO_2 Threatens Human Nutrition, *Nature* 510: 139–142, doi: 10.1038/nature13179.

[200] U.S Global Change Research Program (2016) The Impacts of Climate Change on Human Health in the United States: A Scientific Assessment. A. Crimmins, et al., Washington, DC, 312 pp. http://dx.doi.org/10.7930/J0R49NQX

[201] Professor Corey Bradshaw, a population ecologist at Flinders University, has been quoted as saying "It is nearly impossible to say exactly what Earths human carrying capacity is, because it is entirely dependent on technologies like farming, electricity production and transport—and on how many people we are willing to condemn to a life of poverty or malnutrition." http://www.bbc.com/earth/story/20160311-how-many-people-can-our-planet-really-support

[202] Global Warming and Violent Behavior, http://www.psychologicalscience.org/observer/global-warming-and-violent-behavior

8.8.2 No More Carbon Dioxide?

Since the mid-19th century, Earth has warmed about 1 degree Celsius (1.8 degrees Fahrenheit).[203] Research shows that the chances of keeping global temperature rise below 1.5 degrees Celsius (2.7 degrees Fahrenheit) are essentially nil (see Table 8.1).[204] The National Climate Assessment reports that to stand a good chance (66 percent) of keeping warming below 2 degrees Celsius (3.6 degrees Fahrenheit), future CO_2 emissions must not exceed about 790 billion tons. At current rates, that gives us until about 2033 to 2037 (**Figure 8.21**) to convert to carbon-free power generation, end deforestation, and walk away from the internal combustion engine and other producers of CO_2.[205]

Think back to the year 2000. In the time since, do you believe all of humanity could have given up our addiction to gasoline, coal, beef, dairy products, lumber, airplane travel, cement, and other staples of modern living? The challenge laying before us no less daunting. Our chances of meeting this challenge will swiftly diminish unless the world takes immediate action to cut CO_2 emissions, which totaled about 41 billion tons in 2017.[208]

Yes, there are proven, low-carbon replacements for some of these things. Solar and wind power are replacing fossil fuels as electricity sources at exponential rates. In another decade, renewables will be cheaper than coal—many people hope that marks the point at which we will truly abandon coal. But, we will be more than halfway toward using up our remaining 2 degrees Celsius carbon budget by then.

To stand a good chance of avoiding 3 degrees Celsius (5.4 degrees Fahrenheit) of warming, we have on the order of 50 years if we continue to emit CO_2 at 40 billion tons per year. Continuing to emit CO_2 at this rate is a clear possibility even with the rapid rise in renewables we have witnessed in the past few years. The reason for this, is that economists are projecting strong economic growth in India and China and much of the demand for new energy will be met with fossil fuels.

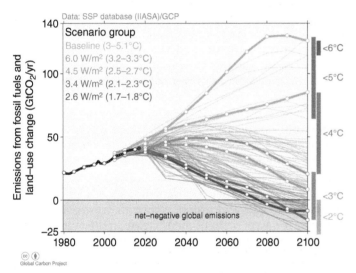

FIGURE 8.21 Scenario pathways to end of century temperature. In preparation for the IPCC Sixth Assessment Report, researchers have developed new modeling scenarios termed Shared Socioeconomic Pathways (SSPs).[206] SSPs describe plausible future changes in demographics, human development, economy, institutions, technology, and environment and when climate policies are included, they lead to different levels of radiative forcing at the end of the century.[207] Red (filled) dot—2017 emissions.

Source: http://folk.uio.no/roberan/GCP2017.shtml

8.8.3 Global Energy Demand

Energy demand accounts for 66 percent of total greenhouse gas emissions and 80 percent of CO_2. Any effort to reduce emissions and mitigate climate change must involve the energy sector. It is the year-to-year demand for new energy by a growing global economy that tends to govern emissions. Human population growth, deforestation, more efficient use of energy, fossil fuel use by the transportation sector, conversions to electricity that can be supplied by clean energy in various parts of the economy (e.g., home heating), and other constantly shifting parameters, also play important roles in governing year-to-year carbon emissions. But it is a growing economy that sets the basic foundation on which emissions rise or fall. A growing global economy requires new sources of energy. At present, and into the near future, only the fossil fuel industry is equipped to respond rapidly to new demands for energy. The year 2017 was a typical example.

2017, Energy Demand Lifted by economic growth concentrated in China and India, global energy demand increased by 2.1 percent in 2017, compared with 0.9 percent the previous year and 0.9 percent on average over the previous five years.[209] Most of this increase was met with oil, gas, and coal. However, renewable forms of energy made notable gains. More than 40 percent of the growth in demand was driven by China and India; 72 percent of the growth

[203] Global Warming Index http://www.globalwarmingindex.org

[204] DeAngelo et al. (2017)

[205] VIDEO: Future Earth Global Carbon Budget 2017 http://futureearth.org/blog/Budget2017

[206] O'Neill, B.C., et al. (2017) The Roads Ahead: Narratives for Shared Socioeconomic Pathways Describing World Futures in the 21st Century, *Global Environmental Change*, v. 42: 169–180.

[207] Description of SSPs https://energiogklima.no/blogg/oil-gas-in-a-low-carbon-world/

[208] Global Carbon Project update for 2017 http://www.globalcarbonproject.org/carbonbudget/index.htm

[209] IEA (2018) *Global Energy & CO2 Status Report 2017, International Energy Agency*: http://www.iea.org/publications/freepublications/publication/GECO2017.pdf

was met by fossil fuels, one-quarter by renewable sources of energy, and the remainder by nuclear power. In 2017:

1. CO_2—Global energy-related emissions grew by 1.4%.
2. Oil—World demand rose by 1.6%, boosted by an increasing share of sport-utility vehicles and light trucks in major economies, and growth in the petro-chemical industry.
3. Natural Gas—Global demand grew by 3% largely due to low-cost and abundant supplies. China alone accounted for almost 30% of this growth.
4. Coal—Global demand rose about 1%, reversing a decline in the past two years. Demand came mainly from Asia, driven by an increase in coal-fired electricity generation.
5. Renewables—The highest growth of any energy source, meeting 25% of new demand.
6. Electricity—Demand increased by 3.1%. China and India account for 70% of this growth.
7. Efficiency—Improvements slowed down dramatically in 2017, because of weaker policies and lower energy prices.

Energy Outlook Meeting the temperature objectives of the Paris Agreement requires urgently reducing greenhouse gas emissions. Rockstrom et al.[210] show that CO_2 reductions equaling 50 percent each decade, beginning in the 2020's, can lead to net-zero emissions around mid-century. This pathway is necessary to limit warming to 2 degrees Celsius (3.6 degrees Fahrenheit). Are projections of near-term energy use consistent with this?

According to projections,[211] global energy demand expands by 30 percent over the next two decades, the equivalent of adding another China and India to today's energy use. Driving this increase are global economy growth at 3.4 percent per year, human population growth from 7.4 billion to more than 9 billion in 2040, and urbanization that adds a city the size of Shanghai to the world's urban population every four months. The largest growth in demand (30 percent) comes from India.

The roles of natural gas, renewable energy, and energy efficiency play major roles in meeting the growth in energy demand. Renewable energy meets 40 percent of the increase. In India, the share of new demand met by coal drops from three-quarters in 2016 to less than half in 2040. Global oil demand continues to grow to 2040 but at a steadily decreasing pace. Natural gas use rises by 45 percent to 2040.

Renewable energy captures two-thirds of global investment in power plants by 2040. They become the most affordable source of new energy. Rapid expansion of solar power, led by China and India, helps it become the largest

source of low-carbon capacity by 2040. By 2040 the share of all renewables in total power generation reaches 40 percent.

However, oil demand continues to rise by 2040: oil use to produce petrochemicals is the largest source of growth, closely followed by rising consumption for trucks (fuel-efficiency policies cover 80 percent of global car sales today, but only 50 percent of global truck sales), for aviation, and for shipping.

According to the International Energy Agency "...global energy-related CO_2 emissions increase slightly to 2040..."[212] although they do not venture how much an increase is projected to occur. Clearly, though, they do not foresee bending the curve of emissions over the next two decades.

The U.S. Energy Information Administration[213] has modeled global energy use to the year 2040. They project the following patterns: coal sustaining a 20-year-long plateau, natural gas plentiful and growing, carbon-free wind and solar growing rapidly in percentage terms but not fast enough to bring emissions down in absolute terms, and petroleum holding its own as the main source of energy for transportation, despite the arrival of electric vehicles.

With populations growing and developing nations getting richer, total energy consumption will keep climbing despite gains in energy efficiency. Fossil fuels hold a 77 percent market share, and as a result carbon emissions will increase in parallel. Worldwide emissions of CO_2 from the burning of fossil fuels is projected to grow 16 percent by the year 2040 (compared to 2015).

The energy company BP provides an annual energy outlook.[214] BP considers a number of different scenarios without favoring any as forecasts of the future. However, their analysis centers on the "Evolving Transition" scenario in which world GDP more than doubles by 2040. This drives an increase in global energy demand, 50 percent of which comes from India and China. Although renewable energy accounts for 40 percent of the increase in demand, carbon emissions continue to rise and increase by around 10 percent by 2040.

In the energy outlook provided by ExxonMobile[215] (2018), a number of global trends drive future energy use. By 2030, the world's economic middle class will expand from 3 billion to more than 5 billion people. As these people develop modern businesses and gain access to cars, appliances and air-conditioned homes, energy use rises in parallel. Despite efficiency gains, global energy demand will increase nearly 25 percent. Strongest growth will be in China and India where demand will increase about 40 percent. Electricity from solar and wind together will

[210] Rockström, J., et al. (2017) A roadmap for rapid decarbonization. *Science*, 355 (6331): 1269 DOI: 10.1126/science.aah3443
[211] IEA (2017) *World Energy Outlook, International Energy Agency*: https://www.iea.org/weo2017/#section-4-5

[212] IEA (2017)
[213] EIA (2017) International Energy Outlook 2017, U.S. Energy Information Administration, https://www.eia.gov/outlooks/ieo/exec_summ.php
[214] BP (2018) *BP Energy Outlook, 2018 Edition*: https://www.bp.com/en/global/corporate/energy-economics/energy-outlook.htm
[215] ExxonMobile (2018) *2018 Outlook for Energy*: http://corporate.exxonmobil.com/en/energy/energy-outlook/a-view-to-2040#/section/1-key-takeaways-at-a-glance

grow about 400 percent. However, oil will continue to play a leading role in the world's energy mix, with growing demand driven by commercial transportation and the chemical industry. Global energy-related CO_2 emissions will likely peak by 2040 at about 10 percent above the 2016 level.

Recall that we opened this section with Rockstrom et al. recommending CO_2 reductions equaling 50 percent each decade in order to meet Paris Accord targets. The outlook of leading energy economists and other market watchers indicates instead, that the world is on a pathway of rising CO_2 emissions by 2040. Exactly how high emissions will rise is somewhat irrelevant as any increase dedicates Earth to a surface temperature that is above the Paris target of 2 degrees Celsius.

Aerosols Limiting global warming to 1.5 or 2.0 degrees Celsius requires strong mitigation of greenhouse gas emissions. At the same time, aerosol emissions (which have a cooling effect by scattering sunlight) tied to these greenhouse gases will decline. Researchers[216] have shown that the net effect of reducing both greenhouse gas emissions and aerosols leads to net warming of 0.5 to 1.1 degrees Celsius (0.9 to 1.98 degrees Fahrenheit). Associated with this are increases in precipitation (2.0 to 4.6 percent) and extreme weather events, especially in the Northern Hemisphere, where most aerosol reductions would occur.

Negative Emissions The Intergovernmental Panel on Climate Change[217] considered 116 potential scenarios for staying below 2.0 degrees Celsius, and 101 of those scenarios (87 percent) included negative emissions in the form of carbon capture and storage.[218] In other words, it's difficult to envision meeting the Paris target by cutting carbon pollution alone.

The term "negative emissions" refers to the use of technology to remove CO_2 from the air. Out of the four scenarios assessed by the IPCC, RCP2.6 is the only one that stands a good chance of limiting global warming to <2 degrees Celsius.[219] Various technology options exist to achieve negative emissions, but none have been proven commercially viable, and some have not even been demonstrated to work at global scale.[220]

One negative emission technology, Bioenergy with Carbon Capture and Storage (BECCS) is widely favored for its potential low cost. BECCS relies on burning biomass crops specifically grown as an energy source, capturing the emissions, and storing them underground in geologic repositories such as former oil and gas reservoirs, saline groundwater, and where the carbon can chemically react with the crust to remain stable. BECCS achieves a "double gain" as the biomass draws carbon dioxide out of the air as it grows, which is then stored in Earth's crust. Of the IPCC model scenarios that included negative emissions in the form of carbon capture and storage, 67 percent concluded that BECCS would provide at least 20 percent of the world's primary energy by 2100.

Although some aspects of BECCS have been shown to be technically feasible for deployment in the U.S.[221] other aspects remain unproven such as: Are the land, water and fertilizer needs sustainable? Is it economically viable? and, is it commercially viable at a scale that would make it useful? There are other important and unanswered questions that challenge the global deployment of BECCS: 1) In order for bioenergy to be globally meaningful, would the land necessary to grow energy crops compete with agriculture or natural ecosystems? 2) Is there sufficient capacity, and in the right locations, to store captured carbon emissions underground? Is long-term storage safe and stable? 3) What is the cost of BECCS and how will it be financed? 4) Are there socioeconomic barriers, such as public acceptance, to deployment? 5) In an era of growing drought, extreme weather, and growing numbers of asylum seekers, where are the optimal lands to support BECCS? What governmental institutions are needed to safeguard BECCS as a global resource? How do all these map onto the geologic repositories and access to them?

Recent Studies Three studies provide perspective on our discussion. One study[222] concluded that there is a 95 percent chance that warming will exceed 2 degrees Celsius (3.6 degrees Fahrenheit) by the end of the century, and only a 1 percent chance that it stays below 1.5 degrees Celsius (2.7 degrees Fahrenheit). The likely range of warming is 2.0 to 4.9 degrees Celsius (3.6 to 8.8 degrees Fahrenheit), and the median forecast is 3.2 degrees Celsius (5.7 degrees Fahrenheit).[223]

A second study[224] concluded that even if humanity stopped burning fossil fuels immediately, global warming would continue to warm 2 more degrees Celsius by the end

[216] Samset, B.H., et al. (2018) Climate impacts from a removal of anthropogenic aerosol emissions. *Geophysical Research Letters*, 45, 1020–1029. https://doi.org/10.1002/2017GL076079

[217] IPCC (2014) Climate Change 2014: Synthesis Report. Contribution of Working Groups I, II and III to the Fifth Assessment Report of the Intergovernmental Panel on Climate Change [Core Writing Team, R.K. Pachauri and L.A. Meyer (eds.)]. IPCC, Geneva, Switzerland, 151 pp.

[218] Fuss, S., et al. (2014) Betting on negative emissions, *Nature Climate Change*, 2014; DOI: 10.1038/nclimate2392

[219] Gasser, T., et al. (2015) Negative emissions physically needed to keep global warming below 2°C, *Nature Communications*, 6, 7958, doi:10.1038/ncomms8958

[220] CarbonBrief, https://www.carbonbrief.org/explainer-10-ways-negative-emissions-could-slow-climate-change

[221] Baik, E. et al. (2018) Geospatial analysis of near-term potential for carbon-negative bioenergy in the United States, *Proceedings of the National Academy of Sciences*, doi:10.1073/pnas.1720338115

[222] Raftery et al. (2017)

[223] Climate Change. http://www.cnn.com/2017/07/31/health/climate-change-two-degrees-studies/index.html

[224] Mauritsen, T. and Pincus, R. (2017) Committed Warming Inferred from Observations, *Nature Climate Change* 7: 652–655, doi: 10.1038/nclimate3357

of the century. The study also calculated that if emissions continue for 15 more years, a very likely scenario, global temperature will rise as much as 3 degrees Celsius.

And let us recall the work of Brown and Caldeira[225] in late 2017, warning that most climate models tend to underestimate the effects of anthropogenic greenhouse gases. They found that models that do the best job of simulating observed climate change predict some of the worst-case scenarios for the future. Using a group of models that perform the best at simulating recent past climate, their conclusion was that if countries stay on a high-emissions trajectory, there's a 93 percent chance the planet will warm more than 4 degrees Celsius by the end of the century. Previous studies placed those odds at 62 percent.

In summary, modeling of allowable carbon release, energy projections for the next two decades, and recent research all indicate we are unlikely to achieve the goals of the Paris Agreement, and the potential for global warming to reach 3 degrees Celsius (5.4 degrees Fahrenheit) is high.

8.8.4 Earth When It Is 3°C Warmer

What will life on planet Earth look like when temperatures are 3 degrees Celsius (5.4 degrees Fahrenheit) warmer? As concluded by the IPCC, many aspects of climate change and associated impacts will continue for centuries, even if anthropogenic emissions of greenhouse gases are stopped. Worse, the risks of abrupt or irreversible changes increase as the magnitude of the warming increases. If we surpass 3 degrees Celsius, it has been estimated by scientists that life on our planet will change as we know it. Rising seas, extinctions, droughts, increased wildfires, intense hurricanes, decreased crops and fresh water and the melting of the Arctic are expected.[226]

The impact on human health will be profound. Rising temperatures and shifts in weather will lead to reduced air quality, food and water contamination, more infections carried by mosquitoes and ticks and stress on mental health.[227] Currently, the World Health Organization estimates[228] that 12.6 million people die globally due to pollution, extreme weather and climate-related disease. It has been estimated that between 2030 and 2050, climate change is expected to cause approximately 250,000 additional deaths per year, from malnutrition, malaria, diarrhea and heat stress.

One study[229] has found that greenhouse gas emissions by 2100 will be enough to trigger an abrupt and unavoidable "catastrophic sixth mass extinction". Another[230] calculates that oxygen production by marine algae, the organisms that produce approximately 70 percent of the air we breathe, may collapse if we continue high rates of carbon burning to the end of the century. Already oxygen levels in the ocean have declined by 2 percent over the past five decades because it is growing more stratified, causing habitat loss for many fish and invertebrate species.[231]

Although carbon emissions have only warmed the surface 1 degree Celsius, the impacts are already severe and widespread, in part because they magnify existing stress that human communities place on natural systems. Researchers[232] found that across 65 percent of the terrestrial surface, land use and related pressures have caused biotic intactness to decline beyond 10 percent, considered a "safe" planetary boundary. Changes have been most pronounced in grassland biomes and biodiversity hotspots.

Sixth Mass Extinction Assembling a census of the biomass of all kingdoms of life, one study[233] found that the world's 7.6 billion people represent just 0.01 percent of all living things. Yet since the dawn of civilization, humanity has caused the loss of 83 percent of all wild mammals and half of plants, while livestock kept for food flourishes. Today, plants constitute 82 percent of all living matter, bacteria 13 percent, and all other creatures make up just 5 percent of the world's biomass. Since the rise of human civilization, the destruction of wild habitat for farming, logging, and development has been enormous. Of all the mammals on Earth, 96 percent are livestock and humans, and only 4 percent are wild. Of all birds, 70 percent are chickens and other poultry, and only 30 percent are wild.

With the addition of climate change, critical systems that support life are being pressed to the very edge of annihilation.[234] Wildlife is dying out due to habitat destruction, overhunting, toxic pollution, invasion by alien species and

---

[225] Brown and Caldeira (2017)

[226] As UN says world to warm by 3 degrees, scientists explain what that means: http://america.aljazeera.com/articles/2015/9/23/climate-change-effects-from-a-3-c-world.html

[227] The Medical Society Consortium on Climate and Health (2017) *Medical Alert! Climate Change is Harming Our Health*, https://medsocietiesforclimatehealth.org/reports/medical-alert/

[228] See WHO page http://www.who.int/mediacentre/news/releases/2016/deaths-attributable-to-unhealthy-environments/en/

[229] Rothman, D.H. (2017) Thresholds of Catastrophe in the Earth System, *Science Advances* 2017, doi: 10.1126/sciadv.1700906

[230] Sekerci, Y. and Petrovskll, S. (2015) Mathematical Modeling of Plankton-Oxygen Dynamics under the Climate Change, *Bulletin of Mathematical Biology*, doi: 10.1007/sl11538-015-0126-0

[231] Schmidtko, S., et al. (2017) Decline in Global Oceanic Oxygen Content During the Past Five Decades, *Nature*, 542: 335–339, 16 February, doi: 10.1038/nature21399

[232] Newbold, T., et al. (2016) Has land use pushed terrestrial biodiversity beyond the planetary boundary? A Global Assessment, *Science*, 15 July, v. 353, Is. 6296, 288-291, doi:10.1126/science.aaf2201.

[233] Bar-On, Y.M., et al. (2018) The biomass distribution on Earth, *PNAS*, May 21, 201711842.

[234] Ceballos, G., et al. (2017) Biological annihilation via the ongoing sixth mass extinction signaled by vertebrate population losses and declines, *PNAS*, 114 (30) E6089-E6096; doi:10.1073/pnas.1704949114

climate change. Species populations of vertebrate animals have decreased in abundance by 58 percent since 1970, largely due to the loss and degradation of habitat.[235] This and other signs have led scientists to conclude that life on Earth has entered the sixth global mass extinction in Earth's 4.6 billion year history.[236]

Climate change effects have now been documented across every ecosystem on Earth.[237] For instance, the average rate of vertebrate species loss over the last century is up to 100 times higher than the background rate. On a fully natural Earth, the number of species that have gone extinct in the last century would have taken, depending on the vertebrate taxon, between 800 and 10,000 years to disappear. Instead, 477 vertebrates have gone extinct since 1900, rather than the 9 that would be expected at natural rates. These estimates reveal an exceptionally rapid loss of biodiversity over the last few centuries, indicating that a sixth mass extinction is underway.

Observed ecological impacts include shifts in species ranges, shrinking body size, changes in predator-prey relationships, new spawning and seasonal patterns, and modifications in the population and age structure of marine and terrestrial species. Deeper changes are afoot however and include disruptions that scale from the gene to the ecosystem. Impacts to entire ecosystems point toward an increasingly unpredictable future for humans including increased pests and disease outbreaks, reduced productivity in fisheries, and decreasing agriculture yields. Dr. Brett Scheffers at the University of Florida has said "Genes are changing, species' physiology and physical features such as body size are changing, species are rapidly moving to keep track of suitable climate space, and there are now signs of entire ecosystems under stress."[238] This trend amounts to a massive anthropogenic erosion of biodiversity and of the ecosystem services essential to civilization.

Biogeochemical Crisis With the addition of climate change, impacts extend beyond the biosphere. In September 2016, the global atmospheric carbon dioxide concentration permanently passed 400 parts per million. This concentration is the highest since the Miocene Epoch (15 million years ago), when global temperature was 3 to 6 degrees Celsius warmer than present.[239]

Today's release of planet-warming carbon dioxide is about ten times faster than the most rapid event of any time in at least the past 66 million years, when an asteroid impact killed the dinosaurs.[240] The magnitude and speed with which we are polluting the troposphere exceeds any recent geologic precedent. In other words, coupled with extinction, we are literally living in the midst of, and are the cause of, a biogeochemical crisis that is planetary in scale.

Warning to Humanity As we discussed in **Chapter 1**, over 15,000 scientists have published a "Warning to Humanity."[241] They said humans have pushed Earth's ecosystems to their breaking point and are well on the way to ruining the planet. The warning listed a series of dire trends over recent decades:

1. A 26% reduction in the amount of fresh water available per person;
2. A drop in the harvest of wild-caught fish, despite an increase in fishing effort;
3. A 75% increase in the number of ocean dead zones;
4. A loss of nearly 300 million acres of forestland, much of it converted for agricultural uses;
5. Continuing significant increases in global carbon emissions and average temperatures;
6. A 35% rise in human population;
7. A collective 29% reduction in the numbers of mammals, reptiles, amphibians, birds, and fish.

The authors conclude "Soon it will be too late to shift course away from our failing trajectory, and time is running out. We must recognize, in our day-to-day lives and in our governing institutions, that Earth with all its life is our only home."[242] In fact, climate change impacts have now been documented across every ecosystem on Earth, despite an average warming so far of only ~1 degree Celsius (1.8 degrees Fahrenheit).[243] Impacts to entire ecosystems point toward an increasingly unpredictable future for humans including increased pests and disease outbreaks, reduced productivity in fisheries, and decreasing agriculture yields.

[235] WWF (2016) Living Planet Report 2016. Risk and resilience in a new era. WWF International, Gland, Switzerland.

[236] Ceballos, G., et al. (2015) Accelerated modern human-induced species losses: Entering the sixth mass extinction, *Science Advances*, 19 June, v.1, no. 5, doi:10.1126/sciadv.1400253.

[237] Scheffers, B.R., et al. (2016) The broad footprint of climate change from genes to biomes to people. *Science*, November, doi: 10.1126/science.aaf7671

[238] Climate change already dramatically disrupting all elements of nature: https://www.sciencedaily.com/releases/2016/11/161110115540.htm

[239] Tripati, A.K., et al. (2009) Coupling of CO_2 and ice sheet stability over major climate transitions of the last 20 million years, *Science*, 326(5958), 1394-1397, http://www.sciencemag.org/cgi/content/abstract/1178296

[240] Zeebe, R.E., et al. (2016) Anthropogenic carbon release rate unprecedented during the past 66 million years, *Nature Geoscience*, doi: 10.1038/ngeo2681

[241] Ripple, W.J., et al. (2017) World Scientists' Warning to Humanity: A Second Notice, *BioScience*, doi: 10.1093/biosci/bix125, http://scientists.forestry.oregonstate.edu/sites/sw/files/Ripple_et_al_warning_2017.pdf

[242] The warning came with steps to reverse negative trends, but the authors suggest that it may take a groundswell of public pressure to convince political leaders to take the right corrective actions. Such activities could include establishing more terrestrial and marine reserves, strengthening enforcement of anti-poaching laws and restraints on wildlife trade, expanding family planning and educational programs for women, promoting a dietary shift toward plant-based foods and massively adopting renewable energy and other "green" technologies.

[243] Scheffers et al. (2016)

8.9 The Socio-Economic Framework of All Humanity Is at Risk

Nearly one-third of the world's population is now exposed to climatic conditions that produce deadly heat waves.[244] In 2003 and 2010, respectively, seasonal extremes in temperature caused fatalities in excess of 70,000 people in Europe and 10,000 people in Russia. High death tolls from numerous other heat waves, now occurring annually across the planet, are stunning proof that extreme heat is already exceeding the capacity of human metabolism. For instance, an increase of 0.5 degree Celsius (0.9 degree Fahrenheit) in summer mean temperatures increases the probability of mass heat-related mortality in India[245] by 146 percent.[246]

If global temperatures increase up to 2 degrees Celsius (3.6 degrees Fahrenheit), the combined effect of heat and humidity will turn summer into one long heat wave. Temperature will exceed 40 degrees Celsius (104 degrees Fahrenheit) every year in many parts of Asia, Australia, Northern Africa, South and North America.[247] But if temperatures rise by 4 degrees Celsius (7.2 degrees Fahrenheit), a new and severe scenario is on the horizon. Scientists predict[248] the emergence of a "super-heatwave" with temperatures peaking at above 55 degrees Celsius (131 degrees Fahrenheit) making large parts of the planet unlivable including densely populated areas such as the U.S. east coast, coastal China, large parts of India and South America.

Climate change has already revealed the political turmoil that results from these conditions. Families from Syria, joined by others from North Africa, flee scarce resources, civil war, and violent extremism, an outgrowth of the deepest drought in 1000 years, and linked to anthropogenic global warming.[249] Far right, sometimes fascist, politicians in Europe, are rising in prominence on a surging tide of voters who object to the influx to refugees.[250] The famous withdrawal of Britain from the European Union, known as "Brexit," is a reaction to this refugee crisis.

Heat waves, drought[251], famine, and violence[252] all rise as the air gets hotter. This is not a distant scenario. Extreme heat waves are projected to cover double the amount of global land by 2020 and quadruple by 2040, regardless of future emissions trends.[253] The political turmoil that results from climate refugees seeking new homes, and the rise of demagogue politicians in response, brings to mind movies of totalitarian governments: Hunger Games, Divergence, and perhaps others.

Is climate change going to make Earth unlivable? The answer is certainly "yes" for parts of the planet, and the early signs are already in play. The United Nations Global Report on Internal Displacement[254] reveals that violence and natural disasters displace twice as many people internally (within their own borders) as those who flee across borders. During 2015, there were 27.8 million new displacements associated with conflict, violence, and disasters in 127 countries. This is roughly equivalent to every man, woman and child in New York City, London, Paris, and Cairo grabbing what they could carry and fleeing their homes in search of safety. Disasters displaced around 19.2 million people across 113 countries in 2015, more than twice the number who fled conflict and violence. Over the past eight years, a total of 203.4 million or an average of 25.4 million displacements have been recorded every year.

Extreme weather including heat waves, drought, and flooding is on the rise as a result of climate change.[255] Humans are more prone to violence and conflict in hotter conditions. Drought leads to food shortages and famine. Conflict and famine displace communities from their homelands. These realities beg certain questions:

- How effective will humans be in reducing greenhouse gas emissions?
- How effective will reduced emissions be in slowing and stopping warming?
- What will the impacts of global warming be in various localities across the globe where humans live?
- How will we adapt to these impacts?

[244] Mora et al. (2017)

[245] Mazdiyasni, O., et al. (2017) Increasing probability of mortality during Indian heat waves. *Science Advances, June, doi:* 10.1126/sciadv.1700066

[246] Im, E.-S., et al. (2017) Deadly heat waves projected in the densely populated agricultural regions of south Asia, *Science Advances*, 02 Aug., v. 3, no. 8, e1603322, doi: 10.1126/sciadv.1603322

[247] Russo, S., et al. (2017) Humid Heat Waves at Different Warming Levels, *Scientific Reports* 7.1: doi: 10.1038/s41598-017-07536-7

[248] Russo et al. (2017)

[249] Kelley, C.P., et al. (2015) Climate Change in the Fertile Crescent and Implications of the Recent Syrian Drought, *PNAS, USA* 112: 3241–3246.

[250] Political Extremism. https://www.bloomberg.com/news/articles/2016-10-20/nationalists-and-populists-poised-to-dominate-european-balloting

[251] Dai, A. (2013) Increasing Drought Under Global Warming in Observations and Models, *Nature Climate Change* 3.1: 52–58, doi: 10.1038/nclimate1633

[252] Raleigh, C., et al. (2014) Extreme Temperatures and Violence, *Nature Climate Change* 4: 76–77.

[253] Coumou, D. and Robinson, A. (2013) Historic and Future Increase in the Global Land Area Affected by Monthly Heat Extremes, *Environmental Research Letters* 8.3: 034018, doi: 10.1088/1748-9326/8/3/034018

[254] GRID (2016) *Global Report on Internal Displacement:* http://www.internaldisplacement.org/globalreport2016/.org/

[255] Medvigy, D. and Beaulieu, C. (2011) Trends in Daily Solar Radiation and Precipitation Coefficients of Variation Since 1984, *Journal of Climate* 25, no. 4: 1330–1339, doi: 10.1175/2011JCLI4115.1; Schleussner, C-F., et al. (2017) In the Observational Record Half a Degree Matters, *Nature Climate Change*, doi: 10.1038/nclimate3320

8.9.1 Heat Stress

While the amount of warming to date may not seem like much, it has already multiplied the number of places experiencing dangerous or extreme heat waves by 50 times.[256] Global warming over the last century means heat extremes that previously only occurred once every 1,000 days are happening four to five times more often. Simply being outside away from air conditioning, during the hottest months of the year, is becoming increasingly unhealthy for some parts of the globe.

Since 1980 more than 1,900 locations worldwide have experienced air temperatures high enough to kill people[257] and it has been estimated that between 2030 and 2050, climate change is expected to cause approximately 250,000 additional deaths per year, from malnutrition, malaria, diarrhea, and heat stress.[258]

In the United States, research[259] shows that nearly 210 million Americans—or two-thirds of the population—live in counties vulnerable to health threats from unexpectedly high summer temperatures. High temperatures can cause heat exhaustion and heatstroke or worsen preexisting cardiovascular and respiratory conditions. An estimated 1,300 excess deaths occurred annually during extreme summer heat from 1975 to 2004, and more than 65,000 people end up in emergency rooms each summer with heat-related illnesses.

If carbon emissions are not reduced, 74 percent of the world's population will be exposed to deadly heat waves by 2100. Even if emissions are aggressively reduced, 48 percent of the world's human population will be affected.[260] Let's say global warming is limited to 2 degrees Celsius (3.6 degrees Fahrenheit), as approved in the 2015 Paris Agreement.[261] A number of equatorial megacities such as Karachi[262] and Kolkata[263] will nonetheless become nearly uninhabitable, annually encountering deadly heat waves like those that crippled them in 2015. With only 1.5 degrees Celsius (2.7 degrees Fahrenheit) of global warming, twice as many megacities (such as Lagos, Nigeria, and Shanghai, China) could become heat stressed than historically,

exposing over 350 million more people to deadly heat by 2050.[264] The results underscore that, even if the Paris targets are realized, there is still a deadly future in store for vulnerable urban populations—triggering large-scale relocations of human communities.

At 4 degrees Celsius (7.2 degrees Fahrenheit), the deadly European heat wave of 2003,[265] which killed more than 70,000 people, will be a normal summer. At 6 degrees Celsius (10.8 degrees Fahrenheit), according to the National Oceanic and Atmospheric Administration,[266] summer labor of any kind will be impossible in the lower Mississippi Valley, and everybody in the United States east of the Rockies would be under more heat stress than anyone, anywhere, in the world today.

8.9.2 Food

Humans get 75 percent of their food—either directly or indirectly as meat—from four crops: maize (corn), wheat, rice, and soybeans. The world's rising population is likely to reach 10.1 billion by the end of the century and has access to food thanks to increasing yields of these four crops. But our preference for burning fossil fuels is cutting crop production and nutritional value.[267] For instance, cereals and grasses will have reduced zinc, iron, and protein levels under the elevated CO_2 conditions predicted for the middle of this century.[268]

Climate change may actually benefit some plants by lengthening growing seasons and increasing carbon dioxide used in photosynthesis. Yet, other effects of global warming, such as more pests, droughts, and flooding, will be less benign. In a warmer world, yields of wheat are expected to decrease by 6 percent, rice by 3.2 percent, maize by 7.4 percent, and soybean by 3.1 percent.[269]

Research suggests that extra CO_2 improved yields for these crops by roughly 3 percent over the past 30 years. Unfortunately, in the case of wheat and maize, that wasn't enough to overcome losses caused by warm temperatures. Temperature problems are overriding CO_2 benefits. Climate-related losses contribute as much as 18.9 percent to the average price of a given crop. Climate change is not

[256] Fischer, E.M. and Knutti, R. (2015) Anthropogenic Contribution to Global Occurrence of Heavy-precipitation and High Temperature Extremes, *Nature Climate Change v.*5, June: 560–565.

[257] Heat waves. https://www.hawaii.edu/news/2017/06/19/rising-deadly-heatwaves/

[258] World Health Organization, Climate Change and Health, fact sheet; http://www.who.int/mediacentre/factsheets/fs266/en/

[259] National Resources Defense Council, Extreme Heat: https://www.nrdc.org/climate-change-and-health-extreme-heat#/map

[260] Mora et al. (2017)

[261] U.N. Framework Convention on Climate Change http://unfccc.int/2860.php

[262] 2015 Pakistan Heat Wave https://en.wikipedia.org/wiki/2015_Pakistan_heat_wave

[263] Kolkata to remain world's most heat stressed city this century: https://www.skymetweather.com/content/climate-change/kolkata-to-remain-worlds-most-heat-stressed-city-this-century/

[264] Matthews, T.K.R., et al. (2017) Communicating the Deadly Consequences of Global Warming for Human Heat Stress, *PNAS* 114, no. 15: 3861-3866, doi:10.1073/pnas.1617526114

[265] 2003 European Heat Wave https://en.wikipedia.org/wiki/2003_European_heat_wave

[266] Dunne, J.P., et al. (2013) Reductions in Labor Capacity from Heat Stress under Climate Warming, *Nature Climate Change*, 3: 563–566, doi:10.1038/nclimate1827

[267] Lobell, D.B., et al. (2013) Climate Trends and Global Crop Production Since 1980, *Science*: 1204531, doi:10.1126/science.1204531

[268] Myers, S.S., et al. (2014) Increasing CO_2 Threatens Human Nutrition, *Nature* 510: 139–142, doi:10.1038/nature13179

[269] Zhao, C., et al. (2017) Temperature Increase Reduces Global Yields of Major Crops in Four Independent Estimates *PNAS* 114, no. 35: 9326–9331; doi:10.1073/pnas.1701762114

disastrous to global food security, but it is a multibillion-dollar-per-year effect.

Loss of crop yield and nutritional value are bad news. But a more insidious food problem is lurking. Research[270] shows that rising carbon dioxide deteriorates the quality of the plants we eat.[271] Every leaf and every blade of grass on Earth makes more sugar as CO_2 levels rise.[272] Mathematical biologist Irakli Loladze has said "We are witnessing the greatest injection of carbohydrates into the biosphere in human history–[an] injection that dilutes other nutrients in our food supply."

Across nearly 130 varieties of plants and more than 15,000 samples collected from experiments over the past three decades, the overall concentration of minerals such as calcium, magnesium, potassium, zinc, and iron had dropped by 8 percent on average. The ratio of carbohydrates to minerals is going up. Plants that humans rely on for basic nutrition are becoming junk food.[273]

There have been marked changes in temperature extremes across the contiguous United States. The number of high temperature records set in the past two decades far exceeds the number of low temperature records. Extreme temperatures in the contiguous United States are projected to increase even more than average temperatures. This trend will challenge food production.

Climate change's effects on global food supply could lead to more than 500,000 deaths by 2050 as people around the world lose access to good nutrition.[274] For every degree of warming, cereal crops decline by 10 percent. Thus, if the planet is 5 degrees Celsius (9 degrees Fahrenheit) warmer at the end of the century, and population growth leads to as many as 50 percent more people to feed, there will about 50 percent less grain to give them.

But the problems start to develop far sooner than the end of the century. Alone, wheat provides about 20 percent of all human protein. However, rising carbon dioxide levels decrease the protein and nutrient content of wheat.[275] By 2050, with 9 billion people, the demand for wheat will increase 60 percent, yet actual yield will decrease 15 percent. This situation sets the stage for food shortages, and exorbitant prices, which further stokes shortages.

Producing certain types of food, such as beef, leads to the emission of huge quantities of greenhouse gas. It takes 16 calories of grain to produce just a single calorie of hamburger meat, butchered from a cow that spent its life polluting the climate with methane because it has been fed corn to make it grow faster, rather than grass, its natural diet.

Reducing heavy red meat consumption—primarily beef and lamb—would lead to per capita food and land use-related greenhouse gas reductions of 15 to 35 percent by 2050.[276] Going vegetarian could reduce those per capita emissions by half. In the average American diet, greenhouse gas emissions resulting from the food they eat are about the same as the emissions resulting from the energy they use.[277]

The tropics are already too hot to efficiently grow grain, and those places where grain is produced today are already at optimal growing temperature—which means even a small warming will push them down the slope of declining productivity.[278] Theoretically, a warmer climate will make it easier to grow food in northern regions. But you can't easily move croplands north a few hundred miles, because yields in places such as remote Canada and Russia are limited by the quality of soil there; it takes many centuries for the planet to produce optimally fertile dirt.

8.9.3 Drought, Water Scarcity

Water is a critical natural resource upon which all human activities and ecosystem processes depend. Globally, human access to clean freshwater is challenged by pollution, lack of water retrieval and delivery infrastructure, changes in precipitation due to climate change (including changing seasonality, snowpack, drought, desertification, and rain characteristics), and poverty. More than 5 billion people could suffer water shortages by 2050 due to climate change, increased demand, and polluted supplies.[279]

Global demand for water has been increasing at a rate of about 1 percent per year as a function of population growth, economic development and changing consumption patterns, among other factors, and it will continue to grow significantly over the next two decades. Industrial and domestic needs for water will increase much faster than agricultural demand, although agriculture will remain the largest overall user. The vast majority of the growing demand for water will occur in countries with developing or emerging economies.

At the same time, the global water cycle is intensifying due to climate change, with wetter regions generally

[270] Myers et al. (2014)

[271] Loladze, I. (2002) Rising CO_2 and Human Nutrition: Toward Globally Imbalanced Plant Stoichiometry? *Trends in Ecology and Evolution*, 17: 457–461, http://www.cell.com/trends/ecology-evolution/abstract/S0169-5347%2802%2902587-9

[272] Loladze, I. (2014) Hidden Shift of the Ionome of Plants Exposed to Elevated CO_2 Depletes Minerals at the Base of Human Nutrition, *eLife*, doi:10.7554/eLife.02245

[273] Politico The Great Nutrient Collapse 9/13/2017, http://www.politico.com/agenda/story/2017/09/13/food-nutrients-carbon-dioxide-000511

[274] Springmann, M., et al. (2016) Global and Regional Health Effects of Future Food Production under Climate Change: A Modeling Study, *The Lancet* 387, no. 10031: 1937–1946.

[275] Myers et al. (2014)

[276] Sustainable Diet http://www.wri.org/blog/2016/04/sustainable-diets-what-you-need-know-12-charts

[277] Red Meat. http://www.climatecentral.org/news/studies-link-red-meat-and-climate-change-20264

[278] Wallace-Wells (2017)

[279] WWAP (2018) *The United Nations World Water Development Report 2018: Nature-Based Solutions for Water*. Paris, UNESCO: http://www.unwater.org/publications/world-water-development-report-2018/

becoming wetter and drier regions becoming even drier. At present, an estimated 3.6 billion people (nearly half the global population) live in areas that are potentially water-scarce at least one month per year, and this population could increase to some 4.8–5.7 billion by 2050.[280]

Higher temperatures and more extreme, less predictable weather conditions are projected to affect availability and distribution of rainfall, snowmelt, stream discharge, groundwater yield, and further deteriorate water quality. Low-income communities, who are already the most vulnerable to any threats to water supply are likely to be the worst affected.[281] Globally, water scarcity already affects 4 out of every 10 people. A lack of water and poor water quality increases the risk of diarrhea, which kills approximately 2.2 million people every year, as well as trachoma, an eye infection that can lead to blindness, and many other illnesses.

Increasing temperatures on the planet and more variable rainfall are expected to reduce crop yield in many tropical developing regions, where food security is already a problem.[282] Large areas of cropland are forecast to experience debilitating drought before the end of the century. Some of the world's most productive farmland is turning to desert. It is a characteristic of a warmer world that rainfall grows more extreme, drought deepens and spreads, and soils retain less moisture. These are exactly the conditions that stress global food production, freshwater resources, and ultimately human health.[283]

Modeling future rainfall patterns is very difficult, yet there is general agreement among modelers that drought will develop with greater frequency nearly everywhere food is produced. Without dramatic reductions in emissions, what are today reliable regions of food production in southern Europe, the Middle East, densely populated parts of Australia, Africa, and South America, and the breadbasket regions of China will all be in permanent extreme drought before the end of the century.[284] The American Plains and Southwest have been forecast[285] to experience "mega-drought," worse than any drought in the past 1000 years.[286]

By 2025, 1.8 billion people are expected to be living in countries or regions with absolute water scarcity, and two-thirds of the world population could be under water stress conditions. Around 700 million people in 43 countries suffer today from water scarcity. By 2025, 1.8 billion people will be living in countries or regions with absolute water scarcity, and two-thirds of the world's population could be living under water stressed conditions. As temperatures continue to rise, the continued release of CO_2 will result in almost half the world's population living in areas of high water stress by 2030. This includes between 75 million and 250 million people in Africa. Worldwide, water scarcity in some arid and semi-arid places will displace between 24 million and 700 million people, with sub-Saharan Africa experiencing the largest number of water-stressed countries of any region.[287]

Water demand globally is projected to increase by 55 percent before the middle of the century. Much of the demand is driven by agriculture, which accounts for 70 percent of global freshwater use. Food production will need to grow by 69 percent by 2035 to feed the swelling human population.[288] Water withdrawal for energy, used for cooling power stations, is also expected to increase by over 20 percent. In other words, the demand for water in the near future is immense **Figure 8.22**.

A NASA-led study[289] reveals that many of the world's freshwater sources are being drained faster than they are being replenished. Of the world's major aquifers, 21 out of 37 are receding, from India and China to the United States and France. Because of population and irrigation demands, the water table in the Ganges Basin in India is lowering by an estimated 6.31 centimeters (2.5 inches) every year. Jay Famiglietti, senior water scientist at NASA, warned[290] "the water table is dropping all over the world. There's not an infinite supply of water."

Approximately 1.1 billion people worldwide lack access to water, and a total of 2.7 billion find water scarce for at least one month of the year. Inadequate sanitation is also a problem for 2.4 billion people—they are exposed to diseases, such as cholera and typhoid fever, and other waterborne illnesses. Two million people, mostly children, die each year from diarrheal diseases alone.[291]

Many of the water systems that keep ecosystems thriving and feed a growing human population have become stressed. Rivers, lakes, and aquifers are drying up or becoming too polluted to use. More than half the world's wetlands have disappeared. Agriculture consumes more water than

[280] WWAP (2018)

[281] World Health Organization (2017) *Climate Change and Health, July, fact sheet*; http://www.who.int/mediacentre/factsheets/fs266/en/

[282] Zhang, X. and Cai, X. (2011) Climate change impacts on global agricultural land availability, *Environmental Research Letters*, 6, 8p., doi:10.1088/1748-9326/6/1/014014

[283] Hatfield, J.G., et al. (2014) Ch. 6: Agriculture. Climate Change Impacts in the United States: The Third National Climate Assessment. Melillo, G.J., et al. (eds.) U.S. Global Change Research Program, 150-174.

[284] Dai (2013)

[285] NASA https://www.nasa.gov/press/2015/february/nasa-study-finds-carbon-emissions-could-dramatically-increase-risk-of-us

[286] Cook, B.I., et al. (2015) Unprecedented 21st century drought risk in the American Southwest and Central Plains. *Science Advances*, 12 February doi: 10.1126/sciadv.1400082

[287] UNDESA (2014) International Decade for Action "Water for Life", United Nations Department of Economic and Social Affairs: http://www.un.org/waterforlifedecade/scarcity.shtml

[288] Global Water Forum, http://www.globalwaterforum.org/2012/05/21/water-outlook-to-2050-the-oecd-calls-for-early-and-strategic-action/

[289] Richey, A.S., et al. (2015) Quantifying Renewable Groundwater Stress with GRACE, *Water Resources Research* 51: 5217–5238, doi:10.1002/2015WR017349

[290] NASA, The World is Running Out of Water: https://www.washingtonpost.com/news/wonk/wp/2015/06/16/new-nasa-studies-show-how-the-world-is-running-out-of-water/?utm_term=.db06b9f872cb

[291] Water Scarcity, World Wildlife Fund: https://www.worldwildlife.org/threats/water-scarcity

FIGURE 8.22 Drought and water scarcity are projected to increase in a warmer world. Some of the world's most productive farmland is turning to desert. It is a characteristic of a warmer world that rainfall grows more extreme, drought deepens and spreads, and soils retain less moisture.

Source: http://www.weathernationtv.com/news/new-noaa-tool-helping-predict-u-s-droughts/

any other source and wastes much of that through inefficiencies. Climate change is altering patterns of weather and water around the world, causing shortages and droughts in some areas and floods in others.[292]

At the current consumption rate, this situation will only get worse. By 2025, two-thirds of the world's population may face water shortages. And ecosystems around the world, which provide human communities with critical goods and services, will suffer even more.

8.9.4 Sea-Level Rise

When the Fifth Assessment Report of the Intergovernmental Panel on Climate Change was released in its final form in 2014[293] the worst-case scenario for potential sea-level rise by the end of the century was a little less than 1 m (3.2 ft). However, recent studies of melting in Antarctica,[294,295,296] Greenland,[297] and Alpine ice systems,[298] and the rate of

ocean warming[299] suggest it is now physically plausible to see more than double this amount.[300,301]

Global warming has reached at least 1 degree Celsius (1.8 degrees Fahrenheit) above the natural background. The last time it was this warm was during the Eemian interglacial (ca. 125,0000 years ago)[302] when global mean sea level was approximately 6.6 meters (21.6 feet) above present.[303] Models indicate that under current warming, mean sea level will eventually stabilize at about 2.3 meters (7.5 feet) above present, although it may take several centuries, depending on the future rate of emissions.[304] Continued warming will increase the height at which sea level eventually stabilizes.

Simulations of sea level rise when the world is 2 to 3 degrees Celsius (3.6 to 5.4 degrees Fahrenheit) warmer, show that melting of the Antarctic ice sheet could raise global sea level by up to 3 meters (9.8 feet) by the year 2300 and continue for thousands of years thereafter.[305]

[292] World Wildlife Fund https://www.worldwildlife.org/threats/water-scarcity

[293] IPCC (2014) *Synthesis Report.*

[294] DeConto, R.M. and Pollard, D. (2016) Contribution of Antarctica to Past and Future Sea-level Rise, *Nature* 531, no. 7596: 591–597.

[295] Khazendar, A., et al. (2016) Rapid Submarine Ice Melting in the Grounding Zones of Ice Shelves in West Antarctica, *Nature Communications* 7: 13243

[296] Scheuchl, B., et al. (2016) Grounding Line Retreat of Pope, Smith, and Kohler Glaciers, West Antarctica, Measured with Sentinel-1a Radar Interferometry Data. *Geophysical Research Letters*, 43, no. 16: 8572, doi: 10.1002/2016GL069287

[297] Tedesco, M., et al. (2016) The Darkening of the Greenland Ice Sheet: Trends, Drivers, and Projections (1981-2100), *The Cryosphere*, 10: 477–496.

[298] Ciraci, E., et al. (2015) Mass Loss of Glaciers and Ice Caps from GRACE During 2002-2015, AGU Poster, Fall.

[299] Glecker, P.J., et al. (2016) Industrial Era Global Ocean Heat Uptake Doubles in Recent Decades, *Nature Climate Change.* doi:10.1038/nclimate2915

[300] Le Bars, D., et al. (2017) A High-end Sea Level Rise Probabilistic Projection Including Rapid Antarctic Ice Sheet Mass Loss, *Environmental Research Letters*, 12, no. 4, April 3.

[301] Sweet, W. V., et al. (2017) *Global and Regional Sea Level Rise Scenarios for the United States.* NOAA Technical Report NOS CO-OPS 083, Silver Spring, 56p. plus Appendices.

[302] Hoffman, J.S., et al. (2017) Regional and global sea surface temperatures during the last interglaciation, *Science*, 355(6322), 276-279, doi: 10.1126/science.aai8464

[303] Kopp, R.E., et al. (2009) Probabilistic assessment of sea level during the last interglacial stage, *Nature*, 462, 863-867, doi: 10.1038/nature08686.

[304] Levermann, A., et al. (2013) The multimillennial sea-level commitment of global warming: *PNAS*, July 15, doi: 10.1073/pnas.1219414110.

[305] Golledge, N.R., et al. (2015) The multi-millennial Antarctic commitment to future sea-level rise: *Nature*, 526 (7573): 421 doi: 10.1038/nature15706.

Research indicates that on multiple occasions over the past three million years, when global temperatures increased 1 to 2 degrees Celsius (1.8 to 3.6 degrees Fahrenheit), melting polar ice sheets caused global sea levels to rise at least 6 meters (19.7 feet) above present levels.[306]

According to the 4th National Climate Assessment,[307] relative to the year 2000, global mean sea level is very likely to rise 9-18 centimeters (0.3–0.6 feet) by 2030, 15-38 centimeters (0.5–1.2 feet) by 2050, and 30-130 centimeters (1.0–4.3 feet) by 2100 (**Figure 8.23**). The report states that there is very high confidence in the lower bounds of these estimates; medium confidence in upper bounds for 2030 and 2050; and low confidence in upper bounds for 2100.

Future pathways of greenhouse gas emissions have little effect on projected sea-level rise in the first half of the century, but significantly affect projections for the second half of the century. Emerging science regarding Antarctic ice sheet stability suggests that, under high future greenhouse

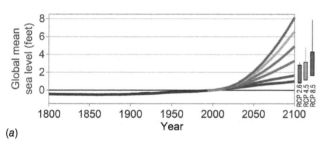

(a)

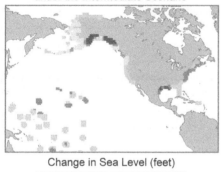

(b)

FIGURE 8.23 (a) Global mean sea level (GMSL) rise from 1800 to 2100. Six Interagency GMSL scenarios, the very likely ranges in 2100 for greenhouse gas emission scenarios from the IPCC-AR5 (boxes), and lines augmenting the very likely ranges on the basis of new research. (b) Relative sea level (RSL) rise (ft) in 2100 projected for the Interagency Intermediate Scenario (1-m [3.3 ft] GMSL rise by 2100).[308]

Source: https://science2017.globalchange.gov/chapter/12/

gas emission scenarios, a rise exceeding 2.4 meters (8 feet) by 2100 is physically possible, although the probability of such an extreme outcome cannot currently be assessed. Regardless of emissions, it is extremely likely that sea-level rise will continue beyond 2100.

It is virtually certain that sea-level rise this century and beyond will pose a growing challenge to coastal communities, infrastructure, and ecosystems from increased (eventually becoming permanent) inundation, more frequent and extreme coastal flooding, erosion of coastal landforms, and saltwater intrusion within coastal rivers and aquifers. To assess community vulnerability to rising sea levels requires an understanding of physical causes, historical evidence, and projections of the future. The 4th National Climate Assessment states that a risk-based perspective on sea-level rise should place emphasis on how changing sea levels alter the coastal zone and interact with coastal flood risk at local scales.

Coastal flooding during extreme high-water events has become more frequent and damaging.[309] Trends in annual flood frequency that surpass local emergency preparedness thresholds for minor tidal flooding (i.e., "nuisance" levels of about 30 to 60 centimeters [1 to 2 feet]) that begin to flood infrastructure and trigger coastal flood "advisories" by NOAA's National Weather Service have increased 5- to 10-fold or more since the 1960s along the U.S. coastline (**Figure 8.24**).

Unmitigated sea level rise is expected to reshape the distribution of human communities, potentially stressing landlocked areas unprepared to accommodate a wave of coastal migrants.[310] By the year 2100, rising sea levels could force up to 2 billion people inland, creating a refugee crisis among one-fifth of the world's population.[311] Worse yet, there won't be many places for those migrants to go.

Researchers have identified three obstacles, or "barriers to entry," that stand in the way of people driven inland from their homes by rising seas.[312] The first is that drought and desertification make some areas uninhabitable at worst, and incapable of sustaining a large influx of migrants at best. Second: if climate refugees flock to cities, the increasing urban sprawl might require land formerly used to grow food. Those communities could lose the ability to feed their inflated populations. Third: regions and municipalities might erect walls and post guards to prevent climate migrants from entering and settling down. This phenomenon is dubbed the "no-trespass zone."

[306] Dutton, A., et al. (2015) Sea-level rise due to polar ice-sheet mass loss during past warm periods, *Science*, 10 Jul., v. 349, Is. 6244, doi: 10.1126/science.aaa4019

[307] Wuebbles et al. (2017)

[308] Sweet et al. (2017)

[309] Sweet, W.V. and Park, J. (2014) From the Extreme to the Mean: Acceleration and Tipping Points of Coastal Inundation from Sea Level Rise, *Earth's Future* 2: 579–600, doi:10.1002/2014EF000272

[310] Hauer, M.E. (2017) Migration induced by sea-level rise could reshape the US population landscape, *Nature Climate Change*, 7, 321-325, doi:10.1038/nclimate3271

[311] Geisler, C. and Currens, B. (2017) Impediments to inland resettlement under conditions of accelerated sea level rise, *Land Use Policy*, v.66 (July), p. 322-330, https://doi.org/10.1016/j.landusepol.2017.03.029

[312] Geisler and Currens (2017)

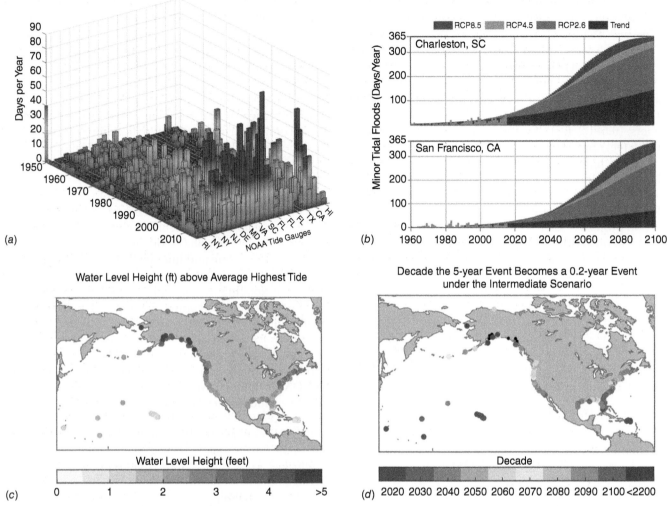

FIGURE 8.24 (a) Tidal floods (days per yr) exceeding NOAA thresholds for minor impacts at 28 tide gauges through 2015. (b) Historical exceedances (orange, light grey), future projections based on continuation of historical trend (blue, darkest), and future projections under median RCP2.6, 4.5, and 8.5 conditions. (c) Water level heights above average highest tide associated with a local 5-year recurrence probability, and (d) the future decade when the 5-year event becomes a 0.2-year (5 or more times per year) event under the Interagency Intermediate scenario; black dots imply that a 5-year to 0.2-year frequency change does not unfold by 2200 under the Intermediate scenario.[313]

Source: https://science2017.globalchange.gov/chapter/12/

This discussion highlights a problem; too much of the conversation around sea level rise adaptation is focused on building sea walls, learning to live with regular flooding, and relocating communities inland. Sea level rise greatly exceeding 1 meter, barriers to entry, and the occurrence of rapid onset events such as hurricanes and extreme flooding that force residents to relocate ahead of planning, mean that these overly simplistic ideas of adaptation could leave human communities unprepared for a mass migration that could dwarf the current refugee crisis in Europe.[314]

8.9.5 Out of Control Climate

Former NASA climate scientist Jim Hansen has modeled the potential impacts of a world that warms to 2 degrees

Celsius (3.6 degrees Fahrenheit) and concludes this level of warming could be dangerous to humanity. Among his findings: rapid melting of ice sheets, extreme heating of the tropics, superstorms, and multimeter sea-level rise this century.[315] There are already signs that the climate system is shifting in this direction: slowed North Atlantic circulation, a stratified Southern Ocean storing warm water that rapidly melts floating ice shelves, Antarctic ice streams that are irreversibly retreating, massive tropical cyclones in the Atlantic, and accelerating global sea-level rise.

Should "multimeter" sea-level rise occur, the human population displaced from the shoreline and forced into the role of climate refugees will number in the hundreds of

[313] Wuebbles et al. (2017)
[314] Geisler and Currens (2017)

[315] Hansen, J., et al. (2016) Ice Melt, Sea Level Rise and Superstorms: Evidence from Paleoclimate Data, Climate Modeling, and Modern Observations that 2°C Global Warming Could be Dangerous, *Atmospheric Chemistry and Physics* 16: 3761–3812, https://doi.org/10.5194/acp-16-3761-2016.

millions. The situation will be "essentially ungovernable" in Hansen's view. In his words[316] "There is a possibility, a real danger, that we will hand young people and future generations a climate system that is practically out of their control. We conclude that the message our climate science delivers to society, policymakers, and the public alike is this: we have a global emergency. Fossil fuel CO_2 emissions should be reduced as rapidly as practical."

This narrative provides little reason for confidence in humanities future. The fact that the current U.S. administration is busy rolling back many of the successes in energy efficiency, renewable power, and decreased emissions made by the previous president, reinforces the basis for a pessimistic attitude.

The scene is indeed bleak, and the trajectory of humankind is into a 3-degree world, higher if poorly understood feedbacks and tipping points materialize. This will likely be a world characterized by a global scale refugee crisis as communities in the tropics find their region no longer habitable, heat waves every summer in the middle latitudes that render the outdoors intolerable, global sea-level rise of several meters essentially dooming the world's major port cities, and extreme weather disasters consisting of massive floods, heat waves, great tropical cyclones, mega-drought, and torrential rainfall.

Ironically, judging by todays rapid, but not rapid enough, progress in sustainable energy, these climate extremes will take place in a world of solar panels, wind mills, electric cars, and cleaner air.

8.9.6 Ominous Signs

Human communities are in a precarious situation. The socio-economic foundation that underpins modern society is at risk. Surface temperatures continue to increase and are already hotter than at any time in the history of civilization. Global sea level rise is accelerating. In the ocean: anoxia is expanding, heat uptake is accelerating, and acidification is intensifying. On land: wildfires are larger, flooding is increasing because of more intense rainfall, winter ends earlier and summer drought is longer and deeper. Desertification, aridity, and water scarcity are expanding.

In unplanned response, human communities in North Africa and the Middle East are moving away from traditional homelands. Political turmoil is growing where they seek refuge. Unusually intense weather events (e.g., Hurricane Harvey), in atypical locations (e.g., Northern California wildfire), create unlivable conditions (e.g., Puerto Rico) that lead to internally displaced migrants. The rapid warming humans have set in motion, coupled with pollution and expanding development, is putting at risk ecological systems that we depend on.

The situation is dire, and with continuation of these trends, it may become desperate.

False Hope Is stabilizing temperature at 1.5 degrees Celsius (2.7 degrees Fahrenheit) truly out of reach? In September 2017 researchers published[317] modeling results indicating that the goal of stabilizing warming to 1.5 degrees Celsius (2.7 degrees Fahrenheit) was still achievable. They showed that limiting carbon dioxide emissions after 2015 to about 200 billion tons would keep additional warming to less than 0.6 degrees Celsius (1.08 degrees Fahrenheit) in 66 percent of CMIP5[318] climate models. The available carbon budget rises to 240 billion tons if communities engaged in ambitious mitigation of short-term greenhouse gases, damaging land-use policies, and other non-carbon dioxide causes of climate change.

Their report was globally extolled as providing a pathway to achieving the goals of the Paris Agreement. However, criticism[319] emerged with a focus on three issues. First: The results rely on a temperature dataset called HadCRUT4, which begins in 1850 and estimates global surface temperatures have warmed about 0.9 degree Celsius (1.6 degrees Fahrenheit) since that time (hence their calculation of a carbon budget leading to an additional 0.6°C warming). However, HadCRUT4 only covers 84 percent of Earth's surface. There are large gaps in its coverage including the Arctic, Antarctica, and Africa. The Arctic is the fastest-warming part of the planet, which means the HadCRUT4 data underestimate global warming.

Second: HadCRUT4 uses measurements of shallow ocean water rather than the air above the ocean surface. The air warms faster than the water, and so once again, HadCRUT4 underestimates global warming.

Third: A previous study[320] has showed that if calculations start earlier than 1850, an additional 0.2 degree Celsius (0.36 degree Fahrenheit) of warming is observed. Again, indicating that HadCRUT4 underestimates global warming, and thus shrinking the required carbon budget.

Combined, these concerns mean that warming has been 0.2 to 0.3 degree Celsius (0.36 to 0.54 degree Fahrenheit) more than assumed in the study, requiring a significantly smaller carbon budget to stop at 1.5 degrees Celsius (2.7 degrees Fahrenheit) and putting it essentially out of reach. Each additional 0.1 degree Celsius (0.18 degree Fahrenheit) of underestimation means the 2.0 degrees Celsius (3.6 degrees Fahrenheit) carbon budget needs to shrink 20

[316] Global Emergency. https://thinkprogress.org/leading-climate-scientists-we-have-a-global-emergency-must-slash-co2-asap-24ae324ecd8a/

[317] Millar, R, et al. (2017) Emission budgets and pathways consistent with limiting warming to 1.5°C. *Nature Geoscience*, 10, 741–747

[318] We discussed the Coupled Model Intercomparison Project, or CMIP, in Chapter 6. It is a United Nations program promoting collaboration among groups who run climate models. CMIP5 provided modeling support for the most recent IPCC assessment, AR5.

[319] Schurer, A.P., et al. (2018) Interpretations of the Paris Climate Target, *Nature Geoscience*, 10.1038/s41561-018-0086-8

[320] Schurer, A.P., et al. (2017) Importance of the pre-industrial baseline for likelihood of exceeding Paris goals, *Nature Climate Change*, 7, 563-567, doi:10.1038/nclimate3345

percent. Thus, allowable carbon emissions actually total 60 percent less than those published in the study. This means the global effort to cut carbon emissions needs to be significantly more urgent, and substantially more restrictive than the study implied.

8.9.7 Paris

The happy news is that all of the world's nations have signed the Paris Agreement.[321] The agreement aligns humanity in a single effort[322] to reduce greenhouse gas emissions and hold global temperature rise this century well below 2 degrees Celsius (3.6 degrees Fahrenheit). Because developing nations may lack capacity to rapidly restrict emissions, assistance is provided by the Green Climate Fund.[323] The Paris Agreement is a truly historic step forward that signifies a unified international commitment to avoiding the worst effects of climate change. Unfortunately, as we have discussed, the ability to achieve the Paris goal of limiting warming to 1.5 degrees is now lost.

The first significant signs of progress have become evident; 2016 was the third year in a row that global carbon emissions remained level, even as the global economy grew.[324] Although more than 40 billion tons of emissions is certainly not good news for future climates, there was cause for optimism within the numbers, as some major economies saw their emissions drop. For example, in just 2 years, China's coal use swung from 3.7 percent growth in 2013 to a decline of 3.7 percent in 2015.[325] In the United States, carbon emissions fell 14 percent over the previous decade.[326]

Despite the plateau in global emissions between 2014 and 2016, November of 2017 brought disappointing news: global emissions of CO_2 rose by 2 percent in 2017—a huge amount. The majority of that rise was attributable to China where a resurgence in the economy led to more coal use fueling a 3.5 percent growth in greenhouse gas emissions.

Also bad, higher natural gas prices in the United States led to a rise in coal burning, for the first time in five years, while oil use also increased. As a result, emissions that had been declining in the United States at about 1.2 percent per year dropped less than half a percent. In the E.U., emissions dropped less than 0.25 percent after a decade of annual declines topping 2 percent per year. In India, emissions growth slowed, rising by just 2 percent in 2017,

compared with an average of 6 percent per year over the past decade. Unfortunately, that good news was tainted because the reduction in India's emissions is almost certain not to last.

The world knows what it has to do to stop warming. It has been shown[327] that if we can reduce carbon dioxide emissions by 50 percent each decade, this can lead to net-zero emissions around mid-century, a path necessary to limit warming to 2 degrees Celsius (3.6 degrees Fahrenheit). Increases in greenhouse gas emissions in 2017 show that, although there are reasons for optimism, humanity still has far to go before it solves the climate problem.[328]

8.9.8 Critical Point

Poised at a critical time, how we behave in the next decade will determine the nature of future human prosperity on Earth. The Paris Agreement translates into a finite planetary carbon budget. Researchers[329] have found that to achieve a 50 percent chance of limiting warming to 1.5 degrees Celsius (2.7 degrees Fahrenheit) by 2100, and a greater than 66 percent probability of meeting the 2 degrees Celsius (3.6 degrees Fahrenheit) target, global CO_2 emissions must peak no later than 2020. Gross emissions must decline thereafter from ~40 billion tons of CO_2 per year in 2020, to ~24 by 2030, ~14 by 2040, and ~5 by 2050. However, a close analysis[330] of worldwide socio-economic patterns found that the likely range of global temperature increase is 2.0 to 4.9 degrees Celsius (3.6 to 8.82 degrees Fahrenheit), with a median 3.2 degrees Celsius (5.76 degrees Fahrenheit), a 5 percent chance that it will be less than 2 degrees Celsius (3.6 degrees Fahrenheit), and only a 1 percent chance that it will be less than 1.5 degrees Celsius (2.7 degrees Fahrenheit).

Literally everything that we depend upon, from the air we breathe, freshwater and food, to transportation, shelter, and security are all contingent on climate. Because of the fast pace at which climate change is happening, in many locations these requirements for human life teeter on the edge of sustainability. They are naturally adapted, or humanly engineered, to a climate that no longer exists.

Climate research has identified unexpected and very worrying trends in both human behavior (e.g., increased conflict resulting from decreased water and food resources)[331] and natural systems (e.g., melting ice releasing unique

[321] UNFCCC page: http://unfccc.int/paris_agreement/items/9485.php
[322] With the temporary exception of the U.S. under the Trump Administration which has announced they will withdraw from the agreement.
[323] Green Climate Fund http://www.greenclimate.fund/home
[324] Carbon emissions. http://www.reuters.com/article/us-carbon-emissions-iea/energy-carbon-emissions-in-2016-flat-for-third-year-iea-idUSKBN16O10B
[325] B. P. Global, *BP Statistical Review of World Energy* (BP Global, ed. 65, 2016).
[326] CarbonBrief, https://www.carbonbrief.org/analysis-why-us-carbon-emissions-have-fallen-14-since-2005?utm_content=bufferf70e6&utm_medium=social&utm_source=twitter.com&utm_campaign=buffer

[327] Rockström et al. (2017)
[328] World's Carbon Emissions Set to Spike by 2% in 2017, *Nature*, 551, 283, (16 November 2017), doi:10.1038/nature.2017.22995
[329] Rogelj, J., et al. (2015) Energy system transformations for limiting end-of-century warming to below 1.5°C, *Nature Climate Change*, 5, 519-527, doi:10.1038/nclimate2572
[330] Raftery et al. (2017)
[331] Hsiang, S.M., et al. (2013) Quantifying the influence of climate on human conflict. *Science*, 13 Sept., v. 341, Is. 6151, doi: 10.1126/science.1235367

forms of bacteria and viruses that humans have not been exposed to)[332] that are a result of global warming. We witness political turmoil already at play,[333] and severe damage to critical ecosystems appears to be unavoidable.[334]

It is important to be aware of these impacts, but at the same time extreme pessimism can paralyze human action.[335] There is abundant evidence that sustainable energy is on the rise,[336] governments on every continent have made significant commitments to fight global warming, and trailblazing corporations[337] are engaging in sustainable practices and providing low-carbon services and products. There appears to be every reason to be optimistic that the worst effects of climate change will never materialize.

Close examination of the facts reveals that we are already on the path to a world that is 3 degrees Celsius (5.4 degrees Fahrenheit) or warmer. This is going to be dangerous for humanity. Forward progress in building an energy system based on clean and renewable resources is remarkably strong. Unfortunately, according to expert projections, it will not be sufficient to offset simultaneous growth in energy demand that will be largely sustained by fossil fuels.

My generation has hesitated on our way to building a sustainable future. The consequences of that hesitation may be existential. Now we must answer to our grandchildren. For the future we have committed them to may drive them to ask "Do I want to bring children into this world?"

Animations and Videos

1. NASA Planet on Fire, computer simulation of global fires and aerosols https://svs.gsfc.nasa.gov/11192

2. NOAA model simulation of surface temperature 1970 to 2100 based on "middle of the road" future emissions https://www.gfdl.noaa.gov/html5-video/?width=940&height=527&vid=gfdlglobe_tref_d4h2x1_1970_2100_30f_1280x720&title=Surface%20air%20temperature%20anomalies

3. Greater Future Warming Inferred from Earth's Recent Energy budget, https://patricktbrown.org/2017/11/29/greater-future-global-warming-inferred-from-earths-recent-energy-budget/

4. Is a Warmer World a Better World? https://www.youtube.com/watch?v=1SAqdG3gJH0&app=desktop

5. Future Earth Global Carbon Budget 2017 http://futureearth.org/blog/Budget2017

6. Time history of atmospheric carbon dioxide from 800,000 years ago until January, 2009, http://www.esrl.noaa.gov/gmd/ccgg/trends/history.html (This marvelous animation by NOAA illustrates the growing concentration of CO_2 in the atmosphere as well as the increase in measurement stations in recent history.)

7. NASA, Flow of Ice Across Antarctica, http://www.jpl.nasa.gov/video/index.cfm?id=1015

8. Witness a Glacier's Staggering Seven-Year Retreat, http://io9.com/5905656/witness-a-glaciers-staggering-seven+year-retreat

Comprehension Questions

1. Describe the Berkeley-based BEST study and explain why it is significant.

2. What are the primary natural processes that contributed to global temperature rise in the past few decades?

3. Describe the trend in greenhouse gas emissions in recent years.

4. Which nations are the largest carbon dioxide contributors?

5. What effect is warming having on the Arctic?

6. Are global warming and the weather related? How?

7. What is drought and why is global warming causing it to change?

8. List the ways urban areas are vulnerable to climate change.

9. How is climate change affecting the world's ecosystems?

10. Describe equilibrium climate sensitivity.

Thinking Critically

1. Where is up-to-date information on climate change available that is reliable?

2. Satellite studies of global temperature have been controversial. Why?

3. What role does the El Niño Southern Oscillation play in the year-to-year climate?

4. Why has it been so difficult to decrease the production of greenhouse gases?

[332] Legendre, M., et al. (2014) Giant DNA virus revived from the late Pleistocene, *PNAS*, Mar, 111 (11) 4274-4279; doi:10.1073/pnas.1320670111

[333] Kelley, C. P., et al. (2015) Climate change in the Fertile Crescent and implications of the recent Syrian drought. *PNAS, USA*, 112, 3241–3246

[334] Heron et al. (2016)

[335] Hecht, D. (2013) The neural basis of optimism and pessimism, *Experimental Neurobiology*, 22(3): 173–199, https://dx.doi.org/10.5607%2Fen.2013.22.3.173.

[336] Financial Times declares a winner in the war for energy's future, and Big Oil won't be happy: https://thinkprogress.org/renewable-energy-is-unstoppable-declares-financial-times-40c222f439bb/

[337] General Motors Is Going All Electric: https://www.wired.com/story/motors general-electric-cars-plan-gm/

5. What are the causes of Arctic amplification? Describe why it is a global concern and not just a regional issue.

6. Describe the stability of the Antarctic ice sheet.

7. Extreme weather has increased in frequency. Why? How is global warming tied to extreme weather?

8. Describe the types of weather changes caused by global warming.

9. How are the world's ecosystems changing as a result of global warming?

10. Why is improving understanding of climate sensitivity relevant to controlling future climate change?

Activities

1. Visit the Intergovernmental Panel on Climate change website http://www.ipcc.ch/ and answer the following questions.
   (a) How is the IPCC organized and what is its purpose?
   (b) Describe the Fifth Assessment Report and what its purpose is.
   (c) What are the major conclusions of AR5?

2. Research climate change in your state.
   (a) What agency is in charge of tracking climate change?
   (b) What conclusions have they reached about threats related to climate change in your state?
   (c) What is being done about these threats?

3. View the animation "Time History of Atmospheric Carbon Dioxide from 800,000 years ago until January 2009" at http://www.esrl.noaa.gov/gmd/ccgg/trends/history.html and answer the following questions.
   (a) Describe what you learned as you watched this animation.
   (b) Watch it a second time; what did you learn this time?
   (c) Has it been valuable to spend taxpayers dollars to track global carbon dioxide concentration? Why?
   (d) What have we learned that is valuable from this effort?

Key Terms

Attribution

Carbon budget

Cascading
  extinction

Ecological
  sensitivity

Equilibrium climate
  ensitivity (ECS)

Flash droughts

Global Carbon Law

Palmer Drought Severity
  Index (PDSI)

Slow modes

Transient climate
  response (TCR)

KEY TERM GLOSSARY

Acidic Having the properties of an acid or containing acid; having a pH below 7.

Adaptation Response to climate change that reduces human vulnerability. Even if emissions are stabilized relatively soon, climate change and its effects will persist for centuries, and adaptation will be necessary in order for communities to be resilient and sustainable.

Aerosol particles Tiny particles in the atmosphere that may scatter sunlight directly back into space and thus reduce global warming, or absorb sunlight and re-radiate heat and promote warming. Aerosols in the lower atmosphere can modify the size of cloud particles, changing how the clouds reflect and absorb sunlight, affecting Earth's energy budget. Aerosols can come from volcanic outgassing, human industrial activity, desert dust, and other sources.

Albedo Proportion of light or radiation (e.g., sunlight) that is reflected by a surface.

Algal blooms Rapid growth of microscopic algae in water, often resulting in a colored froth on the surface. Harmful algal blooms occur when colonies of algae grow out of control. Their death and decay can remove dissolved oxygen from the water leading to a *dead zone.*

Alpine glaciers Located in mountain systems and forming at high elevation where snow accumulates for centuries in bowl-shaped hollows called cirques. Under thick accumulations, snow will recrystallize to ice which slowly flows out of the cirque and down the adjacent valley. Several cirque glaciers can join together to form a single valley glacier. They are all alpine glaciers.

AMOC Atlantic Meridional Overturning Circulation A system of currents in the Atlantic Ocean characterized by the northward flow of warm, salty water in the upper layers of the Atlantic, and a southward flow of denser water in the deep Atlantic.

Annual Greenhouse Gas Index (AGGI) The ratio of the total direct radiative forcing due to long-lived greenhouse gases for any year for which adequate global measurements exist to that which was present in 1990 (the baseline year for the Kyoto Protocol); Calculated annually by NOAA.

Anoxic Areas of sea water, fresh water, or groundwater that are depleted of dissolved oxygen, a severe condition of hypoxia. The U.S. Geological Survey defines anoxic groundwater as having a dissolved oxygen concentration of less than 0.5 milligrams per liter.

Anthropogenic aerosols Solid tiny sulfate (SO_4) particles or liquid droplets of sulfur dioxide (SO_2), similar to the sulfur aerosols produced by volcanic eruptions. The largest amount of anthropogenic aerosols comes from burning coal and oil, mostly in the northern hemisphere where industrial activity is centered.

Anthropogenic climate change Changing climate caused by atmospheric warming due to the increase in greenhouse gases resulting from human activity.

Anthropogenic greenhouse effect Enhancement of the natural greenhouse effect caused by release of greenhouse gases from human activities such as burning fossil fuels (e.g., to produce electricity) and certain types of land use (e.g., deforestation).

Anthropogenic hypothesis Hypothesis that human agriculture (involving deforestation, rice wetland production, and animal husbandry) has controlled global climate for several thousand years (since the early and mid-Holocene) because of its influence on the carbon cycle.

Arctic amplification Extremely rapid warming taking place in the Arctic which is happening two times faster than the global average.

Armoring Practice of building stone or cement walls on eroding shorelines. Leads to beach loss. Also known as coastal hardening.

Arrhenius, Svante August (1859–1927) Nobel-Prize winning Swedish scientist, first to investigate the effect that doubling atmospheric carbon dioxide would have on global climate.

Attribution Research seeking to understand role of global warming in specific weather events.

Back radiation Heat radiated from greenhouse gases to Earth's surface.

Beach loss Destruction of beach as an environment usually because of armoring (coastal hardening).

Biogenic ooze Class of marine sediment collecting on the seafloor due to accumulation of plankton skeletal particles; most commonly consisting of silica (e.g., from Diatoms and Radiolaria) or carbonate (e.g., from Foraminifera and Coccolithophores) composition.

Biosphere Realm of all living things.

Bipolar seesaw Paleoclimatic behavior wherein the North Atlantic and the South Atlantic regions reflect opposite climate trends. While one warms, the other will cool (and vice versa); recorded in ice cores from Greenland and Antarctica; result of alternating shut down and recovery of the Atlantic Meridional Overturning Circulation (AMOC). Vigorous AMOC exports heat from S. Atlantic leading to cooling there and warming in N. Atlantic. Slowing or shut down of AMOC leads to heat build-up in S. Atlantic and cooling in N. Atlantic.

Boundary conditions In Regional Climate Modeling, an existing set of parameters such as water vapor content, winds, temperature, precipitation, or other value that establishes the beginning climate state for a modeling experiment.

Calcification Bicarbonate ion combines with dissolved calcium (Ca^{2+}) in seawater forming calcium carbonate ($CaCO_3$, the mineral calcite), the primary component of limestone.

Calibrated uncertainty language Specific words and phrases used in reports that reflect specific calculations of confidence and probability (e.g., https://science2017.globalchange.gov/chapter/front-matter-guide/).

Callendar, Guy Stewart (1898–1964) English steam engineer and inventor; contributed to theory linking rising carbon dioxide concentrations in the atmosphere to global temperature; earlier proposed by Svante Arrhenius, theory has been called the *Callendar effect*. Callendar thought this warming would be beneficial, delaying a "return of the deadly glaciers."

Carbon budget Quantity of carbon-based emissions (e.g., methane, carbon dioxide) that can be emitted in total over a specified time that will force a specified atmospheric temperature.

Carbon cycle Biological and geological types; conceptual model describing movement and storage of carbon through natural and anthropogenic processes and reservoirs.

Carbon dioxide CO_2; colorless, odorless gas produced by burning carbon and organic compounds and by respiration; naturally present in air (~0.03%); absorbed by plants in photosynthesis; absorbs and reradiates heat (infrared radiation); primary greenhouse gas because of long (up to millennia) residence in atmosphere.

Carbon footprint Amount of carbon dioxide and other carbon compounds emitted due to the consumption of fossil fuels by a particular person, group, or activity.

Carbonate Oxidized carbon buried as carbonate (CO_3) such as limestone; composed of calcium carbonate, $CaCO_3$, the mineral calcite.

Carbonic acid Carbon dioxide reaction with water to form carbonic acid (H_2CO_3); carbonic acid condenses and precipitates as rain and reacts with the most common groups of minerals – the silicates—during *hydrolysis*, an important chemical weathering process.

Cascading extinction Ecological cascade effect involving a series of secondary extinctions triggered by the primary extinction of a key species in an ecosystem.

Central-Pacific El Niño Known by several names: warm-pool El Niño, dateline El Niño, El Niño Modoki; new type of El Niño emerged in recent decades, with warmest waters in the tropical central Pacific Ocean rather than in the tropical eastern Pacific.

Clathrate Gun Hypothesis Idea that ocean warming or sea level lowering can trigger sudden release of frozen methane on the seafloor (i.e., sea bed permafrost); because methane is a powerful greenhouse gas, temperatures rise further, and the cycle repeats. Once started, the runaway process could be as irreversible as the firing of a gun.

Clay Class of minerals (i.e., hydrous aluminum phyllosilicates); formal size class (0.98–3.9 μm [microns]) of particles of natural rock; also a soil type that combines one or more clay minerals with possible traces of quartz (SiO_2), metal oxides (Al_2O_3, MgO etc.) and organic matter.

Climate Long-term (e.g., 30 years) average weather pattern in a particular region; result of interactions among land, ocean, atmosphere, water in all of its forms, and living organisms; described by many weather elements: e.g., temperature, precipitation, humidity, sunshine, wind, and more.

Climate change Change in climate patterns, in particular apparent from the mid to late 20th century onwards and attributed largely to increased levels of atmospheric carbon dioxide and other greenhouse gases produced by anthropogenic fossil fuel use and certain types of land use.

Climate feedbacks Result of changing climate that in turn influences the original change; a positive feedback amplifies climate change, a negative feedback reduces it.

Climate forcing Difference between insolation (sunlight) absorbed by Earth and energy radiated back to space.

Climate model Mathematical relationships constructed by researchers to simulate atmospheric and ocean circulation in order to project future climate changes resulting from increases in atmospheric concentrations of anthropogenic greenhouse gases.

Climate proxies Chemical and other clues in ice and sediment that identify past climate change; central to the field of *paleoclimatology*.

Climate variability Short-term change in climate caused by oceanographic and atmospheric influences. Climate is described as the average, long-term condition of the weather. Even though it is an average condition, it still experiences changes from one state to another and back again.

CMIP United Nations World Climate Research Program sponsors the *Coupled Model Intercomparison Project* or CMIP, a global research endeavor to compare and share data among the scientific groups who build, run, and interpret climate models.

Coal abatement Reduction in the amount of coal burned to generate power for a community.

Coastal erosion Landward retreat of a shoreline due to deficiency in available sediment (e.g., lack of sand on a beach); consequence of sea level rise, short-term wave processes, and human impacts.

Cold front Boundary of an advancing mass of cold air; trailing edge of the warm sector of a low-pressure system.

Concentrated Animal Farming Operations CAFO, U.S. Department of Agriculture; farm in which animals are raised in confinement in large numbers as commercial source of protein.

Conduction Energy transfer directly atom to atom; flow of energy along a gradient.

Continental ice sheets Vast expanse of ice completely covering all underlying terrain; forming on land and spreading outward in all directions; originating in polar regions due to long-term snow accumulation, not related to the high elevation of mountain systems; expansion or retreat signals a significant change in climate.

Control runs Climate model experiments in which climate is held constant under a set of unchanging conditions (e.g., constant CO_2); to compare models that use the same input values.

Convection Fluid movement caused by less dense (i.e., hotter) material to rise, and more dense (i.e., cooler) material to sink under the influence of gravity; convection results in transfer of heat.

Convention to Combat Desertification U.N. program in countries experiencing serious drought and/or desertification, particularly Africa; designed to mitigate the effects of drought and desertification through national action that incorporates strategies supported by international cooperation and partnership arrangements.

Conveyor belt hypothesis Idea that rapid climate change results from a shutdown (or turn-on) in the thermohaline circulation system in the North Atlantic; supported by paleoclimate proxies in ice cores and marine sediments.

Coral bleaching Loss of color in coral due to stress such as rising water temperature, excessive sedimentation, or other negative environmental change. Caused by either loss of pigments by microscopic algae (zooxanthellae) living in symbiosis with their host organisms (coral polyps), or because zooxanthellae have been expelled.

Core Solid innermost section of Earth's interior; located below the mantle; outer boundary is 2,890 km (1,800 mi) beneath Earth's surface.

Coriolis force Force experienced by mass in rotating system acts perpendicular to direction of motion and to axis of rotation; deflects moving objects to the right in northern hemisphere and to left in southern hemisphere; important in formation of cyclonic weather systems.

Coriolis, Gaspard-Gustave (1792–1843) French mathematician, known for work on forces in a rotating frame of reference, leading to the *Coriolis effect*.

Coupled general circulation model Mathematical simulation of atmosphere and ocean circulation and heat transfer; basis for full climate model.

Critical point Height of sea level rise beyond which flooding rapidly accelerates and threatens an entire coastal community; thought of as a *tipping point*, a limit at which some aspect of the climate system irretrievably shifts to a new state

Crust Outermost, rocky layer of Earth, composed of two general types: continental (granite) and oceanic (basalt).

Cryosphere Earth system of ice environments, currently at risk due to global warming.

Crystallization Solidification of atoms or molecules into highly structured form called a crystal.

Cut-off low Closed upper-level low pressure system that is displaced from basic westerly air currents, and moves independently.

Dalton Minimum Period of low sunspot count, representing low solar activity; named after English meteorologist John Dalton, lasting from about 1790 to 1830, or 1796 to 1820.

Dansgaard–Oeschger Periodic sharp spikes in oxygen isotopes in Greenland ice cores suggesting abrupt warming of atmosphere temperature of up to 15°C; climate changes rapidly, within decades, and persists in stable state for 500 yrs to a few millennia; suborbital timescales, cannot be explained by orbital parameters; DO events are *interstadials* that last ca. 500 to 2500 yrs before returning to a colder state – *stadials*; related to Atlantic bipolar seesaw.

Dead zones More common term for hypoxia, a reduced level of oxygen in the water; areas in the ocean of such low oxygen concentration that animal life suffocates and dies.

Deforested Land use involving clearing or removal of a forest typically for agriculture.

Desert dust Wind-blown mineral particles from deserts of North Africa and Asia; absorb sunlight as well as scatter it; warms the layer of the atmosphere where they reside inhibiting formation of storm clouds and suppressing rainfall, thus causing desert expansion.

Desertification Desert expansion; decreased availability of natural resources, especially water and food; spurs political unrest and conflict as one group of people, in order to maintain its own supply level, appropriates resources from another group, or migrates into their resource-rich lands.

Dynamic sea level Tilt of tropical Pacific Ocean surface; correlates to global mean surface temperature; tilt down to the east signals enhanced heat storage in the west Pacific warm pool, associated with decreased global mean surface temperature (consistent with La Niña–like condition); tilt down to the west signals increased global mean surface temperature (consistent with El Niño–like condition).

Dynamical downscaling Climate modeling technique that applies the fundamental governing laws of physics, coupled with parameterizations of various climate processes, to run a regional climate model on a high resolution grid.

Earth's energy imbalance Difference of solar energy absorbed by Earth and energy radiated to space; if positive, more energy coming in than going out, Earth warms; if negative, Earth cools.

Eccentricity Changing shape of Earth's orbit from more circular to more elliptical and back again on 100,000-year and 400,000-year cycles; affects insolation when Earth is farthest from the Sun (aphelion) and when Earth is closest to the Sun (perihelion); shifts seasonal contrast in the Northern and Southern Hemispheres.

Ecological sensitivity Higher CO_2 leads to increasing ecological change and biosphere stress; species face increasing competition and turnover from invasive species and shifting environmental conditions.

Eemian Name given to the warmest phase of the last interglacial period, ca. 125,000 yrs ago.

El Niño Reduction of trade winds and movement of the western Pacific warm pool to the east; appearance of unusually warm, nutrient-poor water off northern Peru and Ecuador, typically in late December.

El Niño Southern Oscillation Irregular variation in wind and sea surface temperatures over tropical eastern Pacific Ocean; affecting global climate; two phases: El Niño and La Niña.

Electromagnetic spectrum Range of wavelengths or frequencies over which electromagnetic radiation extends.

Energy budget Net flow of energy into and out of the Earth system.

Equilibrium climate sensitivity (ECS) Global mean near-surface temperature when it has equilibrated to atmospheric CO_2 concentrations that are double the preindustrial level of CO_2 (~280 ppm), may take centuries to fully develop.

Evaporation Conversion of liquid to vapor (gas).

Evapotranspiration Sum of evaporation and plant transpiration from land and ocean surfaces to the atmosphere. Evaporation accounts for movement of water to the air from sources such as the soil, canopy interception, and water bodies.

Exposure People, assets, or ecosystems, which, when combined with hazards (i.e., dangerous events or trends) and vulnerability (i.e., susceptibility to harm) constitute *risk* from climate change.

Faculae Bright regions surrounding dark spots (i.e., sunspots) on surface of the sun.

Far field Locations (e.g., tropical Pacific) not directly experiencing crustal deformation by ice sheet melting.

Ferrel Cell Atmospheric circulation cell defined by air sinking at 30° latitude and rising at 60° latitude.

Ferrel, William (1817–1891) Demonstrated that owing to Earth's rotation, air and water currents moving distances of tens to hundreds of kilometers are deflected to the right in the Northern Hemisphere and to the left in the Southern Hemisphere (i.e., Coriolis Force).

Findings Determinations of fact reported by the *Intergovernmental Panel on Climate Change*, and other scientific organizations.

Flash droughts Severe drought that develops rapidly.

Fluorinated gases Human-made gases; known to stay in the atmosphere for centuries and contribute to global greenhouse effect; hydrofluorocarbons (HFCs), perfluorocarbons (PFCs), sulfur hexafluoride (SF_6) and nitrogen trifluoride (NF_3); used in refrigeration and variety of consumer products.

Foraminifera Single-celled planktonic animal with a perforated chalky shell through which slender protrusions of protoplasm extend. Most kinds are marine, and when they die, their shells form thick ocean-floor sediments (i.e., biogenic ooze).

Fourier, Jean-Baptiste Joseph (1768–1830) French mathematician and physicist; first to calculate that insolation alone did not sufficiently account for Earth's surface temperature; considered possibility that atmosphere acts as insulator. Today, this is widely recognized as the first proposal of the *greenhouse effect*, although Fourier never called it that.

General circulation models Mathematical relationships reproducing fundamental processes in Earth's climate system: solar radiation, surface heating differences, atmosphere and ocean circulation, and other phenomena linked to the climate.

General circulation of the atmosphere System of winds that transport heat from the equator, where solar heating is greatest, toward the cooler poles, giving rise to *climate zones*.

Geologic proxies Natural materials used to understand some aspect of past climate.

Glacial landforms Erosion and accretion features from glaciers, usually formed in an *ice age*.

Glacial An ice age.

Global biogeochemical cycling Chemical compounds found on Earth's surface move (i.e., cycle) between the air, the water cycle, Earth's crust, and living organisms.

Global carbon law Proposal by Johan Rockstrom of the Stockholm Resilience Institute that with 50% reductions in global carbon emissions per decade, global warming can be limited to 2°C; these reductions scale from individuals to entire globe.

Global mean sea level Global average elevation of the ocean surface.

Global overshoot Degree to which humanity's demand for ecological resources and services in a given year exceeds what Earth can regenerate in that year.

Global warming Gradual increase in overall temperature of Earth's atmosphere due to anthropogenic greenhouse effect; caused by increased levels of CO_2, CFC's, and other pollutants.

Global warming hiatus Also *global warming pause* or *global warming slowdown*; period of relatively little change in global surface temperature 1998-2013.

Global warming potential (GWP) Measure of heat a greenhouse gas traps in the atmosphere compared to heat trapped by a similar mass of CO_2; calculated over a specific time interval (e.g., 20, 100, or 500 yrs); expressed as a factor of CO_2 (i.e., GWP = 1).

Greenhouse effect Trapping of heat radiated from planet surface after it has been warmed by the sun; warming of lower atmosphere due to the greater transparency of the atmosphere to visible radiation from the sun than to infrared radiation emitted from the planet's surface.

Greenhouse gases Gas contributing to greenhouse effect by absorbing infrared radiation, (e.g., carbon dioxide and chlorofluorocarbons).

Groundwater extraction When water is withdrawn from an aquifer more rapidly than it is replenished; as water table drops, may cause subsidence (i.e., lowering of land surface).

Groundwater inundation Flooding by rise in water table due to sea level rise; typical of low lying coastal plain settings.

Hadley Cell Large-scale atmospheric convection cell in which air rises at equator and sinks at medium latitudes, typically about 30° north or south.

Hazards Dangerous event or trend related to climate change, which, with *exposure* of assets, and *vulnerability* (susceptibility to harm), constitute *risk* (potential for consequences) from climate change.

Heat engine Processes in atmosphere and hydrosphere that even out solar heating imbalances through evaporation of surface water, convection, rainfall, winds, and ocean circulation, when distributing heat around the globe.

High air pressure Anticyclone where the atmospheric pressure at Earth's surface is greater than its surrounding environment.

Holocene Epoch Current episode of geologic time, beginning ca. 10,000 yrs ago.

Hydrolysis Process where silicate minerals react with carbonic acid and produce clays.

Hydrosphere All water on (e.g., lakes and rivers), below (e.g., aquifers), and above (e.g., clouds) Earth's surface.

Ice ages Periods of long-term lowering in surface and atmospheric temperature, resulting in expansion of continental and polar ice sheets and alpine glaciers.

Ice shelves Floating sheet of ice permanently attached to a landmass

Ice stream Ice that moves significantly faster than surrounding ice; a type of glacier; significant features of the Antarctic where they account for 10% of total ice volume.

Ice-albedo feedback Positive feedback where a reduction in ice area alters the albedo, this in turn triggers additional reductions in ice.

Insolation Amount of solar radiation reaching an area.

Interglacial cycles For the past 1 million yrs, climate change has been characterized by cycles of alternating ice ages and interglacials spaced approximately 100,000 yrs.

Interglacials Warm episodes that separate two adjacent ice ages (i.e., glacials).

Intergovernmental Panel on Climate Change Scientific reporting body under the auspices of the U.N., dedicated to providing an objective, scientific view of climate change and its political and economic impacts; issues climate change assessment reports on 5 to 7 yr frequency.

Interstadials Minor period of slightly warmer climate during an ice age.

Intertropical Convergence Zone (ITCZ) Known as *doldrums*; area encircling Earth near the Equator, where the northeast and southeast trade winds converge.

Keeling Curve Graph plotting ongoing change in concentration of atmospheric CO_2 since the 1950s; based on continuous measurements taken at NOAA's Mauna Loa Observatory in Hawaii that began under the supervision of Charles Keeling: https://scripps.ucsd.edu/programs/keelingcurve/

Keeling, Charles David (1928-2005) American scientist whose recording of CO_2 at the Mauna Loa Observatory first alerted the world to the possibility of anthropogenic contribution to the greenhouse effect and global warming. The Keeling Curve measures the progressive buildup, and seasonal variability, of carbon dioxide, a greenhouse gas, in the atmosphere.

King tides Popular, non-scientific term used to describe exceptionally high tides; typically occurring during a new or full moon and at its perigee, or during specific seasons.

Köppen, Wladimir (1846–1940) Widely used climate zone classification; based on subdivision of world's climate five major types (tropical, dry, temperate, continental, and polar), each further divided into one or more subcategories.

La Niña Cooling of the eastern and central equatorial Pacific that occurs at irregular intervals and is associated with widespread changes in weather patterns; typified by accelerated trade winds and growth of the western Pacific warm pool.

Latent heat Heat required to convert a solid into a liquid or vapor, or a liquid into a vapor.

Limestone Sedimentary rock, composed mainly of calcium carbonate or dolomite, used as building material and in the making of cement.

Lithification Process in which sedimentary particles are cemented, or compacted under pressure, expel fluids, and gradually become solid rock; Essentially a process of porosity destruction through compaction and cementation.

Lithosphere Rigid, rocky outer layer of Earth, consisting of the crust and the solid outermost layer of the upper mantle; extends to a depth of about 100 km.

Lithospheric flexure Action of bending Earth's lithosphere under the weight of thickening ice, growing volcano, sediment accumulation at a delta, or deformation related to tectonics.

Little Ice Age Period of colder climate that reached its peak during the 17th century.

Local extinction Loss of plants or animals due to migration out of an area where they had previously been naturally abundant; due to environmental changes related to climate change.

Local relative sea level change Localized sea level change that reflects the relative difference between the rate of vertical land motion and the rate of sea-level rise.

Low air pressure System of winds with lower pressure at its center than the area around it. Winds blow towards the low pressure, and the air rises in the atmosphere where they meet. As the air rises, the water vapor within it condenses forming clouds and often precipitation too.

Mantle Part of Earth's interior between the core and the crust; 2,900 km thick, makes up nearly 80% of Earth's total volume. The mantle is composed of solid rock capable of flow.

Marine ice-cliff collapse Calving of floating ice shelves in Antarctica produces a marine ice-cliff that may not be stable under its own weight if its height exceeds 90 m. This could lead to unstoppable collapsing retreat that would drive rapid sea level rise.

Maunder Minimum name used for the period around 1645 to 1715 during which sunspots became exceedingly rare, as was then noted by solar observers.

Medieval Climate Anomaly (MCA) Known also as *Medieval Warm Period, Medieval Climate Optimum,* or *Medieval Climatic Anomaly*; period (approximately 950 to 1250 CE) of warm climate in the North Atlantic region, possibly related to warming events in other regions during that time (e.g., China); other regions, however, were colder, such as the tropical Pacific. Averaged global mean temperatures similar to early-mid 20th century warming. Possible causes include increased solar activity, decreased volcanic activity, and changes in ocean circulation.

Mesoscale eddies Three dimensional ocean surface gyres, representing ocean turbulence, with anomalous salinity, temperature, and sea surface height; horizontal scales of less than 100 km, timescales on the order of a month, and depths extending 1000's m; when intersecting oceanic islands they may create anomalous high tides (i.e., king tides).

Methane Odorless, colorless, flammable gas (CH_4), primary component of *natural gas*; powerful greenhouse gas over 25 times more potent than CO_2; atmospheric residence of approximately one decade.

Mid-latitude cyclones Atmospheric low pressure in mid-latitudes (i.e., the temperate zone, approximately 30° to 60° N or S of the equator); counterclockwise air circulation in the Northern Hemisphere and clockwise circulation in the Southern Hemisphere.

Mid-latitude westerlies Prevailing winds from the west in the middle latitudes between 30° to 60° latitude; originate from descending arm of Hadley Cell at 30° latitude (i.e., horse latitudes).

Mid-Pliocene Warm Period Recent time (ca. 3.3 to 3.0 million yrs ago) with mean global temperatures substantially warmer for sustained period (2°C to 3°C above pre-industrial temperatures); Recent enough that continents and ocean basins had nearly reached present geographic configuration.

Milankovitch cycles Natural climate cycles of ice ages and interglacials caused by three orbital parameters: *eccentricity*, *obliquity*, and *precession*. Named for Milutin Milankovitch, Serbian astrophysicist and engineer.

Milankovitch, Milutin (1879–1958) Serbian astrophysicist and engineer; founded field of planetary climatology by calculating temperatures of Earth's upper atmosphere and planets of the inner Solar system; explained climate changes resulting from shifts in Earth's orientation to the sun, now known as *Milankovitch cycles*.

Miocene Epoch Earliest epoch of Neogene Period, ca. 23 to 5.3 million years ago; time of warmer global climates; first appearances of kelp forests and grasslands.

Mitigation Reducing greenhouse gas emissions in order to slow and stop global warming.

Model ensemble Common practice of running two or more related but different climate models and synthesizing results as a single score (or spread).

Monsoon Seasonal prevailing SW wind bringing rain to SE Asia between May and September.

Monsoon winds SE Asia, seasonal SW wind bringing abundant precipitation in normal years.

Multimodel mean Individual climate models contain random noise (i.e., weather) that limits the value of individual model results. Averaging the results of many models minimizes noise, providing an estimate of how climate may respond to human-induced global warming.

Natural greenhouse effect Natural warming of the troposphere by greenhouse gases that absorb and re-radiate heat from Earth's surface after it is warmed by the sun.

Neap tides Two times each month, when the Moon is in a half-moon phase, the high tides are not very high and the low tides are not very low because the gravitational attraction of the Moon and the Sun are pulling in different directions.

Negative feedback Process that reduces the effect of changing the climate.

Nested modeling When one model is used to define certain values and those values are used by another model to calculate a different set of values, the two models are nested.

Net heating Difference between amount of incoming sunlight and amount of heat radiated back to space.

Neutral year Year that is neither in an El Niño state nor in a La Niña state.

Nitrous oxide Colorless non-flammable greenhouse gas (N_2O) with slight metallic scent and taste; emitted by agricultural and industrial activities, as well as during combustion of fossil fuels and solid waste; has a global warming potential of 265–298 times that of CO_2 for a 100-year timescale. N_2O emitted today remains in the atmosphere for more than 100 years, on average.

Northeast trade winds Prevailing tropical winds blowing from the NE in the Northern Hemisphere developed at the descending limb of the Hadley Cell.

Obliquity In astronomy, angle of tilt of Earth's axis of rotation.

Ocean acidification Ongoing decrease in the pH of Earth's ocean caused by the uptake of anthropogenic CO_2 from the atmosphere.

Oceanic conveyor belt Global, deep-ocean circulation beginning in the North Atlantic as water from the Gulf Stream grows denser until it sinks and moves south as the North Atlantic Deep Water. Most of it eventually upwells in the Southern Ocean, but the oldest water upwells in the Pacific after approximately 1000 yrs.

Orbital parameters Variations in the distribution of sunlight on Earth's surface owing to three configurations of the Earth-Sun geometry: 1) eccentricity, changes in the shape of Earth's orbit; 2) obliquity, changes in the tilt of Earth's axis of rotation, and; 3) precession, a wobble in Earth' axis of rotation. These three control variations in sunlight that closely match the timing of ice ages and interglacials recorded by ice cores and deep-sea sediments. Mathematically described by Milutin Milankovitch.

Orographic effect Meteorological process where warm moist air rises into cooler air due to a topographic barrier. Characterized by condensation, cloud formation, and precipitation.

Ozone Greenhouse gas (O_3) formed in troposphere when pollutants emitted by cars, power plants, industrial boilers, refineries, chemical plants, and other sources chemically react in the presence of sunlight; also forming a layer in the stratosphere which absorbs most of the ultraviolet radiation reaching Earth from the sun.

Pacific Decadal Oscillation Pattern of sea surface temperature in the Pacific Ocean with two characteristic phases: positive or warm, and negative or cold. Occurs as a multi-year, or multi-decadal expression of sea surface temperature variability.

Paleocene–Eocene Thermal Maximum (PETM) Intense, abrupt global warming ca. 56 million years ago, at boundary of Paleocene and Eocene epochs; caused by massive carbon injection into the atmosphere (perhaps lasting no longer than 20,000 yrs); entire warm period lasted for about 200,000 years as global temperatures increased by 5–8°C.

Paleoclimatology Study of past climate; central to the question "Is modern global warming part of a natural climate cycle?"

Palmer Drought Severity Index (PDSI) Standardized index that spans -10 (dry) to +10 (wet) using readily available temperature and precipitation data to estimate relative dryness.

Parameterization Method of replacing processes that are too small-scale or complex to be physically represented in a climate model by a simplified set of equations or empirical coefficient.

Paris agreement International agreement among 195 parties within the United Nations Framework Convention on Climate Change dealing with greenhouse gas emissions mitigation, adaptation, and finance starting in the year 2020.

Permafrost Thick subsurface layer of soil that remains frozen throughout the year, occurring chiefly in polar regions.

Photosynthesis Certain organisms (i.e., green plants) produce carbohydrates (a source of energy to drive metabolism) from sunlight and CO_2 during photosynthesis.

Phytoplankton Plankton consisting of microscopic plants.

Plate tectonics Theory explaining structure of Earth's lithosphere, and many associated phenomena, as resulting from the interaction of rigid lithospheric plates that move slowly over the underlying mantle.

Pleistocene Epoch Geological epoch lasting from 2,588,000 to 11,700 yrs ago, spanning the most recent period of repeated glaciations

Polar Cell Part of global atmospheric circulation, located poleward of Ferrel Cell in both hemispheres.

Polar easterlies Dry, cold prevailing winds blowing from high-pressure areas of the polar highs at the North and South Poles towards low-pressure areas within the Westerlies at high latitudes.

Polar front Boundary of the Ferrel and Polar circulation cells. The polar front is an area of rising air, cloud development, and precipitation that forms at a transitional zone separating Arctic and polar air masses from tropical air masses.

Polar vortex Large area of low pressure and cold air surrounding both poles that weakens in summer and strengthens in winter. The term vortex refers to an organized persistent westerly flow of air that characterizes the system.

Positive feedback Process that amplifies the effect of changing the climate.

Pouillet, Claude Servais Mathias (1790-1868) French physicist, developed first mathematical treatment of greenhouse effect; speculated that H_2O and CO_2 trap infrared radiation in the atmosphere, warming Earth enough to support living organisms.

Precession Gradual shift in the orientation of Earth's axis of rotation in a cycle of approximately 26,000 years

Processes The rates at which elements and compounds move between places where they are temporarily stored (reservoirs) and where they are exchanged (processes) can be measured directly and modeled using computer programs.

Radiative cooling Process by which long-wave (infrared) radiation is emitted to balance the absorption of short-wave (visible) energy from the sun.

Radiative equilibrium Assumes that incoming radiative energy from the Sun is equal to the outgoing radiation emitted by the planet: incoming energy = outgoing energy. That is, the planet is in energy balance. If radiative equilibrium is not achieved, then the temperature of the planet must rise or fall.

Radiative forcing Processes in Earth's climate system that cause an imbalance between the amount of sunlight entering the atmosphere and the amount of energy radiating to space.

Rain shadow Arid or semi-arid region down wind of an orographic barrier.

RCPs *Representative Concentration Pathways*; used by CMIP5 modelers in IPCC AR5, 2014. Four RCPs were used representing greenhouse gas concentration (not emissions) trajectories; RCP2.6, RCP4.5, RCP6, and RCP8.5 are named for possible radiative forcing values in the year 2100 relative to pre-industrial values ($+2.6$, $+4.5$, $+6.0$, and $+8.5$ W/m², resp.).

Regional variability Variations in elevation of the ocean surface from one region to another.

Relative sea-level rise Local sea level rise at a rate representing the difference between vertical land motion and global mean sea level rise.

Representative Concentration Pathways See RCPs.

Reservoirs Natural storage of certain elements and compounds that are otherwise exchanged by natural or anthropogenic processes.

Resilience Capacity to recover quickly from difficulties; toughness.

Respiration Process in living organisms involving production of energy, typically with intake of oxygen and release of CO_2 from oxidation of complex organic substances.

Risk Potential for consequences when something of value is at stake and the outcome is uncertain. Risks from climate change impacts arise from the interaction between three qualities: *hazards*, *vulnerability*, and *exposure*.

Rock cycle Idealized cycle of processes undergone by rocks in the crust, involving igneous intrusion, uplift, erosion, transportation, deposition as sedimentary rock, metamorphism, re-melting, and further igneous intrusion

Rock Solid aggregation of minerals.

Sahel Zone of transition in Africa between the Sahara to the north and the Savanna to the south; semi-arid climate, stretching across south-central latitudes of Northern Africa between the Atlantic Ocean and the Red Sea.

Satellite altimetry Method of mapping the ocean surface with radar emitted from Earth-orbiting satellite; information from tide gauges and satellite altimetry used together improve understanding of sea-level change.

Sea level fingerprints As glaciers melt, ocean waters nearby are less strongly attracted by gravity of the ice and move away causing sea level to rise faster far away from the glacier; resulting pattern of sea level change is known as a sea level fingerprint.

Self-reinforcing cycles Also known as positive feedback.

Self-stabilizing cycles Also known as negative feedback.

Sensible heat Related to changes in temperature of a gas or object with no change in phase.

Silica rocks Rocks in Earth's crust dominated by silica minerals.

Slow modes Paleoclimate studies indicate temperature is more sensitive to CO_2 changes than has been observed; significant levels of warming take more than a century to fully develop (e.g., the Southern Ocean and the Eastern Equatorial Pacific require more than 100 years); these ultimately account for the greatest portion of warming found in climate models and are missed when looking at the relatively brief span of historical observations.

Soil respiration Production of CO_2 when soil organisms respire, includes respiration of plant roots, the rhizosphere, microbes and fauna; from 1989 to 2008 soil respiration increased by about 0.1% per yr. In 2008, the global total of CO_2 released from the soil was about 98 gigatons, 10 times more carbon than humans are now putting into the atmosphere each year by burning fossil fuels.

Southeast trade winds Prevailing tropical atmospheric circulation in Southern Hemisphere related to Hadley Cell.

Spatial resolution Measure of smallest accurate size on a map or photo.

Spreading center Where new oceanic crust is produced from rock in Earth's mantle.

Spring tides When Moon is bright and full or nearly dark, gravitational attraction of the Moon and that of the Sun are both pulling on the ocean in the same direction. During those times of the month, the high tides are higher than average and the low tides are lower than average.

Stadials Short-term, temporary climate stage when temperature falls and more ice is formed.

Statistical downscaling Method of climate modeling using a two-step process 1) development of statistical relationships between local climate variables and global-scale predictors, and 2) application of relationships to output of climate modeling in order to project future climate; accuracy of statistical downscaling relies on important assumptions. Chiefly, have the statistical relationships between variables and predictors been appropriately defined? and do they remain appropriate under the changing climate conditions of the future?

Stefan–Boltzmann Law Total heat emitting power of an ideal black body is proportional to the fourth power of its absolute temperature.

Subduction zone Where oceanic lithosphere is recycled back into the mantle.

Subtropical high-pressure zones Belts of atmospheric high pressure situated at 30°N and 30°S; product of Hadley Cell.

Sulfur aerosols Sulfur-rich particles in the atmosphere.

Sunspot cycle Periodic increase and decrease in the number of sunspots averaging about 11 yrs.

Sustainability Avoidance of resource depletion.

Temporal resolution Precision of a measurement or modeling experiment with respect to time.

Thermal expansion Rise in sea level caused by heating (i.e., expansion) of seawater.

Thermal infrared radiation Heat.

Thermohaline circulation Vertical ocean circulation driven by density of seawater created by differences in heat and salt content.

Tidal range Difference in height between high tide and low tide.

Tide gauge Water-surface measurement device to monitor rise and fall of ocean surface.

Tipping point Point when small changes accumulate to cause a larger, irreversible change.

Total solar irradiance Total amount of solar radiation in W/m² received outside Earth's atmosphere on a surface normal to incident radiation, and at Earth's mean distance from Sun.

Transient climate response (TCR) Change in global air temperature within 20 yrs of CO_2 content doubling over pre-industrial (e.g., 280 ppm) value.

Trewartha, Glenn (1896–1984) American geographer, reworked Koppen classification system in 1968, (updated again in 1980) and reclassified the temperate category into three types (subtropical, temperate and continental, and boreal).

Tropic of Cancer Latitude 23.5°N that is northernmost point reached by the overhead sun.

Tropic of Capricorn Latitude 23.5°S that is southernmost point reached by the overhead sun.

Tropics Geographic band between Tropic of Cancer, 23.5°N, and Tropic of Capricorn, 23.5°S.

Tropospheric ozone See ozone.

Tyndall, John (1820–1893) Prominent 19th-century physicist who made discoveries in the realm of infrared radiation and the physical properties of air.

Uncertainty Margin of error of a measurement, given by a range of values likely to enclose the true value.

United Nations Framework Convention on Climate Change International environmental treaty negotiated at the United Nations Conference on Environment and Development informally known as the Earth Summit, held in Rio de Janeiro from 3 to 14 June 1992.

United States Global Change Research Program (USGCRP) Established by Presidential Initiative in 1989 and mandated by Congress in the Global Change Research Act (GCRA) of 1990 to "assist the Nation and the world to understand, assess, predict, and respond to human-induced and natural processes of global change."

Uplift Weathering Hypothesis Idea that uplift of silica-rich (e.g., continental) crust drives increase in chemical weathering (e.g., hydrolysis) which decreases atmospheric CO_2.

Upwelling Deep, cold water rises toward the surface when surface waters displaced by wind are replaced by cold, nutrient-rich water that "wells up" from below.

Urban heat island Urban area that is warmer than surrounding rural areas due to human activities; most noticeable during the summer and winter.

Variability Although climate is the long-term average condition of the weather, it is variable, meaning that it can change from one season to another, from one year to another, and over even longer time periods such as decades.

Vertical land motion Upward or downward movement of the land; at the coast, vertical land motion causes local relative sea level change.

Volcanic aerosols Following an eruption, large amounts of sulfur dioxide (SO_2), hydrochloric acid (HCl) and ash are spewed into Earth's stratosphere.

Volcanism Volcanic activity.

Vulnerability Susceptibility to harm, which with *hazards* (events) and *exposure* (assets), constitute *risk*.

Warm front boundary of an advancing mass of warm air, in particular the leading edge of the warm sector of a low-pressure system.

Water vapor Atmospheric gas H_2O, rises with air temperature and is a powerful greenhouse gas.

Weather Atmospheric conditions over a short time; defined by variables such as humidity, precipitation, pressure, temperature, and the wind.

Weather fronts Boundary separating two masses of air of different densities; principal cause of meteorological phenomena outside the tropics.

Weathering Chemical, biological, and physical processes that destroy rock and minerals

Western Pacific Warm Pool Region of sea surface temperatures warmer than 28.5°C located in the western tropical Pacific.

Younger Dryas Cold event (ca. 12.9–11.6 thousand yrs ago) seen as classic example of rapid climate change. The North Atlantic region cooled during this interval with a weakening of Northern Hemisphere monsoon strength.

INDEX